Chemiluminescence and Bioluminescence
Past, Present and Future

Chemiluminescence and Bioluminescence
Past, Present and Future

Edited by

Aldo Roda
Department of Pharmaceutical Sciences, University of Bologna-Alma Mater Studiorum, Bologna, Italy

RSCPublishing

ISBN: 978-1-84755-812-1

A catalogue record for this book is available from the British Library

Published by The Royal Society of Chemistry,
Thomas Graham House, Science Park, Milton Road,
Cambridge CB4 0WF, UK

Registered Charity Number 207890

For further information see our web site at www.rsc.org

Preface

The use of chemiluminescence and bioluminescence detection techniques in various fields of analytical and bioanalytical chemistry is continuously expanding thanks to recent achievements that have made available both new chemical or molecular biology tools and advanced instrumentation for light measurement and imaging.

The recent 2008 Nobel Prize in Chemistry awarded to Osamu Shimomura, Martin Chalfie and Roger Y. Tsien for the discovery and development of the green fluorescent protein (GFP) testifies to the importance of luminescence in biosciences.

The observation of bioluminescence in nature has stimulated, in the last 50 years, basic photophysics and biochemical research aimed at explaining different aspects of this fascinating phenomenon, such as unravelling its functions in nature, studying the evolution of bioluminescent species and investigating its biochemical mechanisms. As a result, many bioluminescent systems have been carefully characterized and various luciferins and luciferase enzymes have been isolated and are nowadays available as analytical tools. Nevertheless, much work needs to be done, since many bioluminescent marine and terrestrial systems are still unknown.

Disclosure of the fine mechanism responsible for light emission in bioluminescent systems also enabled scientists to design new synthetic chemiluminescent compounds, exploiting the knowledge of which functional groups are responsible for light emission in nature. All work performed in the 1500–1700s on the observation and accurate classification of bioluminescent organisms and the early attempts to demonstrate the mechanisms of the light emission phenomenon paved the way for subsequent achievements, particularly at the end of the 1800s and during the 1900s. Nowadays, we have extraordinary tools, allowing us to detect a few molecules in a complex sample.

In recent years, tremendous progress in the chemiluminescence field has occurred, thanks to the development of new chemical probes, enhancers and advanced instrumentation. Bio- and chemiluminescence have been employed successfully in biospecific assays, exploiting the high detectability of such

Chemiluminescence and Bioluminescence: Past, Present and Future
Edited by Aldo Roda
© Royal Society of Chemistry 2011
Published by the Royal Society of Chemistry, www.rsc.org

detection systems in combination with high specificity of target recognition offered by antibodies and nucleic acid probes. Furthermore, the intrinsic high sensitivity of bio- and chemiluminescence in small reaction volumes has also allowed the development of high-throughput screening methods, thus boosting the drug discovery process. The possibility of coupling bio- or chemiluminescent reactions with other enzyme reactions has led to amplified enzyme-based assays suitable for automation and miniaturization using immobilized enzymes. Finally, the availability of sensitive and compact light detectors has allowed successful implementation of bio- and chemiluminescence detections in portable analytical devices.

Bioluminescent reporter gene technology has propelled revolutionary advances in many fields of basic and applied science, from drug discovery to environmental monitoring. Thanks to its high sensitivity, bioluminescence has been widely used to monitor cellular events and unravel the molecular mechanisms associated with signal transduction and gene expression. More recently, its potential has been also applied to monitor protein–protein interactions *via* resonance energy transfer processes or split-complementation strategies. These findings in signal transduction and protein "chattering" have been achieved at the cellular and tissue level with *in vitro* bioluminescent assays and also in whole organisms, as in bioluminescence imaging. Non-invasive "*in vivo*" bioluminescence imaging has emerged as a powerful alternative to other imaging techniques and has been applied for the study of tumor metastasis and progression, for monitoring bacterial and viral infections and in studies of stem cell homing. In recent years many xenograft and transgenic bioluminescent animals have been created and proposed as preclinical models for drug development and for pathophysiological studies.

In this book, the history and most recent advances in the applications of bio- and chemiluminescence in analytical chemistry are reported. The first part (Basics of Chemiluminescence and Bioluminescence) reviews the historical development of bio- and chemiluminescence, as well as the fundaments of such phenomena and the most recent advancements in luminescence instrumentation. The second part (Analytical Applications of Chemiluminescence and Bioluminescence) deals with applications of bio- and chemiluminescence in various research fields, such as life sciences, drug discovery, diagnostics, environment, agrofood and forensics.

I hope that this book will become a reference text not only for researchers currently employing such detection techniques in their research activity, but also for those approaching bio- and chemiluminescence for the first time.

Aldo Roda
Bologna, Italy

Contents

Chemiluminescence and Bioluminescence: Past, Present and Future
Edited by Aldo Roda
© Royal Society of Chemistry 2011
Published by the Royal Society of Chemistry, www.rsc.org

Chapter 11 Ultrasensitive Bioanalytical Imaging

Mara Mirasoli, Simona Venturoli, Massimo Guardigli,
Luisa Stella Dolci, Patrizia Simoni, Monica Musiani and
Aldo Roda

Chapter 12 "*In vivo*" Molecular Imaging

Eric L. Kaijzel, Thomas J. A. Snoeks, Ivo Que,
Martin Baiker, Peter Kok, Boudewijn P. Lelieveldt and
Clemens W. G. M. Löwik

Chapter 15   Cell-based Bioluminescent Biosensors   511
Kendrick Turner, Nilesh Raut, Patrizia Pasini,
Sylvia Daunert, Elisa Michelini, Luca Cevenini,
Laura Mezzanotte and Aldo Roda

Part 1
Basics of Chemiluminescence and Bioluminescence

CHAPTER 1

A History of Bioluminescence and Chemiluminescence from Ancient Times to the Present

ALDO RODA

Laboratory of Analytical and Bioanalytical Chemistry, Department of
Pharmaceutical Sciences, Alma Mater Studiorum-University of Bologna,
Via Belmeloro 6, 40126 Bologna, Italy

1.1 Introduction

My interest in the historical aspects of bioluminescence (BL) and chemiluminescence (CL) began in 1988. That year Mario Pazzagli of Florence and I
organized the 5th International Symposium of the International Society for
Bioluminescence and Chemiluminescence (ISBC), held in Florence and
Bologna. During this meeting a day in Bologna was set aside to honor the
distinguished scientist William McElroy for his work on firefly luciferase
machinery. A Degree *ad honorem* in Pharmaceutical Chemistry was conferred
on him by the Faculty of Pharmacy of the University of Bologna. I had became
acquainted with McElroy when, as a visiting scholar in San Diego in 1982, I
worked under the direction of his wife Marlene DeLuca.

It was Marlene's premature death in 1987 that inspired the ISBC board to
have a day of our symposium dedicated to McElroy, during which almost all of
his pupils, including J. W. Hastings, H. H. Selinger, T. O. Baldwin, F. McCapra
and K. V. Wood, presented a lecture in honor of Marlene DeLuca.[1] Several of
the speakers made references to a book by McElroy's mentor, Edmond Newton
Harvey, that piqued my curiosity. I subsequently learned that Harvey's text
A History of Luminescence from the Earliest Times until 1900, published by the

Chemiluminescence and Bioluminescence: Past, Present and Future
Edited by Aldo Roda
© Royal Society of Chemistry 2011
Published by the Royal Society of Chemistry, www.rsc.org

American Philosophical Society, Philadelphia, in 1957 is by far the most complete and definitive work on the subject.[2] Since becoming acquainted with Harvey's book I have had a growing fascination with the history of the study of luminescent phenomena, particularly after learning that the first description of a formal study of luminescence was performed in my home town of Bologna. In 1603, one Casciarolo, a shoemaker by trade and alchemist by passion, accidentally discovered a phosphor that glowed after exposure to light. This substance eventually became known as the Bolognian Stone or the Bolognian Phosphor.

Ten year later, for the 10th International Symposium of the ISBC held in Bologna, I decided to give a presentation on the Bolognian Stone. The story of its discovery is quite interesting, and I enjoyed the opportunity to present some local history as well. Hunting for more detailed information I went to our historical University library to look for the ancient book quoted by Harvey on this subject. To my surprise, the research done by Harvey was so detailed there was little left for me to discover. My recounting of the story of the Bolognian Stone can be found in the Proceedings of the Symposium,[3] as well as on the ISBC web site.[4] While preparing for the present chapter I naturally referred again to Harvey's book, and found myself freshly amazed at how much information it contains – essentially all existing knowledge of the history of the study of luminescence from ancient times until 1900. The scope of this work is all the more impressive when one considers that when it was written in the 1950s our means of communication were much less sophisticated, and collection of all this information must have been a very arduous task. I should add that, since then, very little has been published in the way of an update. I feel very fortunate indeed to have found a copy of Harvey's book through Amazon, used but still in good condition, because, as for many others in our field, it has become a sort of bible for me. Because Harvey did such a thorough job of telling the history of the study of luminescence from ancient times until 1900, I will give only an overview of that period, emphasizing instead the period subsequent to 1900, up to and including the present. In more recent years BL and CL have become fundamental tools in molecular biology and related sciences, with analytical chemistry in particular benefitting from implementation of highly sensitive, rapid, miniaturized methods based on BL and CL. Attesting to the importance of these tools, in 2008 the Nobel Prize for Chemistry was awarded jointly to Osamu Shimomura, Martin Chalfie and Roger Y. Tsien for the discovery and development of green fluorescent protein (GFP).[5]

1.2   Ancient Era

Highlights from this period include texts from China in the East as well as from Greece and Rome in the West.

The oldest known written observations on bioluminescent phenomena in nature were made in China, dating roughly from 1500 to 1000 BCE, regarding fireflies and glow-worms; in these writings, however, there is no mention of any effort being directed at understanding or applying the knowledge of such phenomena.

According to legend, in 1000 BCE, a Chinese emperor possessed a magical painting on which the image of an ox appeared at sunset. This was the first known case of a man-made substance that is capable of storing and emitting light. Unfortunately, the chemical composition of the paint was unknown. In later Chinese literature the BL of sea water is described as a glow that could have been due to Protozoa and minute Metazoa. In the *Hai Neishih Chou Chi*, an account of marvelous sea islands attributed to Tung Fang Shuo (second century BCE), the author states that "if one travels on the sea, one may see fiery sparks when the water is stirred," which was likely due to the presence of dinoflagellates. Mention of the BL of fireflies and glowworms is frequently found in texts from this period. For example, in the *Shin Chung*, a book of odes written in the first century CE, these creatures are referred to as *night travelers*. In some later writings from this period a more methodical approach was taken, with descriptions of the anatomy of bioluminescent animals and their behaviors.

Many folklore-based, artistic works, including poems and paintings, document awareness of bioluminescent phenomena among the peoples of ancient Japan, India, Africa as well as Central and South America. For example, one of the most common insects depicted in Mayan art, especially on ceramic bowls, is the firefly. There are several interesting references to fireflies in Conquest-period documents, most importantly, in the "Popol Vuh," a sacred text that gives the firefly a role in religion and mythology. The reference in the "Popol Vuh" is particularly interesting as it associates fireflies with the practice of cigar smoking (Figure 1.1).[6] Firefly and glowworm are mentioned in

Figure 1.1 Detail of a Maya vase illustrating fireflies with cigars in their mouths. (K8007, © J. Kerr, reproduced with permission.)

the ancient Indian text, the Vedas and in the poem *Mahabharata* (200 BCE). In the Buddhist scripture, the *Dhammapada*, the Pali word *khajjopakana* is used for firefly.

What is probably the first written record of BL in marine fauna has been attributed to Anaximenes of Miletus (585–528 BCE), a Greek philosopher. It was the great Aristotle (384–322 BCE) who wrote the first detailed observations on light from marine species, in addition to insects, describing up to 180 animals. He was also the first to identify *cold* light, *i.e.*, he recognized that the self-luminosity of BL organisms was not accompanied by heat, unlike the light from a candle flame. Aristotle wrote about the luminosity of dead fish, now known to be due to infection by luminescent bacteria, and also about the above-mentioned light that occurs when sea water containing dinoflagellates is agitated by stirring with a rod.[2]

In 215 BCE Titus Livius reported that in Sardinia "the shores were also luminous with frequent fires."

Figure 1.2 Pliny offering his work to Emperor Titus, miniature from *Naturalis Historia* (Books I-XVI), Firenze, Biblioteca Medicea Laurenziana, ms. Plut. 82.1, c. 2v. (Reproduced with permission.)

A few centuries later, a more complete record of BL organisms was made by Pliny the Elder (23–79 CE) in his *Naturalis Historia* (Figure 1.2), in which he gives detailed descriptions of many luminescent animals, from glow-worms and fireflies to a luminous mollusk (*Pholas dactylus* – a Roman delicacy). The specific name was the one given by the Romans, who liked to eat them, as is attested by Pliny in *Naturalis Historia* (IX, 87):

"Concharum e genere sunt dactyli ab humanorum unguium similitudine appellati. His natura in tenebris remoto lumine, alio fulgere claro, et quanto magis humorem habeant, lucere in ore mandentium, lucere in manibus, atque etiam in solo ac veste, decidentibus guttis: ut procul dubio pateat, succi illam naturam esse, quam mir-aremur etiam in corpora."

[The group of shells are dactyls, so named for their similarity to human fingernails. By their nature they are in the dark without light, shine forth clearly, and have as much humor, they shine in the mouth of someone who eats them, gleamed in his hand, and even for land and among his clothes, releasing drops: from which it is obvious that the nature of the juice is that which we admire in the body.]

Pliny also describes the purple luminescent jellyfish *Pelagia nocticula* that are common in the bay of Naples, where Pliny died during the eruption of Vesuvius (79 CE). They were called *Pulmo marinus* by the Romans owing to a slime secreted from the outer surface of the bell that was considered to be a remedy for fevers.

1.3 Middle Ages to the Sixteenth Century

The period extending from the fall of the Roman Empire (around 500 CE) to the Renaissance in Europe is known as the Middle Ages, a time dominated by superstition and belief in magic. All learning and study were controlled by the church, and many studies of natural phenomena such as luminescence were of an "applied" nature, that is, used to maintain belief in spirit or for producing magical potions. During this time, in other parts of the world, such as Japan, Arabia, India, China, South and Central America, civilizations with rich cultures were flourishing, but we have little information about any studies of BL or CL that may have been performed in those areas.

One of the few such references we have is to the two most common firefly species in Japan, the *genji-botaru* and the *heike-botaru*, whose names have an interesting history: it is said that the souls of countless soldiers who were killed in the Battle of Dannoura (1185) turned into fireflies. The larger and more profuse *genji-botaru* fireflies were no doubt named after the winning Genji, and the smaller *heike-botaru* after the defeated Heike (Figure 1.3).

In China, Li Shin-Chen (1518–1593) traveled widely in major Chinese provinces (Honan, Kiangsu, Anhwei), collecting specimens and studying the natural occurrences of minerals, plants and animals. After completion of the final

(A)

(B)

Figure 1.3   (A) "Catching Fireflies" – Kiyohara Hitoshi (Japan, 1896–1956). (B) "The
Actor Kawarazaki Gonjuro Surrounded by Fireflies" – Utagawa Yosh-
itsuya (Japan, 1822–1866). (© Los Angeles County Museum of Art.
Reproduced with permission.)

draft in 1587 he visited Nanking to arrange for publication of his findings. The
first printed edition (the so-called Chin-ling xylograph) did not appear until
1596, after Li's death. In the book he distinguished three main kinds of luminous insects: the common firefly (*Luciola* spp.) and many other genera, the
glowworm and the luminescent flies living in or near water (probably midges, or
mayflies infected with luminous bacteria).

In the text "Tractatus de Simplicibus" Arab writers and the Persian botanist
Ibn-al-Baithar (1197–1248) describe the *Hobaheb*, a beetle with wings that light
up during the night. In 1372 Al-Damiri (1344–1405), a great Arab zoologist of
Cairo, in his zoological dictionary *Hayat-al-Hayawan* (The life of animals)
describes an insect with wings that emits light, as if it were on fire.

In Europe, Isidore, the Archbishop of Seville (560–636) who was later cano-
nized, wrote the "Etymologiae", in which he described luminous stones as well as
the *cincidela* (firefly) a genus of beetles that shines while either walking or flying.

In the thirteenth century Albert Magnus (1206–1280), who came to be regarded as the first naturalist, described and catalogued many luminous species, but not many more than Pliny did several centuries earlier. His work was published only after the invention of the printing press in 1478, in the form of a book that was given the title *De Animalibus*. He made an extract of fireflies, *Liquor lucidus*, that had permanent luminescence.

Dante was one of the few Italians fearless enough to mention BL in his prose. In the Canto XXVI of "Inferno" of *The Divine Comedy*, one of the last poems he penned before his death in 1321, he wrote of peering down into the Eighth Chasm of Hell and seeing "fireflies innumerous spangling o'er the vale."

The Renaissance was a time of tremendous growth in the Arts, and it was also a time of exploration of parts of the world that had been largely unknown to Europeans. Among other novelties, explorers brought back reports of *burning seas*, now known as *milky seas*, a phenomenon that remains little understood. Christopher Columbus in 1492 observed a mysterious light in the sea near San Salvador that was likely due to the presence of a marine worm, *Odontosyllis*, that is known to inhabit those Caribbean waters.

The Spaniard Gonzalo Fernandez Oviedo (1478–1557) who was the official chronicler of Indian affairs during his stay in different parts of the new world, published extensive data on natural history; among other creatures he identified the elaterid beetle (*Pyrophorus*), bioluminescent caterpillar and what eventually came to be known as the railroad worm (*Phenogoides*).

Conrad Gesner (1516–1565), Professor of Natural History and Medicine in Zurich, is credited with writing the first book devoted to luminescence, *De Lunaris*, in which he writes about luminescent animals and plants, as well as luminous stones (Figure 1.4). This book, together with the *Historia Animalium* (also written by Conrad Gesner), probably contains the first mention of luminous sea pens, although the Romans knew BL sea pens (*Pennatulaceans*), referring to them as *Penna marina*.

In 1551 in the Russian Church code called the *Stoglav* there is the curious and poetic pronouncement that on the Summer Solstice, the Feast of Kupala, a fern (in Russian *paparotnik*) blooms on this Mid-Summer Night, and that whoever finds this flower and takes it home will have good fortune ensured for the rest of his life. We know that this fern does not exist, but we also know that there is a mushroom (*Mycena* sp.) that is a bioluminescent mycelium that was used to mark the paths in the forest at night, and perhaps the legend was born here.[7]

1.4 Seventeenth Century

This is the century of the scientific approach as promulgated by the philosophers Sir Francis Bacon (1561–1624) in England and Renee Descartes (1594–1650) in France. Their ideas laid the foundations of Science itself: the philosophical groundwork that permitted a rational and materialistic approach to the study of natural phenomena had been laid and, in the spirit of times, this

(B)

CONRADI GESNERI,

DE

UNARIIS HERBIS,
ET REBUS NOCTU
LUCENTIBUS,

COMMENTARIOLUS.

NARIAM, qvam recentiores
nostrosè describunt, & noctu lucere
ejunt, herbam videri fixisiam.

Onstrosà Lunariæ descriptio-
ne, qvalèm in progressu affere-
mus, videri possunt aliqvi homines
stultos & avaros deludere, ac va-
lactatos ad inqvirendum frustrà inci-
oluisse. Sic Virgilius etiam in Palæmo-
ogâ tertiâ his versibus:
, qvibus in terris, & eris mihi magnus A-
pollo,
patras Cæli spatiam non amplius ulnas;
Gram-

a 2

(A)

Figure 1.4   (A) Frontispiece of *De Lunariis herbis, et rebus noctu Lucentibus, Commentariolus* by Conrad Gesner (1669). (B) C. Gesner, "The Practice of the New and Old Physicke," London, 1599.

approach was applied to finding or making the *Philosopher's Stone*, that which would be capable of turning "ignoble metals" into gold. Intellectual interest expanded to include luminescent phenomena, in living creatures as well as inorganic luminescence from phosphorescent stones. Concerning many new discoveries there was tremendous curiosity and a strong desire for explanations.

Conditions were thus ripe for excitement when, in 1602, one Vincenzo Casciarolo, a cobbler by trade and dilettante alchemist, discovered the Bolognian Phosphorus in Bologna , Italy (Figure 1.5). It was this natural stone, subsequently referred to also as the Bolognian Stone or Litheophosphorus, that became the first object of scientific study of luminescent phenomena.

The most complete text on the subject of Vincenzo Casciarolo and his stone, *Litheosphorus Sive De Lapide Bononiensi*, was written in 1640 by Fortunius Licetus (1577–1657), Professor of Philosophy at the University of Bologna.

The rudimental alchemistic approaches of the cobbler Casciarolo led, by a process of heating and calcination of the stone, to the discovery of its mysterious and magical ability to "accumulate" light when exposed to the sun and to emit it in the darkness. Fortunius Licetus (Figure 1.6) wrote:

"The first way is to reduce it to meal and work it into cakes either with plain water or the white of an egg. After they have dried out they are put in layers with coal in a blast furnace and, after a very hot fire has been made, they are calcinated for

Figure 1.5 Piece of Bolognian Stone, barium sulfate (barite), found on Monte Paderno, Bologna – the same place where Casciarolo collected the stone for his experiments. (Private collection of Aldo Roda).

four or five hours. When the oven has cooled off the cakes are taken out. From this powder various animals are formed in little boxes (pyxidiculum) that shine wonderfully in the dark. The lixivium is prepared in the same way and once it is dry it produces a sulphurous, fetid, sharp, and biting salt."

The recipe and method of preparation of the Bolognian Stone were kept secret only briefly, the first detailed description on the methods being published in 1625 by Pierre Potier (Poterius), physician to the King of France, in his widely used *Pharmacopea Spagirica*, a treatise on inorganic remedies based on the teachings of Paracelsus. Potier lived for some time in Bologna, and he was able to give an acceptable interpretation of the discovery. Potier's version of the recipe is quoted by E.N. Harvey.[2]

Casciarolo showed his *lapis solaris* to many learned men of the time. There was widespread interest in Italy despite the fact that attempts to use it as the *Philosopher's Stone* were unsuccessful. Galileo Galilei (1564–1642) participated in scientific debate regarding the stone, and he also presented it to Giulio Cesare La Galla (1576–1624), Professor of Philosophy of the Collegio Romano, who first reported this phenomenon in the book *De Phenomenis in Orbe Lunae* (1612). La Galla asserted that the untreated stone was not able to emit light, but that it acquired this property only after calcination. He explained this phenomenon, as related to him by Galileo, as a certain quantity of fire and light to which the stone was exposed being trapped in the stone and then slowly released from it, comparing its absorption to that of water by a sponge. Subsequently, Ovidio Montalbani (1601–1671), Professor of Astronomy and Mathematics at the University of Bologna, published a brief report, "De Illuminabili Lapide Bononiensi Epistola" (1634), in which he gave a dissertation on the various colors of light that could be obtained from the stone, and was the first to suggest that the light resulted from a kind of burning. Licetus was detailed and very enthusiastic, so much so that publication of the book led to a famous controversy between himself and Galileo Galilei: whereas Licetus sustained that the faint light of a crescent moon was produced by phosphorescence similar to that of the Bolognian Stone, Galileo believed that it was a reflection of sunlight from the Earth to the Moon.[3] Galileo Galilei wrote of the Bolognian Stone:

"It must be explained how it happens that the light is conceived into the stone, and is given back after some time, as in childbirth".

The debate extended throughout Europe. John Evelyn (1620–1706) was one of the first Englishmen to learn of the stone during his visit to Bologna in 1645. Although he apparently observed the luminescence, confirming that it occurred with various colors of light, he did not take any samples back to England because, as was reported in the *Philosophical Transactions of the Royal Society* in 1666, the recipe for the preparation of the stone had apparently been lost.[3] In 1691, Marsigli wrote "Del fosforo minerale e sia della pietra Bolognese" and dedicated it to his English colleague Robert Boyle, who unfortunately died

(A)

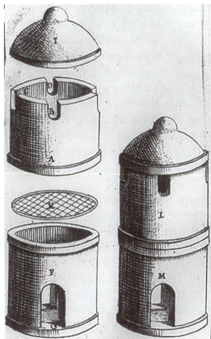

(B)

Figure 1.6 (A) Portrait of Fortunius Licetus and the title-page of his book *Litheosphorus Sive De Lapide Bononiensi* (1640) Biblioteca Comunale dell' Archiginnosio, Bologna. Reproduced with permission. (B) Drawing of the furnace used by Vincenzo Casciarolo to calcinate the Bolognian Stone. Taken from the book *Il Fosforo o vero la Pietra Bolognese* by Marc'Antonio Cellio (1680) (© BUB, Biblioteca Universitaria di Bologna. Reproduced with permission.)

before he was able to read it; because of Boyle's untimely death, the work was put aside and not published until 1698. Marsigli also devoted a special dissertation on the Bolognian Stone at the Academy of Sciences of Paris. Among the treatises of chemistry that have dealt with the stone, the famous *Cours de Chimie* written by Nicolas Lémery (1645–1715) is perhaps the one that has the strongest experimental point of view. This work was published in many editions and translations, the latest of which was published in Italian by Gabriele Hertz in 1719. This version tells how to find the stone, describes in detail how to produce phosphorescence, proposes a theory to explain its brightness and even mentions that it has depilatory properties. Lémery was not very flattering toward his predecessors stating that:

"Poterius, Montalbanus, Maginus, Licetus, Menzelus, and some others have written about this stone, and have given the manner of mortar, but their descriptions are useless, because, continuing, no result is obtained".

The second volume of the *Dictionnaire de Chimie Macquer*, published in Paris by Lacombe in 1769, devotes several pages to the Pierre de Boulogne, interpreting its luminescence in terms of Stahl's phlogiston theory, reflecting the effort of the German chemist S. Maargraf, a strong supporter of this theory. The Bolognian Stone even found its way into French schoolbooks, one example being the *Cours de physique experimentale et de Chimie, Ecole des centrales al'usage, spécialment de l'Ecole Centrale de la Côte d'Or*, published in Dijon and Paris in 1801, in which there are directions for making small phosphorescent cakes. Over the years many other prominent men searched for the Bolognian Stone or visited Monte Paderno, among them Goethe in 1786. Today we know that the stone they were searching is barite (barium sulfate) in the form of heavy silvery concretions with fibrous radial formations that widen toward the periphery of the stone; when pulverized, then mixed with egg white or another binder and calcined with charcoal it turns into phosphorescent barium sulfide.

Digressing again to China, over the centuries there have been few written references to inorganic luminescence. One mentions was a so-called nightshining jewel (*yeh kuang pi*), a mineral brought from Roman Syria to China during the Han Dynasty (206 BCE to 220 CE). This may have been chlorophane, a variety of fluorspar (calcium fluoride), which lights up when heated or scratched. D. J. MacGowan (1814–1893) drew attention long ago to a remarkable story indicating that the preparation of artificial phosphorus may have been known in the Sung Dynasty (940–1280) as reported by the monk Wèn-Jung in the book *Hsiang Shan Yeh Lu* written in the eleventh century.

Interestingly, in the same period as the discovery of the Bolognian Stone a parallel contribution to BL came from Ulisse Aldrovandi (1522–1605), Professor of Natural Sciences at the University of Bologna. In 1602 he wrote the book *De Animalibus Insectis* in which insects of different species from different countries are classified and described.[8]

Aldrovandi also reported the view of various authors concerning the light of fireflies and glowworms in the chapters "De Cincidela" and "De Coccoio". Aldrovandi wrote:

"This kind of insect has been endowed by a wondrous kindness of nature. For it does not harm man when touched or held in his hand. It does not attack him with its bite nor irritate or molest him by its sting when it encounters him . . . Glow worms are said visible at night because they have an internal and inborn light . . . They have this light implanted from fire" *(Figure 1.7)*.

Aldrovandi stated that one could write on paper with firefly juice and that this script could be easily read at night. In his book *De Reliquis Animalibus Exanguibus. De Mollibus, Crustaceis, Testaceis et Zoophytis* he gives minute descriptions of the mollusc *Pholas dactilus* and jellyfish (Figure 1.8).[8]

In the seventeenth century methodological approaches to scientific study were being developed by various thinkers, perhaps all of whom would have agreed the declaration by Robert Boyle (1627–1691) that "Experiment is the interrogation of Nature." There was disagreement, however, on the specifics. Two leading schools of thought were the English Baconian System and the approach of the Frenchman René Descartes.

Robert Boyle (1627–1691) was a leading intellectual of his day. He performed and published the results of many seminal experiments and came to be regarded as the "Father of modern chemistry." He was a strong believer in the Baconian approach, collecting and categorizing both positive and negative results with no intention of formulating any grand theories, in contrast with Descartes' deductive approach. While Boyle is best known for the experiments that resulted in the establishment of Boyle's Law (the inverse relationship between the pressure and the volume of a gas), he had a strong interest in a wide range of topics in chemistry and physics. He produced many papers on luminescent phenomena that can be read online in the *Proceedings of the Royal Society*. Boyle was evidently aware of the numerous studies regarding both inorganic and animal light; he made many of his observations on luminescent phenomena that we now know to be the effects of oxygen, which had not then been identified. He noted that the removal of the air over iron heated to the point of red-hot emission had no effect on the emission whereas for a candle flame or a glowing coal the luminescence was extinguished and did not recover when air was reintroduced. He noted that a live mouse died in the absence of air and did not revive when air was readmitted, but that for a piece of glowing wood or a glowworm the light only dimmed and then the glow was exuberantly re-emitted when air was allowed to re-enter the test chamber. The present day explanation is that oxygen is an essential requirement for both the respiration of living creatures and for BL. In this type of situation the light from most BL systems may only dim and not extinguish, because extremely small amounts of oxygen still support the emission, and when the oxygen is removed for some short period the reaction precursors build up and then produce a burst reaction to luminescence when the oxygen is added.

(B)

(A)

Figure 1.7 Ulisse Aldrovandi's plates with watercolor drawings of animals. (A) Plate showing insects, on the left upper side *Lampyris prona* and *Lampyris supina* (vol. VII, c. 90); (B) plate showing sea animals; in the center, the sea pen (*Penna marina*). (© BUB, Biblioteca Universitaria di Bologna. Reproduced with permission.)

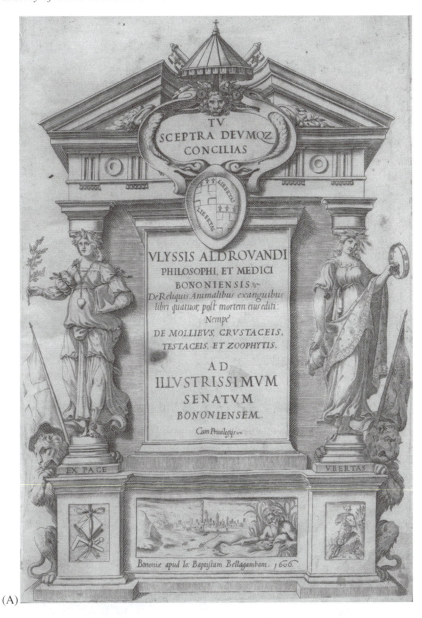

(A)

Figure 1.8 (A) Frontispiece of *De reliquis animalibus exanguibus libri quatuor, post mortem eius editi: nempe de mollibus, crustaceis, testaceis, et zoophytis"* by Ulisse Aldrovandi (Bologna, 1606); (B) drawing of *Pulmo marinus*; and (C) drawing of *Pholas* species. (© Biblioteca del Dipartimento di Biologia Evoluzionistica Sperimentale dell'Università di Bologna. Reproduced with permission.)

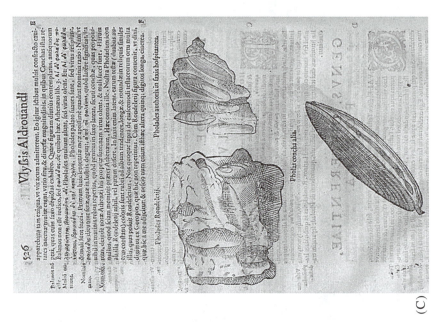

(C)

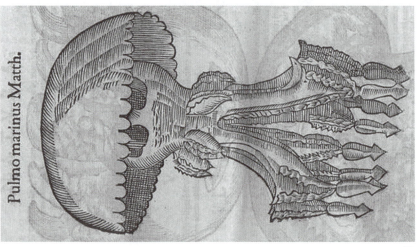

(B)

Figure 1.8 Continued.

In contrast to Boyle's Baconian approach to organizing research, Descartes favored observations followed by deductive reasoning to formulate a theory. In *Principia Philosophia* (1644), regarding light, he proposed that it originated from the friction of the particles rubbing together and that it was transmitted by one particle pushing against the next. This explanation fit for various forms of luminescence, such as light from seawater when it is agitated by the oars of a boat, or the BL of fish that occurs when particles of salt penetrate the pores.

By the late seventeenth century the understanding of light changed with the emergence of two dominant opposing theories: the wave theory of Hooke and Huygens *versus* the corpuscular theory that came to be associated more with Newton than Descartes. Owing mainly to the dominant reputation of Newton, the corpuscular theory remained unchallenged until almost 150 years later when it was clearly incompatible with the results of the diffraction experiments of Young and the work of Fresnel.

There were other important contributors to the study of luminescence in the seventeenth century. Among them, Athanasius Kircher (1602–1680), a German Jesuit priest, was the first to produce a book with a substantial treatment of luminescence. In his text *Ars Magna Lucis et Umbrae* Kircher describes the effect of a wood extract (*lignum nephriticum*) in water and discusses the use of fireflies to illuminate homes.

The most highly regarded and comprehensive book on BL in this period was *De Luce Animalium* (1647) by the Danish physician Bartholin (1616–1680). Although an Arabian alchemist named Alchid Bechil probably discovered phosphorus as early as the twelfth century, the first report of artificial CL occurred in 1669. Like other alchemists of the time (Figure 1.9), Henning Brand, a German alchemist and physician, searched for the *Philosopher's Stone*, a substance that purportedly could transform base metals (like lead) into gold.

He isolated a substance from urine that glowed continuously in the dark and called it *phosphorus mirabilis*, better known today as white phosphorus.

Still another important contributor to the field of BL was Marcello Malpighi of Bologna, a pioneer in the use of the recently invented microscope and famous for the discovery and localization of capillaries in 1688. He was the first to recognize that the light emitting material in fireflies is associated with particular granules in the cell:

(May 30: de Cicindela noctiluca. The end of the body cavity, the last two segments, contains a fluid which is the source of light. In daylight it appears yellowish and contains a milky substance. In the dark it lights sulphur yellow. The fluid contains a mass of small yellow globules in a similar slimy substance, which is half fluid . . . The rhythmic light ceases and became continuous when one cut off the last two segments . . .) (Figure 1.10).[9]

Use of the microscope paved the way for the study of marine organisms such as dinoflagellates. In an interesting story from this period, English explorers

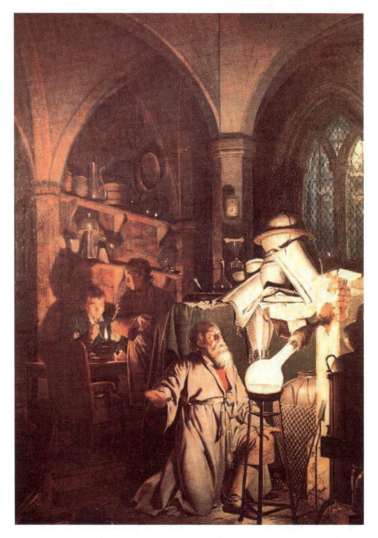

Figure 1.9   *The Alchemist in Search of the Philosopher's Stone* by Joseph Wright, 1771.
(© Derby Museums and Art Gallery. Reproduced with permission.)

apparently mistook the light from firefly beetles for the lights of Spanish campfires and decided not to land in Cuba in 1634, perhaps thus altering the history of the new world.

Domenico Bottoni, an Italian physician famous in his time, in the book *Pyrologia Topographica id est De igne dissertatio juxta loca cum eorum descriptionibus* (1692) discussed the differences between *Lucem*, *Lumen* and *Ignem* with special reference to the firefly (*Nitedula*) (Figure 1.11). Bottoni held that motion was the basis of this luminescence. He was particularly critical of

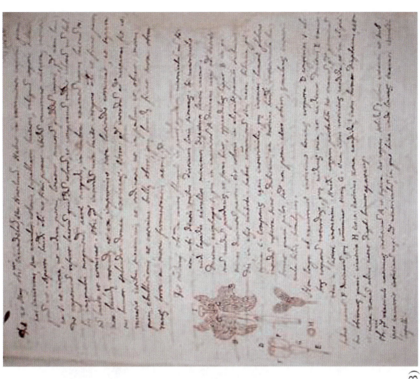

(B)

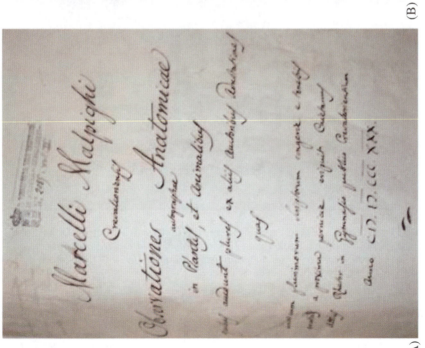

(A)

Figure 1.10 Frontispiece of *Observationes anatomicae in plantis et animalibus* by Marcello Malpighi (sec. XVII) (A) and a hand-written page of the manuscript describing observations of the bioluminescence of fireflies (B). (© BUB, Biblioteca Universitaria di Bologna. Reproduced with permission.)

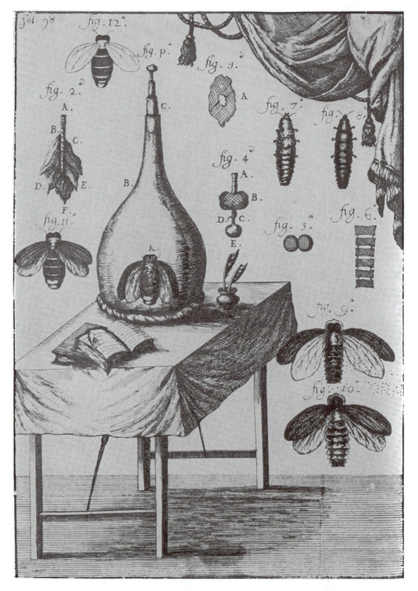

Figure 1.11   Drawing showing experiments on the firefly carried out by Domenico
Bottoni in his book *Pyrologia Topographica, id est, De igne dissertatio
juxta loca cum eorum descriptionibus* (Naples, 1692). (© Biblioteca del
Seminario di Storia della Scienza, Bari.)

Kircher, who had stated that:

*" . . . the nitedulae (fireflies) have obtained this innate light as something
intrinsic by the providence of nature in order that they can see and be seen . . . "*

Bottoni counters that:

" . . . rather we should explain those phenomena by natural reasons, resorting to the world of the senses as much as we can."

His own explanation of luminescence was that:

" . . . when there is motion induced, fiery sparks come forth' seemed particularly applicable to sugar and to the light of sea."

Harvey observed that, if we recall how quickly air in motion will rekindle a smoldering fire and produce a bright light, Bottoni's argument does not seem so ridiculous.

1.5   Eighteenth Century

Science in this century became highly experimental, particularly with investigations of electricity, heat and light and also, most importantly, in the determination of the composition of air and the discovery of oxygen, While not specifically about BL systems, the discovery of oxygen and ideas about the nature of combustion and the associated generation of light emission led to the suspicion that respiration, combustion and BL could be related processes. The air-pump experiments of Boyle where repeated and the results confirmed. In Copenhagen, Forster observed control of the firefly flash on removal or admission of air to the insect.

In 1797 Spallanzani in Naples, Italy, carried out careful experiments on the effect of oxygen and water, showing that parts of luminous medusae gave no light when dried but, if moistened again, would emit light the same as they did before drying. Similarly luminous material from the clam *Pholas* was made into a paste with flour, dried, kept for a year and when reconstituted with water again emitted light, and the same was also found to be true with luminous material extracted from jellyfish.

He thus showed that air (oxygen), water, and some photogenic substance are necessary for light production. Spallanzani's experiment, which has been confirmed for a great many luminous forms, also shows that animal luminescence is not a vital process, in the same sense that for example the conduction of a nerve impulse is a vital process, *i.e.* using the terminology of early physiologists, *living protoplasm* is not necessary for light production. In another experiment Spallanzani found that removal of air dimmed the light from luminous wood or a dead bioluminescent squid, but readmission of oxygen caused it to glow even brighter than it did before the air was removed. He found the light to disappear under nitrogen, hydrogen, and carbon dioxide ("fixed air"), concluding that BL was a process of *slow burning*.[10]

Regarding the frequently observed phenomenon of light emission from seawater, in 1717 Sir Isaac Newton wrote:

" . . . *do not all Bodies . . . emit Light as often as those parts are sufficiently agitated; whether that agitation be made by Heat, or by Friction, or Percussion, or Putrefaction, or by any vital Motion, or any other Cause, as for instance Seawater in a raging Storm."*

Also in 1717, J. J. D. De Mairan described luminous *pulmon marine* (jellyfish) and stated his theory that BL was due to movement of sulfur particles that had became disengaged from surrounding material.

Subsequently, early users of the newly invented microscopes demonstrated that the light emitted from seawater originated from marine microorganisms that were identified many years later, in 1753, as dinoflagellates (*Nocticula miliaris*) by Henry Baker.

At about the same time, in 1747, Benjamin Franklin became the first to state that phosphorescence could be due to electricity. In 1753 he wrote:

"It is indeed possible, that an extremely small animalcule, too small to be visible even by the best glasses, may yet give a visible light".

In 1754 Godeheu de Riville was the first to describe phosphorescent ostracods (crustaceans).[2]

In the late seventeenth century various dictionaries and encyclopaediae began to appear, some of which contained articles on phosphors. One of these was the *Dictionaire Universel* (1690) of Antoine Furetière (1619–1688), compiled under the patronage of the French Academy, that became a reference text for eighteenth-century scientists. Under the entry "Phosphore" Furetière gave particular mention to the Pierre de Boulogne. He distinguished four kinds of phosphorus and gave nine meanings for the world "lumière", but the words luminescence or phosphorescence are not mentioned.[2] The first technical dictionary in English was the *Lexicon Technicum*, a universal English dictionary of arts and sciences published in 1704 by John Harris (1667–1719). In this text "Phosphorus" was defined as a "Chymical Preparation;" there are instructions for making the Bolognian and Baduin's phosphorus, and Robert Boyle's pertinent experiments are presented. In the *Allgemeines Lexicon der Kunsteund Wissenschsften*, published in 1721 by Johann Theodore Jablonski (1654–1731), a sizable entry is devoted to the *Johannis wurmlein* (the glowworm), including instructions for preparing a *liquor lucidus* from the worm. The *French Encyclopédie* completed in 1745 divides the Phosphores into: (1) animal bodies that electric fluid has penetrated and rendered luminous, such as the glowworm, fireflies, the insect of the Canal of Venice, flies of the Antilles, the tongue of the irritated viper and shellfish; and (2) inanimate electric phosphores and other luminous phenomena. In addition, the *Encyclopedia Britannica* (1768–1771) gave a lengthy article devoted to electric light, the main novelty of the time. In the late eighteenth century there was little or no advancement in BL probably because experimental work in luminescent phenomena involved the study of the physical properties of light.

Scientific endeavors of the 1700s were significant for delineating the problems that were about to be attacked by the better equipped scientists of the next century. During the period straddling the eighteenth and nineteenth centuries luminescence changed from being an obscure phenomenon shrouded in mystery to one studied in a rational and enlightened way.[2]

1.6   Nineteenth Century: Modern Science

Charles Darwin (1809–1882) related that during one of his trips, while his ship passed through a bed of jellyfish caught in a storm, the sea was illuminated by debris left after violent thrashing from the agitated water. As with other types of animals he sought an explanation for their peculiarities in his theory of natural selection. However, BL phenomena did not fit easily into this conceptual framework. His controversial opus *On the Origin of Species by Means of Natural Selection*, published in 1859, included his annotations regarding bioelectricity in electric fish as well as brief mention of luminous organs. In Chapter 4, which is devoted to the difficulties encountered with his theory of the natural selection, Darwin wrote:

"The luminous organs which occur in few insects, belonging to widely different families, and which are situated in different parts of the body, offer under our present state of ignorance, a difficulty almost exactly parallel with that of the electric organs."

Adding to his difficulty in assigning a role for natural selection in luminescent phenomena, at that time very little was known about that nature of BL, an ignorance that would persist well into the twentieth century.

During the nineteenth century, as scientific investigation matured, the sciences became organized into different fields such as physics, chemistry and biology, and Newton's corpuscular theory of light was abandoned after the interference and diffraction experiments of Thomas Young, which led to the formulation of wave theory, according to which light consisted of waves occurring in a space-filling ether, a throwback to the *plenum* of Descartes. Another 100 years would pass before the existence of such an ether was disproved and the particle–wave duality theory of the nature of light became the most widely accepted. With the availability of technology that permitted accurate measurements for the investigation of the nature of light, the attention of physicists turned to optics, and the study of animal luminescence became the province of biologists.

After McCartney observed that the BL of dinoflagellates occurred in response to mechanical stimulation of the water in which they were situated, he proposed that the emission of light could have some sort of protective function. Microscopic study of dinoflagellates revealed that their luminescence was localized in discrete particles floating in the cytoplasm that in recent times have been given the name "scintillons." With chemical sciences developing rapidly at this time, it was soon realized that the mixtures reacting to produce luminescence outside of organisms were organic in nature.

In 1877 Radziszewski, a professor of Chemistry at Lemberg in Galicia, discovered that lophine (2,4,5-triphenyl-1H-imidazole) exhibits yellow chemiluminescence in solution after mixing with an alkaline alcoholic solution in the presence of air. In three years he prepared numerous chemiluminescent compounds, and claimed that lophine was as bright as luminous animals. The most import observation from Radziszewski was that lophine does not emit light when heated by itself, which anticipated Wiedermann's distinction of CL from incandescence in 1888.

The idea that the BL was the result of a chemical reaction was verified in 1885 by Raphael Dubois, a scientist in Lyon, France. Dubois made a paste of the luminescent material from the clam *Pholas* then suspended this in cold water, producing a glowing solution that he divided into two parts. One part was heated near boiling, whereupon the glow was extinguished. After the first part of the cold-water extract ceased to glow, the re-cooled hot-water sample was mixed in and light emission started up again. Dubois also showed that extracts of the click beetle (*Pyrophorus*) similarly produced light from a hot-water, cold-water reaction.[2] This experiment subsequently had a major impact on the study of BL. Dubois concluded that the BL was chemical in nature and that the heat stable part was probably an organic molecule, which he named *luciferine*. The active ingredient in the first part of the extract had to be kept cool for an effective reaction (in other words, it was heat labile) and, with the discovery of enzymes, this compound was labeled *luciferase*. These names have persisted to the present time, except for the *e* being dropped from *luciferine*. It is also now realized that luciferin and luciferase are the generic names for substances that are usually specific to a given bioluminescent system, *e.g.* the luciferin and luciferase of the firefly as published by Dubois in 1901. Dubois was also able to produce light by oxidizing luciferin without luciferase but using small crystals of $KMnO_4$, or by adding H_2O_2 (with or without blood-containing hemoglobin), BaO_2, PbO_2 and other oxidizing agents.[2]

By the end of the nineteenth century most BL species inhabiting the top layers in the ocean had already been described; with the advent of deep water trawling, below 200 fathom, it was found that all marine species at those depths are bioluminescent.

In 1883 Heller demonstrated by microscopic examination of *shining wood* that fungal threads (mycelia) growing on the wood were the source of the luminescence. The light of luminous wood was first documented in the early writings of Aristotle in 382 BCE and, many centuries later, more recently in 1667 by Robert Boyle, who noticed glowing soil, also realizing that heat was absent from this form of light. Other early scientists such as Conrad Gesner and Francis Bacon thought that the light was due to small insects or animal interactions. The first mention that the light of luminous wood was due to fungi occurred from a study of luminous timbers used as supports in mines by Bishoff in 1823. This opened the way for further study by many other scientists, including the modern experimental work began by Fabre in 1855 that established the basic parameters of bioluminescent fungi: the light was without heat; the light ceased in a vacuum, in hydrogen and carbon dioxide; the light was independent of humidity, temperature and light, and did not burn any brighter in pure oxygen.[2]

It is well recognized that in the modern era Eilhard Wiedermann was the first, in 1888, to use the term CL to describe a chemical reaction emitting *cold* light.[2] Light is created from sources apart from heat and is distinct from incandescence and black body radiation, or other effects that cause materials to glow at high temperatures. Although cold light clearly presented serious theoretical problems for nineteenth-century physics, practical concerns made these problems especially timely, *e.g.* the emerging incandescent lighting industry needed information about energy relations in light sources. Only in 1888 did Eilhard Wiedemann confront the contradiction that cold light presented to Kirchoff's law, proposing use of the term *luminescence* to designate light emission that was more intense than would be expected from the source's temperature, *i.e.* an emission that does not follow Kirchhoff's law. Wiedemann classified types of luminescence by the means of excitation as follows: phosphorescence, fluorescence, electroluminescence, chemical luminescence, thermoluminescence (produced by heating but whose characteristics were not temperature dependent), triboluminescence (produced by friction) and crystal luminescence (a type of luminescence that accompanies crystallization). He also pointed out that some emissions, such as light from electrical discharge tubes, are probably composed of both luminescence and temperature-dependent radiation. Wiedemann's classification system and his definition of luminescence were quickly adopted by other scientists.

The glowworm *Arachnocampa luminosa* (a crude translation being glowing spider bug) has been studied in the Wellington Botanic Garden since the end of the nineteenth century. Early entomologists believed that this New Zealand glowworm is a relative of the European firefly.

The Maori have named these insects *titiwai* (meaning projected over water), which describes their general habitat along streams. The name *pura toke* (one eyed worm or blind worm) is also used. The true nature of these insects was first described by George Vernon Hudson, an 18-year-old Englishman living in Karori Wellington, only a short distance from the botanic garden. On arrival in Wellington he commenced studying them, and in 1886 noted they were the larva of a two-winged fly, which he called *fungus gnat*.

1.7   Twentieth Century: the Brightest Era

In the twentieth century the "glow" of luminescence was considerably brightened by the discovery of many new CL compounds, as well as by the results of many studies that provided a scientific explanation for what was once considered magical phenomena, including the biochemical mechanisms of BL systems in terrestrial and marine organisms. In addition, numerous chemical partners involved in BL systems were identified, and the complex structure of luciferin was elucidated. With the maturing of protein chemistry and molecular biology, the structure of many luciferase enzymes were determined and, more recently, these proteins have been cloned.

The basic studies on these naturally occurring luciferins, carried out to understand their fine chemical mechanisms, stimulated research in synthetic chemistry to develop new molecules that carry groups similar to those present in luciferin. These studies aimed to produce molecules able to generate BL through a chemically generated excited state mimicking those of naturally occurring luciferins.

In 1905 Trautz published a review of known CL and BL reactions, attributing them to the presence of "active oxygen." He also introduced the concept of ultraweak or cellular CL, a phenomenon that is not only widespread among oxidations of organic substances in non-aqueous solvents in which short-lived free radicals are present but can also be observed during the interaction of oxygen with stable radicals or other oxygen-sensitive radicals in living organisms.[11]

Interesting studies were performed by Coblentz in 1912 regarding the relationship between evolution and specific BL features in a given BL species.[12] For example, insect BL has evolved to allow insects to signal their sexual availability to potential mates at night. Coblentz also tried to explain the BL intensity as a function of wavelength, and showed that *Photinus pyralis* has the greatest recorded light intensity, 1/50 that of a sperm candle. He used a photograph taken with a spectrograph and a photographic plate that was most sensitive at 590 nm.

Another important discovery in the twentieth century was the surprising ability of luminol (5-amino-2,3-dihydro-1,4-phthalazinedione) to yield a CL reaction in basic solution in the presence of an oxidant such as hydrogen peroxide – first reported in 1928 by Albrecht[13] and its CL properties first used at a crime scene in 1937 in Germany.[13] Luminol was first synthesized by Teichmann in 1853[14] but was used just as a dye for blood detection until the discovery of its CL. Luminol is probably the most popular CL compound studied to date, and it appears to have applications in many fields of analytical bioscience (see below). The *luminol test* for blood detection is still used in forensic science. It has become even more important recently, thanks to the use of ultrasensitive CCD cameras that allow imaging of even tiny traces of blood. In the same period the first synthesis of an acridinium salt was reported by K. Lehmstedt and E.W. Wirth.[15] Later, in 1935, K. Gleu and W. Petsch described the synthesis of lucigenin (bis-*N*-methylacridinium nitrate).[16]

The extraordinary growth in luminescence research that characterized the period after World War II was mainly due to the development of new instrumentation, such as photomultipliers and Vidicon imaging tubes (both around 1950), as well as lasers (about 1960). Progress in microelectronics in the early 1970s and the availability of the first laboratory computers were extremely advantageous (around 1975). In the second half of the twentieth century, following the discoveries and intuitions of the early 1900s, the main basic aspects of BL and CL have been reinvestigated and reinterpreted.

1.7.1 Bio- and Chemiluminescence in the United States and the Americas

In the United States in the first 20 years of the twentieth century the main contributions to the study of BL were almost exclusively those of E. N. Harvey at Princeton University. Harvey's first paper "On the chemical nature of the luminous material of the firefly" was published in *Science* in 1914.[17] He continued his studies, usually publishing alone, until 1929. He worked on different BL systems, including those of *Cypydina* luciferase/luciferin, fish with a luminous organ designed for the growth of luminous bacteria, and the West Indian elaterid beetle *Pyrophorus*. In 1928, Harvey in a paper with K. P. Stevens[18] wrote:

"The West Indian elaterid beetle, Pyrophorus, is one of the brightest of luminous organisms. Pickering, by comparing its light with stars of various magnitudes, determined its intensity to be .004 candle."

He measured the basic photophysical properties with a modified *Macbeth illuminometer*. Harvey is considered to be the father of the modern BL. After World War II, work on BL and CL was carried out mainly by his students or collaborators. Chronologically, Harvey's "inheritors" were F. H. Johnson, who began his activity in 1949, followed by W. D. McElroy in 1951 and then M. J. Cormier and F. I. Tsuji in 1954.

F. H. Johnson continued independent work on *Cypridina* until 1961, when O. Shimomura joined his laboratory. In 1962 they published a paper on the extraction, purification and properties of aequorin, a bioluminescent protein from the hydromedusa *Aequorea*.[19] Before joining Harvey's laboratory, Shimomura was a fellow of professor Hirata at Nagoya University in Japan, where he crystallized the *Cypridina* luciferin. In 1965 Shimomura went to New Zealand to study two kinds of bioluminescent organisms, the cave worm *Arachnocampa* and the freshwater limpet *Latia*. Aequorin was named a *photoprotein* by Shimomura in 1971 since it required only Ca^{2+} and not oxygen for light emission.[20] In 1974 Shimomura and H. Morise published an important work on intermolecular energy transfer in the bioluminescent system of *Aequorea*.[21] When Johnson retired from Princeton in 1977, Shimomura moved to a marine laboratory, but before leaving Princeton he elucidated the structure of the chromophore of GFP. He also studied dinoflagellate luciferin and luminous scale worms. In 1981 he moved to the Marine Biological Laboratory (MBL) in Woods Hole, Massachusetts. In 1992, the cDNA of GFP was cloned by Douglas Prasher, who was then at the Woods Hole Oceanographic Institution. At that time it was commonly believed that expressing the GFP cDNA in living organisms would not produce fluorescent GFP because formation of its chromophore requires condensation and dehydrogenation, reactions that are not expected to occur spontaneously. In 1994, however, when Martin Chalfie of Columbia University tried to express the cDNA in *E. coli* and in a nematode worm, much to his surprise the expressed GFP fluoresced. The results suggested that GFP and similar proteins could be expressed in living organisms. Chalfie's research attracted the interest of many people, triggering rapid development of applications of GFP. Roger Tsien of the University of California, San Diego, engineered GFP by modifying the amino acid residues surrounding the chromophore, thus producing many different fluorescent proteins that emit various colors, from blue to red. Today, GFP is widely used as a fluorescent marker of proteins and cells, and it has become an essential tool in the study of biology, physiology and medicine. The range of applications of the fluorescent proteins is beyond imagination. O. Shimomura, M. Chalfie and R. Tsien were awarded the Nobel Prize for chemistry in 2008 for the discovery and development of GFP (Figure 1.12).

W. McElroy (Figure 1.13), another great collaborator of Harvey, first became interested in BL as a graduate student at Stanford University's Hopkins Marine Station at Monterey, where he took a course under the microbiologist C. B. Van Niel.[22] His interest was further stimulated when he went to Princeton University to finish work on his doctorate. There he met Harvey, an acknowledged giant in biochemistry and BL, who was to become his mentor. After a joint publication with Harvey, he carried out independent work and many distinguished scientists joined his laboratory over the years. Perhaps the two most significant modifications of the basic luciferin/luciferase reaction were

Figure 1.12   Osamu Shimomura (left), Martin Chalfie (middle) and Roger Y. Tsien (right) at their interview with Nobelprize.org in Stockholm, 6 December 2008. (Photograph Merci Olsson "© The Nobel Foundation.")

Figure 1.13 At the 5th International Symposium of the International Society for Bioluminescence and Chemiluminescence (ISBC), held in Florence and Bologna, the University of Bologna awarded the Laurea Honoris Causa in Pharmaceutical Chemistry to the distinguished scientist William McElroy for his work on firefly luciferase machinery.

identified by McElroy, one even before he had been made full professor at Johns Hopkins University. In his landmark 1947 paper entitled "The energy source for BL in an isolated system" McElroy demonstrated that adenosine triphosphate (ATP) is an essential requirement for the *in vitro* reaction of firefly BL.[23] In 1951 McElroy reported the observation that adding adenosine triphosphate (ATP), a high-energy compound found in all living cells, to samples of ground fireflies caused a brilliant flash of light to appear immediately, then persisting for a considerable time, the length of which depended upon the concentration of the ATP. In 1953, McElroy along with Hastings and colleagues found that in bacterial luminescence luciferase catalyzes the oxidation of a reduced form of luciferin, namely, flavin mononucleotide ($FMNH_2$) – a very important compound in cellular respiration – and a long-chain aldehyde.[24] Virtually every application of BL in industry and research depend on the fact that ATP and $FMNH_2$ are the limiting factors for the amount and intensity of light radiated from a sample. Thus, each application owes at least some debt to McElroy and Hastings for their discoveries. In 1955 McElroy's group isolated crystalline firefly luciferase after purification.[25]

In the late 1950s, H. H. Seliger joined McElroy's laboratory and, as explained in a shared paper in 1960, by measuring the quantum yield they demonstrated that the emission efficiency of the oxidation of firefly luciferin was near unity.[26] Selinger's contributions to McElroy's group were particularly outstanding. In 1961 he, in collaboration with W. D. McElroy, E. H. White and G. F. Field, carried out interesting studies on the stereo-specificity of the firefly BL reaction by comparing the reactivity of natural and synthetic luciferins.[27] In 1962, he developed an underwater photometer for day and night measurements of the light emission from marine dinoflagellates.[28]

E. H. White continued to work with Seliger on the spectroscopic properties of firefly luciferin and related compounds, in addition to excited state formation mechanisms.[29] The structure and synthesis of firefly luciferin was reported by E. H. White, F. McCapra, G. Field, and W. D. McElroy in 1961.[30] White, as an organic chemist, continued his work on synthetic analogues of firefly luciferin.[31] Later, in 1975, B. Branchini began to work with him on luciferin analogues with the aim of switching the light emission colors.[32] He continued to work on BL until 1991.[33] Branchini continued his independent studies of luciferin analogues and their reaction mechanisms, identifying a firefly luciferase active site peptide in 1997.[34] In 1999 he proposed a model for explaining how BL color is determined, based on his site-directed mutagenesis studies of firefly luciferase active site amino acids.[35] Red- and green-emitting firefly luciferase mutants for bioluminescent reporter applications have been reported recently.[36] More recently, he cloned luciferase from the Italian firefly *Luciola italica* in collaboration with A. Roda's group in Bologna.[37]

Seliger continued to work on the spectral distribution of firefly light and, in 1965, was joined by J. Lee, who developed a phototube-based system to measure the absolute quantum yields of CL and BL reactions.[38] Lee continued to work on bacterial BL and in 1975 published a study on equilibrium association measurements, quantum yields, reaction kinetics and overall reaction schemes. He also carried out some interesting work on lumazine, starting with a study published in 1985 in collaboration with D. J. O'Kane and V. A. Karle.[39] Subsequently, Lee performed extensive studies on lumazine protein and excitation mechanisms in bacterial BL.[40] In 1970 Lee and Seliger also made important contributions regarding the spectral characteristics of the excited states of luminol CL products.[41]

W. W. Ward began work on BL with Seliger's group on calcium-activated photoproteins in 1974.[42] In collaboration with Cormier and Seliger, Ward carried out important research on *Renilla* and *Renilla*-like luciferases.[43] He subsequently studied *Aequorea* green fluorescent protein and as part of a collaborative work determined the primary structure of the *Aequorea victoria* green-fluorescent protein in 1992.[44]

J. W. Hastings is another BL pioneer who worked with E. Newton Harvey, beginning in 1948. Under Harvey's tutelage he developed techniques to quantify the oxygen requirement in the luminescent reaction of various species.[45] In 1951 Hastings joined the laboratory of William McElroy just after his discovery that light emission in firefly extracts required ATP. With purified luciferase Hasting obtained evidence that oxygen gating is the mechanism for firefly flashing.[46] In McElroy's laboratory, he also investigated luminous bacteria, discovering that a flavin is a substrate for their luciferase.[47] Bacterial luciferases would later become one of the major subjects of Hastings' research. In 1953, he accepted a faculty position in the Department of Biological Sciences at Northwestern University, Evanston, IL. Here, in collaboration with B. M. Sweeney, he demonstrated that *Gonyaulax* had a circadian rhythm of BL.[48] In 1963 he published a work about intermediates in the BL oxidation of reduced flavin mononucleotide.[49] In 1966, Hastings joined the faculty of Harvard University as Professor of Biology.

Continuing his work at Woods Hole, he carried out studies on coelenterate luminescence systems with a graduate student, J. Morin. They observed that in these systems some species emit green light *in vivo* but blue light in isolated enzyme system. In *Obelia* sp., they found that a green fluorescent protein, previously discovered in *Aequorea* sp. and which they dubbed GFP, served as a secondary emitter by virtue of energy transfer from the luciferase-bound excited state.[50] Work on the dinoflagellate *Gonyaulax* continued and, in 1970, Hastings' research provided some of the first evidence for communication among bacteria by means of a process that he called quorum sensing.[51] According to Hastings:

" . . . *the bacteria were producing and releasing into the medium a substance that turned on transcription of specific genes that had been repressed.*"

This occurred only when the concentration of this substance, which they called an *autoinducer*, reached a critical level. At that time, however, the existence of this process was widely disbelieved or ignored, and sometimes derided. Hastings' work with bacterial systems continued throughout the 1980s and by the early 1990s his concept of quorum sensing gained acceptance. More recently, Hastings has been studying another dinoflagellate species, the heterotroph *N. scintillans*, for which he successfully cloned its luciferase gene. Alerted to a red tide of *Noctiluca* in the Gulf of Mexico, his team immediately went to harvest them: "Wow, did we get organisms" exclaims Hastings:

"In a very few minutes, the nets were teeming with Noctiluca and almost nothing else. We brought the frozen cells back to Harvard, but the cloning was not straightforward."

Hastings *et al.* reported that *Noctiluca* luciferase has only a single catalytic domain, lacking residues responsible for pH control. Interestingly, a second domain of the gene codes for a luciferin-binding protein-like sequence, which in *Gonyaulax* occurs as a separate gene. Thus, a single protein in *Noctiluca* appears to possess both catalytic and substrate binding properties that in other species occur as separate proteins.[52]

In 1970 the Canadian E. A. Meighen joined Hastings' laboratory and started to work on the structure of bacteria luciferase subunits,[53] participated in its purification and defined the structural requirements of the flavine substrate of the luciferase.[54] Later, he returned to Canada, where he continued his studies on the chemistry of bacterial luciferase[55] and eventually succeeded in expressing the subunits of bacterial luciferase in *Escherichia coli*.[56] Meighen's work was fundamental for the use of the *lux* gene in molecular biology. He is still active, and in recent years he has reported the development of bright stable BL yeast using bacterial luciferase. He has suggested that bacterial luciferase could be the light-emitting sensor of choice in eukaryotic organisms.[57]

In 1974, T. O. Baldwin joined Hastings' laboratory and began to work on the inactivation of bacterial luciferase mechanisms by various substrates.[58,59] In 1983 he reported the cloning of the luciferase structural genes from *Vibrio harveyi* and its expression in *E. coli*.[60,61] Baldwin continued his intense scientific activity along with his wife M. Ziegler, trying to elucidate the role of lysine residue folding on luciferase activity. He had an important collaboration with the N. Ugarova's group in Moscow on the cloning of luciferase from the firefly *Migrelica* in 1993.[62]

In 1970 K. H. Nealson joined Hastings' laboratory and worked with Hastings on the cellular control of the synthesis and activity of the bacterial luminescent system.[63] Subsequently, in 1977, he started independent work on bacterial luciferase autoinduction[64,65] and later on the cloning of the *Vibrio harveyi* luciferase genes.[66]

In 1986 A. A. Szalay begin to work on gene technology in Baldwin's laboratory, using bacterial luciferase.[67] In 1996 he joined Cormier's group, where he studied *Renilla reniformis* luciferase,[68] and, in more recent years, he has focused his research activity on BL molecular imaging.[69]

Another Harvey fellow is M. J. Cormier, who in collaboration with B. L. Strehler demonstrated in 1954 the requirement for long-chain aldehyde (luciferin) in bacteria luminescence.[70] Cormier continued to work on bacterial BL and in 1973 published an important paper on the mechanism and biochemistry of coelenterate BL.[71] A more important contribution from Cormier was the cloning and expression of the cDNA coding for aequorin, which he did in collaboration with D. Prasher and R. O. McCann. This discovery paved the way for the recent Nobel Prize for Chemistry awarded in 2008 to Shimomura, Chalfie and Tsien. As the laureates reported in their lecture, Prasher in 1992 was the first to delineate the primary structure of the *Aequorea victoria* green fluorescent protein.[72] However, the debate with Cormier and other scientists involved GFP is still active, as reported recently by Ward, and therefore it is very difficult to write the true story.[73] In 1996 Cormier made another important contribution in terms of *Renilla reniformis* luciferase gene expression in mammalian cells.[74]

A distinguished scientist from Japan, F. I. Tsuji, worked in Harvey's laboratory from 1954 to 1966, studying the luminescence of *Cypridina* luciferin,[75] in particular the mechanisms and the kinetic of the reaction; he also purified the enzyme and determined its molecular weight.[76] In the 1970s Tsuji started to work on other bioluminescent organisms; with Y. Haneda he studied light production in the luminous fish *Photoblepharon* and *Anomalops*,[77] and the luminescent system in a myctophid fish, *Diaphus elucens Brauer*, among other animals. In 1972 in collaboration with J. F. Case he studied BL in the marine teleost.[78] During his career Tsuji collaborated with M. J. Cormier, J. E. Wampler and M. Deluca. In 1985 he published the cloning and sequence analysis of cDNA for the luminescent protein aequorin.[79] In the 1970s J. F. Case carried out important studies on the mechanism of New Guinea firefly flash light time and synchronism connected with brain oscillators;[80] he also worked on the anatomy of the luminous organ of various BL fish.[81] In 1980 E. A. Widder and M. I. Latz joined his laboratory, adding their relevant contributions to the study of marine BL.[82] The most comprehensive web page on bioluminescence, from basic principles to an accurate description of BL in marine organisms, is managed by J. F. Case at UCSB in collaboration with S. Haddock of the Monterey Bay Aquarium Research Institute.[83]

After returning to Japan, F. I. Tsuji continued intense and fruitful scientific activity, including a study published in 1991 with Y. Ohmiya on the structural requirements for BL in aequorin.[84] Tsuji and Haneda both worked with *Quantula striata*, regarded as the only land gastropod in the world capable of true BL. This snail can be found fairly easily in Singapore, where it was first discovered in September 1943 by Kumazawa (an entomologist) at the Good-wood Park Hotel, Scotts Road. This snail is normally encountered in secondary forests, lawns, rubbish dumps, under concrete slabs and also in crevices along sidewalks, especially after rain. Because of its curious and unique luminous character, it is probably the most studied land snail species.[85] The luminous organ of this snail, known as the *organ of Haneda*, sends flashes of yellow-green light from beneath the mucous fold of its head in juvenile stage. Another feature is that the mature snail sometimes loses the organ permanently. A social role that has been suggested for this phenomenon is that it may help young snails to find sources of food.[86] Tsuji's group reported recently the redshifted emission of marine copepod *Chiridius poppei*.[87] In 1979, M. Maeda joined the Tsuji's laboratory, helping to develop many applications of BL systems as labels for immunoassays and related techniques.[88,89]

In the late 1970s R. Hart joined Cormier's laboratory, where he studied the mechanism of the enzyme-catalyzed BL oxidation of coelenterate-type luciferin and the luciferase-catalyzed production of non-emitting excited states from luciferin analogues; in addition, he contributed to the elucidation of the excited-state species involved in energy transfer to *Renilla* green fluorescent protein.[90,91]

In 1960, F. McCapra carried out his first postdoctoral year at Johns Hopkins University, working in McElroy's laboratory with Emil White on delineating the structure of firefly luciferin as well as its synthesis.[92] Later, at the University of British Columbia, he performed many studies on new CL molecules, discovering in 1964 the CL of acridinium esters.[93] These compounds were also independently synthesized by Rauhut.[94] Their CL mechanism is also the basis for the mechanism of light emission from firefly luciferin: the reduced form of acridinium esters, an acridan, provides an excellent model for the luciferin. Various analogues have been synthesized, particularly by the group of scientists at Cyanamid Company, for the development of oxalate ester series.[95] In 1967 McCapra studied many luciferin analogues to better elucidate the BL mechanism.[96] After returning to England he became Professor of Organic Chemistry, in 1980, at the University of Sussex. He concentrated his research on mechanisms of CL and BL, determining the most effective route to light emission in organic compounds, the reactions of dioxetanes. McCapra provided the now fully accepted mechanism of light emission in the better-known luminescent organisms.

Dioxetanes were first described by Blitz in 1897.[97] During the 1960s, White, McCapra and Rauhut predicted that they were the critical intermediates in the CL reactions of lophine, acridinium salt, indoles, peroxylates and several luciferins. However, 1,2-dioxetanes until recently could not be implemented into the design of bioassays because they were too thermally labile.[98] Since the first report of the chemical synthesis of a stable 1,2-dioxetane[99] numerous derivatives have been synthesized, offering a wide range of thermal stabilities.[100,101] The most thermostable 1,2-dioxetanes described so far are derived from sterically hindered adamantylidenadamantane.[102,103] Enzyme-cleavable dioxetanes that produced light upon reaction with enzymes were described in 1987,[104] and their use in immunoassays for the detection of enzymatic labels (alkaline phosphatase and β-galactosidase) was described by P. Schaap and I. Bronstein.[105,106] Subsequently, the commercial potential of the technology was shown for nucleic acid probing and sequencing as well as in both nucleic acid and protein blotting. Nowadays, 1,2-dioxetanes have widespread application in molecular biology, immunology and biotechnology.[107,108]

M. DeLuca joined McElroy's laboratory in 1964, beginning a period of intense and productive scientific activity during which she collaborated with almost all of the scientists present in the laboratory. She initially worked with McElroy, then with Cormier until 1971, on the mechanism of *Renilla reniformis* BL.[109] This was followed by work with Selinger on the dehydroluciferin–luciferase complex.[110] She collaborated with Tsuji on the mechanism of the enzyme-catalyzed oxidation of *Cypridina* luciferin[111] until he moved to the University of California in San Diego. In 1980, taking advantage of the availability of the first optimized BL reagents, she began to work on the application of BL in diagnostics and analytical chemistry, in collaboration with J. Wanlund,[112] L. Kricka and A. Roda.[113] The most important studies carried out in Marlene DeLuca's laboratory regarded the cloning of firefly luciferase, which paved the way to modern molecular biology using report gene technology. In particular, in 1984 in collaboration with K. Wood, she synthesized an active firefly luciferase by *in vitro* translation of RNA from their lanterns.[114] One year later, the cloning of firefly luciferase in *E. coli* was reported.[115] In 1986 her laboratory obtained a transient and stable expression of the firefly luciferase gene in plant cells and transgenic plants (Figure 1.14),[116] and in 1987 the firefly luciferase gene was also expressed in mammalian cells.[117] In the November of

Figure 1.14 A luminescent tobacco plant bearing the firefly luciferase gene. It was the first picture of a transgenic multicellular organism expressing bioluminescence and has become an iconic image of genetic engineering. (Reproduced from ref. 116, with permission.)

1987 Marlene DeLuca died prematurely, just after she had made her most important discoveries, those that truly revolutionized the study of molecular biology. In subsequent years the *luc* gene became a fundamental tool in gene reporter technology, superior even to GFP. As mentioned above, the ISBC dedicated a Symposium to her memory in 1988, held in Florence and Bologna, and given the value of her prolific studies a special award in her memory for young scientists has been made during the biennial ISBC Symposium, sponsored by F. Berthold Company; this honor was well-deserved.

Other scientists, such as M. I. Latz from the San Diego Scripps Institution of Oceanography and E. A. Widder from Santa Barbara, have made important contributions to marine BL.[82] More recently, Latz has developed experimental approaches for the interpretation of dolphin-stimulated BL,[118] and he has studied the hydrodynamic stimulation of dinoflagellate BL.[119]

The use of CL labels such as isoluminol or acridinium ester derivatives (ABEI) for the development of sensitive immunoassays was extensively studied by H. R. Schroeder, who already in the 1970s developed the first immunoassay using isoluminol as a label to monitor serum biotin.[120–122] In those years, researchers also investigated *low level* (or *ultraweak*) *luminescence*. Such emission is due to biochemical pathways inside biological structures and organelles that generate light, *i.e.*, enzymatic systems and free-radical mediated reactions producing electronically excited species. G. Cilento from Brazil was particularly active in this area of study.[123–125] Ultraweak CL is involved, for example, in oxidative radical reactions and is associated with phagocytosis by polymorphonuclear leukocytes. Pioneers in this field were E. Cadenas[126–128] and R. C. Allen.[129,130] During the past 20–30 years, the International Society for Bioluminescence and Chemiluminescence has devoted important conference sessions in its Symposia to clinical applications of ultraweak CL.

Peroxyoxalate CL was first reported in 1967 by M. M. Rauhut for the reaction of diphenyl oxalate.[131] This CL reaction, in which the emission is generated by an oxalate ester in the presence of hydrogen peroxide and a suitably fluorescent energy acceptor, is used for example in glow sticks. The three most common aryloxalates used in peroxyoxalate CL are bis(2,4,6-trichlorophenyl)oxalate (TCPO), bis(2,4,5-trichlorophenyl-6-carbopentoxyphenyl)oxalate (CPPO) and bis(2,4-dinitrophenyl)oxalate (DNPO). Other aryloxalates have been synthesized and evaluated with respect to their possible analytical applications.[132]

In the late 1990s P. R. Contag and G. H. Contag introduced the use of *in vivo* BL imaging of gene expression, employing different BL mammalian cells to study tumor growth, tumor metastases and various metabolic processes.[133–135]

Important advancements came from S. Daunert regarding the production of new BL enzymes, analytical application of coupled BL reactions, use of aequorin as a label in immunoassays and the development of BL cell-based biosensors.[136,137] She is the editor of a recent book dealing with BL and photoproteins that includes many important contributions to this field.[138]

In the late 1990s M. Adamczyk from Abbott Laboratories synthesized many acridinium ester analogues and derivatives to be used for direct DNA labeling and in immunoassays.[139,140]

1.7.2   Bio- and Chemiluminescence in Europe

In Europe, the study of BL and CL became particularly active in 1980, when F. McCapra returned to England from the United States. McCapra and his collaborators synthesized many CL compounds and studied their application in bioanalytical chemistry. In 1979, in collaboration with A. K. Campbell, McCapra proposed the use of acridinium esters as sensitive labels in immunoassays. Campbell too was very active in the field of applied CL and BL.[141] In 1979, J. Simpson reported the first application of a CL labeled antibody in immunoassays.[142] Also in England, P. J. Herring of the Institute of Oceanographic Sciences in Surrey carried out much work on the distribution of BL in living organisms, and in particular in marine BL animals, including study of the filters in the ventral photophores of mesopelagic animals.[143,144] He also investigated the far-red BL from two deep sea fish in collaboration with

E. A. Widder, M. I. Latz, and J. F. Case.[145] In 1979, T. P. Whitehead, in collaboration with L. J. Kricka, T. J. Carter and G. H. Thorpe, was the first to introduce BL and CL methods into the field of clinical chemistry.[146] In 1983, L. Kricka went back to England and, in collaboration with Whitehead, G. H. G. Thorpe, T. J. Carter and C. Groucutt, discovered *enhanced chemiluminescence*, *i.e.*, they found the enhancing action of firefly luciferin on the luminol/H_2O_2 CL reaction catalyzed by peroxidase.[147] Two years later they reported that phenol derivatives can be also used as enhancers.[148] Thanks to the use of enhancers, which allow detection of horseradish peroxidase with high sensitivity and stable light signal, this enzyme became the most used CL label in immunoassays and gene probe assays. Enhanced CL is the basis for many popular CL applications still in use today, from sensitive imaging of gels to miniaturized analytical devices based on CL detection. Important advancements in BL and CL also came from P. E. Stanley, who was involved in the field much earlier: in 1969 he worked on instrumentation for the determination of ATP utilizing a luciferase BL system[149] and then, in 1971, he was involved with the determination of sub-picomole levels of NADH and FMN using bacterial luciferase and a liquid scintillation spectrometer.[150] He is the author of many reviews on BL and CL related topics, including instrumentation and commercial kits, and in general he comments on the BL and CL literature. Since 1978, Stanley was co-editor of almost all the Proceedings of the ISBC Symposia,[4] and he also organized some of the Symposia held in Cambridge.

The group of molecular biologists from Nottingham, composed of G. S. Stewart and P. J. Hill, was also very active. In 1987, Stewart carried out important studies on the use of bacterial luciferase as a promoter probe for real-time analysis of gene expression in *Bacillus* spp.[151] He then studied BL quorum sensing and other techniques until 1999, when he suddenly died at 48 years of age. On his side, Hill continued the activity, working on *lux* gene technology, in particular on the engineering of the *luxCDABE* gene.[152,153]

In Sweden, in the late 1970s S. E. Brolin developed several BL amplified enzymatic methods for different analytes.[154] A. Lundin performed many careful studies on ATP BL detection and measurement, including the development of an original luminometric ATP assay. In 1987 he founded a company for the commercialization of BL reagents and ATP assay kits.

In Belgium, E. Schram studied the application of BL and CL in the clinical laboratory.[155] In collaboration with the group of A. M. García-Campaña in Granada, Spain, W. Baeyens investigated the use of CL in separation techniques.[156] A book, jointly edited by these later two authors in 2001, presents the most relevant CL methods and techniques in analytical chemistry.[157] Important achievements in analytical applications of CL were also achieved by J. De Boever, F. Kohen and J. B. Kim, who reported the use of CL tracers such as isoluminol and ABEI.[158,159]

The French researchers J. Mallefet and F. Baguet have been working in Louvain since 1984 on the mechanisms of BL in the luminous organs of mesopelagic fishes from the Strait of Messina and other places around the world.[160] Important contributions have also came from J. C. Nicolas in

Montpellier regarding the development of amplified BL methods for steroids: in 1983 he reported a BL assay for the determination of estrone and estradiol at femtomolar levels.[161] In subsequent years, Nicolas continued his research activity, working on immunoassays, BL gene probe technology and low light imaging.[162] Finally, P. R. Coulet and L. J. Blum in Lyon have worked on highly sensitive CL biosensors and related devices.[163]

In The Netherlands, G. Zomer developed in 1984 CL immunoassays for steroids, employing a new enzyme label system that was superior to all CL labels used in immunoassays reported to that time.[164]

In Germany, in the early 1980s, at the Institute of Organic Chemistry of the University of Wurzburg, W. Adam carried out many studies on the CL electron-transfer mechanisms of dioxetanes and related molecules.[165] Collaboration between the scientific community and the instrumentation company directed by F. Berthold was very profitable in terms of an exchange of ideas that resulted in the commercialization of several sensitive instruments for light measurement. The company was very active in the promotion and support of BL and CL sciences and particularly the ISBC Symposia.

In Italy, research on BL and CL has mainly involved three groups. In Florence, the group headed by M. Pazzagli developed, in collaboration with G. Messeri, sensitive CL immunoassays using different labels such as isoluminol and acridinium esters.[166,167] At the University of Bologna, A. Roda and S. Girotti achieved important advancements in the field of bioanalytical BL and CL, working on flow-based assays for the determination of several analytes using co-immobilized BL enzymes, CL enzyme immunoassays, as well as in the application of BL and CL imaging techniques in immunohistochemistry and *in situ* hybridization.[168–170] In Rome, since 1982, P. De Sole has studied the resting and stimulated CL of polymorphonuclear leukocytes and has investigated the potential clinical and diagnostic relevance of such phenomenon.[171]

In Greece, at the University of Athens A. Calokerinos developed CL based analytical technologies[172] and T. K. Christopoulos in Thessaloniki worked on BL hybridization assays and other BL-based molecular biology techniques.[173]

In Israel, S. Ulitzur was particularly active in the field of environmental monitoring based on BL and CL, reporting several toxicity assays using BL marine bacteria and genetically modified BL *E. coli*.[174,175]

1.7.3 Bio- and Chemiluminescence in Russia

Research on BL and CL in Russia has been very active since the early 1960s, but – at least at that time – most publications were in Russian and therefore were not easily accessible to the international scientific community. A. Yu. Vladimirov was a pioneer in the study of the ultraweak luminescence originating from biochemical reactions.[176] In 1962 R.F. Vasil'ev studied the mechanisms of CL excitation and oxidation.[177] In more recent times, the research group of the Moscow University headed by N. N. Ugarova, including L. I. Brovko, O. V. Lebedeva and I. V. Berezin, have made important advancements in both BL and CL, including the study of the firefly *Luciola migrelica* and the development of BL and CL applications in bioanalysis and microbiology.[178,179] In 2000, L. I. Brovko moved to Canada, were she continued to work on the application of BL in microbiology and in food analysis.[180] A. Egorov in Moscow studied peroxidase-catalyzed CL reactions and in 1994 he cloned the enzyme. Afterwards, he worked on CL immunoassays and biosensors.[181–183] Another important research group was founded in Krasnoyarsk by J. I. Gitelson. This group was particularly active in the isolation and characterization of BL bacteria, especially from marine environments. Since the late 1970s, V. A. Kratasyuk and N. S. Kudryasheva have reported many interesting applications of these BL bacteria.[184–186]

1.7.4 Bio- and Chemiluminescence in Japan

In Japan research on BL began in 1900, when Sakyo Kanda, a true pioneer of BL science, described this phenomenon for the first time in his country. Many of the early publications were in Japanese and thus are not easily accessible to most potential readers, as was true of early Russian studies. I am grateful to Y. Ohmiya, who sent me a list of the principal milestones in BL and CL research in Japan in the twentieth century (Table 1.1). The list does not include the scientific activity of F. I. Tsuji, which has been reported already in Section 1.7.1.

1.7.5 Bio- and Chemiluminescence in China

The ancient Chinese involvement in BL and CL has been reported above; we have little information on the scientific activities in recent years, probably because most scientific reports in Chinese were never translated into English. In contrast, at present, and particularly in the last 20 years, much research activity carried out in China has been documented in many scientific publications in international journals, for both BL and CL.

1.8 Suggested Further Reading

In addition to Harvey's book,[2] another important comprehensive book that covers historical aspects of BL and CL is *Chemiluminescence: Principles and Applications in Biology and Medicine* by A. K. Campbell.[141] A large section of this volume is devoted to BL phenomena, with discussion of the most important milestones in its discovery and application. A more recent text edited by W. R. G. Baeyens and A. M. Garcia-Campaña in 2001[157] gives a detailed history of the most important CL molecules as well as a review of their principal applications in analytical chemistry. Another source of data regarding BL and CL is the Proceedings of the biennial ISBC Symposia. Since 1978, these Symposia have been the most important international conferences for researchers in this field; updated information on both basic aspects and recent

Table 1.1   Principal milestones of the BL and CL research in Japan in the twentieth century.

Year	Milestone
1905	S. Watase named the firefly-squid in Toyama bay *Watasenia scintillans* and identified the luminous organism
1916	E. N. Harvey visited Toyama to study the BL mechanism of firefly-squid and the sea-firefly *Cypridina hilgendorfii*
1926	Y. Yasaki published his study "Bacteriological studies on bioluminescence II. On the nature of the new luminous bacteria, *Microspira phosphorctim Yasaki*"
1928	Y. Yasaki identified the luminous bacterium in *Knight fish*
1935	S. Kanda published "Firefly", in which he reported biology, chemistry, ecology and literature concerning Japanese firefly
1939	Y. Haneda described the luminous mushroom in the South Pacific
1957, 1965	F. H. Johnson visited Japan to study the sea-firefly *Cypridina hilgendorfii* and tried to identify its luciferin
1957	O. Shimomura and Y. Hirata first succeeded in the isolation of crystalline *Vargula* luciferin
1962	O. Shimomura and F. H. Johnson succeeded in the isolation of a photoprotein, named aequorin. They found that aequorin generates blue light on activation with Ca^{2+} cations. They also isolated concurrently green fluorescent protein, GFP. Later, GFP was found to be the true light emitter in the jellyfish body via a sensitized BL process
1966	T. Goto, in collaboration with Y. Kishi, O. Shimomura and Y. Hirata, determined the structure of luciferin and oxyluciferin
1968	T. Goto isolated and identified the Japanese firefly luciferin, which was found to be identical to the luciferin of the North American firefly
1969	Y. Haneda and American researchers traveled to New Guinea to search for luminous organisms ("Alpha Helix New Guinea Expedition")
1971	T. Goto isolated the proposed light emitter oxyluciferin from the lantern of the Japanese firefly and determined its structure by chemical synthesis in collaboration with his associate N. Suzuki
1974	T. Goto and K. Okada found that the firefly luciferin could be generated from the nitrile, a decomposition product of oxylyciferin, by reaction with cysteine. They isolated ^{14}C-labeled luciferin from fireflies in which the ^{14}C-labeled nitrile compound was injected
1975	O. Shimomura proposed an imidazopyrazinone compound named coelenterazine as the BL substrate of aequorin
1975	T. Goto and S. Inoue isolated a new imidazopyrazinone compound, named *Watasenia* luciferin, from the liver of the squid *Watasenia scintillans* and studied its structure by spectroscopic methods coupled with chemical synthesis. On the basis of their results, they proposed that *Watasenia* luciferin is coelenterazine disulfate
1976	T. Goto and K. Okada demonstrated by incorporation experiments of labeled compounds that in fireflies the firefly luciferin was biosynthesized from benzoquinone or hydroquinone. Almost at the same time, F. McCapra carried out similar incorporation experiments using labeled cysteine
1978	T. Goto and S. Inoue isolated *Watasenia* luciferin from the photophores of the squid and established its structure by chemical synthesis

applications of BL and CL can be found in the Proceedings. The complete list of the Symposium volumes, along with those of earlier conferences, starting with the "Faraday Society's Discussion on Luminescence" held in Oxford in 1938 and the "Conference on Luminescence" held in Monterey, CA, in 1954, can be downloaded from the ISBC web page.[4] Notable books that focus mainly on CL are also the "Critical reviews" edited by Knox Van Dyke, dealing with topics such as instrumentation and application,[187] cellular luminescence[188] and immunoassays and molecular biology-based assay.[189] Other information sources are the three volumes of *Methods in Enzymology* completely devoted to BL and CL.[190–192] *Luminescence: The Journal of Biological and Chemical Luminescence* (formerly *Journal of Bioluminescence and Chemiluminescence*), published by John Wiley, publishes scientific papers on fundamental and applied aspects of all forms of luminescence, including BL and CL.[193] Over the years, L. J. Kricka (editor-in-chief of the journal) and P. E. Stanley have written many articles dealing with the overall BL and CL literature, reporting the new advancements in basic and applied sciences.

1.9 Conclusions

For millennia mankind has been fascinated by the eerily beautiful light of fireflies, glowworms, marine animals and other bioluminescent creatures. Primitive understanding of these phenomena was limited to terms of magic and superstition. Paralleling the course of development in other scientific fields, methodological study of luminescence began in the 1400s, with subsequent gradual advancement until nearly exponential growth in our fund of knowledge in the last century and a half or so. Today, while we are still exploring the intricacies of the mechanisms involved in the production of cold light, the practical applications of BL and CL seem endless. Fortunately, we can still be charmed by these sometimes beautiful but always fascinating phenomena.

References

1. M. Pazzagli, E. Cadenas, L. J. Kricka, A. Roda and P. E. Stanley, *Bioluminescence and Chemiluminescence: Studies and Applications in Biology and Medicine*, John Wiley and Sons, Chichester, 1989.
2. E. N. Harvey, *A History of Luminescence from the Earliest Times until 1900*, American Philosophical Society, Philadelphia, PA, 1957.
3. A. Roda, in *Bioluminescence and Chemiluminescence: Perspectives for the 21ˢᵗ Century*, ed. A. Roda, M. Pazzagli, L. J. Kricka and P. E. Stanley, John Wiley and Sons, Chichester, 1998, p. 3.
4. http://www.ISBC.unibo. it (last accesses March 2010).
5. http://nobelprize.org/nobel_prizes/chemistry/laureates/2008 (last accessed March 2010).
6. http://www.mesoweb.com/features/lopes/Fireflies.pdf (last accessed March 2010).

7. http://www.absoluteastronomy.com/topics/Ivan_Kupala_Day (last accessed March 2010).

8. http://amshistorica.cib.unibo.it/diglib/collection.php?set = aldr%3Astamp (last accessed March 2010).

9. M. Malpighi, *Observationes anatomicaes in plantis and animalibus*, 1687.

10. L. Spallanzani, *Mem. Soc. Ital. Verona*, 1794, **VII**, 271.

11. M. Trautz, *Z. Phys. Chem.*, 1905, **53**, 1.

12. W. W. Coblentz, *A Physical Study of the Firefly*, Gibson Bros., Washington, D.C., 1912.

13. H. O. Albrecht, *Z. Phys. Chem.*, 1928, **136**, 321.

14. L. Teichmann, *Z. Ration. Med.*, 1853, **3**, 375.

15. K. Lehmstedt and E. W. Wirth, *Ber. Dtsch. Chem. Ges.*, 1928, **61**, 2044.

16. K. Gleu and W. Petsch, *Angew. Chem.*, 1935, **58**, 57.

17. E. N. Harvey, *Science*, 1914, **40**, 33.

18. E. N. Harvey and K. P. Stevens, *J. Gen. Physiol.*, 1928, **12**, 269.

19. O. Shimomura, F. H. Johnson and Y. Saiga, *J. Cell. Comp. Physiol.*, 1962, **59**, 223.

20. O. Shimomura and F. H. Johnson, *Biochem. Biophys. Res. Commun.*, 1971, **44**, 340.

21. H. Morise, O. Shimomura, F. H. Johnson and J. Winant, *Biochemistry*, 1974, **13**, 2656.

22. W. D. McElroy and R. Ballentine, *Proc. Natl. Acad. Sci. U.S.A.*, 1944, **30**, 377.

23. W. D. McElroy, *Proc. Natl. Acad. Sci. U.S.A.*, 1947, **33**, 342.

24. W. D. McElroy, J. W. Hastings, V. Sonnefeld and J. Coulombre, *Science*, 1953, **118**, 385.

25. A. A. Green and W. D. McElroy, *Biochim. Biophys. Acta*, 1955, **20**, 170.

26. H. H. Seliger and W. D. McElroy, *Biochem. Biophys. Res. Commun.*, 1959, **1**, 21.

27. H. H. Seliger, W. D. McElroy, E. H. White and G. F. Field, *Proc. Natl. Acad. Sci. U.S.A.*, 1961, **47**, 1129.

28. H. H. Seliger, W. G. Fastie, W. R. Taylor and W. D. McElroy, *J. Gen. Physiol.*, 1962, **45**, 1003.

29. W. D. McElroy, H. H. Seliger and E. H. White, *Photochem. Photobiol.*, 1969, **10**, 153.

30. E. H. White, F. McCapra, G. Field and W. D. McElroy, *J. Am. Chem. Soc.*, 1961, **85**, 337.

31. E. H. White, *J. Org. Chem.*, 1966, **31**, 1484.

32. E. H. White and B. R. Branchini, *J. Am. Chem. Soc.*, 1975, **97**, 1243.

33. E. H. White, M. Li and D. F. Roswell, *Photochem. Photobiol.*, 1991, **53**, 125.

34. B. R. Branchini, R. A. Magyar, K. M. Marcantonio, K. J. Newberry, J. G. Stroh, L. K. Hinz and M. H. Murtiashaw, *Biol. Chem.*, 1997, **272**, 19359.

35. B. R. Branchini, R. A. Magyar, M. H. Murtiashaw, S. M. Anderson, L. C. Helgerson and M. Zimmer, *Biochemistry*, 1999, **38**, 13223.

36. B. R. Branchini, T. L. Southworth, N. F. Khattak, E. Michelini and A. Roda, *Anal. Biochem.*, 2005, **345**, 140.
37. B. R. Branchini, T. L. Southworth, J. P. DeAngelis, A. Roda and E. Michelini, *Comp. Biochem. Physiol. B Biochem. Mol. Biol.*, 2006, **145**, 159.
38. J. Lee and H. H. Seliger, *Photochem. Photobiol.*, 1965, **4**, 1015.
39. D. J. O'Kane, V. A. Karle and J. Lee, *Biochemistry*, 1985, **24**, 1461.
40. J. Lee, *Biophys. Chem.*, 1993, **48**, 149.
41. J. Lee and H. H. Seliger, *Photochem. Photobiol.*, 1970, **11**, 247.
42. W. W. Ward and H. H. Seliger, *Biochemistry*, 1974, **13**, 1491.
43. W. W. Ward and M. J. Cormier, *Proc. Natl. Acad. Sci. U.S.A.*, 1975, **72**, 2530.
44. D. C. Prasher, V. K. Eckenrode, W. W. Ward, F. G. Prendergast and M. J. Cormier, *Gene*, 1992, **111**, 229.
45. J. W. Hastings, *J. Cell. Comp. Physiol.*, 1952, **39**, 1.
46. J. W. Hastings, W. D. McElroy and J. Coulombre, *J. Cell. Comp. Physiol.*, 1953, **42**, 137.
47. W. D. McElroy, J. W. Hastings, V. Sonnenfeld and J. Coulombre, *Science*, 1953, **118**, 385.
48. J. W. Hastings and B. M. Sweeney, *J. Cell. Comp. Physiol.*, 1957, **49**, 209.
49. J. W. Hastings and Q. H. Gibson, *J. Biol. Chem.*, 1963, **238**, 2537.
50. J. G. Morin and J. W. Hastings, *J. Cell. Physiol*, 1971, **77**, 313.
51. K. Nealson, T. Platt and J. W. Hastings, *J. Bacteriol.*, 1970, **104**, 313.
52. L. Liu and J. W. Hastings, *Proc. Natl. Acad. Sci. U.S.A.*, 2006, **104**, 696.
53. E. A. Meighen, L. B. Smillie and J. W. Hastings, *Biochemistry*, 1970, **9**, 4949.
54. E. A. Meighen and R. E. MacKenzie, *Biochemistry*, 1973, **12**, 1482.
55. J. Cousineau and E. A. Meighen, *Biochemistry*, 1976, **15**, 4992.
56. J. F. Evans, S. McCracken, C. M. Miyamoto, E. A. Meighen and A. F. Graham, *J. Bacteriol.*, 1983, **153**, 543.
57. R. Szittner, G. Jansen, D. Y. Thomas and E. Meighen, *Biochem. Biophys. Res. Commun.*, 2003, **309**, 66.
58. T. O. Baldwin and M. M. Ziegler, *Proc. Natl. Acad. Sci. U.S.A.*, 1979, **76**, 4887.
59. C. A. Reeve and T. O. Baldwin, *J. Bacteriol.*, 1981, **146**, 1038.
60. D. H. Cohn, R. C. Ogden, J. N. Abelson, T. O. Baldwin, K. H. Nealson, M. I. Simon and A. J. Mileham, *Proc. Natl. Acad. Sci. U.S.A.*, 1983, **80**, 120.
61. T. O. Baldwin, T. Berends, T. A. Bunch, T. F. Holzman, S. K. Rausch, L. Shamansky, M. L. Treat and M. M. Ziegler, *Biochemistry*, 1984, **23**, 3663.
62. J. H. Devine, G. D. Kutuzova, V. A. Green, N. N. Ugarova and T. O. Baldwin, *Biochim. Biophys. Acta*, 1993, **1173**, 121.
63. K. H. Nealson, T. Platt and J. W. Hastings, *J. Bacteriol.*, 1970, **104**, 313.
64. K. H. Nealson, *Arch. Microbiol.*, 1977, **112**, 73.
65. E. G. Ruby and K. H. Nealson, *Science*, 1977, **196**, 432.

66. D. H. Cohn, R. C. Ogden, J. N. Abelson, T. O. Baldwin, K. H. Nealson, M. I. Simon and A. J. Mileham, *Proc. Natl. Acad. Sci. U.S.A.*, 1983, **80**, 120.
67. R. P. Legocki, M. Legocki, T. O. Baldwin and A. A. Szalay, *Proc. Natl. Acad. Sci. U.S.A.*, 1986, **83**, 9080.
68. W. W. Lorenz, M. J. Cormier, D. J. O'Kane, D. Hua, A. A. Escher and A. A. Szalay, *J. Biolumin. Chemilumin.*, 1996, **11**, 31.
69. Y. A. Yu, T. Timiryasova, Q. Zhang, R. Beltz and A. A. Szalay, *Anal. Bioanal. Chem.*, 2003, **377**, 964.
70. B. L. Strehler and M. J. Cormier, *J. Biol. Chem.*, 1954, **211**, 213.
71. M. J. Cormier, K. Hori, Y. D. Karkhanis, J. M. Anderson, J. E. Wampler, J. G. Morin and J. W. Hastings, *J. Cell. Physiol.*, 1973, **81**, 291.
72. D. C. Prasher, V. K. Eckenrode, W. W. Ward, F. G. Prendergast and M. J. Cormier, *Gene*, 1992, **111**, 229.
73. W. W. Ward, *BioProcess. J.*, 2008, **7**, 26.
74. W. W. Lorenz, M. J. Cormier, D. J. O'Kane, D. Hua, A. A. Escher and A. A. Szalay, *J. Biolumin. Chemilumin.*, 1996, **11**, 31.
75. E. N. Harvey and F. I. Tsuji, *J. Cell. Physiol.*, 1954, **44**, 63.
76. F. I. Tsuji and R. Sowinski, *J. Cell. Comp. Physiol.*, 1961, **58**, 125.
77. Y. Haneda and F. I. Tsuji, *Science*, 1971, **173**, 143.
78. F. I. Tsuji, A. T. Barnes and J. F. Case, *Nature*, 1972, **237**, 515.
79. S. Inouye, M. Noguchi, Y. Sakaki, Y. Takagi, T. Miyata, S. Iwanaga, T. Miyata and F. I. Tsuji, *Proc. Natl. Acad. Sci. U.S.A.*, 1985, **82**, 3154.
80. F. E. Hanson, J. F. Case, E. Buck and J. Buck, *Science*, 1971, **174**, 161.
81. M. Anctil and J. F. Case, *Am. J. Anat.*, 1977, **149**, 1.
82. E. A. Widder, M. I. Latz, P. J. Herring and J. F. Case, *Science*, 1984, **225**, 512.
83. http://lifesci.ucsb.edu/ ~ biolum (last accessed March 2010).
84. M. Nomura, S. Inouye, Y. Ohmiya and F. I. Tsuji, *FEBS Lett.*, 1991, **295**, 63.
85. http://shell.kwansei.ac.jp/ ~ shell/life/e_striata.html (last accessed March 2010).
86. F. I. Tsuji and I. Frederick, in *The Mollusca, Vol. 2: Enviromental Biochemistry and Physiology*, ed. P. W. Hochachka, Academic Press, New York, 1983, p. 257.
87. K. Suto, H. Masuda, Y. Takenaka, F. I. Tsuji and H. Mizuno, *Genes Cells*, 2009, **14**(6), 727.
88. H. Arakawa, M. Maeda and A. Tsuji, *Anal. Biochem.*, 1979, **97**, 248.
89. H. Ohkuma, K. Abe, Y. Kosaka and M. Maeda, *Luminescence*, 2000, **15**, 21.
90. R. C. Hart, K. E. Stempel, P. D. Boyer and M. J. Cormier, *Biochem. Biophys. Res. Commun.*, 1978, **81**, 980.
91. R. C. Hart, J. C. Matthews, K. Hori and M. J. Cormier, *Biochemistry*, 1979, **18**, 2204.
92. E. H. White, F. McCapra, G. F. Field and W. D. McElroy, *J. Am. Chem. Soc.*, 1961, **83**, 2402.

93. F. McCapra and D. G. Richardson, *Tetrahedron Lett.*, 1964, **43**, 3167.
94. M. M. Rauhut, D. Sheehan, R. A. Clarke, B. G. Roberts and A. M. Semsel, *J. Org. Chem.*, 1965, **30**, 3587.
95. L. J. Bollyky, R. H. Whitman, B. G. Roberts and M. M. Rauhut, *J. Am. Chem. Soc.*, 1967, **89**, 6523.
96. F. McCapra and Y. C. Chang, *J. Chem. Soc., Chem. Commun.*, 1967, 1011.
97. H. Blitz, *Justus Liebigs Ann. Chem.*, 1897, **29**, 238.
98. F. McCapra, *Q. Rev.*, 1970, **20**, 485.
99. K. R. Kopecky and C. Mumford, *J. Can. Chem.*, 1969, **47**, 709.
100. W. Adam, in *The Chemistry of Peroxides*, ed. S. Patai, John Wiley & Sons, New York, 1983, p. 829.
101. T. Wilson, in *Singlet Oxygen Vol 2*, ed. A. A. Frimer, CRC Press, Boca Raton, FL, 1985Ch. 2.
102. J. H. Wieringa, J. Stratling, H. Wynberg and W. Adam, *Tetrahedron Lett.*, 1972, **13**, 169.
103. G. Schuster, N. J. Turro, H.-C. Steinmetzer, A. P. Schaap, G. Faier, W. Adam and J. C. Lui, *J. Am. Chem. Soc.*, 1975, **97**, 7110.
104. A. P. Schaap, R. S. Handley and P. B. Girl, *Tetrahedron Lett.*, 1987, **28**, 935.
105. A. P. Schaap, M. D. Sandison and R. S. Handley, *Tetrahedron Lett.*, 1987, **28**, 1159.
106. I. Bronstein, B. Edwards and J. C. Voyta, *J. Biolumin. Chemilumin.*, 1989, **4**, 99.
107. C. E. Olesen, J. Mosier, J. C. Voyta and I. Bronstein, *Methods Enzymol.*, 2000, **305**, 417.
108. C. S. Martin and I. Bronstein, in *Nonisotopic Probing, Blotting and Sequencing*, ed. L. J. Kricka, 2nd Ed., Academic Press, San Diego, CA, 1995, p. 494.
109. M. DeLuca, M. E. Dempsey, K. Hori, J. E. Wampler and M. J. Cormier, *Proc. Natl. Acad. Sci. U.S.A.*, 1971, **68**, 1658.
110. M. DeLuca, L. Brand, T. A. Cebula, H. H. Seliger and A. F. Makula, *J. Biol. Chem.*, 1971, **246**, 6702.
111. F. I. Tsuji, M. DeLuca, P. D. Boyer, S. Endo and M. Akutagawa, *Biochem. Biophys. Res. Commun.*, 1977, **74**, 606.
112. J. Wannlund, J. Azari, L. Levine and M. DeLuca, *Biochem. Biophys. Res. Commun.*, 1980, **96**, 440.
113. A. Roda, L. J. Kricka, M. DeLuca and A. F. Hofmann, *J. Lipid. Res.*, 1982, **23**, 1354.
114. K. V. Wood, J. R. de Wet, N. Devji and M. DeLuca, *Biochem. Biophys. Res. Commun.*, 1984, **124**, 592.
115. J. R. de Wet, K. V. Wood, D. R. Helinski and M. DeLuca, *Proc. Natl. Acad. Sci. U.S.A.*, 1985, **82**, 7870.
116. D. W. Ow, J. R. DE Wet, D. R. Helinski, S. H. Howell, K. V. Wood and M. DeLuca, *Science*, 1986, **234**, 856.

117. J. R. de Wet, K. V. Wood, M. DeLuca, D. R. Helinski and S. Subramani, *Mol. Cell. Biol.*, 1987, **7**, 725.

118. J. Rohr, M. I. Latz, S. Fallon, J. C. Nauen and E. Hendricks, *J. Exp. Biol.*, 1998, **201**, 1447.

119. M. I. Latz, A. R. Juhl, A. M. Ahmed, S. E. Elghobashi and J. Rohr, *J. Exp. Biol.*, 2004, **207**, 1941.

120. H. R. Schroeder, P. O. Vogelhut, R. J. Carrico, R. C. Boguslaski and R. T. Buckler, *Anal. Chem.*, 1976, **48**, 1933.

121. H. R. Schroeder, C. M. Hines, D. D. Osborn, R. P. Moore, R. L. Hurtle, F. F. Wogoman, R. W. Rogers and P. O. Vogelhut, *Clin. Chem.*, 1981, **27**, 1378.

122. H. R. Schroeder, R. C. Boguslaski, R. J. Carrico and R. T. Buckler, *Methods Enzymol.*, 1978, **57**, 424.

123. G. Cilento, *J. Theor. Biol.*, 1975, **55**, 471.

124. G. Cilento, *J. Biolumin. Chemilumin.*, 1989, **4**, 193.

125. G. Cilento and W. Adam, *Free Radical Biol. Med.*, 1995, **19**, 103.

126. K. Kakinuma, E. Cadenas, A. Boveris and B. Chance, *FEBS Lett.*, 1979, **102**, 38.

127. A. Boveris, E. Cadenas and B. Chance, *Fed. Proc.*, 1981, **40**, 195.

128. E. Cadenas, *Photochem. Photobiol.*, 1984, **40**, 823.

129. R. L. Stjernholm, R. C. Allen, R. H. Steele, W. W. Waring and J. A. Harris, *Infect. Immun.*, 1973, **7**, 313.

130. R. C. Allen, *Adv. Exp. Med. Biol.*, 1982, **141**, 411.

131. M. M. Rauhut, L. J. Bollyky, B. G. Roberts, M. Loy, R. H. Whitman, A. V. Iannotta, A. M. Semsel and R. A Clarke, *J. Am. Chem. Soc.*, 1967, **89**, 6515.

132. K. Nahashima, K. Maki, S. Akiyama, W. H. Wang, Y. Tsukamoto and K. Imai, *Analyst*, 1989, **114**, 1413.

133. P. R. Contag, I. N. Olomu, D. K. Stevenson and C. H. Contag, *Nat. Med.*, 1998, **4**, 245.

134. C. H. Contag, S. D. Spilman, P. R. Contag, M. Oshiro, B. Eames, P. Dennery, D. K. Stevenson and D. A. Benaron, *Photochem. Photobiol.*, 1997, **66**, 523.

135. C. H. Contag and M. H. Bachmann, *Annu. Rev. Biomed. Eng.*, 2002, **4**, 235.

136. W. Huang, A. Feltus, A. Witkowski and S. Daunert, *Anal. Chem.*, 1996, **68**, 1646.

137. S. Ramanathan, M. Ensor and S. Daunert, *Trends Biotechnol.*, 1997, **15**, 500.

138. S. Daunert and S. K. Deo, *Photoproteins in Bioanalysis*, Wiley-VCH, Weinheim, 2006.

139. M. Adamczyk, P. G. Mattingly, J. A. Moore and Y. Pan, *Org. Lett.*, 1999, **1**, 779.

140. M. Adamczyk, P. G. Mattingly, J. A. Moore, Y. Pan, K. Shreder and Z. Yu, *Bioconjugate Chem.*, 2001, **12**, 329.

141. A. K. Campbell, *Chemiluminescence: Principles and Applications in Biology and Medicine*, Ellis Horwood, Chichester, 1988.

142. J. S. A. Simpson, A. K. Campbell, M. E. T. Ryall and J. S. Whitehead, *Nature*, 1979, **279**, 646.
143. P. J. Herring, *Symp. Soc. Exp. Biol.*, 1985, **39**, 323.
144. P. J. Herring, *J. Biolumin. Chemilumin.*, 1987, **1**, 147.
145. E. A. Widder, M. I. Latz, P. J. Herring and J. F. Case, *Science*, 1984, **225**, 512.
146. T. P. Whitehead, L. J. Kricka, T. J. Carter and G. H. Thorpe, *Clin. Chem.*, 1979, **25**, 1531.
147. T. P. Whitehead, G. H. G. Thorpe, T. J. N. Carter, C. Groucutt and L. J. Kricka, *Nature*, 1983, **305**, 158.
148. G. H. G. Thorpe, L. J. Kricka, S. B. Moseley and T. P. Whitehead, *Clin. Chem.*, 1985, **31**, 1335.
149. P. E. Stanley and S. G. Williams, *Anal. Biochem.*, 1969, **29**, 381.
150. P. E. Stanley, *Anal. Biochem.*, 1971, **39**, 441.
151. G. S. Stewart, *FEMS Microbiol. Lett.*, 1998, **163**, 193.
152. P. J. Hill, S. Swift and G. S. Stewart, *Mol. Gen. Genet.*, 1991, **226**, 41.
153. M. K. Winson, S. Swift, P. J. Hill, C. M. Sims, G. Griesmayr, B. W. Bycroft, P. Williams and G. S. Stewart, *FEMS Microbiol. Lett.*, 1998, **163**, 193.
154. S. E. Brolin, G. Wettermark and H. Hammar, *Strahlentherapie*, 1977, **153**, 124.
155. F. Gorus and E. Schram, *Clin. Chem.*, 1979, **25**, 512.
156. A. M. García-Campaña, W. R. Baeyens, X. R. Zhang, E. Smet, G. Van Der Weken, K. Nakashima and A. C. Calokerinos, *Biomed. Chromatogr.*, 2000, **14**, 166.
157. A. M. Garcia-Campaña and W. R. G. Baeyens, *Chemiluminescence in Analytical Chemistry*, Marcel Dekker, New York–Basel, 2001.
158. J. De Boever, F. Kohen and D. Vandekerckhove, *Clin. Chem.*, 1983, **29**, 2068.
159. F. Kohen, J. De Boever and J. B. Kim, *Methods Enzymol.*, 1986, **133**, 387.
160. J. Mallefet and F. Baguet, *Comp. Biochem. Physiol. C*, 1987, **87**, 233.
161. J. C. Nicolas, A. M. Boussioux, A. M. Boularan, B. Descomps and A. Crastes de Paulet, *Anal. Biochem.*, 1983, **135**, 141.
162. J. C. Nicolas, *J. Biolumin. Chemilumin.*, 1994, **9**, 139.
163. L. J. Blum, S. M. Gautier and P. R. Coulet, *J. Biolumin. Chemilumin.*, 1989, **4**, 543.
164. E. H. Jansen, G. Zomer, R. H. Van den Berg and R. W. Stephany, *Vet. Q.*, 1984, **6**, 101.
165. W. Adam, F. Vargas, B. Epe, D. Schiffmann and D. Wild, *Free Radical Res. Commun.*, 1989, **5**, 253.
166. F. Kohen, M. Pazzagli, J. B. Kim and H. R. Lindner, *Steroids*, 1980, **36**, 421.
167. M. Pazzagli, G. Messeri, R. Salerno, A. L. Caldini, A. Tommasi, A. Magini and M. Serio, *Talanta*, 1984, **10**, 901.
168. A. Roda, S. Girotti, S. Ghini, B. Grigolo, G. Carrea and R. Bovara, *Clin. Chem.*, 1984, **30**, 206.

169. A. Roda, S. Girotti, A. L. Piacentini, S. Preti and S. Lodi, *Anal. Biochem.*, 1986, **155**, 346.
170. A. Roda, P. Pasini, M. Musiani, S. Girotti, M. Baraldini, G. Carrea and A. Suozzi, *Anal. Chem.*, 1996, **68**, 1073.
171. P. De Sole, S. Lippa and G. P. Littarru, *Adv. Exp. Med. Biol.*, 1982, **141**, 591.
172. A. Calokerinos and L. P. Palilis, in *Chemiluminescence in Analytical Chemistry*, ed. A. M. Garcia-Campaña and W. R. G. Baeyens, Marcel Dekker, New York–Basel, 2001, p. 321.
173. B. Galvan and T. K. Christopoulos, *Anal. Chem.*, 1996, **68**, 3545.
174. R. Bar and S. Ulitzur, *Appl. Microbiol. Biotechnol.*, 1994, **41**, 574.
175. S. Ulitzur, *J. Biolumin. Chemilumin.*, 1998, **13**, 365.
176. A. Yu. Vladimirov, *Ultra weak Luminescence during Biochemical Reactions*, Nauka, Moscow, 1965 (in Russian).
177. R. F. Vassil'ev and A. A. Vichutinskii, *Nature*, 1962, **194**, 1276.
178. N. I. U. Filippova and N. N. Ugarova, *Biokhimiia*, 1979, **44**, 1899(in Russian).
179. N. N. Ugarova, L. I. U. Brovko, O. V. Lebedeva and I. V. Berezin, *Vestn. Akad. Med. Nauk. SSSR*, 1985, **7**, 88(in Russian).
180. L. Y. Brovko and M. W. Griffiths, *Sci. Prog.*, 2007, **90**, 129.
181. B. B. Kim, V. V. Pisarev and A. M. Egorov, *Anal. Biochem.*, 1991, **199**, 1.
182. A. M. Egorov, I. G. Gazaryan, B. B. Kim, V. V. Doseyeva, J. L. Kapeljuch, A. N. Veryovkin and V. A. Fechina, *Ann. N. Y. Acad. Sci.*, 1994, **721**, 73.
183. M. Yu. Rubtsova, G. V. Kovba and A. M. Egorov, *Biosens. Bioelectron.*, 1998, **13**, 75.
184. V. A. Kratasyuk, E. V. Vetrova and N. S. Kudryasheva, *Luminescence*, 1999, **14**, 193.
185. B. A. Illarionov, V. M. Blinov, A. P. Donchenko, M. V. Protopopova, V. A. Karginov, N. P. Mertvetsov and J. I. Gitelson, *Gene*, 1990, **86**, 89.
186. V. A. Kratasyuk and J. I. Gitelson, *Uspekhi Mikrobiologii*, 1987, **21**, 3(in Russian).
187. K. Van Dyke, C. Van Dyke and K. Woodfork, *Luminescence Biotechnology: Instruments and Applications*, CRC Press, Boca Raton, FL, 2001.
188. K. Van Dyke and V. Castranova, *Cellular Chemiluminescence*, CRC Press, Boca Raton, FL, 1987.
189. K. Van Dyke and R. Van Dyke, *Luminescence Immunoassay and Molecular Applications*, CRC Press, Boca Raton, FL, 1990.
190. *Methods Enzymol.*, 1978, **57**.
191. *Methods Enzymol.*, 1986, **133**.
192. Methods Enzymol., 2000, **305**.
193. http://www.wiley.com/bw/journal.asp?ref = 1522-7235 (last accessed March 2010).

CHAPTER 2

The Nature of Chemiluminescent Reactions

GIJSBERT ZOMER

Netherlands Vaccine Institute, P.O. Box 457, 3720AL Bilthoven, The Netherlands

2.1 Introduction

The chemiluminescence phenomenon has attracted attention ever since the first observation of bioluminescence. The mystical beauty, together with the how and why questions, have inspired poets and scientists, in fact all beholders of this "cold light". My first encounter with chemiluminescence was in the 1970s at the University of Groningen, where I witnessed an experiment when it was attempted to purify adamantylidenadamantane-1,2-dioxetane by bulb to bulb distillation. During the process beautiful blue light was emitted. At that time I fell in love with chemiluminescence and it still gives me pleasure to drop acridinium ester to a basic hydrogen peroxide solution, and just admire the flashes of light.

There has been renewed interest in the subject of the chemistry of chemiluminescent reactions since the last review by McCapra[1] appeared. Therefore, although numerous reviews[2-15] on the application of chemiluminescence have appeared, there seems to be a need to address the chemistry of some of these chemiluminescence reactions that find use in applications. This chapter tries to fill this gap. The main focus will be on results obtained during the last ten years or so on the chemistries of 1,2-dioxetanes, peroxyoxalates, luminol and acridinium/ acridan esters.

The phenomenon of chemiluminescence can be defined as the emission of light resulting from a chemical reaction. For a chemical reaction to produce light, some essential requirements should be met:

Chemiluminescence and Bioluminescence: Past, Present and Future
Edited by Aldo Roda
© Royal Society of Chemistry 2011
Published by the Royal Society of Chemistry, www.rsc.org

1. To populate an electronically excited (singlet) state the reaction has to be sufficiently exothermic. The free energetic requirement can be calculated using:

$$-\Delta G \geq \frac{hc}{\lambda_{ex}} = \frac{28600}{\lambda_{ex}} \qquad (2.1)$$

   Therefore, chemiluminescence reactions producing photons in the visible (400–750 nm) range require around 40–70 kcal mol^{-1}.

2. This electronically excited state has to be accessible on the reaction coordinate.

3. Photon emission from the excited state has to be a favourable energy release route. This means that either the product of the reaction has to be fluorescent or – if by energy transfer – an excited state can be populated (this energy transfer can occur intra- or intermolecularly).

The chemiluminescence quantum yield, defined as the number of photons emitted per reacting molecule, can be expressed as:

$$\Phi_{CL} = \Phi_R \times \Phi_{ES} \times \Phi_F \qquad (2.2)$$

where Φ_R reflects the chemical yield of the reaction, Φ_{ES} is the fraction of the product entering the excited state and Φ_F is the fluorescent quantum yield. If energy transfer occurs from the excited state to an acceptor fluorescent molecule, secondary emission from the acceptor results. The efficiency of this energy transfer (Φ_{ET}) and the fluorescence efficiency of the acceptor (Φ_F') have to be taken into account:

$$\Phi_{CL} = \Phi_R \times \Phi_{ES} \times \Phi_F' \times \Phi_{ET} \qquad (2.3)$$

The kinetics, *i.e.* the relationship between chemiluminescence intensity and time, is expressed by equations incorporating reaction rate constants and concentrations of substrates and reactants. Often, relatively simple expressions can be used to describe the chemiluminescence intensity–time profile of a chemiluminescent reaction. As an example consider the horseradish peroxidase (HRP) catalysed enhanced chemiluminescent peroxidation of an acridan ester (Scheme 2.1).[16]

In this scheme HRP can be considered as an Fe^{3+} species that by hydrogen peroxide becomes oxidized to an Fe^{5+} species (Compound I). The enhancer plays a double role: on the one hand it is used to reduce Compound I to a Fe^{4+} species (Compound II) and Compound II back to the resting ferric state of HRP. On the other hand, the enhancer radical formed in these reactions can oxidize the acridan ester to the acridinium ester, reforming the enhancer. This peroxidation reaction is therefore catalytic both in HRP and in enhancer.

This sequence of reactions can adequately be described by assuming the formation of the acridinium ester as the rate-determining step. This acridinium ester subsequently reacts with hydrogen peroxide in the light-generating step.

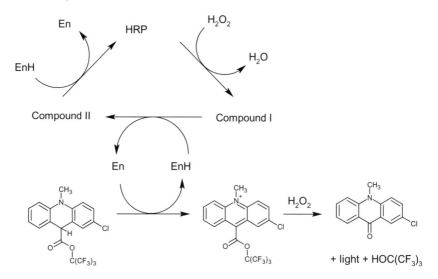

Scheme 2.1 Simplified mechanism of enhanced chemiluminescent peroxidation of acridan ester.

These sets of reaction can be described as two unimolecular irreversible consecutive reactions with rate constants k_1 and k_2:

$$A \xrightarrow{k_1} B \xrightarrow{k_2} C \qquad (2.4)$$

Assuming concentrations for A, B and C as x, y and z, respectively, the differential equations are:

$$\frac{-\mathrm{d}x}{\mathrm{d}t} = k_1 x \qquad (2.5)$$

$$\frac{-\mathrm{d}y}{\mathrm{d}t} = -k_1 x + k_2 y \qquad (2.6)$$

$$\frac{\mathrm{d}z}{\mathrm{d}t} = k_2 y \qquad (2.7)$$

Assuming an initial concentration of a for A the differential equations can be integrated, giving:

$$x = a\mathrm{e}^{-k_1 t} \qquad (2.8)$$

$$y = \mathrm{e}^{-k_2 t} \left[(k1 a \mathrm{e}^{(k_2 - k_1)t}) / (k_2 - k_1) - k_1 a / (k_2 - k_1) \right] \qquad (2.9)$$

$$z = a\left[1 - (k_2 \mathrm{e}^{-k_1 t}) / ((k_2 - k_1) + k_1 \mathrm{e}^{-k_2 t}) / (k_2 - k_1) \right] \qquad (2.10)$$

Using Microsoft Excel the formulas for x, y and z can be used to calculate and plot the concentrations at different time points. An example of such a plot, assuming $A = 2 \times 10^6$, $k_1 = 0.006$, and $k_2 = 0.0002$, is shown in Figure 2.1. The observed chemiluminescence response can be fitted to the calculated one using Microsoft Excel solver add-in. By employing this procedure values for k_1 and k_2 can be calculated.

Of course this approach disregards other (side) reactions that take place (a full mathematical model for the peroxidase catalysed luminol chemiluminescent reaction has been described)[17] but which do not lead to light emission. For instance, it is assumed that the reactions involved are (pseudo) first order (in HRP), *i.e.* the amounts of acridan ester and hydrogen peroxide are sufficient and no deterioration of HRP or enhancer occurs. Nevertheless, this approach describes chemiluminescence–time curves rather well, and is also applicable to flash like chemiluminescence reactions, *e.g.* from acridinium ester or peroxyoxalates.[18,19] Using this type of analysis valuable information can be deduced from the effects of compounds on the shape of these curves, *e.g.* in antioxidant research.[20]

Besides the kinetics of the reaction, mechanistic information can be obtained from product analysis studies, fluorescence efficiencies, deuterium isotope effects, quantum mechanical calculations, *etc.*

Highly exothermal reactions often involve oxidation because of the strong bonds between carbon and oxygen that are formed. Thermodynamically, the oxidation of glucose yields about $700 \, \text{kcal} \, \text{mol}^{-1}$ of energy (2.11):

$$C_6H_{12}O_6 + O_2 \xrightarrow{\text{oxidation}} 6CO_2 + 6H_2O + \text{energy} \qquad (2.11)$$

$$6CO_2 + 6H_2O + \text{energy} \xrightarrow{\text{photosynthesis}} C_6H_{12}O_6 + O_2 \qquad (2.12)$$

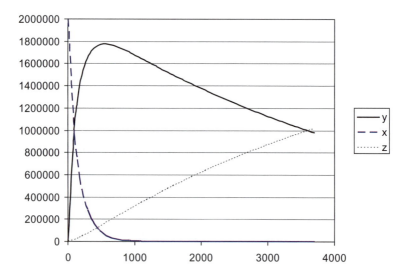

Figure 2.1   Kinetic concentration–time profiles for a consecutive $A \rightarrow B \rightarrow C$ reaction.

In the reverse direction (2.12) this reaction can be seen as a way energy can be stored by means of photosynthesis. In this way chemiluminescence as a result of an oxidation reaction can be viewed as the reverse of photosynthesis.[21]

A well-known example of a chemiluminescence reaction involves the base-catalysed chemiluminescent autoxidation of acridan ester (Scheme 2.2).[22]

During this reaction the base abstracts the acridan C9 proton. The carban-ion, in an autoxidation reaction, forms the peroxide. Ring closure with expelling of the phenolate anion gives an unstable highly strained dioxetanone. Light emission occurs from decomposition of the dioxetanone. This remarkable sequence of events was envisioned and put into practice by Frank McCapra,[23] who studied this system as model for the firefly bioluminescence (Scheme 2.3; for a recent mechanistic proposal see Lung *et al.*[24]).

As pointed out by McCapra,[22] there are some striking similarities between the two systems: the pK_A at the CH alpha to the ester group is about 20 in both cases; both systems show a deuterium isotope effect of about 3, indicating that the C–H bond is broken in the rate-determining step. Moreover, both reactions proceed best in a somewhat hydrophobic surrounding. The pK_A of the leaving group in the acridan case is about 10 (for unsubstituted phenyl) while AMP is somewhat more acidic with a pK_A of about 6. Another important aspect of both systems is the electron donating ability of the dioxetanone substituent, facilitating the population of the excited state. In firefly luciferin the phenolate anion is essential for light production since the methylated analogue is virtually non-luminescent. Thus, although the dioxetanone intermediates have never been isolated for these cases, there seems to be good evidence for their invol-vement in the excitation step. As will become obvious in the rest of this chapter, dioxetanones and dioxetanes play a pivotal role in the oxidative chemilumi-nescence phenomenon.[25]

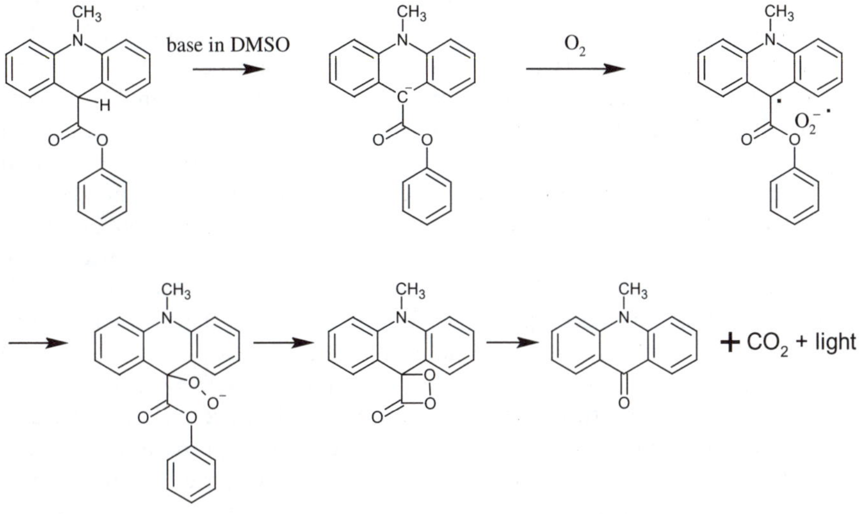

Scheme 2.2 Chemiluminescent autoxidation of acridan ester.

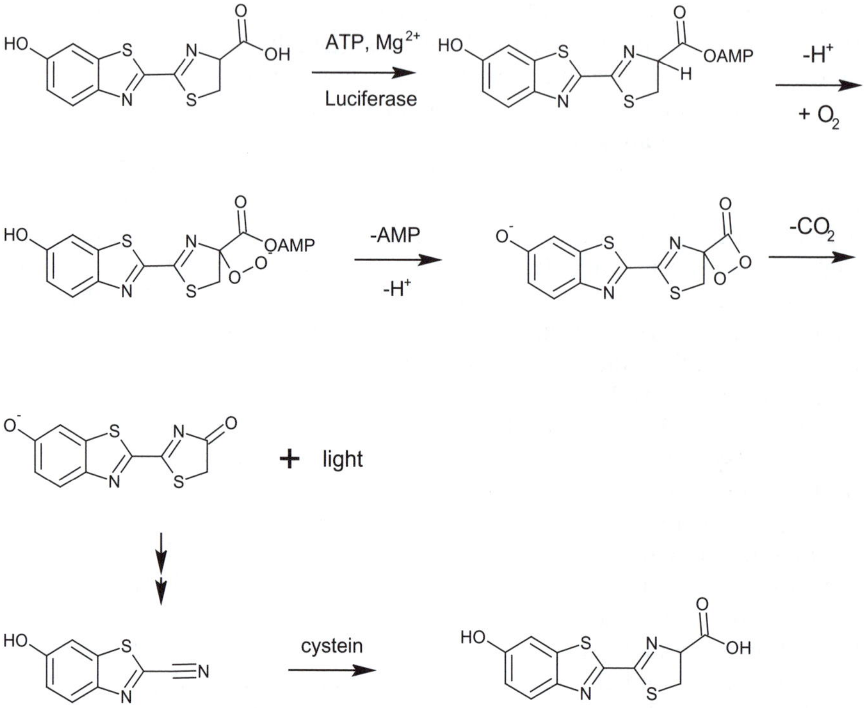

Scheme 2.3 Simplified mechanism of firefly bioluminescence.

2.2 1,2-Dioxetane Chemiexcitation

Ever since the discovery of their chemiluminescence properties[26] there has been much interest in, and research devoted to, unravelling the mechanism of the 1,2-dioxetane excitation step. Two mechanisms have been advocated (Scheme 2.4).

The thermal fragmentation of a dioxetane leads to two carbonyl fragments (in this example acetone and acetaldehyde), a fraction of which is in excited triplet (major) and singlet (minor) states. The synchronous, concerted mechanism which is a symmetry forbidden pericyclic rearrangement has been suggested by McCapra[27] and Turro[28] while O'Neal and Richardson[29] describe a mechanism that proceeds *via* a biradical structure. This biradical mechanism is supported by kinetic[30] and *ab initio* calculation[31] results. More recent calculations on dioxetane decomposition[32,33] seem to suggest that neither extreme is in accordance with the actual mechanism. The calculations indicate that the O–O bond is broken first and that, after the biradical is formed, there is no energy barrier left for further fragmentation (see, however, C. Tanaka and J. Tanaka,[32] who suggest that a little activation energy still is required to break the C–C bond). This means that although not really concerted the reaction of these simple dioxetanes takes place in a way that does not involve two consecutive steps with a biradical as an intermediate. This study failed, however, to

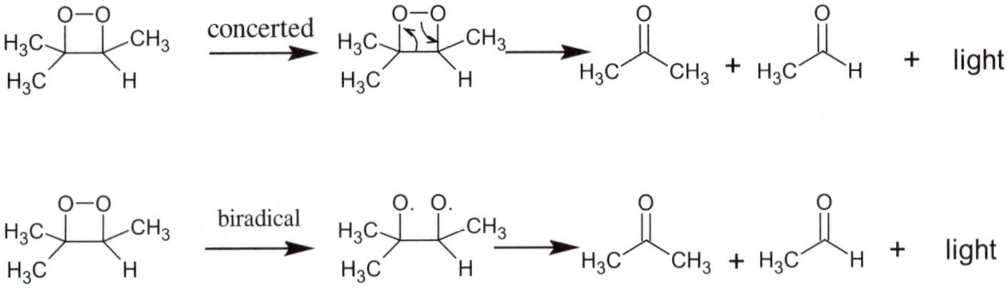

Scheme 2.4 Mechanisms of 1,2-dioxetane decomposition.

correctly reproduce the activation barrier or identify the entropic trap that facilitates chemiluminescence. Careful product studies from the decomposition of 3,3-dimethyldioxetane in 1,4-cyclohexdiene as the solvent also support an intermediate mechanism (asynchronous, concerted).[34] Theoretical work[35] on the isoelectronic 1,4 tetramethylene biradical indicates that the biradical species actually is trapped entropically, with this trapping being related to the number of degrees of freedom. This idea was further elaborated by De Vico *et al.*[36] who calculated at the multistate multiconfigurational second-order perturbation level of theory the thermal decomposition of the parent 1,2-dioxetane and the associated activation energy and singlet *versus* triplet excited state yields. According to their results the 1,2-dioxetane decomposition can be viewed as beginning with O–O' and C–C' stretching, followed by O–C–C'–O' dihedral torsion, asymmetric O–C/O'–C' bonds stretching, and asymmetric C/C' pyramidalization. An alternative minimum energy path (MEP) was computed from the S_0 transition state involving only the torsional mode and that along this path S_0, S_1 and T_1 potential energy surfaces are (almost) degenerate, allowing the molecule to change state. This entropic trapping increases the lifetime of the molecule in a region where state crossing is possible. Using this result it is easy to see that increasing substituents on the 1,2-dioxetane ring increases the number of degrees of freedom and, therefore, the entropic trapping, giving a higher yield of excited state products. Recently, the 1,4 biradical of 1,2-dioxetanedione has been experimentally observed by electron spin resonance (ESR) spectroscopy.[37]

At the same level of quantum mechanical calculation the firefly bioluminescence mechanism has also been studied.[38]

The most stable 1,2-dioxetane synthesized so far is adamantylidenadamantane-1,2-dioxetane. The compound was synthesized by Wieringa[39] in 1972 from adamantylidenadamantane by sensitized oxygenation (Scheme 2.5).

The crystal structure of adamantylidenadamantane-1,2-dioxetane was obtained[40] and indicated a twisted four-ring structure with a dihedral angle of about 21°. The two bulky substituents prevent elongation of the O–O and C–C bonds, leading to a somewhat twisted structure where the O–O bond becomes longer while the C–C bond is unaltered. This subject has been discussed in more detail by Baader.[41] As indicated before, electron-donating substituents on the dioxetane are important for high yield chemical excitation to the singlet state. This was elegantly shown by Schaap and Gagnon[42] (Scheme 2.6) using a phenol substituted dioxetane, 1-(4-hydroxyphenyl)-6-phenyl-2,5,7,8-tetraoxobicyclo [4.2.0]octane (HPTBO).

By removing the phenolic proton, the rate of decomposition was increased more than a million fold. Moreover, the chemiexcitation changed from a triplet to a singlet state. This important finding forms the base of the notion that by changing substituents on a dioxetane efficient chemiluminescence could be triggered chemically or enzymatically. By combining the stability enhancing adamantane substituent with a protected phenolic substituent stable dioxetanes could be devised and synthesized that, dependent on the protecting group, could be made to chemiluminesce by an external trigger (Scheme 2.7). With X = *tert*-butylsilyl and using tetrabutyl fluoride in DMSO as the trigger a very bright blue chemiluminescence could be observed.[43,44]

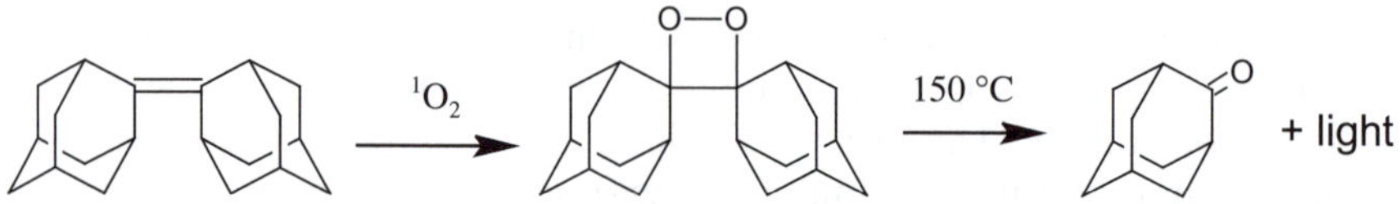

Scheme 2.5 Thermochemiluminescent decomposition of adamantylideneadamantane-1,2-dioxetane.

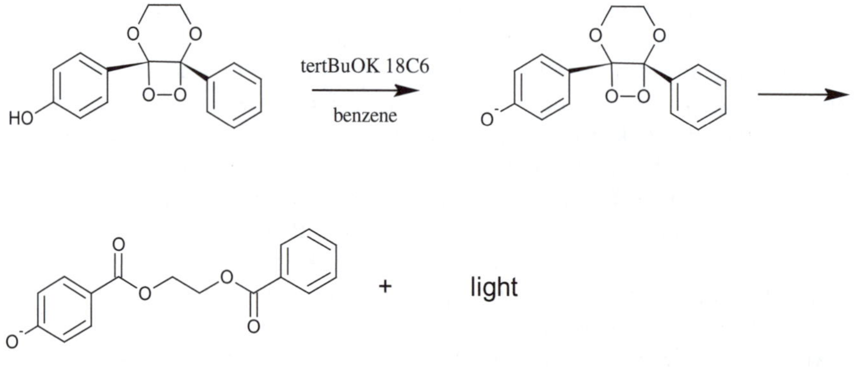

Scheme 2.6 Base-induced chemiluminescent decomposition of HPTBO.

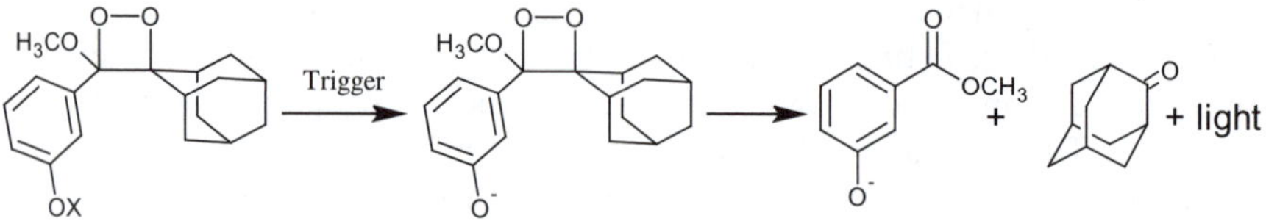

Scheme 2.7 Triggerable 1,2-dioxetanes.

The idea of triggering a chemiluminescence signal as desired from a protected dioxetane was further expanded by using acetate (with base trigger) or phosphate (with alkaline phosphatase enzyme). This very efficient chemiexcitation has been rationalized in terms of an intramolecular chemically initiated electron exchange luminescence (CIEEL) mechanism. The intermolecular CIEEL mechanism was originally proposed by Koo and Schuster[45] while studying the decomposition of diphenoyl peroxide in the presence of fluorescers. It turned out that the catalytic effect of the fluorescers was proportional to their ease of oxidation. Later, this mechanism was refined by Wilson[46] and McCapra[47] in such a way that a partially developing charge transfer replaced a full one-electron transfer during the chemiexcitation. Recent theoretical studies by Isobe *et al.*[48] indicate that indeed charge transfer from a donor moiety in the molecule to the O–O bond of the peroxide provokes the breakdown of the dioxetane into two radicals within a solvent cage. This is immediately followed by C–C bond breakage with simultaneous back charge transfer to chemically excite the donor containing resulting carbonyl group.

Mass spectrometric analysis has also been used to elucidate the mechanism of dioxetane chemiexcitation. Matsumoto *et al.*[49] have studied the matrix-assisted laser desorption ionization time-of-flight mass spectra (MALDI-TOF-MS) of certain dioxetanes and their thermolysis decomposition products (Scheme 2.8).

Under the used conditions the dioxetane showed full scan fragments associated with elimination of 2-methyl-1-propene (56u) and pivaldehyde (86u). These fragments were not observed in the full scan spectra of the keto ester. When MS-MS spectra were recorded of the deprotonated ion of the dioxetane and of the keto ester, fragments originating from the loss of 56u and 86u were observed for both ions. It was concluded that intramolecular charge transfer induced decomposition of the dioxetane takes place to afford excited species while the keto ester is excited vibrationally in MS-MS. The same group[50] recently described the solvent promoted chemiluminescent decomposition of the same dioxetane. They showed that the decomposition of 1,2-dioxetanes containing a 4-benzothiazol-2-yl-3-hydroxyphenyl substituent is increased in various aprotic polar solvents without the addition of base. Measurement of the activation parameters for the base induced and solvent promoted decomposition revealed a large negative entropy term in the latter case, indicative of a transition state with considerable less disorder, than for the base induced reaction. This was rationalized in terms of hydrogen bonding of the phenolic hydrogen with the solvent.

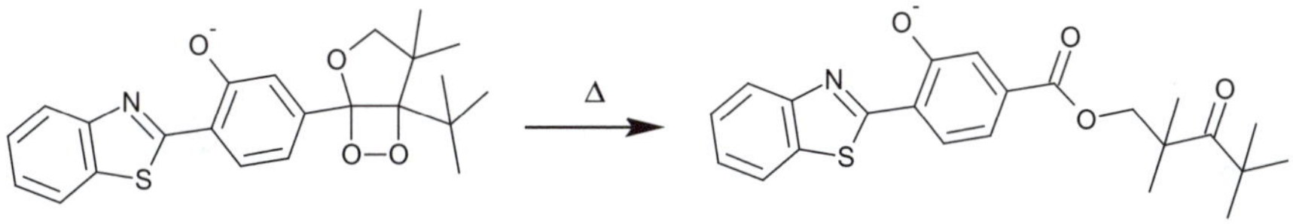

Scheme 2.8 Thermal decomposition of dioxetane as studied by MALDI-TOF mass spectrometry.

Another remarkable finding, first described by Edwards and Bronstein,[51] relates to the position of the aromatic hydroxyl group relative to the attachment point of the dioxetane. As an example consider the following case (Scheme 2.9).

Dioxetanes substituted with a naphthalen-2-yl group bearing a silylated hydroxyl at the 4, 5 or 7 position ("odd" pattern) show much more efficient chemiluminescence upon base treatment than analogues hydroxylated at the 3, 6 or 8 position ("even" pattern) upon fluoride-catalysed decomposition.[52] By measuring the fluorescence quantum yields of the ester products it was concluded that the significant differences in chemiluminescent efficiency are mainly attributable to the more efficient chemiexcitation step. *Ab initio* calculations in an analogue case[48] also indicate that the activation energy is dramatically reduced from 19.4 to 3.8 kcal by the deprotonation at the odd position.

The next example from the Matsumoto group[53] showed that even with an even pattern of substitution the efficiency of chemiluminescence could be increased up to a factor of 210 by lithium ion catalysis. The finding that the order of ionic radius of the alkali metal was inversely proportional to the chemiluminescence efficiency strongly suggested a tight metal coordination (Scheme 2.10).

Most of the hitherto described triggerable 1,2-dioxetanes show high chemiluminescence efficiencies in aprotic solvents like DMSO and acetonitrile. Generally, they are much less active in aqueous solution. Also, the rate of decomposition in aqueous solution is much lower in water compared to aprotic solvents. This can be rationalized by the fact that in aqueous medium the phenolate anion becomes hydrated and, therefore, is less easily oxidized.

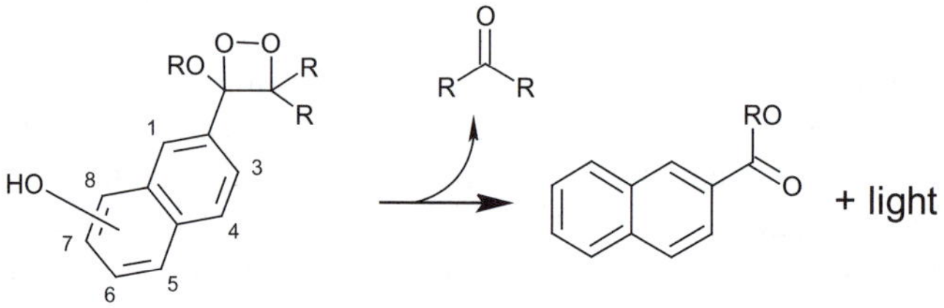

Scheme 2.9   Numbering of substituted 1,2-dioxetane, indicating odd and even positions.

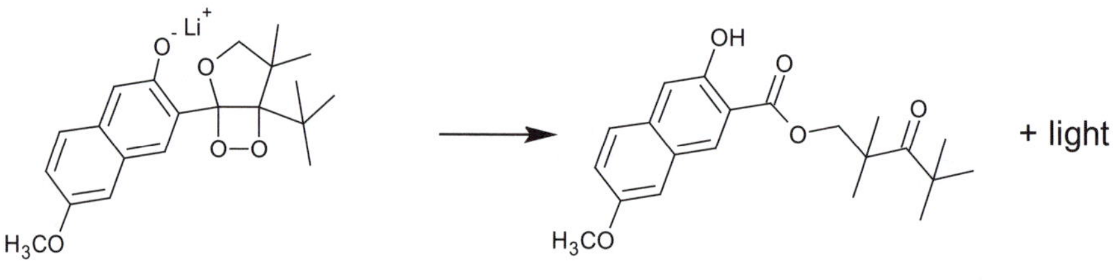

Scheme 2.10   Tight metal coordination explains the observed chemiluminescence.

Recently,[53] several benzoxazole (**2.1**, X = O) and benzothiazole (**2.1**, X = S) dioxetanes (R=H, OCH₃), bearing a 4-4-(benzoxazol-2-yl)-3-hydroxyphenyl or (benzothiazol-2-yl)-3-hydroxyphenyl, have been studied with respect to decomposition rate and chemiluminescence efficiency in aqueous solution and in acetonitrile.

2.1

These 1,2-dioxetanes show efficient chemiluminescence after reaction with tetrabutylammonium fluoride in acetonitrile and also in basified water–acetonitrile mixtures with fast decomposition rates. Careful study of the decomposition rate at different water–acetonitrile mixtures revealed different behaviour between the benzoxazole and benzothiazole series. The benzoxazole derivative showed the fastest reaction in pure acetonitrile, while the benzothiazole derivative showed an increase in decomposition rate constant going to a more aqueous solvent mixture.

The study of solvent effects during enzymatic triggering of dioxetanes is obviously very important in realizing commercial applications of the compounds. Advances in this field have been the topic of many patent applications[54–57] and mostly involve the use of surfactants (*e.g.* quaternary ammonium salts) and 1,2-dioxetanes with improved aqueous solubility.

An interesting chemiluminescence reaction involving epoxides (Scheme 2.11) as the starting material has been reported by Imanishi *et al.*[58]

The epoxides, prepared from the corresponding olefins, showed light emission when treated with basic hydrogen peroxide followed by acid. The epoxide can be considered to be present as a neutral or zwitterionic species. In the zwitterionic state the acridinium moiety is prone to nucleophilic attack by hydroperoxide anion, forming the *anti* adduct. Rotation around the central C–C bond gives part of the perhydroxylated adduct in the *syn* state. The proposed mechanism involves the formation from the *syn* adduct of an intermediate 1,2-dioxetane that decomposes *via* a CIEEL type mechanism. It turned out that the use of different acids has an effect on the observed chemiluminescence, both in terms of chemiluminescence intensity and time profile. Malonic acid gave the highest chemiluminescence quantum yield (about 0.2%). This was interpreted as a more favourable interaction of malonic acid with the *syn* adduct.

2.2.1 Synthesis of 1,2-Dioxetanes

Dioxetanes can be prepared in several ways (Scheme 2.12).

The first 1,2-dioxetane was synthesized by ring closure of the bromo hydroperoxide by Kopecky and Mumford.[26]

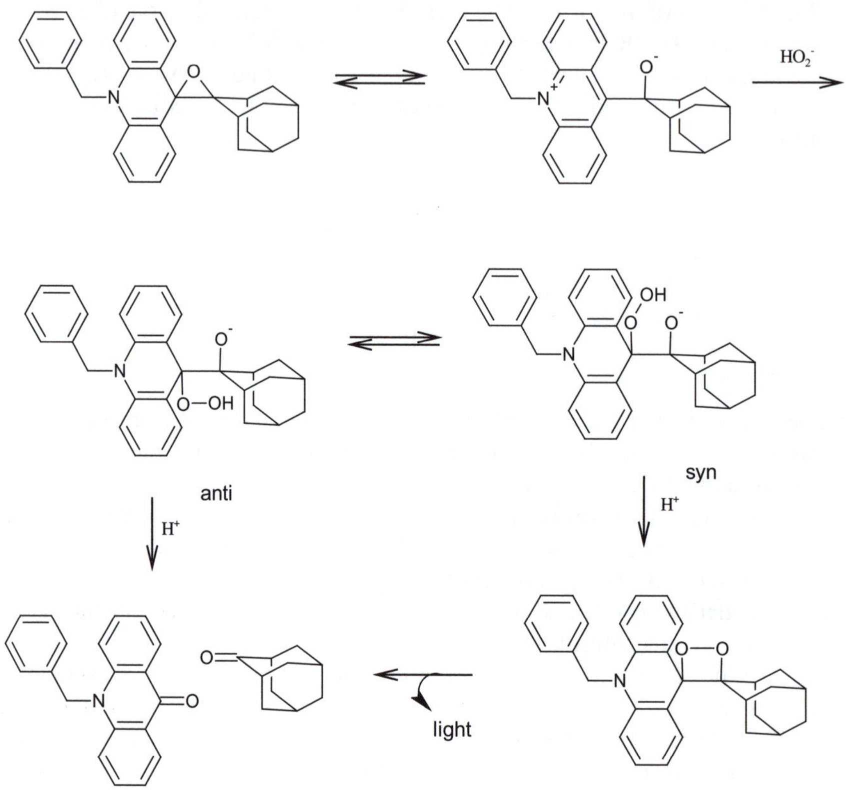

Scheme 2.11   Proposed mechanism for chemiluminescent decomposition of an epoxide.

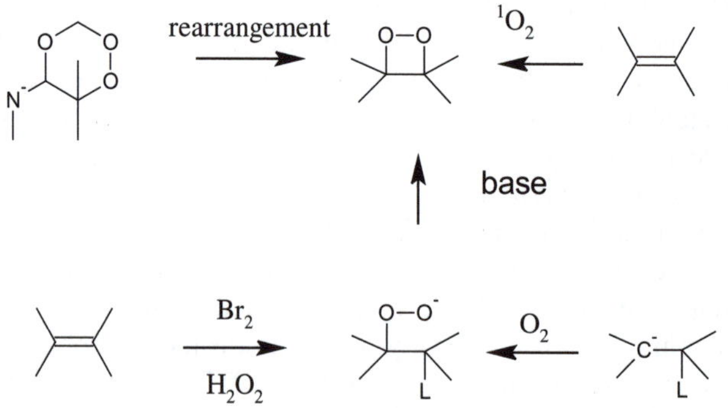

Scheme 2.12   Synthetic routes toward 1,2-dioxetanes.

The easiest way to prepare simple 1,2 dioxetanes is by direct oxygenation of an olefinic bond using singlet oxygen.[59–80]

Other types of activation can also be used, *e.g.* epoxide ring opening as shown by Leclercq *et al.* in 1982 (Scheme 2.13).[81]

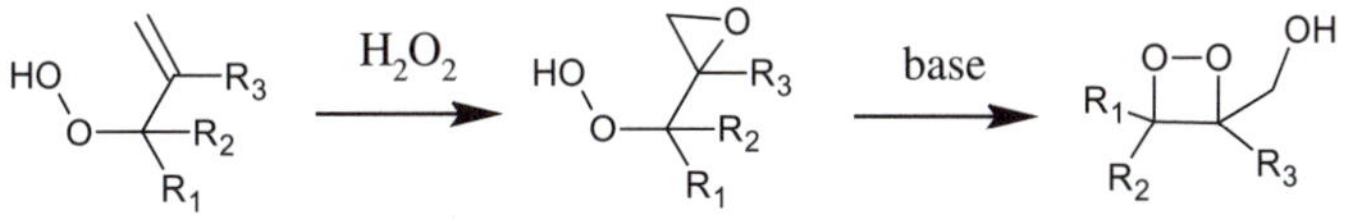

Scheme 2.13 Synthesis of a functionalized 1,2-dioxetane.

2.3 Peroxyoxalates

Apart from peroxyoxalates, other oxalic acid derivatives are known to give the chemiluminescence reaction.

Actually, the reaction of oxalyl chloride with hydrogen peroxide in the presence of a fluorescent molecule, *e.g.* diphenylanthracene, constitutes the first example of this chemiluminescence reaction (Scheme 2.14).[82] During the reaction in ether solvent, carbon dioxide, carbon monoxide and a little oxygen gas was formed. Addition of a radical inhibitor caused a delay in light emission, but not when the reaction was carried out in dimethyl phthalate.

Reaction of oxalyl chloride with *tert*-butyl hydroperoxide in the presence of diphenylanthracene did not produce light. Addition of water to the reaction mixture produced moderate light emission. At first, these reactions were thought to afford applications in the field of emergency light sources requiring no external power. The well-known light stick is an example. Later, applications in analytical chemistry became important. These are reviewed elsewhere in this book.

Other leaving groups, *e.g.* certain sulfonamides (**2.2** and **2.3**) result, in very efficient (>30%) chemiluminescence.[83]

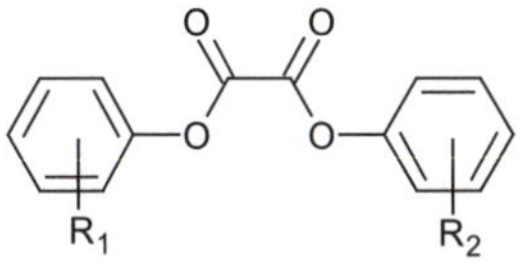

2.2

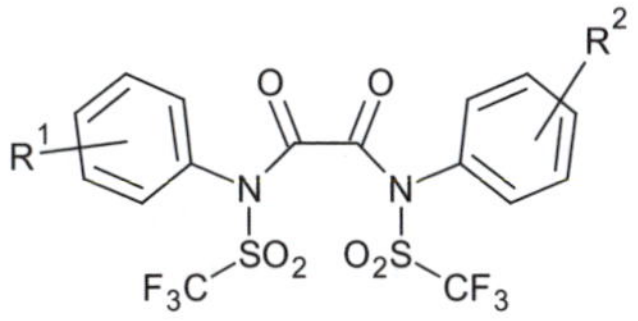

2.3

The chemical products formed as a result of this reaction are carbon dioxide and the leaving group (in the case of oxalyl chloride this is HCl). The emitted light stems from the excited state of the fluorescer. Already in the early days, 1,2-dioxetanedione (**2.4**) was suggested to be a high-energy intermediate in the reaction.[84]

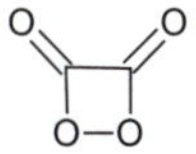

2.4

The fact that light emission could be delayed by delayed addition of the fluorescer to the reaction mixture containing peroxyoxalate and hydrogen

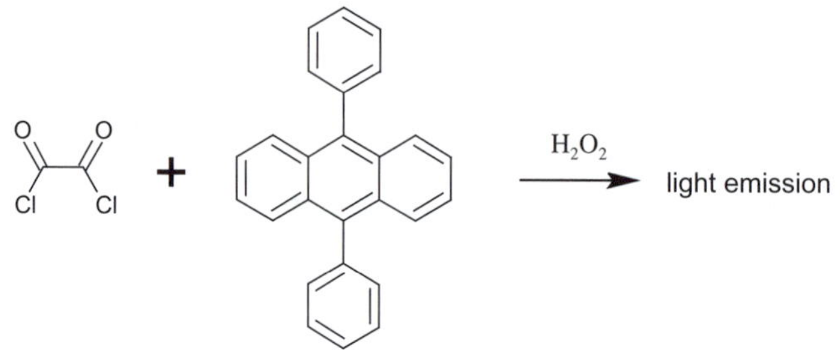

Scheme 2.14 Chemiluminescent reaction of oxalyl chloride with hydrogen peroxide in the presence of 9,10-diphenylanthracene.

peroxide was an important observation. One of the intermediates, therefore, has to have a certain lifetime. Addition of the fluorophore then evidently catalysed in some way the chemiluminescence reaction. To accommodate these observations a charge-transfer complex between the high-energy intermediate and the fluorescer was proposed. The high efficiency of the chemiluminescence reaction was thought to occur *via* the CIEEL (chemically initiated electron exchange luminescence) mechanism[85] or at least some form of charge transfer interaction.

Another high-energy intermediate candidate is **2.5**, a 1,2-dioxetanone still bearing one of the leaving groups.

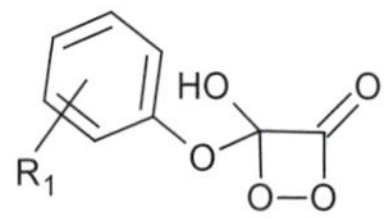

2.5

In an elegant experiment Motoyoshiya *et al.*[86] studied the chemiluminescence reaction of a hybrid form of activated oxalic acid containing both an aromatic phenolate and a sulfonamide leaving group (**2.6**).

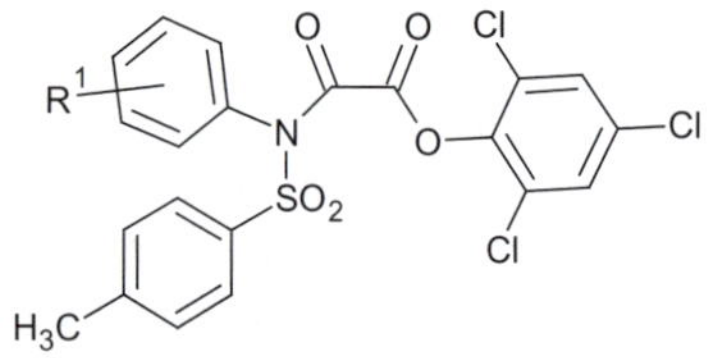

2.6

When these compounds were treated with aqueous hydrogen peroxide in tetrahydrofuran in the presence of diphenylanthracene (DPA) chemiluminescence was observed. An NMR study of this reaction revealed a gradual decrease of starting material with concomitant formation of 2,4,6-trichlorophenol and the tosylanilides. Importantly, no signals due to intermediates were observed. Because the parent bis-sulfonamide oxalate shows virtually no reaction the initial nucleophilic attack expels 2,4,6-trichlorophenolate. From the kinetic results obtained through NMR product formation studies with different R^1 substituents ($R^1 = $ H, p-CH$_3$, m-Cl, p-Cl, m-OCH$_3$, p-OCH$_3$) a Hammett relationship was found with a ρ-value of $+1.75$. Light emissions under neutral and basic (sodium carbonate) conditions also revealed a Hammett relationship with ρ-values of $+2.66$ and $+1.20$, respectively.

From these observations – a larger ρ for the light emission than for the elimination of the sulfonanilide – it was concluded that the interaction of the 1,2-dioxetanone with DPA is the most favourable pathway for light emission. This interaction therefore takes place before the elimination of the second leaving group (Scheme 2.15).

The same paper[86] describes experiments that shed further light (*sic*) on the intermediacy of the 1,2-dioxetanone (instead of 1,2-dioxetanedione). When the mixed oxalyl derivative (Scheme 2.16) was allowed to react with aqueous hydrogen peroxide in THF only very feeble chemiluminescence was observed.

When *N*-2-naphthyl-*N*-tosylamide was added to the reaction mixture enhanced chemiluminescence was observed with an emission spectrum in good

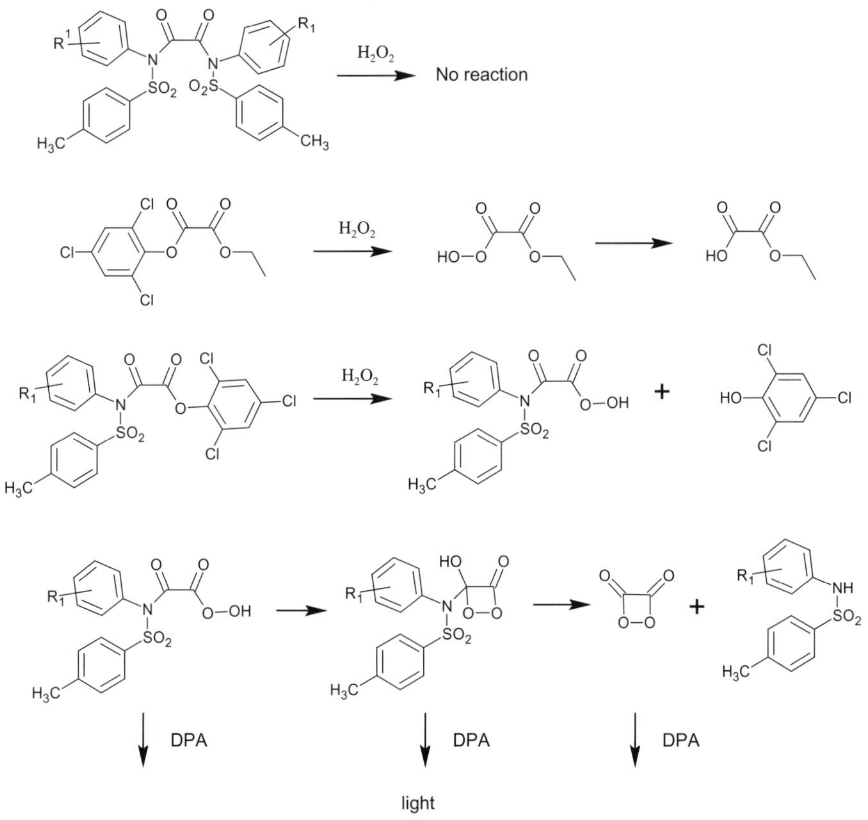

Scheme 2.15 Proposed routes to high-energy intermediates.

agreement with the fluorescence spectrum of *N*-2-naphthyl-*N*-tosylamide. From the linearity of the double reciprocal plot of concentration of *N*-2-naphthyl-*N*-tosylamide and light intensity this was shown to be a bimolecular reaction of *N*-2-naphthyl-*N*-tosylamide and the high-energy intermediate. Therefore, the light emission does not involve the 1,2-dioxetanedione intermediate. In that case chemiluminescence would have been observed also without the addition of *N*-2-naphthyl-*N*-tosylamide.

The group of Baader[87] has performed kinetic studies on the chemiluminescent decomposition of a peracid intermediate of the peroxyoxalate reaction. In the absence of base and the presence of DPA no chemiluminescence is observed, ruling out the peracid as the high-energy intermediate. In the presence of base light emission corresponding to the fluorescence of DPA is observed. Different base systems were studied. With oxygen bases (hydroxide, *tert*-butoxide and *p*-chlorophenolate) deprotonation of the peracid occurred. Because of a competing dark reaction involving nucleophilic substitution of the trichlorophenolate the quantum efficiency was rather low (Scheme 2.17).

When using a nitrogen base such as imidazole, the efficiency of the chemiluminescence reaction increased by a 100-fold. The decay rate constant in this case is dependent on the imidazole concentration: at low concentrations (<20 mM) a linear relation is observed while at high concentrations (100–250 mM) a second-order relation is obtained. This was explained by assuming a dual role for imidazole, as a basic and nucleophilic catalyst. The structure for

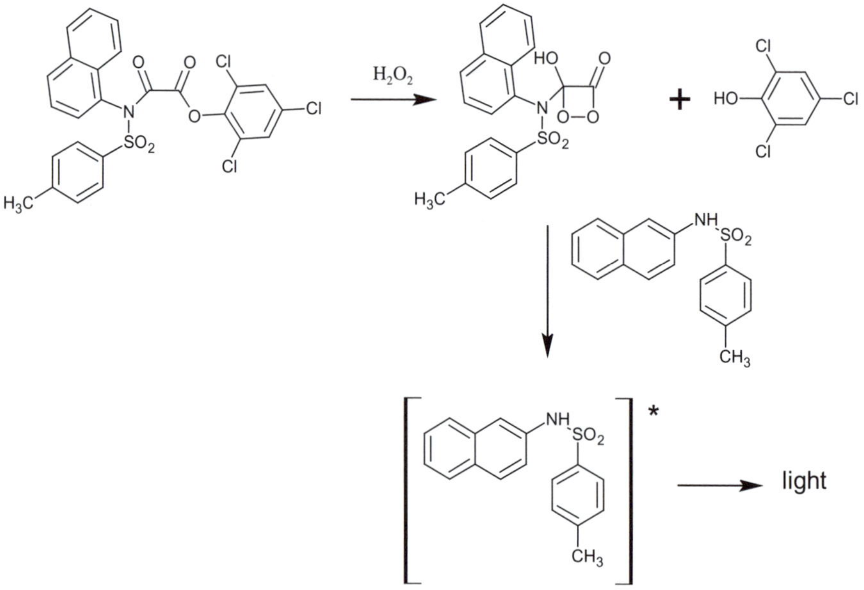

Scheme 2.16 Hybrid oxalic acid derivative containing fluorescer does not emit light until fluorescer is added, pointing to a high-energy intermediate that still contains a leaving group.

the high-energy intermediate was assumed to be 1,2-dioxetanedione. The 1,2-dioxetanone structures were considered less likely candidates, because the same cyclization step should be involved at low and high imidazole concentrations but in the latter case the much better leaving group imidazole is expelled and this step is not observed kinetically (Scheme 2.18). The chemiluminescence efficiency decreased with increasing imidazole concentration. This was interpreted as indicative of an interaction between imidazole and the high-energy intermediate in competition with the interaction of the high-energy intermediate with DPA.

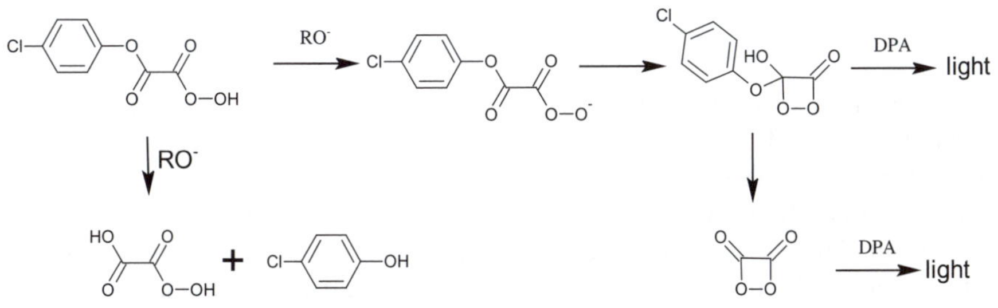

Scheme 2.17 Decomposition of peracid derivative with alkoxide.

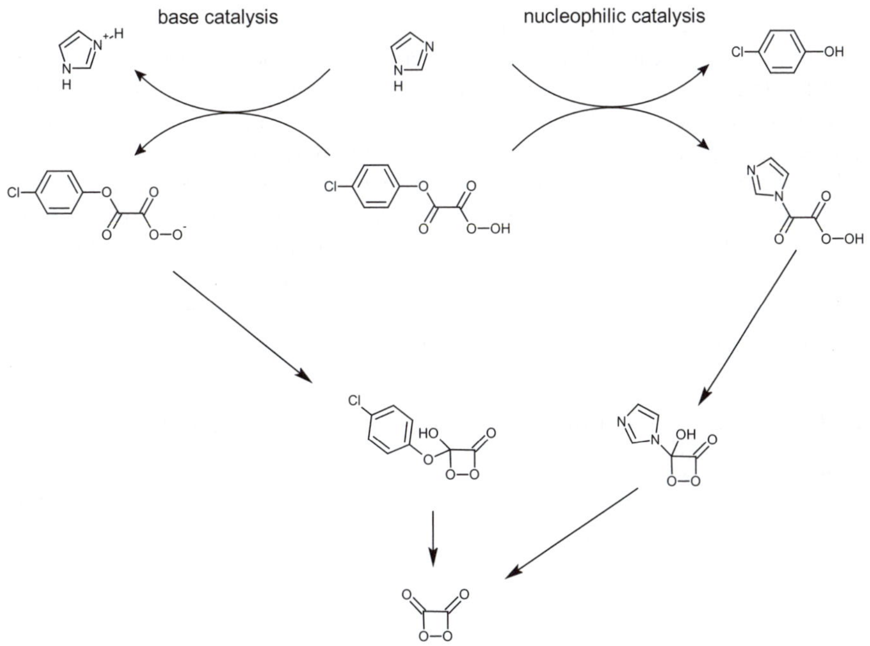

Scheme 2.18 Decomposition of oxalic peracid derivative with imidazole, showing general base and nucleophilic catalysis.

Peroxyoxalates (**2.7**) with electron-donating substituents or fluorescent groups behave at first sight rather unexpectedly (Table 2.1).[88] For electronegatively substituted aromatic peroxyoxalates there is a linear relation in the double reciprocal plot of the chemiluminescence quantum yield and the DPA concentration. When electron-donating substituents are used, this plot is no longer linear for all cases.

2.7

This deviation from linearity means that the excess luminescence for peroxyoxalates **2.7b, c, f** and **g** is not caused by excited DPA. Because the substituted phenol products are essentially non-fluorescent, the results were explained by assuming that the production of excess chemiluminescence is due to excimer formation. These excimers could be formed as the result of an intramolecular interaction in the dioxetanone intermediate still bearing the phenolic moiety. However, no definitive proof of this mechanism could be given, but it seems important to note that electron-donating substituents of peroxyoxalates in some cases can lead to interesting chemiluminescence behaviour of peroxyoxalates.

Instead of the well-known fluorescent organic compounds, certain nanomaterials also are capable of accepting energy from peroxyoxalate treatment with hydrogen peroxide. This was shown by the group of Cui,[89] who prepared gold nanoparticles of different sizes (about 3 and 6 nm) and studied the chemiluminescence emission of these particles upon reaction with bis(2,4,6-trichlorophenyl) oxalate (TCPO) and hydrogen peroxide. During the reaction blue light was emitted with a quantum yield of about 0.003% (for the 6 nm particles). They showed that the light emission originates from the gold nanoparticles. The proposed mechanism involves chemical excitation of the gold particles by a high-energy intermediate formed during the TCPO–hydrogen peroxide reaction. The efficiency of the chemiluminescence depends

Table 2.1 Effect of substitution pattern on linearity of bis reciprocal plot of quantum yield *versus* DPA concentration.

Structure	R^I	*Linear (L) or nonlinear (NL)*
2.7a	H	L
b	$4\text{-}CH_3$	NL
c	$4\text{-}OCH_3$	NL
d	$3,5\text{-}(OCH_3)_2$	L
e	$3,4,5\text{-}(OCH_3)_3$	L
f	$2,4\text{-}(OCH_3)_2$	NL
g	$4\text{-}(CH_3)_2CH$	NL

on the size of the gold particle, with a size of 3 nm being less efficient. According to the authors, the quantum yield can be improved substantially through optimizing the size and surface of the gold nanoparticles.

2.4 Luminol

The oxidation of luminol is one of the oldest described chemiluminescence reactions. It shows under optimum conditions (pH 11–13, hydrogen peroxide) a chemiluminescence excitation quantum yield of about 4% in aqueous solution, which is raised to about 9% in aprotic solvents.[90] Oxidation in aqueous systems gives rise to blue light emission, while in aprotic solvents this shifts to yellow-green.

Luminol (LH_2) can be considered a diprotic acid with pK_As of 6 and 13, respectively. During the chemiluminescence reaction under basic conditions the prevalent luminol anion (LH^-) is oxidized to luminol radical anion ($LH^{\cdot}$) (Scheme 2.19).

This can be achieved by a great number of one-electron oxidants (HRP, metals like cobalt, copper, manganese, iron, *etc.*). Also, many free radical

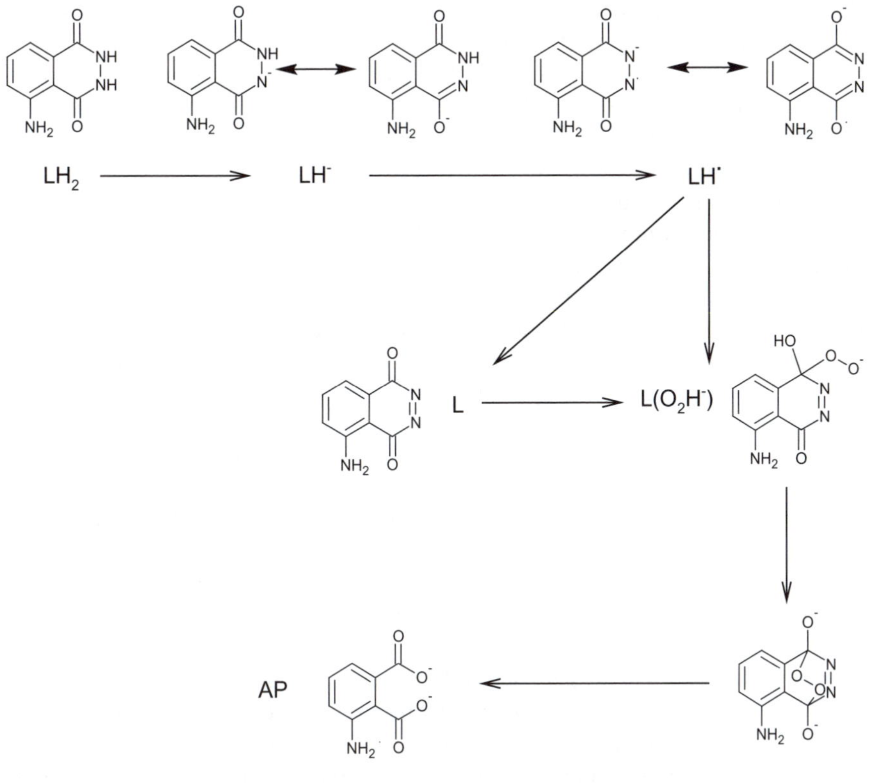

Scheme 2.19 Mechanism of chemiluminescent luminol oxidation.

species like N_3^-, $CO_3^{\cdot -}$ and ClO_2^- are effective oxidants for this step. Chelators such as EDTA have a positive effect in certain cases, *e.g.* the Co^{2+}-catalysed oxidation.[91] The enhancement effect of carbon dioxide on the chemiluminescence reaction has been explained by the proposed formation of the peroxycarbonate radical ($^{\cdot}CO_4^-$).[92] Superoxide has been shown to be ineffective.[93] In a second oxidation step $LH^{\cdot}$ is further oxidized to either aminodiazaquinone (L) or directly (by superoxide anion) to hydroperoxide adduct (LO_2H^-). Another reaction that takes place is the dismutation of two molecules of $LH^{\cdot}$, forming L and LH^-. Addition of hydrogen peroxide to L can also give the hydroperoxide adduct (LO_2H^-). From this adduct an endoperoxide species can be formed, from which molecular nitrogen is expelled, generating excited state 3-aminophthalate dianion (AP). Gold nanoparticles with sizes from 6 to 99 nm can also oxidize luminol in the presence of base (0.01 M) and hydrogen peroxide (0.15 M).[94] The highest chemiluminescence signals were observed with 38 nm particles. The luminescent compound was shown to be AP, and the gold nanoparticles acted catalytically. A mechanism was proposed with a role for the gold particle as a matrix for cleaving the O–O bond of hydrogen peroxide with the generation of hydroxyl radicals that might be stabilized by the particle. Dissolved oxygen played no role in the chemiluminescence reaction.

Gold particles in the presence of silver nitrate[95] are also effective. In this case the gold nanoparticles serve as nucleation centres to catalyse the oxidation of luminol with concomitant reduction of silver nitrate to silver atoms. The produced luminol radicals can react with dissolved oxygen to give LO_2H^-, ultimately forming excited state AP. Reductive compounds interfering with this reaction can be determined using this system.[96] Platinum nanoparticles also catalyse the luminol oxidation.[97] The role and influence of dissolved oxygen on the hemin-catalysed luminol chemiluminescence reaction has been studied by Baj *et al.*[98] They compared the effect of degassing using a helium sparge on the linear calibrations between the amount of oxidant and chemiluminescence peak area. Two groups of oxidants were used, one containing OOH (hydrogen peroxide, *n*- and *t*-butyl hydroperoxide, 3-chloroperbenzoic acid and dibenzoyl peroxide) and the other containing iodosobenzene diacetate and iodosobenzene. To their surprise they found that by using 3-chloroperbenzoic acid or dibenzoyl peroxide as the oxidant the chemiluminescence intensity increased after sparging with helium (with the other oxidants light intensity decreased, as expected). This effect could be tentatively explained by assuming a competitive pathway, depicted in Scheme 2.20, leading to excited state AP that does not involve oxygen.

The addition of surfactants, *e.g.* Triton-X100, has a favourable effect on the chemiluminescence signal from the luminol–hydrogen peroxide reaction in basic solution.[99]

When HRP and hydrogen peroxide are used as the oxidizing system the luminol radical ($L^{\cdot}$) can be formed from the reaction of compound I and compound II (Scheme 2.21).[100] Luminol reacts about 70-fold faster with compound I than with compound II. With excess hydrogen peroxide compound III may be formed.[101] Compound III reacts slowly with luminol to give

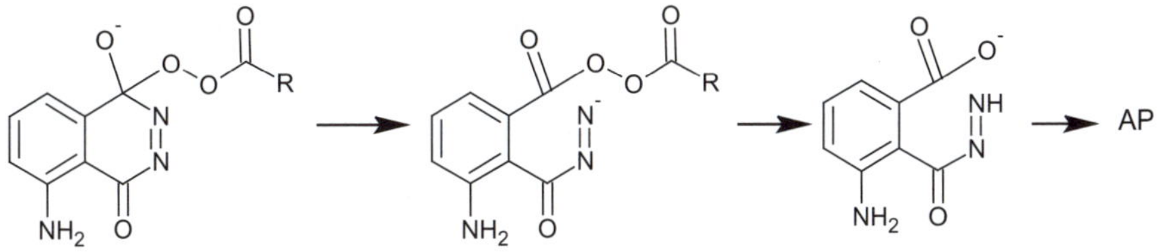

Scheme 2.20 Possible competitive pathway leading to chemiluminescence in the absence of oxygen.

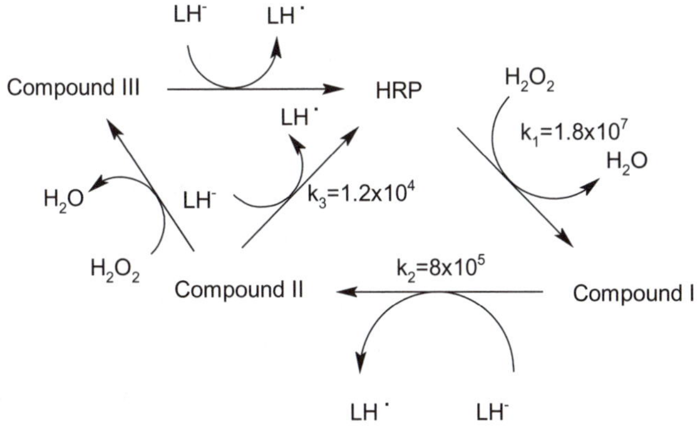

Scheme 2.21 HRP-catalysed oxidation of luminol.

HRP. The inactivation of HRP by excess hydrogen peroxide could be diminished by the addition of imidazole.[102]

When the chemiluminescent oxidation of luminol by hydrogen peroxide catalysed by HRP is performed in a liposome environment encapsulating HRP, the chemiluminescence signal increased seven-fold compared to the reaction in bulk solution,[103] proving that luminol and hydrogen peroxide permeate into the inner phase of the liposome. Liposomes can contain up to 1200 molecules of HRP.[104] They can be used to label antibodies,[105,106] potentially increasing the detectability of antibodies.

The light output of the HRP reaction can be increased by so-called enhancers, as was serendipitously discovered by Thorpe *et al.*[107] The mechanism[108] of this enhanced chemiluminescence involves the very efficient generation of enhancer radicals (En·) increasing HRP turnover. Figure 2.2 shows the structures and reaction rate constants of several enhancers with compounds I and II. In general, compound I is more reactive towards enhancer than compound II although their reduction potentials are estimated to be very similar.[109] Because the pK_As of the enhancer phenol radical cations are several units lower than the parent phenols (p$K_A \sim 10$) this implies that at pH 7–8 oxidation is accompanied by deprotonation. These enhancer radicals proved to be very efficient oxidizers of luminol by electron transfer (Scheme 2.22).

	k_{cpd-I}/L mol^{-1}s^{-1}	k_{cpd-II}/L mol^{-1}s^{-1}
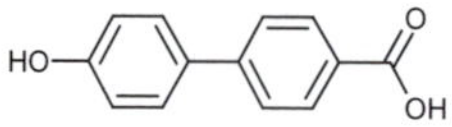	1.6×10^6	3.8×10^5
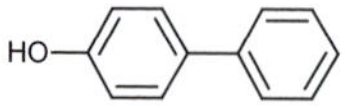	3.9×10^7	2.6×10^6
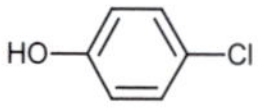	2.4×10^6	3.6×10^4
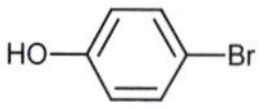	8.3×10^6	5.7×10^5
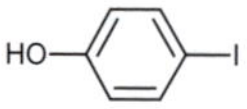	2.8×10^7	3.4×10^6
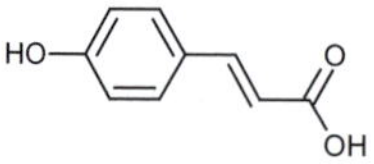	6.0×10^7	1.9×10^7
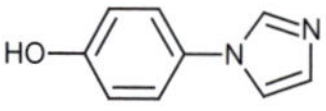	1.2×10^6	1.8×10^5
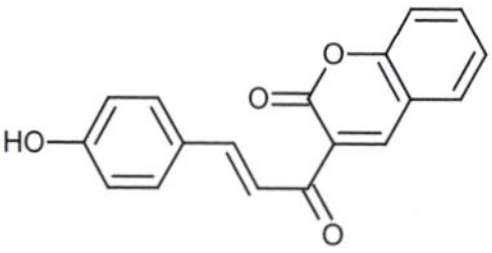	5.0×10^6	8.3×10^5

Figure 2.2   Structures of substituted phenol enhancers along with reaction rate constants for HRP turnover.

$$En^\cdot + LH^- \quad \rightleftharpoons \quad EnH + LH^\cdot$$

Scheme 2.22   Reversible hydrogen atom transfer between enhancer and luminol.

The position of this equilibrium is determined by the difference between the reduction potential of the redox couples, $L^\cdot/LH^-$ and $En^\cdot/EnH$. It was found that the rate of reaction with compound II is also determined by the reduction potential of the $En^\cdot/EnH$ couple.

For these enhancers, a bell-shaped dependence of the luminescence efficiency on the reduction potential of $En^\cdot$ was observed,[108] with its maximum[(for the enhancer 4-(1-imidazoyl)phenol] at approximately 0.8 V *versus* normal hydrogen electrode (NHE), which is roughly the same potential of luminol, 0.87 V, *versus* NHE.[93] Therefore, the enhancement effect is due to a much more

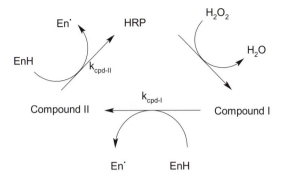

Scheme 2.23   HRP turnover by enhancer oxidation.

efficient oxidation of luminol. The two effects, increased rate of HRP turnover (Scheme 2.23) and reversible electron transfer between enhancer radicals and luminol (Scheme 2.22), both play important roles in explaining the enhancement effect. From this it might be inferred that the concentration of enhancer is proportional to the chemiluminescent signal.

This is not entirely true, for most enhancers there is an optimum concentration. This can be explained by assuming secondary reactions of enhancer radicals with HRP resulting in deactivation of the enzyme.[110] This loss of enzyme activity can be partially prevented by the addition of, for example, serum albumin or polyelectrolytes,[111,112] and skimmed milk and detergents.[113] The last study, using 4-[4-(2-methyl)thiazolyl]phenol as the enhancer, reached an HRP detection limit of 6×10^{-18} moles. The same limit of detection was obtained using 3-(10-phenothiazinyl)propane-1-sulfonate together with an acylation catalyst (4-dimethylaminopyridine or 4-morpholinopyridine).[114] Substituted phenylboronic acids can also be used as enhancers, either alone or in a synergistic combination with 4-substituted phenols.[115,116] The enhancer effect of substituted phenylboronic acids is probably due to the *in situ* formation of the corresponding (enhancer) phenols from reaction with hydrogen peroxide, as pointed out by Akhavan-Tafti.[117] High salt concentration, *e.g.* 3 M potassium chloride, also has an enhancing effect.[118] From this study it was concluded that the enhancing effect was mainly due to the non-enzymatic steps of the luminol chemiluminescence reaction rather than a more efficient enzymatic process.

Direct detection of HRP encapsulated in liposomes could be improved 150-fold by addition of the enhancer 4-iodophenol,[119] showing that the enhancer was also capable of crossing the liposomal membrane.

2.5 Acridinium and Acridine Compounds

Acridinium esters with leaving groups R_1 having a $pK_A < 10$ (*i.e.* a pK_A that is smaller than that of hydrogen peroxide) can generate efficient light emission upon reaction with basic hydrogen peroxide. The efficiency of the light generation is related to the pK_A of the conjugated acid of the leaving group, as pointed out by McCapra[120,121] and Nelson *et al.*[122]

Nelson *et al.* have studied the chemiluminescence and hydrolysis properties of several substituted aromatic acridinium esters (**2.8**). The chemiluminescence profiles (signal intensity *versus* time) depended very much on the substitution. pK_A-Lowering substituents, *e.g.* bromo and fluoro, resulted in a faster reaction and shorter signal duration, while electron-donating groups (methyl and methoxy) caused a slower kinetics (Table 2.2). The optimal detection pH followed the same trend (more acidic leaving groups required lower pH). Hydrolysis rates at pH 7.6 were governed more by the steric effects of the substituents.

2.8

The mechanism of this reaction (Scheme 2.24) is thought to proceed by initial attack of the hydroperoxide anion at the C9 position of the acridinium nucleus, followed by an intramolecular nucleophilic attack of the resulting hydroperoxide at the ester carbonyl group. During this process a four-membered highly strained dioxetane structure (D, E or F) is formed that breaks down with the generation of excited state acridone (G), carbon dioxide and the conjugated acid of the leaving group.

More advanced calculations (for $R_1 = OPh$)[123] indicate that the formation of excited state acridone (G) is most likely to occur from structure D. In earlier work it was thought that the chemiluminescence occurred from decomposition of E (see, however, also McCapra[120]). The rate-determining step (C→D)

Table 2.2 Effect of substituents on kinetics of acridinium ester (**2.8**) reaction with hydrogen peroxide.

X	Y	Z	pH	Peak (s)	Duration (s)	$t_{1/2}$ pH 7.6
H	H	H	11.9	0.42	4.6	0.67
Br	Br	H	10.2	0.22	1.8	2.68
CH_3	H	H	11.9	0.75	14.0	2.00
OCH_3	H	H	13.0	0.60	8.4	2.10
CH_3	H	F	11.3	0.22	1.2	2.24

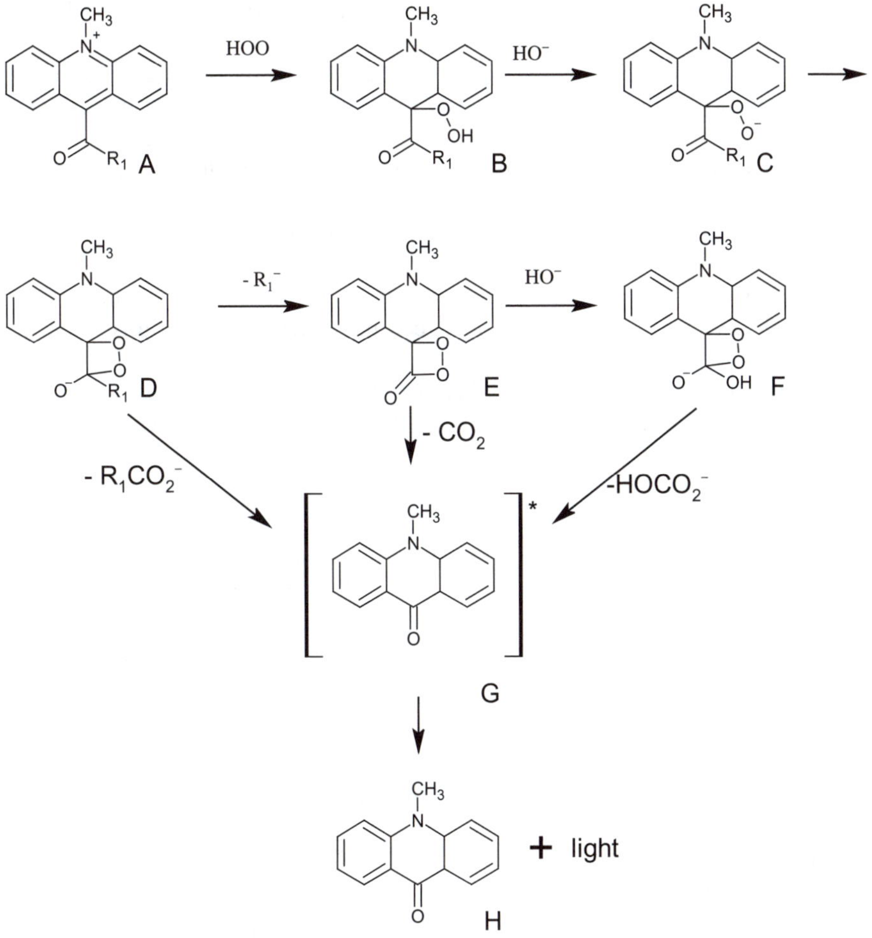

Scheme 2.24 Mechanism of acridinium ester chemiluminescent reaction with basic hydrogen peroxide.

involves leaving group R_1. Therefore, the properties of R_1, be it electronic or steric, will influence the chemiluminescence properties.[124]

As indicated by McCapra, the acidity of the leaving group is proportional to the chemiexcitation yield of the reaction. Kinetic analysis of the acridinium ester chemiluminescence reaction[125] has corroborated the mechanism and provided reaction rate constants. In the absence of hydrogen peroxide acridinium esters hydrolyse rapidly at high pH. This process starts with the formation of the so-called pseudo-base of the acridinium compound (Scheme 2.25).

Pseudo-base formation is a reversible reaction of water (or other nucleophile) at the C9 position of the acridinium moiety. Further reaction with base and oxygen then affords *N*-methylacridone and *N*-methyl-acridinium-9-carboxylic acid.

With nucleophiles other than water (*e.g.* sulfite, methanol, *etc.*) acridinium ester can be protected.[126]

Pseudo-base formation also seems to be dependent on the substitution pattern on the acridinium with, for example, 2,7-dimethoxy substituents being less prone to pseudo-base formation.[127] The same holds for acridinium compounds having sulfonamide leaving groups.[128] Sulfonamide leaving groups were introduced[124,129] as more stable alternatives to the phenyl esters.

The use of certain surfactants[130] helps increase the detectability of acridinium ester labelled proteins.

The corresponding acridine esters also show chemiluminescence upon reaction with hydrogen peroxide under certain reaction conditions (Scheme 2.26).[131]

In mixtures of tetrahydrofuran (THF) and water (67/33 mol%) either bright blue or bright yellow-green chemiluminescence is observed, depending on the

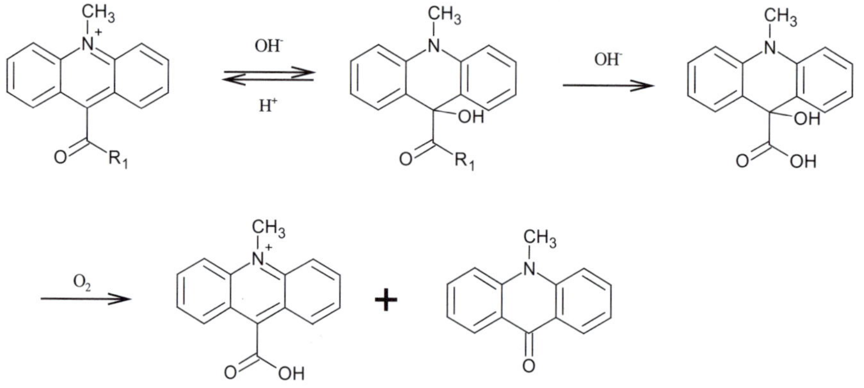

Scheme 2.25   Pseudo-base formation leading to a "dark" reaction of acridinium ester.

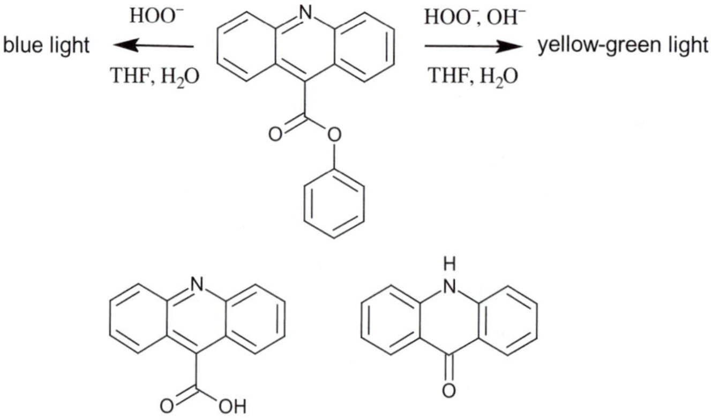

Scheme 2.26   Chemiluminescent oxidation of acridine ester.

conditions. Acridine-9-carboxylic acid and acridone are the isolated products of the reaction. Acridine-9-carboxylic acid is essential non-fluorescent under the reaction conditions. The fluorescence of the product mixture is blue under most conditions (up to $pH \sim 12$) although the chemiluminescence emitted during the reaction is yellow-green. It was shown that the fluorescence of neutral acridone matched the blue chemiluminescence while the fluorescence of the acridone anion coincided with the yellow-green chemiluminescence. The authors showed that in the product mixture specific excitation of the acridone anion could be accomplished at 442 nm, while neutral acridone could be excited at 376 nm. They conclude that the excited states of neutral and anionic acridone do not achieve equilibrium under the reaction conditions and that excited state acridone and acridone anion can be generated separately during the reaction.

This reaction does not involve a dioxetanone as an intermediate.

Acridine-9-carbonylimidazole (**2.9**) is another example of an acridine compound capable of chemiluminescence upon slightly basic hydrogen peroxide treatment.[132]

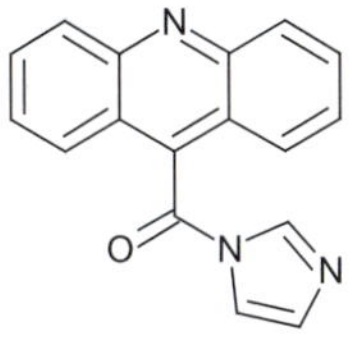

2.9

This compound can be used to quantify hydrogen peroxide concentrations over a wide pH range. The resulting chemiluminescence lasts for many hours and levels of hydrogen peroxide down to 0.4 µM can be quantified.

Acridinium compounds are very versatile in all kinds of applications. This is partly due to the flexibility (adaptability) of the acridinium system. If an intermediate with the general structure **2.10** can be formed during a chemical reaction, this reaction will be chemiluminescent.

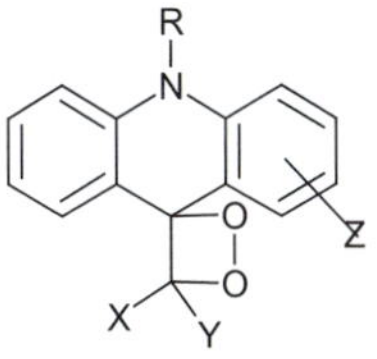

2.10

This has been exploited over the last 40 years or so in many different ways. One way to prepare **2.10** involves the oxygenation of double bonds using singlet oxygen (from sodium hypochlorite and hydrogen peroxide in dimethylformamide). An example with a chemiluminescence quantum yield of 1.3% is shown in Scheme 2.27.[133]

The oldest example uses chloride as the leaving group[134] followed by a phenyl ester (**2.11** and **2.12**).[135] These compounds are triggered by the addition

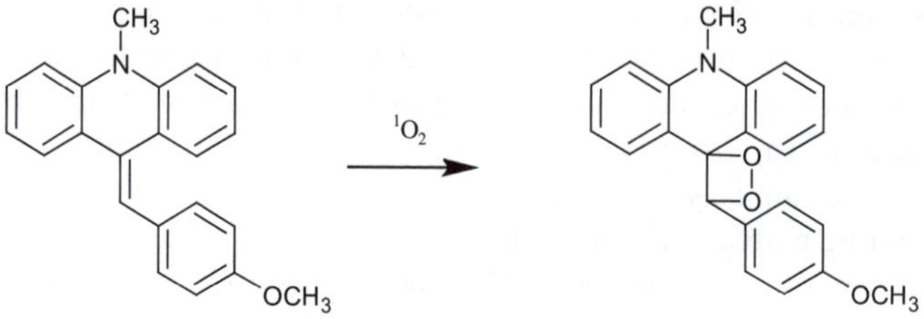

Scheme 2.27 Chemiluminescent oxidation of *N*-methylacridinyl-9-(4-methoxy) benzylidene.

of basic hydrogen peroxide.

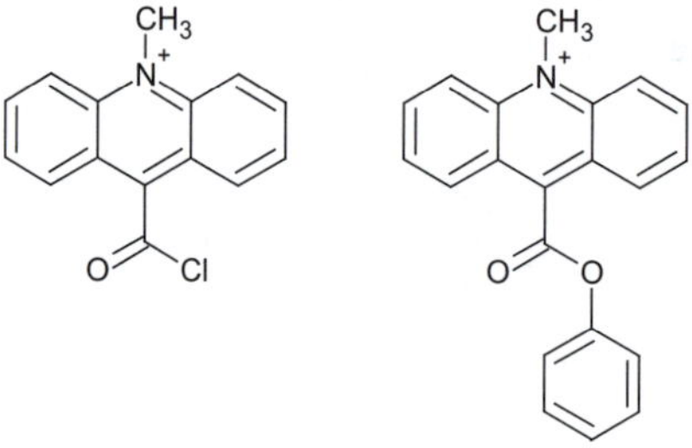

2.11   **2.12**

In 1979 Simpson *et al.* reported the first use of acridinium esters as labels in immunoassay.[136] A few years later Weeks *et al.*[137] used an activated ester substituted acridinium ester (**2.13**) as a label. Although useful this compound suffers from reduced stability at pH > 5.

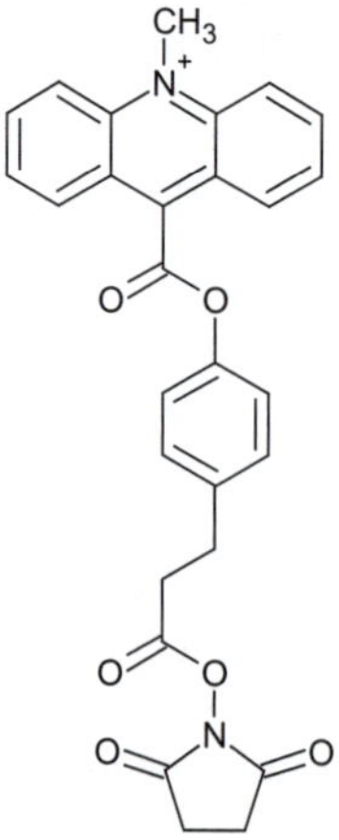

2.13

The evolution of acridinium ester based labelling compounds continued with more stable analogues. This was accomplished by adding extra substituents at the *ortho*-positions of the phenolate leaving group (**2.14**).[138] This increased the hydrolytic stability, probably by steric protection of the ester carbonyl group, or by making the acridinium ester less prone to pseudo-base formation.

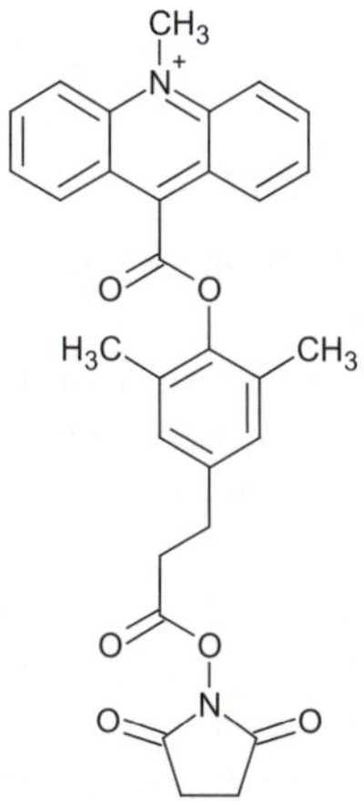

2.14

Another approach to stabilize the acridinium label was used by two groups[124,129] who used sulfonamide leaving groups (**2.15** and **2.16**). The linker group can be attached to the sulfonyl moiety or to the nitrogen.

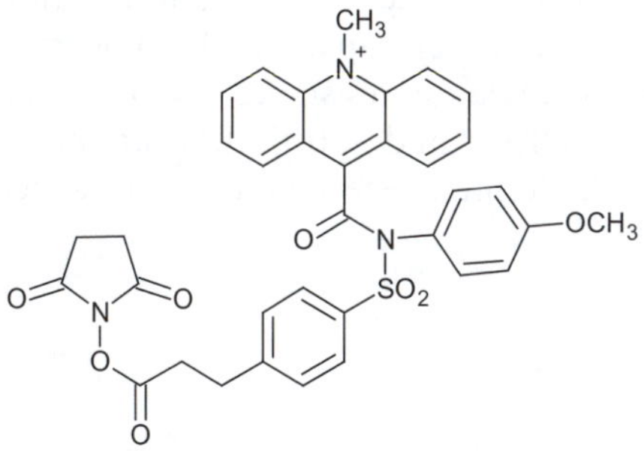

2.15 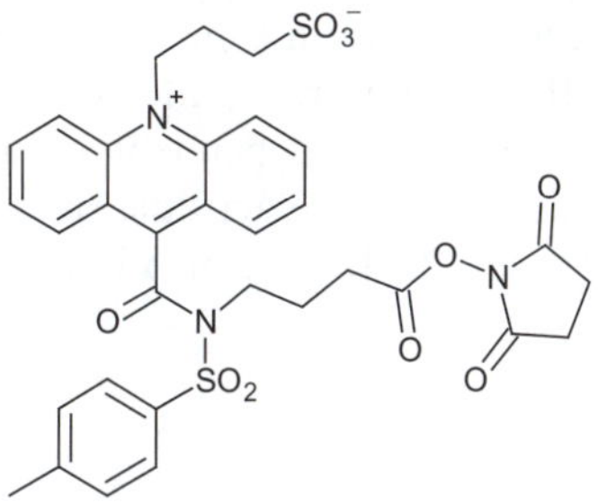   **2.16**

The first N-functionalized acridinium ester was reported by Zomer and Stavenuiter.[139] This approach was later used by Sato in developing immunoassays.[140,141] Apart from phenolate and sulfonamide leaving groups, hydroxamic acids (**2.17**) also proved successful.[142]

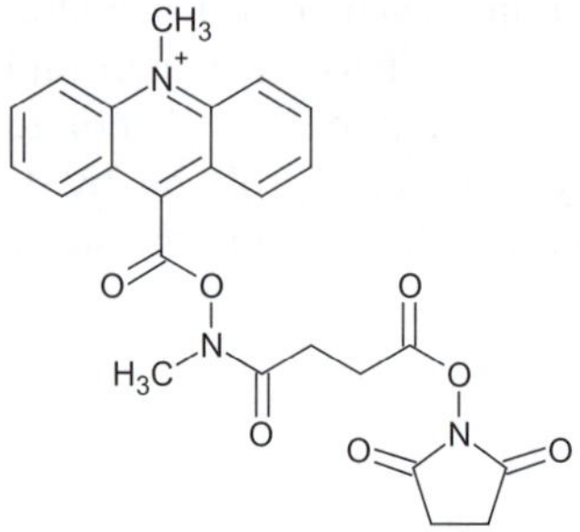

2.17

The acridinium nucleus has also been substituted. Razavi and McCapra synthesized the compounds shown in Figures 2.3 and 2.4.[143,144]

The acridinium ester shown in Figure 2.3 (O–O–C–C torsion angle 60°) has the added flexibility that it disconnects the linking part from the leaving group part. The parent acridinium ester, with an unsubstituted phenyl ester, has a corresponding torsion angle of about 72°. Changes in the leaving group that improve stability or kinetics of the chemiluminescence reaction can be introduced without affecting the linking part. The acridinium shown in Figure 2.4, although sterically congested (O–O–C–C torsion angle of 90°), is less stable. The peri-substituents force the acridine-CO system to deviate from planarity more, thereby perhaps facilitating hydrolysis or pseudo-base formation.[145]

Addition of methoxy-groups to the 2- and 7-positions of the acridine was reported to increase the chemiluminescence quantum yield two- to three-fold (patent application US2008/0014660). Moreover, these substituents were claimed to decrease pseudo-base formation (US2009/0029349), allowing for triggering of the chemiluminescence reaction at relatively low pH.

In conclusion it can be stated that the chemiluminescence efficiency is related to pK_A of the leaving group and substituents on the acridine ring, the kinetics of the light emission is influenced by the leaving group capability, by substitution of the acridine rings and by steric effects, while the resistance towards hydrolysis is mostly dependent on the steric effects and substitution at the 2(7) position(s) of the acridine ring system. Acridinium esters emitting in the near-infrared (**2.18**) have also been reported.[146]

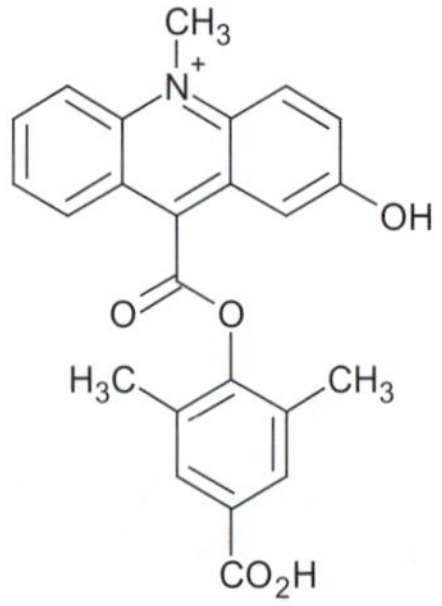

2.18

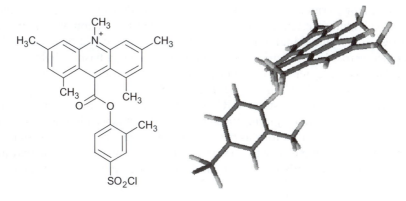

Figure 2.3   Example of a peri-substituted acridinium ester.

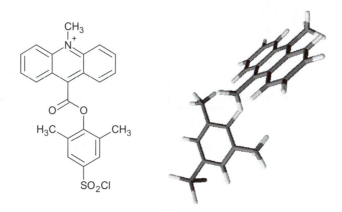

Figure 2.4   Example of a di-ortho substituted acridinium ester.

2.6   Acridan Esters

Acridan esters have been used as model systems for firefly bioluminescence by McCapra.[22] Interest in acridan esters revived when Akhavan-Tafti and Schaap reported their use as signal reagent for horseradish peroxidase.[147] They discovered that acridan esters could be converted into acridinium esters by an enhanced HRP-catalysed oxidation. The formed acridinium ester subsequently reacts with hydrogen peroxide under basic conditions, expelling the leaving group to yield (*via* an intermediate dioxetanone) the excited state of the corresponding acridone. Depending on the substitution on the acridan ring system the reaction either gave a continuous emission of light (acridan **2.19a**, **c** and **d**) under the slightly basic (pH 8) conditions or no signal could be detected (acridan **2.19b** and **e**) (Table 2.3). In both cases it was shown that the corresponding acridinium esters were formed, but that under the conditions used the acridinium esters **2.20b** and **e** were resistant to reaction with hydrogen peroxide. These acridinium esters, in a subsequent step, could be reacted with hydrogen peroxide under more basic conditions, giving a flash of light. In this way, the enzymatic step resulting in the formation of acridinium ester could be separated from the detection step (Scheme 2.28).

Using these acridan esters, HRP could be detected down to 0.1 amol (10^{-19} mol), which is a major improvement over the enhanced luminol reaction. The proposed mechanism for this reaction involves the same enzymatic steps as in the enhanced luminol reaction (*vide supra*). The enhancer radicals that are formed react with acridan ester give, ultimately, acridinium ester. Under the

Table 2.3 Effect of substitution pattern of acridan esters (**2.19**) on chemiluminescence properties.

Structure	R^1	R^2	R^3	Max RLU
a	3-OCH$_3$	H	F	6500
b	1-OCH$_3$	6-OCH$_3$	F	Undetectable
c	H	H	H	55400
d	H	H	F	56880
e	1-OCH$_3$, 4-CH$_3$	6-OCH$_3$	F	Undetectable

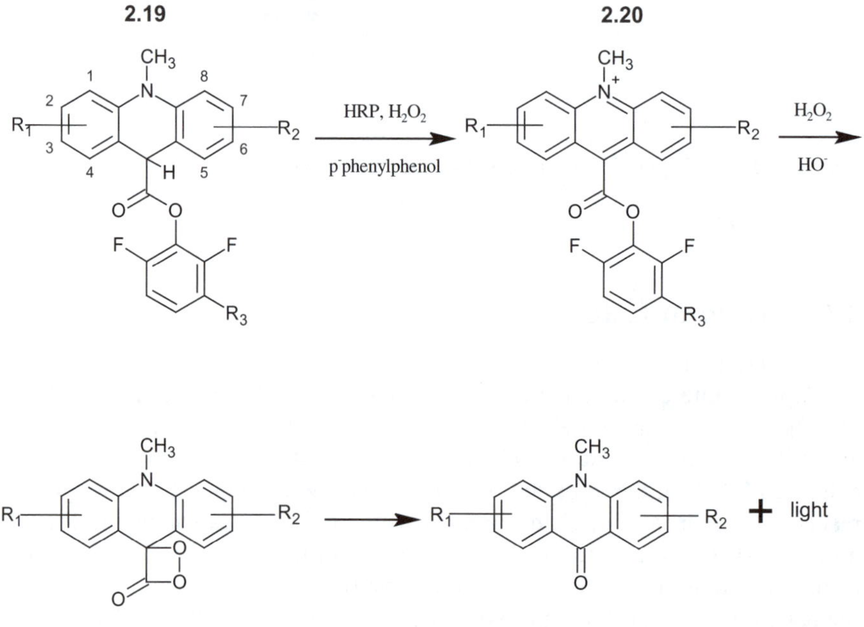

Scheme 2.28 Proposed mechanism of HRP-catalysed enhanced chemiluminescent peroxidation of acridan esters.

reaction conditions the acridinium ester is converted into exited state acridone *via* an intermediate peroxyacridan, which cyclizes to the dioxetanone.

Other acridan bases signal reagents (Scheme 2.29) were developed for, for example, the detection of alkaline phosphatase.[148] Certain cationic aromatic compounds enhanced this reaction.

The same reagents could also be used – after adding suitable linker groups – as protein labelling compounds (Scheme 2.30).[149]

These could be triggered by hydrogen peroxide by subsequently acidifying and adding base to the labelled proteins. The labelled proteins could also be detected on gels.

These acridan compounds can also be oxidized electrochemically[150] *via* two one-electron steps to form an acridinium ester (Scheme 2.31) that in the presence of hydrogen peroxide gives light emission. In the absence of hydrogen peroxide and the presence of dissolved oxygen the intermediate radical can be trapped, forming a peroxy radical adduct. This adduct is reduced forming the

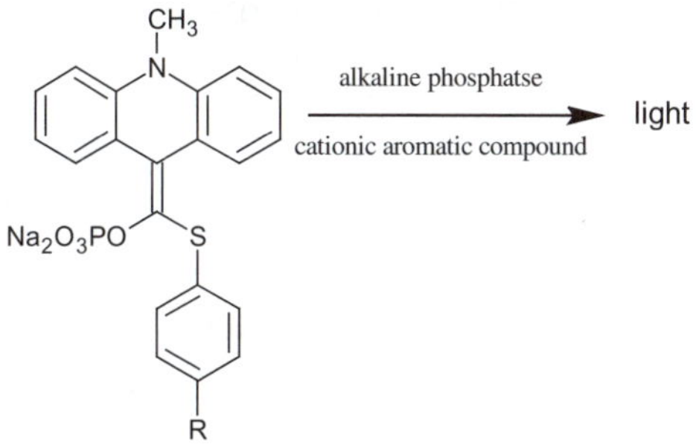

Scheme 2.29   Acridan-based substrate for alkaline phosphatase.

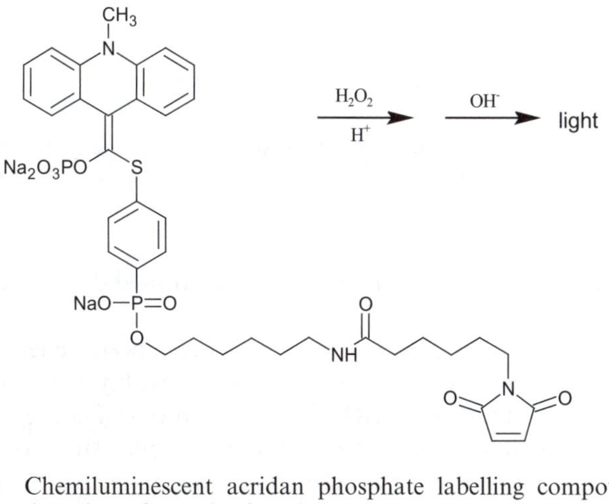

Scheme 2.30   Chemiluminescent acridan phosphate labelling compound used for detection of proteins in gels.

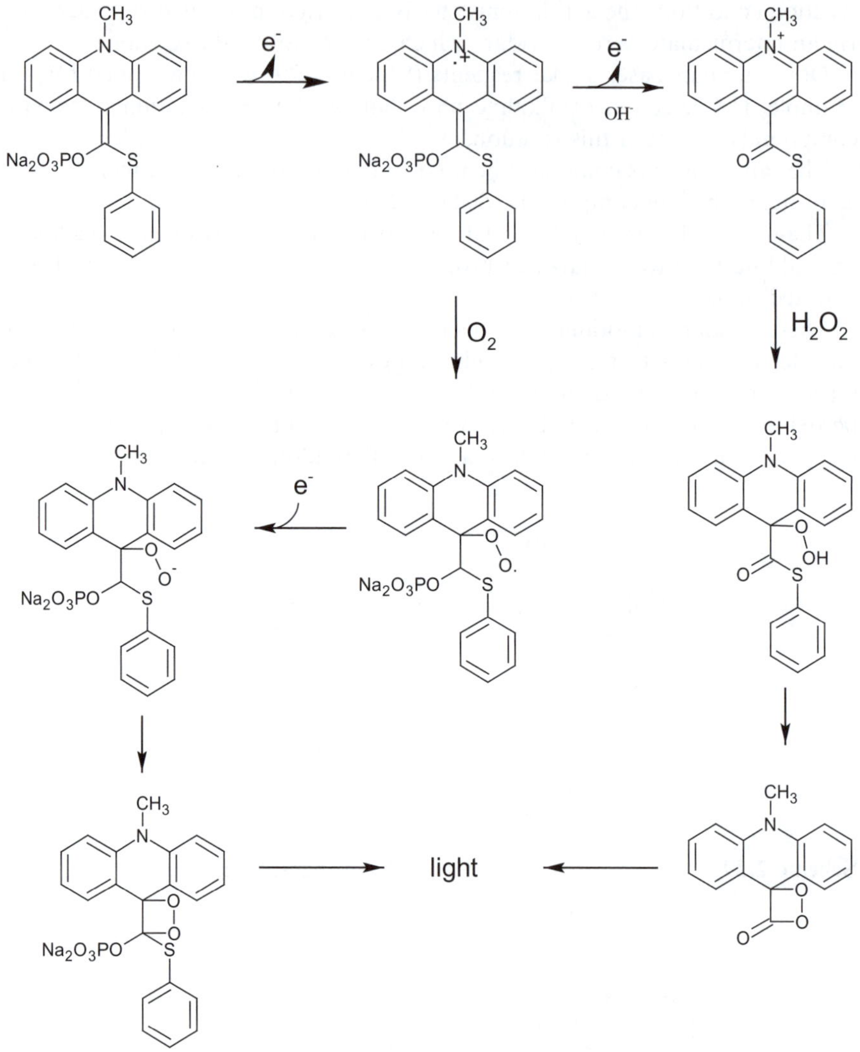

Scheme 2.31 Electrochemiluminescent oxidation of 9-phenylthio(phosphoryloxy) methylidene-10-methylacridan in the presence and absence of hydrogen peroxide.

hydroperoxide anion. Ring closure gives the substituted dioxetane, which upon decomposition emits light.

The use of microwave-triggered metal-enhanced chemiluminescence (MT-MEC) has been developed recently as a possibly very promising new method to detect and quantify HRP.[151-154] As a proof of concept, biotinylated bovine serum albumin was incubated with HRP-labelled streptavidin on a glass surface coated with silver nanoparticles. As a control, uncoated glass was used. When acridan ester and enhancer were added chemiluminescence could be observed. The use of low power microwave pulses increased the chemiluminescence significantly. The silver nanoparticles coated glass surface was three times more effective in increasing the chemiluminescence, indicating the metal enhancement. Very recently, a review has appeared on this subject.[155]

2.7 Concluding Remarks

Chemiluminescence has evolved from being an interesting fact, *via* a promising analytical technique, to a fairly well-understood phenomenon. Research during the last two decades has resulted in a much better understanding of the chemistry of chemiluminescence. This has led to exciting (*sic*) new systems that are not only interesting from a basic scientific perspective but have also has resulted in commercial exploitation of chemiluminescence in all kinds of applications. Nevertheless, as in 1996, when Frank McCapra gave the opening lecture of the 9th International Symposium on Bioluminescence and Chemiluminescence held at Woods Hole, Mass. Entitled "Mechanisms in chemiluminescence and bioluminescence – unfinished business," there still is unfinished business, and I am glad for that. It means that there is still much more to be discovered in this fascinating field.

References

1. F. McCapra, *Meth. Enzymol.*, 2000, **305**, 3.
2. J. L. Adcock, P. S. Francis and N. W. Barnett, *Anal. Chim. Acta*, 2007, **601**, 36.
3. A. R. Bowie, M. G. Sanders and P. J. Worsfold, *J. Biolumin. Chemilumin.*, 1996, **11**, 61.
4. A. J. Brown, P. S. Francis, J. L. Adcock, K. F. Lim and N. W. Barnett, *Anal. Chim. Acta*, 2008, **624**, 175.
5. A. Fan, Z. Cao, H. Li, M. Kai and J. Lu, *Anal. Sci.*, 2009, **25**, 587.
6. P. Fletcher, K. N. Andrew, A. C. Calokerinos, S. Forbes and P. J. Worsfold, *Luminescence*, 2001, **16**, 1.
7. L. Gamiz-Gracia, A. M. Garcia-Campana, J. F. Huertas-Perez and F. J. Lara, *Anal. Chim. Acta*, 2009, **640**, 7.
8. F. Li, C. Zhang, X. Guo and W. Feng, *Biomed. Chromatogr*, 2003, **17**, 96.
9. Y. M. Liu, J. K. Cheng, Y. M. Liu and J. K. Cheng, *J. Chromatogr.*, 2002, **A. 959**, 1.
10. C. A. Marquette and L. J. Blum, *Anal. Bioanal. Chem.*, 2006, **385**, 546.
11. K. Mervartova, M. Polasek, C. J. Martinez, K. Mervartova, M. Polasek and J. Martinez Calatayud, *J. Pharm. Biomed. Anal.*, 2007, **45**, 367.
12. K. Nakashima, R. Ikeda, M. Wada, K. Nakashima, R. Ikeda and M. Wada, *Anal. Sci.*, 2009, **25**, 21.
13. H. A. H. Rongen, R. M. W. Hoetelmans, A. Bult and W. P. van Bennekom, *J. Pharm. Biomed. Anal.*, 1994, **12**, 433.
14. M. Tsunoda and K. Imai, *Anal. Chim. Acta*, 2005, **541**, 13.

15. R. Zhu and W. T. Kok, *J. Pharm. Biomed. Anal.*, 1998, **17**, 985.
16. A. M. Osman, G. Zomer, C. Laane and R. Hilhorst, *Luminescence*, 2000, **15**, 189.
17. L. Li, M. A. Arnold and J. S. Dordick, *Biotechnol. Bioeng.*, 1993, **41**, 1112.
18. M. Hosseini, S. D. Abkenar, M. J. Chaichi and M. Shamsipur, *Acta Chim. Slov.*, 2008, **55**, 562.
19. J. H. Lee, J. Je, J. Hur, M. A. Schlautman and E. R. Carraway, *Analyst*, 2003, **128**, 1257.
20. For an example see: www.zomerbloemen.com.
21. R. F. Vasil'ev, *High Energy Chem.*, 2002, **36**, 170.
22. K. D. Gundermann and F. McCapra, *Chemiluminescence in Organic Chemistry*, Springer-Verlag, Berlin, 1987.
23. F. McCapra, *Biochem . Soc. Trans.*, 1979, **7**, 1239.
24. W. C. Lung, S. Hayashi, M. Lundberg, T. Nakatsu, H. Kato and K. Morokuma, *J. Am. Chem. Soc.*, 2008, **130**, 12880.
25. J. W. Hastings, *J. Mol. Evol.*, 1983, **19**, 309.
26. K. R. Kopecky and C. Mumford, *Can. J. Chem.*, 1969, **46**, 709.
27. F. McCapra, *J. Chem. Soc., Chem. Commun.*, 1968, 155.
28. N. J. Turro and A. Devaquet, *J. Am. Chem. Soc.*, 1975, **97**, 3859.
29. H. E. O'Neal and W. H. Richardson, *J. Am. Chem. Soc.*, 1970, **92**, 6553.
30. W. H. Richardson, F. C. Montgomery, M. B. Yelvington and H. E. O'Neal, *J. Am. Chem. Soc.*, 1974, **96**, 7525.
31. L. B. Harding and W. A. Goddard III, *J. Am. Chem. Soc.*, 1977, **99**, 4520.
32. C. Tanaka and J. Tanaka, *J. Phys. Chem. A*, 2000, **104**, 2078.
33. S. Wilsey, F. Bernardi, M. Olivucci, M. A. Robb, S. Murphy and W. Adam, *J. Phys. Chem. A*, 1999, **103**, 1669.
34. S. Murphy and W. Adam, *J. Am. Chem. Soc.*, 1996, **118**, 12916.
35. N. W. Moriarty, R. Lindh and G. Karlstrom, *Chem. Phys. Lett.*, 1998, **289**, 442.
36. L. D. Vico, Y. J. Liu, J. W. Krogh and R. Lindh, *J. Phys. Chem. A*, 2007, **111**, 8013.
37. R. Bos, S. A. Tonkin, G. R. Hanson, C. M. Hindson, K. F. Lim and N. W. Barnett, *J. Am. Chem. Soc.*, 2009, **131**, 2770.
38. L. W. Chung, S. Hayashi, M. Lundberg, T. Nakatsu, H. Kato and K. Morokuma, *J. Am. Chem. Soc.*, 2008, **130**, 12880.
39. J. H. Wieringa, J. Strating, H. Wynberg and W. Adam, *Tetrahedron Lett.*, 1972, **13**, 169.
40. H. Numan, J. H. Wieringa, H. Wynberg, J. Hess and A. Vos, *J. Chem. Soc., Chem. Comm.*, 1977, 591.
41. E. L. Bastos and W. J. Baader, *Arkivoc*, 2007, 257.
42. A. P. Schaap and S. D. Gagnon, *J. Am. Chem. Soc.*, 1982, **104**, 3504.
43. A. P. Schaap, M. D. Sandison and R. S. Handley, *Tetrahedron Lett.*, 1987, **28**, 1159.
44. A. V. Trofimov, K. Mielke, R. F. Vasil'ev and W. Adam, *Photochem. Photobiol.*, 1996, **63**, 463.

45. J. Y. Koo and G. B. Schuster, *J. Am. Chem. Soc.*, 1978, **100**, 4496.
46. L. H. Catalani and T. Wilson, *J. Am. Chem. Soc.*, 1989, **111**, 2633.
47. F. McCapra, *J. Photochem. Photobiol., A Chem.*, 1990, **51**, 21.
48. H. Isobe, Y. Takano, M. Okumura, S. Kuramitsu and K. Yamaguchi, *J. Am. Chem. Soc.*, 2005, **127**, 8667.
49. H. K. Ijuin, M. Yamada, M. Ohashi, N. Watanabe and M. Matsumoto, *Eur. J. Mass Spectrom.*, 2008, **14**, 17.
50. M. Matsumoto, M. Tanimura, T. Akimoto, N. Watanabe and H. K. Ijuin, *Tetrahedron Lett.*, 2008, **49**, 4170.
51. B. Edwards, A. Sparks, J. C. Voyta and I. Bronstein, *J. Biolumin. Chemilumin.*, 1990, **5**, 1.
52. N. Hoshiya, N. Fukuda, H. Maeda, N. Watanabe and M. Matsumoto, *Tetrahedron*, 2006, **62**, 5808.
53. M. Matsumoto, F. Kakuno, A. Kikkawa, N. Hoshiya, N. Watanabe and H. K. Ijuin, *Tetrahedron Lett.*, 2009, **50**, 2337.
54. H. Akhavan-Tafti and Z. Arghavani, *US Pat.* 5650099.
55. B. Edwards, T. G. Geiser and S. M. Menchen, *US Pat.* 7368296.
56. K. Kitaoka, M. Yamada, M. and S. Kawaguchi, *US Pat.* 7091051.
57. A. P. Schaap, *US Pat.* 6107024.
58. K. Miyashita, M. Minagawa, Y. Ueda, Y. Tada, N. Hoshino and T. Imanishi, *Tetrahedron*, 2001, **57**, 3361.
59. W. Adam, R. Fell and M. H. Schulz, *Tetrahedron*, 1993, **49**, 2227.
60. W. Adam, M. Balci, O. Oakmak, K. Peters, C. R. Saha-Muller and M. Schulz, *Tetrahedron*, 1994, **50**, 9009.
61. D. Heindl, H. P. Josel, R. Beckert, D. Weiss and W. Adam, *Eur. J. Clin. Chem. Clin. Biochem.*, 1997, **35**.
62. N. Hoshiya, N. Watanabe, H. K. Ijuin and M. Matsumoto, *Tetrahedron*, 2006, **62**, 12424.
63. T. Imanishi, Y. Ueda, R. Tainaka and K. Miyashita, *Tetrahedron Lett.*, 1997, **38**, 841.
64. C. W. Jefford and M. F. Deheza, *Heterocycles*, 1999, **50**, 1025.
65. M. Matsumoto, N. Watanabe, H. Kobayashi, H. Suganuma, J. Matsubara, Y. Kitano and H. Ikawa, *Tetrahedron Lett.*, 1996, **37**, 5939.
66. M. Matsumoto, N. Watanabe, H. Kobayashi, M. Azami and H. Ikawa, *Tetrahedron Lett.*, 1997, **38**, 411.
67. M. Matsumoto, T. Hiroshima, S. Chiba, R. Isobe, N. Watanabe and H. Kobayashi, *Luminescence*, 1999, **14**, 345.
68. M. Matsumoto, M. Kawahara and N. Watanabe, *Luminescence*, 1999, **14**, 341.
69. M. Matsumoto, J. Murayama, M. Nishiyama, Y. Mizoguchi, T. Sakuma and N. Watanabe, *Tetrahedron Lett.*, 2002, **43**, 1523.
70. M. Matsumoto, T. Sakuma and N. Watanabe, *Tetrahedron Lett.*, 2002, **43**, 8955.
71. M. Matsumoto, *J. Synth. Org. Chem.*, 2003, **61**, 595.
72. M. Matsumoto, K. Yamada, N. Watanabe and H. K. Ijuin, *Luminescence*, 2007, **22**, 420.

73. E. W. Meijer and H. Wynberg, *J. Chem. Educ.*, 1982, **59**, 1071.
74. A. I. Voloshin, G. L. Sharipov, V. P. Kazakov and G. A. Tolstikov, *Bull. Acad. Sci. USSR, Div. Chem. Sci.*, 1988, **36**, 2634.
75. N. Watanabe, H. Kobayashi, M. Azami and M. Matsumoto, *Tetrahedron*, 1999, **55**, 6831.
76. N. Watanabe, H. Suganuma, H. Kobayashi, H. Mutoh, Y. Katao and M. Matsumoto, *Tetrahedron*, 1999, **55**, 4287.
77. N. Watanabe, K. Nagamatsu, T. Mizuno and M. Matsumoto, *Luminescence*, 2005, **20**, 63.
78. H. Wynberg and H. Numan, *J. Am. Chem. Soc.*, 1977, **99**, 603.
79. W. Adam and A. A. Luis, *Chem. Ber.*, 1982, **115**, 2592.
80. W. Adam and J. B. Wilhelm, *Angew. Chem., Int. Ed.*, 1984, **23**, 166.
81. D. Leclercq, J. P. Bats, P. Picard and J. Moulines, *Synthesis*, 1982, 778.
82. E. A. Chandross, *Tetrahedron Lett.*, 1963, **4**, 761.
83. S. S. Tseng, A. G. Mohan, L. G. Haines, L. S. Vizcarra and M. M. Rauhut, *J. Org. Chem.*, 1979, **44**, 4113.
84. M. M. Rauhut, L. J. Bollyky, B. G. Roberts, M. Loy, R. H. Whitman, A. V. Iannotta, A. M. Semsel and R. A. Clarke, *J. Am. Chem. Soc.*, 1967, **89**, 6515.
85. J. Motoyoshiya, N. Sakai, M. Imai, Y. Yamaguchi, R. Koike, Y. Takaguchi and H. Aoyama, *J. Org. Chem.*, 2002, **67**, 7314.
86. R. Koike, J. Motoyoshiya, Y. Takaguchi and H. Aoyama, *Chem. Commun.*, 2003, **9**, 794.
87. C. V. Stevani and W. J. Baader, *J. Phys. Org. Chem.*, 1997, **10**, 593.
88. R. Koike, Y. Kato, J. Motoyoshiya, Y. Nishii and H. Aoyama, *Luminescence*, 2006, **21**, 164.
89. H. Cui, Z. F. Zhang, M. J. Shi, Y. Xu and Y. L. Wu, *Anal. Chem.*, 2005, **77**, 6402.
90. J. Lee and H. H. Seeliger, *Photochem. Photobiol.*, 1972, **15**, 227.
91. I. Parejo, C. Petrakis, P. Kefalas, I. Parejo, C. Petrakis and P. Kefalas, *J. Pharm. Toxicol. Meth.*, 2000, **43**, 183.
92. C. Xiao, D. A. Palmer, D. J. Wesolowski, S. B. Lovitz and D. W. King, *Anal. Chem.*, 2002, **74**, 2210.
93. G. Merenyi, J. Lind, X. Shen and T. E. Eriksen, *J. Phys. Chem.*, 1990, **94**, 748.
94. Z. F. Zhang, H. Cui, C. Z. Lai and L. J. Liu, *Anal. Chem.*, 2005, **77**, 3324.
95. H. Cui, J. Z. Guo, N. Li and L. J. Liu, *J. Phys. Chem. C*, 2008, **112**, 11319.
96. N. Li, J. Guo, B. Liu, Y. Yu, H. Cui, L. Mao and Y. Lin, *Anal. Chim. Acta*, 2009, **645**, 48.
97. S. L. Xu and H. Cui, *Luminescence*, 2007, **22**, 77.
98. S. Baj, T. Krawczyk and K. Staszewska, *Luminescence*, 2009.
99. X. Liu, A. Li, B. Zhou, C. Qiu and H. Ren, *Chem. Cent. J.*, 2009, **3**, 7.
100. S. B. Vlasenko, A. A. Arefyev, A. D. Klimov, B. B. Kim, E. L. Gorovits, A. P. Osipov, E. M. Gavrilova and A. M. Yegorov, *J. Biolumin. Chemilumin.*, 1989, **4**, 164.

101. M. B. Arnao, M. Acosta, J. A. Del Rio, R. Varon and F. Garcia-Canovas, *Biochem. Biophys. Acta, Prot. Struct. Mol. Enzym.*, 1990, **1041**, 43.
102. O. Nozaki, H. Kawamoto, O. Nozaki and H. Kawamoto, *Luminescence*, 2003, **18**, 203.
103. T. Kamidate, K. Komatsu, H. Tani, A. Ishida, T. Kamidate, K. Komatsu, H. Tani and A. Ishida, *Anal. Sc.*, 2008, **24**, 477.
104. T. Suita, T. Kamidate, M. Yonaiyama and H. Watanabe, *Anal. Sci.*, 1997, **13**, 577.
105. T. Suita and T. Kamidate, *Anal. Sci.*, 1999, **15**, 349.
106. T. Suita, H. Tani and T. Kamidate, *Anal. Sci.*, 2000, **16**, 527.
107. T. P. Whitehead, G. H. G. Thorpe and T. J. N. Carter, *Nature*, 1983, **305**, 158.
108. P. M. Easton, A. C. Simmonds, A. Rakishev, A. M. Egorov and L. P. Candeias, *J. Am. Chem. Soc.*, 1996, **118**, 6619.
109. B. He, R. Sinclair, B. R. Copeland, R. Makino, L. S. Powers and I. Yamazaki, *Biochemistry*, 1996, **35**, 2413.
110. Y. L. Kapeluich, M. Y. Rubtsova and A. M. Egorov, *Luminescence*, 1997, **12**, 299.
111. E. L. Gorovits, E. M. Gavrilova, V. A. Izumrudov, A. M. Egorov and A. B. Zezin, *Dokl. Biochem.*, 1989, **307**, 229.
112. E. L. Gorovits, V. A. Izumrudov, V. V. Pisarev, E. M. Gavrilova and A. M. Egorov, *Biotechnol. Appl. Biochem.*, 1995, **22**, 249.
113. R. Iwata, H. Ito, T. Hayashi, Y. Sekine, N. Koyama and M. Yamaki, *Anal. Biochem.*, 1995, **231**, 170.
114. E. Marzocchi, S. Grilli, L. la Ciana, L. Prodi, M. Mirasoli and A. Roda, *Anal. Biochem.*, 2008, **377**, 189.
115. L. J. Kricka and X. Ji, *J. Biolum. Chemilum.*, 1996, **11**, 137.
116. L. J. Kricka, M. Cooper and X. Ji, *Anal. Biochem.*, 1996, **240**, 119.
117. H. Akhavan-Tafti, R. A. Eickholt, K. S. Lauwers and R. S. Handley, *US Pat.* 7390670.
118. A. B. Collaudin and L. J. Blum, *Photochem. Photobiol.*, 1997, **65**, 303.
119. T. Kamidate, M. Maruya, H. Tani and A. Ishida, *Anal. Sci.*, 2009, **25**, 1163.
120. F. McCapra, *Acc. Chem. Res.*, 1976, **9**, 201.
121. F. McCapra, *Proc. Royal Soc. London, Ser. B Biol. Sci.*, 1982, **215**, 247.
122. N. C. Nelson, A. B. Cheikh, E. Matsuda and M. M. Becker, *Biochemistry*, 1996, **35**, 8429.
123. J. Rak, P. Skurski and J. Blazejowski, *J. Org. Chem.*, 1999, **64**, 3002.
124. P. G. Mattingly, *J. Biolumin. Chemilumin.*, 1991, **6**, 107.
125. D. W. King, W. J. Cooper, S. A. Rusak, B. M. Peake, J. J. Kiddle, D. W. O'Sullivan, M. L. Melamed, C. R. Morgan and S. M. Theberge, *Anal. Chem.*, 2007, **79**, 4169.
126. P. W. Hammond, W. A. Wiese, A. A. Waldrop 3rd, N. C. Nelson and J. Arnold, *J. Biolum. Chemilum.*, 1991, **6**, 35.
127. A. Natrajan, T. Sells, H. Schroeder, G. Yang, D. Sharpe, Q. Jiang, H. Lukinsky, and S. Law, *US Pat.* 7319041.

128. M. Adamczyk, J. C. Gebler, P. G. Mattingly and J. Wu, *Rapid Commun. Mass Spectrom.*, 2000, **14**, 2112.
129. T. Kinkel, H. Lubbers, E. Schmidt, P. Molz and H. J. Skrzipczyk, *J. Biolumin. Chemilumin.*, 1989, **4**, 136.
130. F. J. Bagazgoitia, J. L. Garcia, C. Diequez, I. Weeks and J. S. Woodhead, *J. Biolumin. Chemilumin.*, 1988, **2**, 121.
131. E. H. White, D. F. Roswell, A. C. Dupont and A. A. Wilson, *J. Am. Chem. Soc.*, 1987, **109**, 5189.
132. A. A. Waldrop III, J. Fellers and C. P. Vary, *Luminescence*, 2000, **15**, 169.
133. G. Perkizas and J. Nikokavouras, *Monatsh. Chem.*, 1986, **117**, 89.
134. M. M. Rauhut, D. Sheehan, R. A. Clarke, B. G. Roberts and A. M. Semsel, *J. Org. Chem.*, 1965, **30**, 3587.
135. F. McCapra and D. G. Richardson, *Tetrahedron Lett.*, 1964, **5**, 3167.
136. J. S. A. Simpson, A. K. Campbell, M. E. T. Ryall and J. S. Woodhead, *Nature*, 1979, **279**, 646.
137. I. Weeks, I. Beheshti, F. McCapra, A. K. Campbell and J. S. Woodhead, *Clin. Chem.*, 1983, **29**, 1474.
138. S. J. Law, T. Miller, U. Piran, C. Klukas, S. Chang and J. Unger, *J. Biolumin. Chemilumin.*, 1989, **4**, 88.
139. G. Zomer and J. F. C. Stavenuiter, *Anal. Chim. Acta*, 1989, **227**, 11.
140. H. Sato, H. Mochizuki, Y. Tomita and T. Kanamori, *Clin. Biochem.*, 1996, **29**, 509.
141. N. Sato, K. Shirakawa, Y. Kakihara, H. Mochizuki and T. Kanamori, *Anal. Sc.*, 1996, **12**, 853.
142. R. Renotte, G. Sarlet, L. Thunus and R. Lejeune, *Luminescence*, 2000, **15**, 311.
143. Z. Razavi and F. McCapra, *Luminescence*, 2000, **15**, 239.
144. Z. Razavi and F. McCapra, *Luminescence*, 2000, **15**, 245.
145. J. W. Bunting, V. S. F. Chew, S. B. Abhyankar and Y. Goda, *Can. J. Chem.*, 1983, **62**, 351.
146. A. Natrajan, Q. Jiang, D. Sharpe and S. Law, Sayjong, *US Pat.* 7611909.
147. H. Akhavan-Tafti, K. Sugioka, Z. Arghavani, R. Desilva, R. S. Handley, Y. Sugioka, R. A. Eickholt, M. P. Perkins and A. P. Schaap, *Clin. Chem.*, 1995, **41**, 1368.
148. H. Akhavan-Tafti, Z. Arghavani and R. Desilva, *US Pat.* 6045727.
149. H. Akhavan-Tafti, R. Desilva, K. Sugioka, R. S. Handley and A. P. Schaap, *Luminescence*, 2001, **16**, 187.
150. F. Mirkhalaf and R. Wilson, *Electroanalysis*, 2005, **17**, 1761.
151. M. J. Previte, K. Aslan, S. N. Malyn and C. D. Geddes, *Anal. Chem.*, 2006, **78**, 8020.
152. M. J. Previte, K. Aslan, S. Malyn and C. D. Geddes, *J. Fluoresc.*, 2006, **16**, 641.
153. M. J. Previte and C. D. Geddes, *J. Am. Chem. Soc.*, 2007, **129**, 9850.
154. M. J. Previte, K. Aslan and C. D. Geddes, *Anal. Chem.*, 2007, **79**, 7042.
155. K. Aslan and C. D. Geddes, *Chem. Soc. Rev.*, 2009, **38**, 2556.

CHAPTER 3

Progress and Perspectives on Bioluminescence: from Luminous Organisms to Molecular Mechanisms

J. WOODLAND HASTINGS

Department of Molecular and Cellular Biology, Harvard University, 16 Divinity Avenue, Cambridge, MA, 02138, USA

3.1 Introduction

When I started in the field in 1948, much was known about the organisms that emit light, but very little about the chemistry or enzymology of bioluminescence. McElroy's then recent discovery[1] of the requirement for ATP in firefly light emission provided the first identification of a biochemical component in a luciferase system, and marked the onset of many successful biochemical investigations of luminous systems over the subsequent decades.

After reviewing some highlights of discoveries over those years, I will discuss the definitions of luciferin and photoprotein, two terms that are not well defined or fully accepted in the field. In addition, in different groups of organisms there are many different luciferins and luciferases, but there is no agreed guideline for the selection of identifiers, so I discuss how that identifier might be best selected.

Harvey's 1952 monograph comprehensively reviewed the extensive knowledge of the many different luminous organisms then known. This led him to a conclusion concerning the evolutionary origins, namely that the ability to emit light arose independently many different times in the course of evolution, artfully articulated as "...cropping up here and there as if a handful of damp sand has

Chemiluminescence and Bioluminescence: Past, Present and Future
Edited by Aldo Roda
© Royal Society of Chemistry 2011
Published by the Royal Society of Chemistry, www.rsc.org

been cast over the names of various groups written on a blackboard, with luminous species appearing wherever a mass of sand stuck."[2] Despite this, and a later paper,[3] it seems that the important conclusion did not really stick; Harvey and others continued to refer to luciferin as a single substance. The hand of God threw the sand, it might appear, even while Darwinism was not denied.

Indeed, the first and only major advance in understanding the chemical basis for light emission prior to 1947, the luciferin–luciferase test of Dubois,[4] appears to have had a constraining influence on later research. Even though systems in different organisms were recognized as different, it was evidently assumed that all would conform to some similar biochemical mechanism. One of my enduring memories of a conference on bioluminescence, held in Asilomar, California in 1954, is of the night that many participants crowded into a dark room to watch Harvey and Haneda test for the luciferin–luciferase reaction in several organisms brought for that purpose by both of them and others. All results were negative.

Dubois described the chemical nature of bioluminescence in a way that put it in the category of other biochemical reactions then known. From simple but elegant experiments with luminous beetles, and later with the clam *Pholas*, he showed that light emission not only occurred in aqueous extracts of the luminous organ, but that the activity had two components, one heat stable and the other heat labile. These were obtained as hot and cold water extracts, and suitably named luciferin (light bearing) and luciferase (light enzyme), respectively.

If failure to think outside the envelope of Dubois stymied progress, so did the absence of knowledge concerning the structures of luciferins. Organic chemistry was vigorous and successful in the first decades of the twentieth century, and the oxidation of many organic compounds, such as pyrogallol and the Grignard reagent, were known to emit light, along with the even brighter chemiluminescent emissions from the oxidation of luminol (phthalcyclohydrazides) and biacridium salts.[5] But luciferins other than *Cypridina* (now cypridinid luciferin) were studied little if at all over the period before McElroy, and their chemical natures remained unknown. The failure to use the firefly (beetle) over this long period, especially given its brightness and worldwide abundance, is difficult to understand. Thus, from Dubois to McElroy, studies stayed inside the Dubois envelope, as though all systems would (or should) conform to one model.

A final comment about the Dubois luciferin–luciferase test is that the original postulate was incorrect in the case of the beetle; the limiting activity remaining in the hot water extract was ATP, a co-substrate, and not luciferin. This in no way diminishes the importance of the experiment or the conclusions, but it does show that conclusions may be useful but off the mark.

3.2   Isolation of Many Different Bioluminescence Systems

In the first decades after 1947, many luminous species were characterized biochemically, including the bacteria,[6,7] fungi,[8] dinoflagellates,[9] cypridinids,[10] coelenterates[11–14] and *Pholas*.[15,16]

3.2.1   Insects: Beetles and Diptera

3.2.1.1   Beetles

The discovery of the ATP requirement in the firefly reaction came at a time when protein purification techniques were being perfected, and pure firefly luciferase was soon obtained and crystallized,[17] but its crystal structure would not be determined until some 40 years later.[18] It shows two distinct regions, a large (436 aa) N-terminal and the smaller (110 aa) C-terminal domain, with a small flexible linker peptide.

Firefly luciferin (LH_2) was crystallized and its structure determined to be a fluorescent benzothiazoyl thiazole.[19–21] Biochemical studies revealed that ATP reacts with the luciferin to give an intermediate luciferyl-AMP (LH_2-AMP), the substrate for the subsequent reaction with oxygen;[22] synthetic LH_2-AMP is active for light emission and its subsequent reaction with oxygen was later postulated to result in the formation of an intermediate cyclic peroxide,[23,24] and proof of its structure came some years later.[25] Breakdown of this peroxide leads to the excitation of the product oxyluciferin, whose structure was later shown to be the product of both the chemiluminescent and bioluminescent reactions.[26,27]

The very slow turnover was traced to an inhibitor formed in the reaction, identified as dehydroluciferyl (L-AMP); earlier believed to be the emitter in the reaction, it was shown to be the product of the dark oxidation of LH_2-AMP to yield L-AMP and H_2O_2.[28] The discovery that coenzyme A stimulated light emission[29] remained an enigma until it was shown by Fraga[30] to be due to its reaction with L-AMP to form the less inhibitory dehydroluciferyl-CoA (L-CoA). The involvement of CoA was explained when it was discovered that other enzymes that also activate the carbonyl group by adenylation, namely fatty acid acyl-CoA synthetases (FACSs), are homologous to firefly luciferase, which also suggests the evolutionary origin of luciferase.[31]

Indeed, firefly luciferase is a bifunctional enzyme; it is both an oxygenase and a fatty acid acyl-CoA synthetase.[32] Likewise, non-luminous organisms (*e.g.*, mealworms) have weak luciferase activity with added luciferin,[33,34] attributable to FACSs. Indeed, FACSs have been shown to be functionally convertible into firefly luciferase by site-directed mutagenesis of one of the putative residues for luciferin binding to a serine;[35] they suggest that this serine might be key in the divergence of luciferase from FACS in insects, and that it might be necessary for luminescence in all beetle luciferases.

The presence of a firefly luciferase-like gene in a sponge has been reported;[36] expressed heterologously, the protein emits light with added ATP and firefly luciferin. It could be due to a FACS; neither the presence of luciferin in nor bioluminescence of the sponge has been demonstrated. The same authors also found a gene homologous with a reported luciferin regenerating enzyme.[37–39] As these reports have been questioned,[40,41] the sponge report must also await a clarification in this respect.

During the early years it was well recognized, notably by Strehler, that the specificity of the firefly reaction for ATP meant that it could be used to determine ATP quantitatively,[42] and thus the presence of living organisms. Such practical uses are now manifold and very important commercially; after its cloning[43] many applications have been developed in which luciferase is used as a molecular marker.[44] Even more striking, perhaps, is the use of beetle luciferase in DNA sequencing, which is based on the release of pyrophosphate in the DNA polymerase reaction and its conversion into ATP.[45,46]

3.2.1.2   Diptera

The light organs of the North American dipteran *Orfelia* (formerly *Platyura*) present unusual and intriguing cytological features.[47] After a conversation with Jean-Marie Bassot, Therese Wilson initiated a biochemical study of this glowworm and of the other dipteran larva that shares some of its outward characteristics, that of the fungus gnat *Arachnocampa*, a tourist attraction in New Zealand and Australian caves.

The luciferin–luciferase reaction, a requirement for oxygen, and ATP stimulation had been demonstrated in *Arachnocampa*.[48,49] *Orfelia* turned out to be very different;[50] results with crude luciferin indicated that active material could only be obtained by including DTT or ascorbic acid in the extract before heating, and that it is bound. This was confirmed by gel filtration, which revealed two active peaks, luciferase at $M_r \sim 140$ kDa (probably a dimer) and luciferin in the void volume, thus bound to a much larger protein. The two give light in the assay with no other additions, but the addition of DTT results in a strong stimulation, with the light continuing to increase over a period of at least 30 min, suggesting a continuing slow release of luciferin.

In studies with *Arachnocampa* extracts it was found that light emission in crude extracts is not increased by DTT or ascorbic acid; stimulation by ATP was confirmed.[50] The active luciferin fraction was located on TLC plates and found to emit fluorescence peaking at 415 nm. Addition of luciferin and ATP to the column-purified luciferase results in a rapid rise of light and a slow decay ($t_{1/2} \sim 1$ min), but the kinetics indicate slow, or no, enzyme turnover, as in firefly extracts.

The sequence of a peptide from the *Arachnocampa* luciferase was used to synthesize the corresponding polynucleotide and then to clone and sequence the putative luciferase gene. However, protein expressed heterologously lacked activity (T. Wilson, personal communication). The cloned gene could be a FACS; its nucleotide sequence is similar to both firefly luciferase and FACS enzymes.

3.2.2   Bacterial Luminescence

3.2.2.1   In vitro; a Peroxide Enzyme Intermediate

Over the years there were many negative luciferin–luciferase tests reported but no biochemical studies. However, from *in vivo* studies much had been inferred about the source of electrons for the reduction of luciferin, and Strehler thereby perceived the correct experiment; he added reduced pyridine nucleotide (NADH; DPNH at the time) to a cell extract and obtained light emission.[6] It was soon discovered that FMN stimulated and that its reduced form is the luciferin;[7,51] a second substrate, a long-chain saturated aliphatic aldehyde, was also identified.[52] All such aldehydes are active;[53] tetradecanal was identified as naturally functioning in one bacterial species.[54]

A key early achievement was the demonstration and subsequent isolation of a long-lived intermediate able to emit light in the absence of oxygen, but requiring oxygen for its formation.[55] Its existence was inferred because light emission continued for many seconds or even minutes after all reduced substrate ($FMNH_2$) had been oxidized non-enzymatically (and very rapidly) without emission, thus revealing a single enzyme turnover *in vitro*.

This quasi-stable intermediate ($t_{1/2} \sim 1$ h at 2 °C), postulated to be luciferase-bound flavin peroxide, was isolated by low-temperature chromatography[56] and its structure confirmed to be the 4a-peroxy FMN.[57] All bioluminescence reactions are postulated to involve peroxide intermediates, but the bacterial peroxide is the only one known to be linear; its reaction with aldehyde and then breakdown to emit light may proceed *via* a Baeyer–Villiger type mechanism.[58]

In the absence of accessory emitters, the light peaks at about 495 nm. The fact that the emitter is a flavin was demonstrated by the finding that the color is shifted by reduced flavin analogs; for example, with 5,6-dimethyl FMN the peak emission is at 470 nm, and with 2-thio FMN at 534 nm.[59] The emitter itself is postulated to be the 4a-hydroxy FMN, detected spectrally in the reaction of the peroxide with decanal at 2 °C.[60] We now know that accessory emitters occur in some species, resulting in emissions at different wavelengths, shorter in some[61] and longer in others.[62,63]

The heterodimeric luciferase,[64] in which the alpha subunit was identified as catalytic,[65] was crystallized and shown to have a mobile sequence,[66] in which two lysine residues are critical for the stabilization of reaction intermediates.[67] Recently, the structure of the protein with bound (oxidized) flavin was determined, but neither the structure with reduced flavin nor that of the luciferase-peroxy flavin has been determined.[68]

3.2.2.2 In vivo *Control of Synthesis: Autoinduction*

An unanticipated bonus from studies of luminous bacteria was the discovery of autoinduction. At concentrations less than $\sim 10^6$ cells ml^{-1} in liquid culture, cells grow but do not synthesize luciferase until the cells reach a higher density, at which point luciferase synthesis is five or ten times more rapid than cell growth.[69] This was shown to be due to transcriptional regulation of luciferase genes by a substance produced by the bacteria themselves, released into the medium, freely diffusible and thereby affecting all cells. The substance acts as a specific inducer and can be viewed as a pheromone responsible for chemical communication in bacteria, which was not previously considered to occur. The mechanism was later dubbed "quorum sensing" because the genes are turned on only at higher cell concentrations,[70] as in the light organ of a host such as a fish or squid.[71,72]

The autoinducer was shown to be a homoserine lactone,[73] and synthetic autoinducer was demonstrated to turn on luciferase synthesis in cells at a low density.[74] Later studies revealed that such a mechanism occurs in many different bacteria, involving many different autoinducers, for many different specific purposes, where turning on a specific gene at high cell densities is functionally important.[75] For example, an infectious toxin-producing bacteria may restrict toxin synthesis until the bacterial population is high enough to release a lethal dose in a short time, such that the organism being attacked is unable to raise defenses, as it might do if exposed to lower concentrations over a longer time.[76–78]

3.2.3   Fungi: Similar but not the Same Mechanism as Bacteria

Fungal luminescence is similar to the bacterial system in that the light emission is continuous and persistent, and its color is quite close to the fluorescence of flavin. So the possibility that the biochemistry of the fungal system might be similar to the bacterial reaction seemed reasonable. Indeed, Airth discovered that, as in bacteria, NADH would stimulate light in extracts.[8,79]

However, he found that neither flavin nor aldehyde stimulated, but discovered that the activity could be separated into soluble and particulate fractions. Addition of NADH to the soluble fraction alone resulted in the time-dependent formation of a presumed reduced compound – possibly the luciferin – that emitted light when mixed later with the particulate fraction, which was identified with enzyme activity. Thus, the system could be classed as a luciferin–luciferase reaction.

Airth died prematurely and the work did not continue in his laboratory. But in the Netherlands Kuwabara and Wassink[80] noted that extracts made with hot water (thus, protein denatured) would give light upon the addition of H_2O_2. They considered this to be a chemiluminescence but thought that the emitter might be, or be related to, the luciferin. Purifying the activity from 15 kg of fungal mass, they obtained crystals of the compound but did not identify its structure. They stated that it gave light when mixed with Airth's luciferase fraction, but provided no supporting information and published it only in a non-refereed symposium volume.

They also did not continue the studies, but some years later Shimomura took a similar approach. From a number of different fungal species he identified several related chemiluminescent substances that are able to emit light in the presence of H_2O_2 and Fe^{2+}, naming the parent compound panal; all were assumed to be functional fungal luciferins. Unable to isolate a fraction with luciferase activity, he concluded that fungal luminescence *in vivo* is actually a chemiluminescence.[81]

Many new species of luminous fungi have been discovered in recent years in Brazil,[82,83] providing material for a reinvestigation. Following the Airth protocol, researchers in Sao Paulo were able to duplicate the results and confirm his conclusions,[84] thus opening the way for the characterization of the two fractions, still to be accomplished.

It may be that the *in vivo* emission from fungi comes from both chemi- and bioluminescence, but if so this may be difficult to demonstrate. I believe fungi do emit a true bioluminescence; in some species the emission is very bright (D. Desjardin, personal communication).

3.2.4   Dinoflagellates: Scintillons and Novel Regulation by pH

At the conference on bioluminescence held in Asilomar, California in 1954, mentioned above, Beatrice Sweeney and Francis Haxo reported on the circadian expression of bioluminescence of the dinoflagellate *Gonyaulax polyedra*, brought into unialgal culture by Sweeney at the Scripps Institution of Oceanography in La Jolla.

Excited, I asked Haxo, whom I knew from the time we were both at Johns Hopkins University (1951–53), to allow me to spend a summer in his laboratory to attempt the isolation of the biochemical components of this system. So in 1955 I went in early June with my 6-months-pregnant wife, along with my home-made photomultiplier photometer. Away during that summer, Haxo allowed us to live in his cottage, a stone's throw away on the cliff above the laboratory, and I worked with Sweeney, with whom I had an extended, productive and very enjoyable collaboration.

3.2.4.1   Soluble Extracts

With her help I grew cultures in flasks on shelves constructed in front of north-facing windows, since light for photosynthesis during the day was essential. With cells extracted in water, I could detect a weak emission and a slight stimulation by heated extract. Thinking that dried powder might be active and stable, I plunged harvested cells into acetone at $-25\,°C$; the light emission was blinding, and I was worried that there would be no substrate remaining. But aqueous extracts of the acetone powder gave a respectable light emission, and it was stimulated substantially by heated extracts, indicative of a luciferin. The pH optimum ($\sim$6.5), requirement for molecular oxygen and other features were readily determined.[9]

The luciferin was found to be highly unstable due to autoxidation. Purified under an inert atmosphere, it was found to be a tetrapyrrole related to (and possibly derived from) chlorophyll.[85] The structures of dinoflagellate luciferin and its oxidation product have since been determined,[86] but the energy-rich intermediate has not yet been identified, nor has the emitter.

In *Gonyaulax*, and in several other (but not all) species, luciferin is bound to and sequestered by a protein distinct from luciferase.[87] Its binding is pH dependent, bound at pH 8 and above and not at pH 6 and below. Luciferase activity is similarly pH dependent, active at pH 6 and not at pH 8.[88] It is postulated that a rapid change in pH in scintillons (see below) triggers the rapid flash, with a rise time of less than 10 ms and a duration of ~100 ms.

3.2.4.2   *A Luciferase with Three Active Sites*

Even if the system seemed simple enough in its biochemical components, it turned out to have several unique features, elucidated over subsequent years.[89] The luciferase has three catalytic sequences in a single molecule (M_r, ~140 kDa) and, as mentioned above, the luciferin is similar to chlorophyll and bound to a different protein (LBP, M_r, ~73 kDa). Both are regulated by pH by a novel mechanism, and both are localized in small (0.4 mμ) vesicles, called scintillons (see below). In addition, the 100 ms flashes are triggered by a conducted action potential in the vacuolar membrane, postulated to open membrane channels in the scintillons, allowing protons to enter and effect a pH change. These and other features of the system are described in detail online.[90]

The crystal structure of domain 3 of luciferase (located at its C-terminal end) has been determined.[91] A major part forms a barrel, inside of which the luciferin can bind, with an opening through a channel constricted at pH 8, thus preventing luciferin entry, but open at pH 6 by movement of three alpha helices that form it. Four histidines located in these sequences were shown by site-directed mutagenesis to be responsible for the pH-dependent change; if substituted by alanines, the luciferase is fully active at pH 8, which is attributed to an opened channel.[92]

3.2.4.3   *A Cell Organelle: The Scintillon*

It was found that the pellet from cells extracted at pH 8 gives an enormous additional amount of light simply upon acidification.[93,94] This led to the discovery of a light-emitting organelle responsible for light emission, for which we coined the term scintillon (flashing unit). Scintillons occur as cortically located vesicles, hanging in the vacuole with narrow necks connecting them to the cytoplasm, visualized by immunolocalization.[95,96] An explanation for activity in both soluble and particulate fractions is that some scintillons are ruptured during extraction, while others are pinched off and can be isolated as vesicles. Rapid acidification of isolated scintillons produces a flash kinetically similar to the flash of the living cells; after flashing scintillons can be "recharged" by adjusting to pH 8, and incubating with fresh luciferin for some minutes.[97]

3.2.4.4   *Phylogeny of Dinoflagellate Luciferases*

The structures of dinoflagellate luciferase genes are very similar in seven photosynthetic species; all have three contiguous domains of similar sizes and sequences.[98] However, the intergenic sequences are highly divergent, indicating an absence of selection for sequence conservation in those regions.[99] An eighth luminous species, the heterotroph *Noctiluca miliaris*, which is the most primitive (based on ribosomal RNA), has a single gene for both the enzymatic and luciferin binding functions, expressed as a single protein. The LBP sequence is full-length, but only a single luciferase domain is present, and it lacks a major part of the N-terminal region and three of the histidines found it the other species.[100]

3.2.5 Cypridinids: Luminous Ostracod Crustacea

Harvey learned about *Cypridina hilgendorfii* (now *Vargula hilgendorfii*) during his honeymoon in Japan.[101] The reaction was shown to conform to the two-component (luciferin and luciferase) Dubois model, and it became a model organism for studying the chemistry of bioluminescence. Although progress was made,[102] the structure of its luciferin eluded determination prior to its crystallization in the Goto laboratory in Nagoya, Japan.[10]

3.2.5.1 Function of the Luminescence

The animal ($\sim$2 mm) injects its luciferin and luciferase into the sea water from two separate glands, forming small luminous spots ($<$2 mm) that hang motionless in the water and may maintain their integrity and emission for many seconds or minutes. The light may serve as a decoy to divert predators, or as an aposematic signal, or as a burglar alarm if attacked. However, Morin and colleagues[103,104] have described many Caribbean species where the male creates species-specific structured trains of such light spots (called pulses by them) that serve in courtship. Such behavior has not been reported in species elsewhere, including Japanese ones.

3.2.5.2 Cypridinid Luciferin and Luciferase

The structure of the luciferin (and its total synthesis) were also reported from the Goto laboratory.[105,106] The luciferin, earlier referred to by the genus name of the animal, is now preferably called by the more inclusive term "cypridinid" luciferin (since the same molecule occurs in members of several genera[107]). It has a tripeptide-like structure with an amino-pyrazine skeleton. At neutral pH without luciferase it is highly unstable due to autoxidation and emits a weak chemiluminescence, possibly *via* the same cyclic dioxetanone intermediate that occurs in the enzymatic reaction.[108]

Cypridinid luciferases, cloned and sequenced from both *Vargula hilgendorfii* and *Cypridina noctiluca*, are very similar, monomers with molecular masses of about 68 kDa, some 6 kDa of which is carbohydrate.[109,110] Although EDTA strongly inhibits the reaction,[111] the protein lacks an EF-hand type calcium binding site[109] such as found in apoaequorin, and is not homologous to the coelenterate luciferases (see below).

The luciferase-bound oxyluciferin product is postulated to be the emitter; the light peaks at about 460 nm *in vitro*, close to that emitted by the living organism. Emission is stimulated non-specifically five-fold or more by salt, peaking at concentrations of about 50 mM[112] and can be described kinetically at a neutral pH by two concurrent first-order reactions, light and dark. As with many luciferases, the turnover is not rapid (30 s^{-1}), but is effectively 100-times less than that due to the very slow hydrolysis of the luciferase-bound inhibitory oxyluciferin product.[113]

3.2.6 Coelenterates: Aequorin and GFP

After what must have been many frustrated attempts to demonstrate a luciferin–luciferase reaction in extracts of the jellyfish *Aequorea*, Shimomura discovered that calcium ion was responsible for triggering the reaction.[14] Crude aqueous extracts resulted in bright and long-lived luminescence, yet all attempts to separate the components had failed. The clue, as described by Shimomura,[81] came when he discarded a still-emitting prep into the sink and its brightness increased greatly. He soon traced this to the effect of calcium.

3.2.6.1 Aequorin

By using a chelating agent to exclude calcium during the extraction, Shimomura then showed the active material to be a protein, which by itself would emit light upon the addition of calcium, with no need for any other factor, including oxygen. He referred to this protein as "aequorin," believing that this was an entirely new kind of bioluminescence system, not involving a luciferin–luciferase reaction, later including it in a newly coined term "photoprotein."

Studying the bacterial reaction at the same time (see above), I discovered a quasi-stable intermediate formed after the reaction of luciferase with reduced FMNH$_2$ and oxygen, and postulated it to be peroxy-flavin luciferase.[55] The similarities between this intermediate and aequorin were noted; both were proteins and rather stable; neither required oxygen for the subsequent steps leading to light emission, aequorin being triggered by calcium ion while the bacterial flavin-peroxide required a long-chain aldehyde. While aequorin was very stable in the absence of calcium, the flavin peroxide broke down slowly to give oxidized flavin and H$_2$O$_2$ without light emission.[114] Viewed as an intermediate stored in the cell ready to flash, the peroxy-coelenterazine finally explained the well-known ability of many luminous coelenterates to emit light in the complete absence of dissolved oxygen.[115]

Peroxy-coelenterazine was later confirmed to be the prosthetic group in both aequorin and obelin, a sister protein from the hydrozoan *Obelia*.[116,117] Apoaequorin was shown to function as the luciferase; active aequorin could be regenerated by incubation of coelenterazine with apoaequorin in the presence of oxygen and absence of Ca^{2+}.[118] The reaction is very slow, as with the formation of luciferyl-AMP in the firefly. Thus, calcium does not need to be removed for the flash to be extinguished.

3.2.6.2 Green Fluorescent Protein (GFP)

While aequorin revealed a different type of luciferin–luciferase reaction, the associated GFP was the first protein shown to be responsible for a shift in the wavelength of bioluminescence emission from a luciferase reaction. Shimomura noted green protein fluorescence in *Aequorea* extracts and suggested that this might be responsible for the emission *in vivo*.[14] In further studies, GFP was so named and shown to be localized in photogenic cells of the hydromedusan *Obelia*, and to function in several different luminous coelenterates by radiationless energy transfer.[119,120]

Further developments awaited the cloning of the GFP gene by Prasher,[121] who had carried the project from his graduate work in Milton Cormier's laboratory at the University of Georgia to be completed at his position at the Woods Hole Oceanographic Institution. Although he did not continue with a career as a research scientist, his work was crucial and he provided many others with the cloned gene. Martin Chalfie introduced it heterologously into the nematode, showing its value in determining the temporal course and cellular expression and localization of a protein in development.[122] Roger Tsien mutagenized GFP and developed an array of proteins with different spectral emissions, thereby allowing different proteins to be tracked separately and compared in a single experiment.[123] Its widespread applications are now legend and have effectively revolutionized biological studies in many different areas.[124]

3.2.7 Pholas

The clam *Pholas dactylus* inhabits holes that it drills into the soft rock along the Atlantic and Mediterranean coasts; it emits a cloud of luminescence when disturbed, much like and perhaps functionally similar to the tube-worm *Chaetopterus*.

3.2.7.1 A Different Substrate Binding Protein

Although extracts were found by Dubois to be positive in the luciferin–luciferase reaction, there was no report of further studies until almost a century later by Michelson's group in France.[15,125] They confirmed Dubois' work, and were able to separate the luciferase and luciferin activities; luciferin is associated with a glycoprotein ($M_r \sim 34\,600$ Da), while the much larger luciferase ($M_r \sim 310$ kDa), also a glycoprotein, binds two copper molecules within a subunit structure. The enzymatic reaction is reported to have a quantum yield of 0.09.[16]

The protein-bound luciferin was renamed pholasin[126] but the identity of the luciferin prosthetic group remains unknown; it is reported not to be related to coelenterazine or cypridinid luciferin.[127] Pholasin will emit light without luciferase upon addition of any of several reagents, including ferrous ions, H_2O_2, peroxidases, superoxide anions and hypochlorite, but the reactions can be classed as chemiluminescences. The much brighter reaction *in vivo* is most certainly enzymatic.

3.3   Discussion

3.3.1   Milky Seas

The recently reported imaging of a continuous luminescence in the ocean over an area the size of the state of Connecticut[128] merits renewed study. Previously described repeatedly in logs of merchant ships and dubbed "milky seas," it had been thought but never shown (and still not) that the light is due to luminous bacteria.[129,130] In all studied luminous bacteria, including "free-living" plank-tonic forms, regulation has been shown to involve autoinduction, and sea water does not generally have the requisite free-living bacterial concentration for autoinducer to accumulate,[131] so the candidacy of such bacteria as the source of light is questionable. It may be that they are growing on (possibly decompos-ing) abundant micro filamentous algae. The results of well-designed on-site studies of the phenomenon will be of great interest.

3.3.2   Beetle Luciferase Structure and Color of Emission

The last decade has seen intensive investigations concerning the color of light in different beetles, where peaks *in vivo* range from yellow-green to red (540–620 nm), attributable to the structure of the luciferase[34,132,133] (http://www.photo-biology.info/#Biolum).

In vitro emissions within this spectral range are obtained with the different luciferases but the same luciferin. Specific regions and residues of luciferases in different species (not all the same) have been identified as determinants; an understanding of how the protein structure affects the spectrum is incomplete.

3.3.3   Evolutionary Origins and Phylogeny of Luciferases; Lateral Gene Transfer

Some 25 years ago I estimated that, in considering extant luminous species, the ability to emit light could have originated as many as 30 different times in the course of evolution.[134] This has been upped to 40 or more in a recent review,[135] but both numbers may have to be revised (up or down) as the genes and bio-chemical mechanisms become known. Since no fossil record of luminous organisms exists, there is no way to know of extinct forms, so the total number of evolutionary origins may have been greater. The fact that most luminous species have many closely related non-luminous species is most readily explained by loss, which might not be lethal to a line. In cases where a specific luciferase occurs in a very distant species lateral gene transfer is a possibility; no clear cases of this are known at present.

Although the luciferases of dinoflagellates and krill cross react,[136] it is not known if they are homologous, as the krill luciferase has not been cloned. If they are, it could be a very interesting case of lateral gene transfer.

3.3.4 Luciferins: Biosynthesis and Nutritional Transfer

In addition to the origin of luciferases, the source of the luciferin in different species is of interest. The widespread occurrence and utilization of coelenterazine as luciferin is well known but not explained,[137,138] and knowledge about its synthesis[139] and function[140] in non-luminous species is nowhere near complete. It has been stated recently that luciferins are highly conserved across phyla[135] this is *not* correct. There are many luciferins other than coelenterazine.

Other cases of cross-phylogenetic molecular similarities are those in which the same or a similar luciferin functions in distantly related groups. Cypridinid luciferin serves in the luminous midshipman fish *Porichthys*, and is obtained nutritionally,[141] and is thus analogous to a vitamin in mammals.

Another case is that of euphausid shrimp, which utilizes a luciferin very similar to dinoflagellate luciferin.[85,142] Shrimp feed on luminous dinoflagellates, so nutritional transfer might occur, but this has not been demonstrated. Even if luciferin is not in the diet of the shrimp, it is possible that the shrimp synthesize their own luciferin from ingested chlorophyll; if so, this would free them from the strict diet of luminous species and be another example of convergent evolution.

Heterotrophic luminous dinoflagellates such as *Noctiluca miliaris* lack chloroplasts and chlorophyll yet their bioluminescence system appears to utilize the same luciferin, and they have the luciferin binding protein gene embedded in their luciferase gene.[100] As with the shrimp, it has been speculated that they might obtain luciferin nutritionally, but this appears not to be so; cells grown on a diet lacking luciferin or chlorophyll retain bioluminescence.[35] A cautionary note: *Porichthys* injected with a quantity of luciferin can continue to emit for months, with the total number of photons emitted estimated to be many times greater than the molecules of luciferin, suggesting that it is recycled by the animal.[143] The same might be true in dinoflagellates.

3.3.5 Biochemical Basis for Flashing in Luminous Organisms; the Off Rate

Rapid flashing, ranging from less than 100 ms to about 1 s, is a feature of numerous bioluminescent systems. Many biological processes that are controlled on such a rapid time scale are based on membrane channels, mediated by ion-controlled openings and closures. But the mechanism of the luciferase reaction itself is also relevant to the kinetics in some cases.

The onset of the firefly flash is due to a pulse of oxygen, whose entry is triggered by a nerve impulse that results in the opening of tracheoles and oxygen entry.[144] The flash decay time *in vivo*, however, is not dependent on oxygen removal; if it were, the flash decay times would be affected by physiological conditions that could result in flashes with different kinetics, which they are not. Flash kinetics are species-specific, functioning for mate identification. Instead, the kinetics are defined by the rate constant for the decay of the cyclic peroxide intermediate to give light. Because the formation of LH_2-AMP by reaction of fresh luciferin with ATP is a very slow reaction, very little is formed during or before the removal of oxygen. Subsequent removal of any excess oxygen and the maintenance of anaerobiosis in photocytes are achieved by the activity of the abundant mitochondria, which also produces ATP. This proposed mechanism is supported by studies with the luciferase reaction *in vitro*; in the absence of oxygen the enzyme-luciferyl-AMP intermediate accumulates; upon the subsequent admission of oxygen it is rapidly oxidized to give a flash, with a rapid decay even though excess oxygen is present.[145]

The kinetics of the flashing in *Aequorea* may be explained by an analogous mechanism; luciferase-peroxy-coelenterazine (aequorin), accumulated in cells in the absence of Ca^{2+}, reacts with injected Ca^{2+} to give an excited product in a reaction that is much, much faster than that of its formation. The kinetics of the decay are thus determined by the rate constant of that terminal step.[146,147] The onset of the flash is likely to be a membrane controlled mobilization of Ca^{2+} in the photocytes; the removal of the Ca^{2+} is not needed for the light to decay.

Evidence concerning the flash of dinoflagellate cells indicates that a membrane-mediated entry of H^+ into the scintillons is responsible. Whether an accumulated (peroxy?) intermediate is the substrate is not known; this is another area for future investigation.

3.3.6   Basic and Applied Research – the Positives and Negatives

The onset of a focus on applications of bioluminescence can be marked by the first symposium on applications in 1978;[148] such symposia have been held every two years since. Whereas earlier studies of bioluminescence were mostly driven simply by a desire to gain new understanding of a phenomenon, the prospect of applications has attracted many new workers to the field, some of whom might previously have viewed the study of biological light emission as being outside the main stream.

Luciferase systems, notably firefly, bacterial and coelenterate, were and are used extensively for analytical purposes, some with important commercial applications (*e.g.*, water quality, microbial contamination, DNA sequencing, cellular calcium concentrations). But the now widespread use of luciferase genes and proteins as reporters for gene expression and molecular tags has taken on a life of its own, and in the case of GFP has been recognized by a Nobel Prize. Such developments would not have been possible in the absence of knowledge from basic research, from both within and without the field.

With the increased focus on applications there is also a perceptible and regrettable decrease in the exchange of experimental results and open discussions of new ideas; this is probably attributable to concerns that others will make inappropriate use of information that might be discussed. A likely result is a loss of new ideas and insights that might have emerged from such exchanges.

On the positive side, applied studies have resulted in new insights and understanding of bioluminescence itself, mostly at the molecular level. This is, of course, in addition to the applications themselves, some of which have revolutionized studies in certain areas, notably the subcellular localization of specific molecules. At the same time, the concomitant decrease in truly basic research can be expected to have a long-term negative effect on applications, which are so dependent on basic knowledge. Despite its clear long-range importance, basic research cannot be overtly supported by an applied program.

3.4 Definitions and Terms

Definitions are important and contribute heavily to an understanding of a subject. In the field of bioluminescence, two terms are not well defined or well accepted in the field: luciferin and photoprotein. Also, luciferin and luciferase are generic terms and must be further defined to indicate the organism where they occur. But identification with an individual species may not be satisfactory because different species within a group (which may be narrow or wide) may share the same or similar luciferins and luciferases.

3.4.1 Luciferin

It has been widely, but not universally, accepted that the term luciferin refers to the light-emitting molecule in the reaction; this is based on its etymology from the Latin: *luci-*, light; *fero-*, to bear, carry. In assigning this name, Dubois referred to the substance from the hot water extract prior to its reaction with oxygen, but it is the product that is the actual emitter. So strictly speaking what we call luciferin is a precursor to the emitter in the bioluminescence reaction. Thus, in the bacterial reaction, $FMNH_2$ is the luciferin, but neither it nor FMN, the final oxidized product, is the emitter, which is luciferase-bound hydroxyflavin, a reaction intermediate. Likewise, firefly luciferin is a benzothiazole, but the actual emitter is an oxidized form, electronically excited in the penultimate step.

It is recognized that the oxidation of luciferin provides energy used for the excitation of the product emitter molecule, and it has been proposed that this function should be used to define the word luciferin, instead of the above.[81] I consider this to be insufficient and incorrect; in the bacterial reaction the concomitant oxidation of a long-chain aldehyde also provides part of the energy, but should not be called a luciferin, as it does not function in the light emission. Likewise, in the firefly reaction, ATP is a co-substrate in the reaction but does not contribute to the light emission. Luciferins should thus be defined as substrates in luciferase reactions which, after reaction, give rise to the species that is responsible for light emission.

3.4.2 Photoprotein

The catchy but often confusing term "photoprotein" was introduced in 1966 as a name for proteins that would emit light upon the addition of a substance other than a luciferin. A further criterion used was that the total light is proportional to the amount of protein, which is not so for an enzyme or catalyst. Thus, photoproteins were believed not to turn over as enzymes do, meaning that maximally only a single photon will be emitted per protein molecule.

The molecular nature of the several now-named photoproteins evidently differs. Aequorin was the first to be isolated, and it is now established to be a luciferase-coelenterazine peroxy reaction intermediate (not simply a complex of the three), and to be capable of turning over if additional luciferin is added in the absence of calcium.[118] Thus, after aequorin has emitted, the protein (apoaequorin) can react with oxygen and coelenterazine to form aequorin again, so that aequorin might be referred to as being "precharged."[149]

But the next protein to be classified as a photoprotein was different; isolated from the tube-worm *Chaetopterus*, it emits light upon the addition of Fe^{2+} and peroxide.[150] Another, similarly triggered protein, is from the scaleworm, where the name polynoïdin was given to the protein.[151] Since superoxide, peroxide and Fe^{2+} are known to result in chemiluminescence when added to many different organic compounds, these photoproteins are most likely cellular proteins with bound fluorescent molecules that are not necessarily involved in the reaction pathway in the bioluminescence of the animal.

The protein moiety of a photoprotein might be a luciferin binding protein and the bound fluorescent molecule the luciferin. This appears to be the case in *Pholas*, where (in some other systems also) a special protein sequesters the luciferin.[15,16] This protein was named pholasin and classed as a photoprotein, able to emit light upon the addition of any of several oxidants, but requiring molecular oxygen.[81] In a later study its light emission was determined to be at least one million times less than could be obtained when reacted with luciferase and oxygen, and called a chemiluminescence.[152] Here again the reaction is quite evidently not involved in the pathway of light emission in the animal.

A protein isolated from the millipede *Luminodesmus*, where ATP and Mg^{2+} stimulate the reaction,[153] has also been called a photoprotein.[81] However, there seems to be no reason to exclude the possibility that it could be a luciferin–luciferase reaction somewhat like the firefly. More studies are needed.

Thus, none of the later-named photoproteins appears to be a luciferase reaction intermediate like aequorin, and in none has the identity or structure of the light-emitter been determined. Efforts to characterize such systems should go forward; in my opinion, all *in vivo* bioluminescence reactions probably involve a luciferin and a luciferase in some kind of enzymatic reaction.

3.4.3 A More Inclusive Terminology for Luciferin and Luciferase

It is well appreciated that the terms luciferase and luciferin are generic and must be preceded by an identifier that relates to the organism. Morin[107] proposed that, rather than using species' names, more inclusive terms might be used. Thus "beetle" would be preferable to "*Photinus*" or "firefly" to identify a luciferase or luciferin, since it is more inclusive, and the luciferases within the beetles are homologous and the luciferins believed to be the same. The same kind of terminology could be used for bacterial and dinoflagellate systems; however, there are certainly instances where the species needs to be specified, especially for luciferases.

References

1. W. D. McElroy, *Proc. Natl. Acad. Sci. USA*, 1947, **33**, 342.
2. E. N. Harvey, *Bioluminescence*, Academic Press, New York, 1952.
3. E. N. Harvey, *Fed. Proc.*, 1953, **12**, 597.
4. R. Dubois, *Compt. Rend. Soc. Biol.*, 1885, **37**, 559.
5. R. S. Anderson, in *Bioluminescence*, New York Academy of Sciences, New York, 1948, p. 337.
6. B. L. Strehler, *J. Am. Chem. Soc.*, 1953, **75**, 1264.
7. W. D. McElroy, J. W. Hastings, V. Sonnenfeld and J. Coulombre, *Science*, 1953, **118**, 385.
8. R. L. Airth and W. D. McElroy, *J. Bacteriol.*, 1959, **77**, 249.
9. J. W. Hastings and B. M. Sweeney, *J. Cell. Comp. Physiol.*, 1957, **49**, 209.
10. O. Shimomura, T. Goto and Y. Hirata, *Bull. Chem. Soc. Jpn.*, 1957, **30**, 929.
11. M. J. Cormier, *J. Biol. Chem.*, 1962, **237**, 2032.
12. M. J. Cormier and J. R. Totter, *Annu. Rev. Biochem.*, 1964, **33**, 431.
13. O. Shimomura, F. H. Johnson and Y. Saiga, *J. Cell. Comp. Physiol.*, 1962, **59**, 223.
14. O. Shimomura, F. H. Johnson and Y. Saiga, *J. Cell. Comp. Physiol.*, 1963, **62**, 1.
15. J. P. Henry and A. M. Michelson, *Biochim. Biophys. Acta*, 1970, **205**, 451.
16. A. M. Michelson, *Methods Enzymol.*, 1978, **57**, 385.
17. A. A. Green and W. D. McElroy, *Biochim. Biophys. Acta*, 1956, **20**, 170.
18. E. Conti, N. P. Franks and P. Brick, *Structure*, 1996, **4**, 287.
19. B. Bitler and W. D. McElroy, *Arch. Biochem. Biophys.*, 1957, **72**, 358.
20. E. H. White, G. F. Field, W. D. McElroy and F. McCapra, *J. Am. Chem. Soc.*, 1961, **83**, 2402.
21. E. H. White, G. F. Field and F. McCapra, *J. Am. Chem. Soc.*, 1963, **85**, 337.
22. W. C. Rhodes and W. D. McElroy, *J. Biol. Chem.*, 1958, **233**, 1528.
23. T. A. Hopkins, H. H. Seliger, E. H. White and M. W. Cass, *J. Am. Chem. Soc.*, 1967, **89**, 7148.

24. F. McCapra, Y. C. Chang and V. P. Francois, *J. Chem. Soc., Chem. Commun.*, 1968, **22**.

25. O. Shimomura, T. Goto and F. H. Johnson, *Proc. Natl. Acad. Sci. USA*, 1977, **74**, 2799.

26. E. H. White, M. G. Steinmetz, J. D. Miano, P. D. Wildes and R. Morland, *J. Am. Chem. Soc.*, 1980, **102**, 3199.

27. J. da Silva, J. Magalhaes and R. Fontes, *Tetrahedron Lett.*, 2001, **42**, 8173.

28. H. Fraga, *Photochem. Photobiol. Sci.*, 2008, **7**, 146.

29. R. L. Airth, W. C. Rhodes and W. D. McElroy, *Biochim. Biophys. Acta*, 1958, **27**, 519.

30. H. Fraga, D. Fernandes, R. Fontes and J. da Silva, *FEBS J.*, 2005, **272**, 5206.

31. K. V. Wood, *Photochem. Photobiol.*, 1995, **62**, 662.

32. Y. Oba, M. Ojika and S. Inouye, *FEBS Lett.*, 2003, **540**, 251.

33. V. R. Viviani and E. J. H. Bechara, *Photochem. Photobiol.*, 1996, **63**, 713.

34. V. R. Viviani, *Cell. Mol. Life Sci.*, 2002, **59**, 1833.

35. Y. Oba, K. Iida and S. Inouye, *FEBS Lett.*, 2009, **583**, 2004.

36. W. E. G. Müller, M. Kasueske, X. Wang, H. C. Schroder, Y. Wang, D. Pisignano and M. Wiens, *Cell. Mol. Life Sci.*, 2009, **66**, 537.

37. K. Gomi and N. Kajiyama, *J. Biol. Chem.*, 2001, **276**, 36508.

38. K. Gomi, K. Hirokawa and N. Kajiyama, *Gene*, 2002, **294**, 157.

39. J. C. Day, L. C. Tisi and M. J. Bailey, *Luminescence*, 2004, **19**, 8.

40. S. Inouye, *Cell. Mol. Life Sci.*, 2010, **67**(3), 387.

41. S. M. Marques and J. C. Esteves da Silva, *IUBMB Life*, 2009, **61**, 6.

42. B. L. Strehler and W. D. McElroy, *Methods Enzymol.*, 1957, **3**, 871.

43. D. W. Ow, K. V. Wood, M. DeLuca, J. R. deWet, D. R. Helinski and S. H. Howell, *Science*, 1986, **234**, 856.

44. J. W. Hastings and C. H. Johnson, in *Biophotonics, Part A Methods in Enzymology*, ed. A. G. Mariott and I. Parker, Elsevier Academic, San Diego, 2003, pp. 75–104.

45. B. Gharizadeh, M. Akhras, N. Nourizad, M. Ghaderi, K. Yasuda, P. Nyren and N. Pourmand, *J. Biotech.*, 2006, **124**, 504.

46. M. Ronaghi, S. Karamohamed, B. Pettersson, M. Uhlen and P. Nyren, *Anal. Biochem.*, 1996, **242**, 84.

47. J. M. Bassot, *C. R. Acad. Sci. Paris, Ser. D*, 1978, **286**, 623.

48. O. Shimomura, F. H. Johnson and Y. Haneda, in *Bioluminescence in Progress*, ed. F. H. Johnson and Y. Haneda, Princeton University Press, Princeton, 1966, p. 487.

49. J. Lee, *Photochem. Photobiol.*, 1976, **24**, 279.

50. V. R. Viviani, J. W. Hastings and T. Wilson, *Photochem. Photobiol.*, 2002, **75**, 22.

51. B. L. Strehler, E. N. Harvey, J. J. Chang and M. J. Cormier, *Proc. Natl. Acad. Sci. USA*, 1954, **40**, 10.

52. B. L. Strehler and M. J. Cormier, *J. Biol. Chem.*, 1954, **211**, 213.

53. J. W. Hastings, J. A. Spudich and G. Malnic, *J. Biol. Chem.*, 1963, **238**, 3100.

54. S. Ulitzur and J. W. Hastings, *Proc. Natl. Acad. Sci. USA*, 1978, **75**, 266.
55. J. W. Hastings and Q. H. Gibson, *J. Biol. Chem.*, 1963, **238**, 2537.
56. J. W. Hastings, C. Balny, C. Le Peuch and P. Douzou, *Proc. Natl. Acad. Sci. USA*, 1973, **70**, 3468.
57. J. Vervoort, M. Ahmad, D. J. Okane, F. Muller and J. Lee, *Biophys. J.*, 1985, **47**, A332.
58. A. Eberhard and J. W. Hastings, *Biochem. Biophys. Res. Commun.*, 1972, **47**, 348.
59. G. Mitchell and J. W. Hastings, *J. Biol. Chem.*, 1969, **244**, 2572.
60. M. Kurfürst, S. Ghisla and J. W. Hastings, *Proc. Natl. Acad. Sci. USA*, 1984, **81**, 2990.
61. J. Lee, *Biophys. Chem.*, 1993, **48**, 149.
62. E. Ruby and K. Nealson, *Science*, 1977, **196**, 432.
63. J. W. Eckstein, K. W. Cho, P. Colepicolo, S. Ghisla, J. W. Hastings and T. Wilson, *Proc. Natl. Acad. Sci. USA*, 1990, **87**, 1466.
64. J. M. Friedland and J. W. Hastings, *Proc. Natl. Acad. Sci. USA*, 1967, **58**, 2336.
65. T. W. Cline and J. W. Hastings, *Biochemistry*, 1972, **11**, 3359.
66. A. J. Fisher, F. M. Raushel, T. O. Baldwin and I. Rayment, *Biochemistry*, 1995, **34**, 6581.
67. Z. T. Campbell and T. O. Baldwin, *J. Biol. Chem.*, 2009, **284**, 32827.
68. Z. T. Campbell, A. Weichsel, W. R. Montfort and T. O. Baldwin, *Biochemistry*, 2009, **48**, 6085.
69. K. H. Nealson, T. Platt and J. W. Hastings, *J. Bacteriol.*, 1970, **104**, 313.
70. W. C. Fuqua, S. C. Winans and E. P. Greenberg, *J. Bacteriol.*, 1994, **176**, 269.
71. J. W. Hastings and G. Mitchell, *Biol. Bull.*, 1971, **141**, 261.
72. J. W. Hastings and E. P. Greenberg, *J. Bacteriol.*, 1999, **181**, 2667.
73. A. Eberhard, A. L. Burlingame, C. Eberhard, G. L. Kenyon, K. H. Nealson and N. J. Oppenheimer, *Biochemistry*, 1981, **20**, 2444.
74. R. A. Rosson and K. H. Nealson, *Arch. Microbiol.*, 1981, **129**, 299.
75. C. Fuqua, S. C. Winans and E. P. Greenberg, *Annu. Rev. Microbiol.*, 1996, **50**, 727.
76. B. L. Bassler and R. Losick, *Cell*, 2006, **125**, 237.
77. E. P. Greenberg, *J. Clin. Invest.*, 2003, **112**, 1288.
78. W. L. Ng and B. L. Bassler, *Annu. Rev. Genet.*, 2009, **43**, 197.
79. R. L. Airth and G. E. Foerster, *Arch. Biochem. Biophys.*, 1962, **97**, 567.
80. S. Kuwabara and E. C. Wassink, in *Bioluminescence in Progress*, ed. F. H. Johnson and Y. Haneda, Princeton University Press, Princeton, 1966, p. 233.
81. O. Shimomura, *Bioluminescence: Chemical Principles and Methods*, World Scientific Publishing Co., Singapore, 2006.
82. D. E. Desjardin, M. Capelari and C. V. Stevani, *Fungal Diversity*, 2005, **18**, 9.
83. D. E. Desjardin, A. G. Oliveira and C. V. Stevani, *Photochem. Photobiol. Sci.*, 2008, **7**, 170.

84. A. G. Oliveira and C. V. Stevani, *Photochem. Photobiol. Sci.*, 2009, **8**, 1416.
85. J. C. Dunlap, J. W. Hastings and O. Shimomura, *FEBS Lett.*, 1981, **135**, 273.
86. H. Nakamura, Y. Kishi, O. Shimomura, D. Morse and J. W. Hastings, *J. Am. Chem. Soc.*, 1989, **111**, 7607.
87. M. Fogel and J. W. Hastings, *Arch. Biochem. Biophys.*, 1971, **142**, 310.
88. D. M. Morse, A. M. Pappenheimer and J. W. Hastings, *J. Biol. Chem.*, 1989, **264**, 11822.
89. T. Wilson and J. W. Hastings, *Annu. Rev. Cell Dev. Biol.*, 1998, **14**, 197.
90. J. W. Hastings, W. Schultz and L. Liu, in Photobiological Sciences Online, 2009; http://www.photobiology.info/Hastings.html.
91. L. W. Schultz, L. Liu, M. Cegielski and J. W. Hastings, *Proc. Natl. Acad. Sci. USA*, 2005, **102**, 1378.
92. L. Li, L. Liu, R. Hong, D. Robertson and J. W. Hastings, *Biochemistry*, 2001, **40**, 1844.
93. R. DeSa and J. W. Hastings, *J. Gen. Physiol.*, 1968, **51**, 105.
94. M. Fogel, R. Schmitter and J. W. Hastings, *J. Cell Sci.*, 1972, **11**, 305.
95. M.-T. Nicolas, G. Nicolas, C. H. Johnson, J.-M. Bassot and J. W. Hastings, *J. Cell Biol.*, 1987, **105**, 723.
96. L. Fritz, P. Milos, D. Morse and J. W. Hastings, *J. Phycol.*, 1991, **27**, 436.
97. M. Fogel and J. W. Hastings, *Proc. Natl. Acad. Sci. USA*, 1972, **69**, 690.
98. L. Liu, T. Wilson and J. W. Hastings, *Proc. Natl. Acad. Sci. USA*, 2004, **101**, 16555.
99. L. Liu and J. W. Hastings, *J. Phycol.*, 2006, **42**, 96.
100. L. Y. Liu and J. W. Hastings, *Proc. Natl. Acad. Sci. USA*, 2007, **104**, 696.
101. E. N. Harvey, *Am. J. Physiol.*, 1917, **42**, 318.
102. A. M. Chase, in *Bioluminescence*, New York Academy of Sciences, New York, 1948.
103. T. J. Rivers and J. G. Morin, *J. Exp. Biol.*, 2008, **211**, 2252.
104. A. C. Cohen and J. G. Morin, *J. Crustacean Biol.*, 1990, **10**, 184.
105. Y. Kishi, T. Goto, Y. Hirata, O. Shimomura and F. H. Johnson, *Tetrahedron Lett.*, 1966, 3427.
106. Y. Kishi, T. Goto, S. Inoue, S. Sugiura and H. Kishimot, *Tetrahedron Lett.*, 1966, 3445.
107. J. G. Morin, *Luminescence*, 2009.
108. F. McCapra and Y. C. Chang, *J. Chem. Soc., Chem. Commun.*, 1967, 1011.
109. E. M. Thompson, S. Nagata and F. I. Tsuji, *Proc. Natl. Acad. Sci. USA*, 1989, **86**, 6567.
110. Y. Nakajima, K. Kobayashi, K. Yamagishi, T. Enomoto and Y. Ohmiya, *Biosci. Biotech. Biochem.*, 2004, **68**, 565.
111. O. Shimomura, F. H. Johnson and Y. Saiga, *J. Cell. Comp. Physiol.*, 1961, **58**, 113.
112. R. V. Lynch, F. I. Tsuji and D. H. Donald, *Biochem. Biophys. Res. Commun.*, 1972, **46**, 1544.

113. O. Shimomura, F. H. Johnson and T. Masugi, *Science*, 1969, **164**, 1299.
114. J. W. Hastings and C. Balny, *J. Biol. Chem.*, 1975, **250**, 7288.
115. E. N. Harvey and I. M. Korr, *J. Cell. Comp. Physiol.*, 1938, **12**, 319.
116. J. F. Head, S. Inouye, K. Teranishi and O. Simomora, *Nature*, 2000, **405**, 372.
117. J. Lee, J. N. Glushka and E. S. Vysotski, *Biophys. J.*, 2000, **78**, 2831.
118. O. Shimomura, Y. Kishi and S. Inouye, *Biochem. J.*, 1993, **296**, 549.
119. J. G. Morin and J. W. Hastings, *J. Cell. Physiol.*, 1971, **77**, 313.
120. J. G. Morin and J. W. Hastings, *J. Cell. Physiol.*, 1971, **77**, 303.
121. D. C. Prasher, V. K. Eckenrode, W. W. Ward, F. G. Prendergast and M. J. Cormier, *Gene*, 1992, **111**, 229.
122. M. Chalfie, Y. Tu, G. Euskirchen, W. W. Ward and D. C. Prasher, *Science*, 1994, **263**, 802.
123. R. Y. Tsien, *Annu. Rev. Biochem.*, 1998, **67**, 509.
124. *Green Fluorescent Protein: Properties, Applications, and Protocols*, ed. M. Chalfie and S. R. Kain, Wiley-Interscience, Hoboken, 2006.
125. J. P. Henry, M. F. Isambert and A. M. Michelson, *Biochimie*, 1973, **55**, 83.
126. P. A. Roberts, J. Knight and A. K. Campbell, *Anal. Biochem.*, 1987, **160**, 139.
127. T. Muller and A. K. Campbell, *J. Biolum. Chemilum.*, 1990, **5**, 25.
128. S. D. Miller, S. H. D. Haddock, C. D. Elvidge and T. F. Lee, *Proc. Natl. Acad. Sci. USA*, 2005, **102**, 14181.
129. P. J. Herring and M. Watson, *Mar. Observer*1993, **63**, 22.
130. K. H. Nealson and J. W. Hastings, *Appl. Environ. Microbiol.*, 2006, **72**, 2295.
131. E. G. Ruby, E. P. Greenberg and J. W. Hastings, *Appl. Environ. Microbiol.*, 1980, **39**, 302.
132. B. R. Branchini, T. L. Southworth, M. H. Murtiashaw, R. A. Magyar, S. A. Gonzalez, M. C. Ruggiero and J. G. Stroh, *Biochemistry*, 2004, **43**, 7255.
133. T. Nakatsu, S. Ichiyama, J. Hiratake, A. Saldanha, N. Kobashi, K. Sakata and H. Kato, *Nature*, 2006, **440**, 372.
134. J. W. Hastings, *J. Mol. Evol.*, 1983, **19**, 309.
135. S. H. D. Haddock, M. A. Moline and J. F. Case, *Annu. Rev. Mar. Sci.*, 2010, **2**, 293.
136. J. Dunlap, O. Shimomura and J. W. Hastings, *Proc. Natl. Acad. Sci. USA*, 1980, **77**, 1394.
137. C. M. Thomson, P. J. Herring and A. K. Campbell, *Mar. Biol.*, 1995, **124**, 197.
138. O. Shimomura, S. Inoue, F. H. Johnson and Y. Haneda, *Comp. Biochem. Physiol. B*, 1980, **65**, 435.
139. C. M. Thomson, P. J. Herring and A. K. Campbell, *J. Mar. Biol. Assoc. U. K.*, 1995, **75**, 165.
140. J. F. Rees and F. Baguet, *J. Exp. Biol.*, 1988, **135**, 289.
141. F. I. Tsuji, A. T. Barnes and J. F. Case, *Nature*, 1972, **237**, 515.
142. O. Shimomura, *FEBS Lett.*, 1980, **116**, 203.

143. E. M. Thompson, B. G. Nafpaktitis and F. I. Tsuji, *Photochem. Photobiol.*, 1987, **45**, 529.

144. G. S. Timmins, F. J. Robb, C. M. Wilmot, S. K. Jackson and H. M. Swartz, *J. Exp. Biol.*, 2001, **204**, 2795.

145. J. W. Hastings, W. D. McElroy and J. Coulombre, *J. Cell. Comp. Physiol.*, 1953, **42**, 137.

146. J. W. Hastings, G. W. Mitchell, P. H. Mattingly, J. R. Blinks and M. Van Leeuwen, *Nature*, 1969, **222**, 1047.

147. J. W. Hastings and J. G. Morin, *Biochem. Biophys. Res. Commun.*, 1969, **37**, 493.

148. *Proceedings of the International Symposium on Analytical Applications of Bioluminescence and Chemiluminescence: September 1978*, ed. E. Schram and P. Stanley, State Printing and Publishing, Inc., Westlake Village, 1979.

149. J. W. Hastings, *Annu. Rev. Biochem.*, 1968, **37**, 597.

150. O. Shimomura and F. H. Johnson, *Science*, 1968, **159**, 1239.

151. M.-T. Nicolas, J.-M. Bassot and O. Shimomura, *Photochem. Photobiol.*, 1982, **35**, 201.

152. S. L. Dunstan, G. B. Sala-Newby, A. B. Fajardo, K. M. Taylor and A. K. Campbell, *J. Biol. Chem.*, 2000, **275**, 9403.

153. J. W. Hastings and D. Davenport, *Biol. Bull.*, 1957, **113**, 120.

CHAPTER 4

Instrumentation for Chemiluminescence and Bioluminescence

FRITZ BERTHOLD,[a] MANFRED HENNECKE[b] AND JÜRGEN WULF[c]

[a] Weissenburgstrasse 36, D-75173 Pforzheim, Germany; [b] Berthold Technologies, Calmbacher Str. 22, D-75323 Bad Wildbad, Germany; [c] Sun-Ray-Optics, Langgasse 22, D-88662 Überlingen, Germany

4.1 Introduction

Basically, two classes of instrumentation are available for the measurement of chemiluminescence (CL): luminometers and low-level-light imagers.

Luminometers measure the light emission from discrete samples, either in single tubes or – in most applications – from microplates. Since an earlier review on the same subject,[1] there is an increasing trend away from classical luminometers to multimode readers in research applications, where CL measurement is only one technology, together with absorption and different types of fluorescence measurements.

Low-level-light imagers capture complete and quantitative images of light emission from small animals, plants, *etc.*

There is a certain overlap of applications between luminometers and imagers, since imagers are also used for simultaneous measurements of light emission from all wells of a microplate, allowing high-sample throughput.

Chemiluminescence and Bioluminescence: Past, Present and Future
Edited by Aldo Roda
© Royal Society of Chemistry 2011
Published by the Royal Society of Chemistry, www.rsc.org

4.2 Luminometers

Luminometers are used in such different areas as research, clinical diagnostics or industrial applications. The following discussion will outline the basic principles that are common to all types of luminometers, pointing out features that are specific for certain applications.

We begin with a description of a microplate luminometer, the most widely used instrument type. A microplate luminometer basically consists of the following components:

1 light detector,
2 x-y table carrying the microplate and presenting each well successively to the detector,
3 transfer optics between sample and detector,
4 reagent injector(s) – optional,
5 light-tight housing,
6 electronics and software.

4.2.1 Detector

The standard detector device in luminometers is the photomultiplier. Different types of semiconductor devices are also available, like silicon diodes or avalanche diodes, but these do not match photomultipliers in terms of sensitivity. If limited space is available, they can be used as in portable instrumentation, and where high sensitivity is not required. One might say that the photomultiplier is one of the few vacuum tube devices that has not (yet?) been replaced by semiconductors.

Figure 4.1 shows a photomultiplier as it is normally used in high-sensitivity luminometry.

Photons hitting the transparent photocathode release photoelectrons due to the photoelectric effect, for the discovery of which Albert Einstein was awarded the Nobel Prize in Physics.

The percentage of photoelectrons released in relation to the number of photons hitting the photocathode is called quantum efficiency (QE) and is dependent on the wavelength (λ).

Figure 4.2 shows QE as a function of λ for three different photomultipliers that are used in luminometry, and also the major wavelengths emitted by the most commonly used CL systems.

Multipliers with a standard bialkali photocathode are sensitive over a wavelength range of 360–600 nm, covering all CL chemistries of interest, albeit with low QE in the green range of only about 3% for firefly luminescence at 565 nm.

About 6% QE at 565 nm is achieved with green-enhanced bialkali cathodes.

Multipliers with extended blue or red sensitivity are also available, the latter being required for the light emission from reactive oxygen species, and then

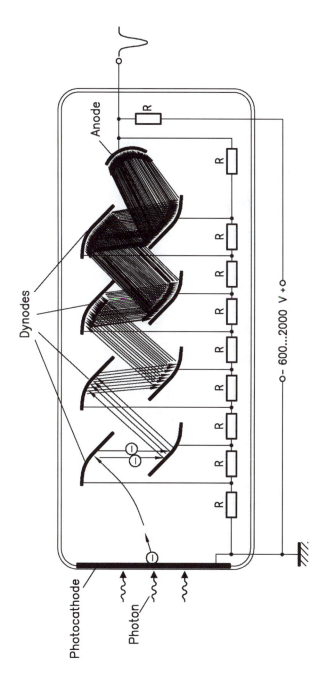

Figure 4.1 Photomultiplier principle. Photons hitting a photocathode release photoelectrons that are amplified by dynodes until they hit the anode. The current flow through a resistor (R) generates a pulse signal for further processing.

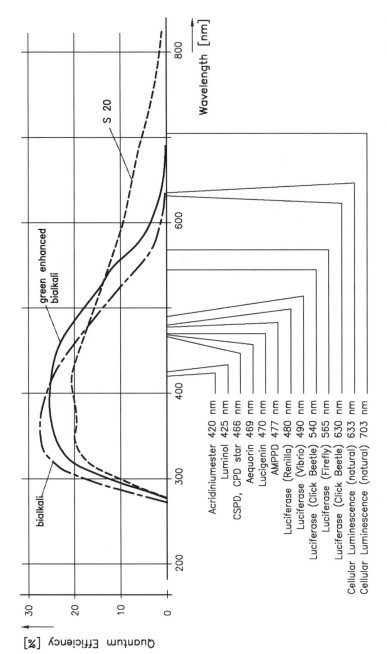

Figure 4.2 Quantum efficiency for different photocathodes and emission from major CL systems. Major wavelengths emitted by CL systems, and typical quantum efficiency, are shown for a multiplier with bialkali photocathode, a multiplier with extended green sensitive bialkali cathode, and a multiplier with S20 cathode.

using S20 type photocathodes, extending sensitivity up to 1000 nm. Since noise increases drastically with red sensitivity, multiplier cooling will be required.

Within the vacuum of the photomultiplier, the photoelectrons are accelerated by an appropriate electrostatic field until they hit the first of a series of successive dynodes.

Practically all of the photoelectrons released hit the first dynode, where a number of secondary electrons – typically between 4 and 6 – are generated. These are again accelerated to the next dynode and so on, forming an avalanche of electrons, which finally hits the anode from where the charges are picked up for electronic processing.

An electron multiplication factor of 10^6 is typical. The signals at the anode are shown in Figure 4.3.

One might either just integrate the electric charges for charge or current measurement, or convert each pulse with an amplitude greater than a set threshold into a digital pulse, which is called photon counting (although photoelectron counting would be more correct). One beauty of photon

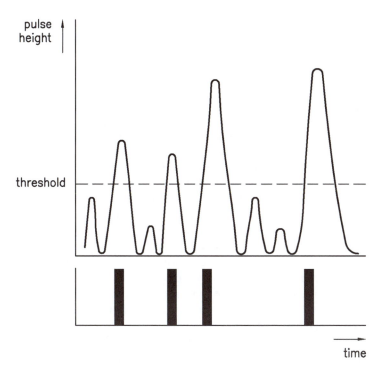

Figure 4.3 Principle of photon counting. The upper diagram shows the analog pulses picked up from a preamplifier connected to anode. Only pulses exceeding the threshold value are converted into logic pulses – lower diagram. The threshold is set so that only signals due to electrons originating from the cathode are registered; smaller pulses are rejected as noise.

counting is that it demonstrates the quantum nature of both light and electricity in a single device.

The sequence of pulse counts follows Poisson statistics, as in a Geiger counter, but at the rates normally occurring Gaussian statistical analysis is fully adequate. For example, if the signal rate is 5000 counts per second (cps), with a measuring time of 2 s, the total number of counts would be 10 000 and the one-sigma error is $\sqrt{10\ 000}$, or 1%.

Most modern instruments use photon counters because of higher long-term stability and somewhat higher analytical detection sensitivity, but in the current-measuring mode the dynamic range can be higher. Therefore, some luminometers combine both technologies, using photon counting at low signal rates, and automatically switching over to current-measuring mode at high signal rates.

Background is mostly due to the spontaneous release of electrons from the photocathode and increases rapidly with temperature, requiring cooling in some cases.

Photomultipliers with a side-window and opaque photocathode are sometimes used in portable instruments, thanks to their compact dimensions and low cost, but the resulting efficiency is far below that of head-on types.

A special type of photomultiplier uses a so-called channeltron with a continuous dynode for electron multiplication, and is distinguished by low background values and improved single electron spectrum.

4.2.2   Reagent Injectors

On-board reagent injection is mandatory for flash-type luminescence, as for aequorin, acridinium ester or luminol without enhancer.

Typically, a flash-luminescence signal reaches its peak in less than 1 s after beginning injection of the starter reagent, and more than 90% of the light is emitted in less than 2 s. To monitor light emission from the beginning, the starter reagent has to be injected into the well in the measurement position. As a consequence, the injector tip blocks some of the emitted light, which cannot reach the detector, requiring careful design to minimize light loss.

Reagent injection in a luminometer must also be designed to ensure good mixing of sample and added reagent, which is a prerequisite for precise results. A special challenge is CL measurement in the presence of magnetic particles. This requires a relatively forceful so-called jet injection, capable of quickly resuspending magnetic particles.

Besides volume, injectors should allow to set injection speed, allowing optimum mixing but avoiding splashing that will occur at too high speeds. To effect mixing over a longer time, most luminometers can also shake the samples.

All materials in the reagent duct must be compatible with the widest range of reagents in order not to be damaged by aggressive chemicals, but also not to impair reagent function, like deactivating enzymes or cells.

Teflon® and Peak® are suitable materials.

4.2.3 Transfer Optics

Transfer optics refers to the parts between the sample and the detector. Their function is to guide as much light as possible from the sample to the detector, and to avoid undesired light from wells other than the one in the measuring position reaching the detector. This effect, called cross-talk, can be minimized by a design where the entrance aperture rests, during measurement, on the microplate well with a light barrier against all other samples. In some luminometers the aperture opening can be optimized for measuring 96-, 384- or even 1536-well microplates.

4.2.4 Multimode Readers

A multimode reader is designed to accommodate several measuring principles in a single instrument. Besides CL, it may be capable of measuring prompt fluorescence, time-resolved fluorescence (TRF), fluorescence resonance energy transfer (FRET), time-resolved fluorescence resonance energy transfer (TR-FRET), fluorescence polarization (FP) and absorption. Besides using optical filters, advanced multimode readers are also equipped with monochromators, which, however, do not have sufficient sensitivity for most CL applications.

The design of most multimode readers was originally based on fluorometers. In principle, every fluorometer with the light-source switched off can measure CL, but normally lacks sensitivity and suffers from high cross-talk levels. Some multimode readers overcome this problem by implementing a separate and optimized optical path for CL measurement. An advantage of a multimode reader in CL measurements is the availability of optical filters, allowing bioluminescence resonance energy transfer (BRET), or other color-discriminating analyses – however, it would be wise to verify that the sensitivity is high enough.

4.2.5 High-throughput Luminometers

Light intensities, and consequently signal rates, are generally lower in CL measurements than in fluorescence. Longer measuring times are therefore required to obtain sufficient statistical precision, and this limits sample throughput.

True high-throughput systems are built around imaging detectors and will be discussed later. An intermediate solution is provided in the form of multi-detector systems, allowing parallel measurement of several wells simultaneously. Examples are the SpecraMax L luminometer from MOLECULAR DEVICES with up to six detectors, or the PerkinElmer TopCount and MicroBeta LumiJET models, each with up to twelve detectors.

4.2.6 Tube Luminometers

Single-tube luminometers for samples in typically 12×75 mm tubes have been widely replaced by microplate instruments. However, they are still available as portable devices, *e.g.*, in industrial hygiene applications, as low-cost diagnostic systems or as low-cost laboratory instruments. Tubes may also be chosen when relatively high sample volumes are required, as in urine testing. There is even an automatic luminometer for 180 sample tubes available, used for chemiluminescence immunoassays for detection of growth hormone doping,[2] or for bacterial screening in milk.[3]

4.2.7 Temperature Control

Most luminometers offer temperature control as an option, sometimes up to about 40 °C, some have incubation chambers for temperatures up to 60 °C. Cooling might also be available, for the samples, the reagents or the photomultiplier so as to reduce background noise.

4.2.8 Relative Light Units

Raw data from luminometric measurements is normally reported as "relative light units" (RLU).

While in absorption measurements one can directly obtain a physically meaningful result, *i.e.*, optical density, luminometric measurements have to undergo a calibration procedure to afford quantitative values for analyte mass or concentration. Of course, the same is true in fluorometric measurements.

For a photomultiplier in current mode, the output depends strongly on the applied high voltage, which in turn determines the electron multiplication factor of the dynodes.

Typically, the internal gain of a photomultiplier may increase by a factor of two for a voltage increase of 120 V. Outside the tube, amplifier gain and the properties of the analog-to-digital-converter determine the RLU value, meaning that this a totally arbitrary unit.

For photon counters, the rate of digital output pulses, or counts per second (cps), is practically identical to the rate of photoelectrons released from the cathode, so that by taking the quantum efficiency into account the rate of photons hitting the photocathode may be calculated.

Both QE and background show great individual variations, even for the same type of photomultiplier, typically about 30–40% for QE, and more than a factor of two for background. Since it is desirable that different individual luminometers of the same model show identical readings for the same sample, the raw data, *i.e.*, cps in the case of photon counting, can be multiplied with an individual so-called RLU factor.

4.2.9 Sensitivity and Dynamic Range

Both sensitivity and dynamic range are important specifications of a luminometer.

Sensitivity means the minimum detectable analyte quantity, and the range up to the maximum amount of analyte that is still measurable, without serious nonlinearity and saturation effects, is the dynamic range of the assay (the dynamic range of the instrument alone would be the range between background and maximum signal measureable).

Brochures of commercial instruments sometimes state rather ambitious values for sensitivity and dynamic range, but frequently do not specify how these data are derived.

The minimal detectable analyte quantity is not only an instrument specification. For example, the minimum detectable amount of ATP depends on the specific type of reagent, and also on the optical properties of the microplate used. A white microplate might produce a ten-times higher signal than a black one.

We recommend the following procedure to determine sensitivity:

1 Measure the response S to a calibrated amount of analyte Q, which should be about in the middle of the linear range. For example: $Q = 1000$ ng ATP, $S = 100000$ RLU s^{-1} and, therefore, 1 RLU s^{-1} corresponds to 10 pg ATP.
2 Measure the background of a statistically sufficient number of samples (wells) and calculate the standard deviation. For example: average background $= 50$ RLU s^{-1}, measured in 10 wells; standard deviation $= 8$ RLU s^{-1}. To be conservative, use the three-sigma criterion, then the sensitivity is $3 \times 8 \times 10 = 240$ pg ATP.

4.2.10   Clinical Applications

The biggest area of CL applications is in clinical diagnostics, used in major fully automatic immunoassay or DNA/RNA probe systems, and including everything from sample preparation to administrative tasks. Examples are ABBOTT's ARCHITECT system, SIEMENS BAYER's ACS 180, GENPROBE's DTS and the SIEMENS DPC Immulite, with several thousand installations in total.

Interestingly, most of these systems use acridinium ester labels. Detailed descriptions are beyond the scope of this chapter.

Recently, small luminometers for clinical applications with extended immunoassay software have become available, approaching the cost of ELISA-readers. These do not need on-board reagent injection, because the label is usually horseradish peroxidase, and with suitable light-generating substrates a long-lasting glow-type signal is produced, so that the starter reagents can be added outside the instrument.

By adding certain basic sample preparation capabilities, semi-automatic CL analyzers at moderate cost are becoming available.

4.2.11   Practical Hints

1 Microplates or sample tubes, like most plastic material, tend to exhibit phosphorescence, leading to elevated background levels after exposure to

light. It is therefore advisable to always store them in the dark. Remember that a luminometer is about the most sensitive instrument to measure light.

2 The right choice of microplates is important to obtain good results. White opaque plates are normally used, but there are considerable differences in quality. If the walls are not completely opaque, cross-talk will occur. Some plates can show strong phosphorescence. Relatively good choices are PORVAIR model 204003 and GREINER LumiTrac 200 plates.

3 Reagent ducts, from reagent container through injector pump up to the injector tip, should never become dry, as during stand-by. They should be filled with wash solution or buffer when not in operation with starter reagents. In the case of ATP measurement, an antimicrobial solution is recommended.

4 During priming of an originally air-filled system, there is always the danger of splashing air–liquid foam around the injector tip area. Therefore, this area should be regularly inspected, and eventually cleaned with a soft Q-tip or similar. To avoid contamination of the tip area, a prime cycle with deionized water is recommended before actually priming the lines with reagent.

5 Care must be taken since the injector tip is quite delicate – even smallest scratches can lead to irregular injection patterns. It is a great advantage when the luminometer design allows easy access to this area, as some instruments do.

4.3 Imagers

Low-level-light imagers have been developed to image CL emission from small animals, plants, blots, *etc*. They are also used for high-throughput systems imaging entire microplates. Some images allow detection even in the single-photon counting regime, allowing the most sensitive measurement of CL emission. The detector of choice presently is the charge-coupled device (CCD), for the invention of which Kao and Boyle were awarded the Nobel Prize in Physics in 2009. A scientific CCD camera is based on the same principle as that used in every digital camera, but differs in certain aspects that will be discussed below. In the future, C-MOS (complementary metal oxide semiconductor) cameras may be expected to compete with CCDs.

Most imagers are also equipped with a fluorescence option, using an excitation light source and appropriate filter combinations for excitation and emission.

An imager consists of the following components:

1 low level light camera
2 transfer optics
3 sample illumination
4 sample holder

5 light tight cabinet containing items 1–4
6 electronics and software.

Several accessories are available for imagers, allowing, besides fluorescence detection, temperature control as well as gas or liquid supply. These accessories are introduced in Section 4.3.7. Finally, *in vivo* imaging including three-dimensional imaging, real-time and multimodal imaging will be covered in Sections 4.3.8 and 4.3.9.

4.3.1 Camera Principles and Types

The heart of a camera is the CCD chip, which is a rectangular array of individual light-sensitive elements called pixels. Contrary to a photomultiplier, which releases a quasi-continuous signal flow, CCD image acquisition is performed in two discrete steps. During the first step, photons impinging upon the pixels create photoelectrons, which are stored and integrated in the pixels. In a second step, after a definite time of integration, all pixels are read out, and the pixels are reset to their initial state. Following this, the next image can be acquired.

The read-out process is performed by applying a definite sequence of pulses to so-called gate electrodes at each pixel, causing a transfer of pixel charges to its immediate neighboring pixel. Because of this process of moving or coupling the charges from one pixel to the next, this device is called a charge coupled device (CCD). A thorough review of CCD function/structure is given by MacKay.[4]

Read-out, also called clocking, is performed in the following steps, as illustrated in principle in Figure 4.4. The method is known as progressive scan read-out and covers the following steps (Figure 4.4):

1 Status of a 2×2 pixel array at the end of an exposure.
2 The charge contents of all rows are transferred down one row, thereby shifting the lowest row into the readout register.
3 The charges in the read-out register are sequentially transferred into the direction of the charge to voltage converter (CVC), where they are amplified and converted into a voltage (see also Figure 4.5 below).
4 Shows the state where the last pixel content in the read-out register has been shifted into the CVC.
5 The contents of the next pixel row is shifted downwards into the read-out register.
6 The contents of the read-out register is shifted towards the CVC, as in step 3.
7 Corresponds to step 4, where the last pixel content in the read-out register is shifted into CVC. Following step 7, the CCD is cleared of all charges and ready for the next exposure.

Each voltage corresponding to a certain pixel as it is generated by the CVC is converted into a discrete number by an analog-to-digital converter

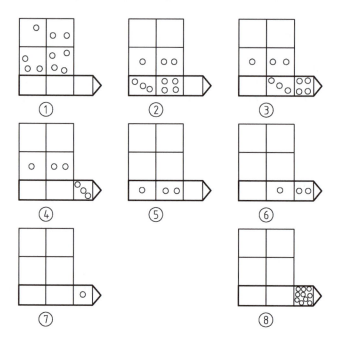

Figure 4.4 Simplified CCD read-out sequence (clocking) for CCD with only four
pixels, plus output register (lowest row) (see text for explanation). Item 8
shows the result of binning where the charge contents of all pixels are
concentrated before read-out (see Section 4.3.2.3, Binning).

(ADC). This number is usually called counts, or digital units (DU), and is
stored in memory.

The following sections describe different detectors used in CL imaging.

4.3.1.1 Full Frame Transfer Chip (FFT)

The read-out principle of a FFT chip has been described above. The architecture of this chip is depicted in Figure 4.5(a). Since all pixels of this type of
chip are light sensitive even during the transfer process, the light exposure
should be interrupted by a mechanical shutter during read-out. This prevents a
vertical smearing of the image. The frame rate (frames per second, fps), which
determines the acquired number of images per second, is typically 2 fps for
1000×1000 pixel (1 Mb) chips. It is an excellent low noise chip for slow scan
cooled camera systems.

4.3.1.2 Frame Transfer Chip (FT)

The FT chip (Figure 4.5b) is equipped with a second array of light insensitive
pixels of equal size and number into which the charges of the first array are

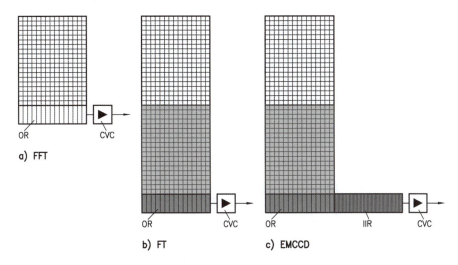

Figure 4.5 Layout of CCD devices used in CL imaging. (a) Full-frame transfer (FFT), with output register (OR) and charge-to-voltage converter (CVC). (b) Frame transfer (FT) with additional array of light-insensitive pixels into which pixel contents are transferred before read-out. (c) Like (b), with impact-ionization register (IIR) for electron multiplication before read-out.

transferred in a single step of less than 1 ms duration. While the second area is clocked out, the first is light sensitive again. Therefore, typical frame rates could be higher, as is listed in Table 4.1. A mechanical shutter is usually not required.

4.3.1.3 Electron Multiplying CCD Chip (EMCCD)

With the chips described above, sensitivity is limited, among other sources, by read-out noise, generated in the CVC and in the ADC. This noise is proportional to the clocking frequency of the read-out process. To avoid excessive read-out noise, the clocking frequency is therefore kept below 3 MHz. To obtain high sensitivity, a typical clocking frequency is only 150 kHz, which gave rise to the name slow scan systems. With a 1 MB pixel chip the read-out time for one image is about 7 s. For many applications this frame rate is too slow.

Higher frame rates combined with high sensitivity are achievable with so-called EMCCD chips (Figure 4.5c). The architecture is similar to a FT chip, but with an extended output register. The charge contents in the output register pass a sequence of so-called multiplication registers (IIR), where the charges are amplified due to impact ionization processes, before they are fed into the CVC and ADC.[5] By this method the signal is increased to a level far above the read-noise, making the read-noise negligible. Multiplication gains of up to 1000 are achievable. Frame rates of 31 fps for 1 MB chips can be attained.

Table 4.1 Technical data of different cameras.

Type	Camera type manufacturer	Pixel size (μm^2)	Max fps @clocking rate	Read noise	Dark current (electron pixel^{-1} s^{-1})	QE % (max.)
FFT	Andor Ikon M-934	13×13	2.2 @ 2.5 MHz	2.5 el. @ 50 kHz	1.2×10^{-4}	95
FT	Hamamatsu C8800	8×8	30 @ 40 MHz	25 el.	25	50
EMCCD	Andor Luca R604	8×8	12.4 @13.5 MHz	<1	0.17	65
ICCD	Andor iStar DH720	26×26	91	N.a.	N.a.	40

EMCCD's are about the most sensitive and versatile chips used in luminescence imagers.[6]

4.3.1.4 Intensifier Technology (ICCD)

This technology uses an intensifier in front of a conventional camera. Modern intensifiers use a microchannel plate (MCP) with millions of very thin, conductive glass capillaries (4–25 µm in diameter) fused together and sliced into a thin plate. Each capillary or channel works as an independent secondary-electron multiplier to form a two-dimensional multiplier array. Incident photons hit a semi-transparent photocathode in front of the MCP, where photoelectrons are generated. These electrons are amplified during the passage of the MCP, and finally hit a phosphorescent screen behind the MCP. The intensified screen is imaged either by a so-called relay optic to a standard CCD camera or the screen is coupled *via* a tapered fiber optic to the CCD chip (Figure 4.6).

Since the light intensity is considerably amplified prior to read-out, an ICCD camera does not necessarily have to be cooled. Fast imaging with high frame rates is possible. An intensified camera may be used in real time observation of mice *in vivo* or in fast calcium signaling.[7,8]

Compared to FT and FFT chips, the QE is typically lower (Figure 4.7). The spatial resolution of about 50 µm and the dynamic range are limited by the MCP-phosphor system.

Table 4.1 summarizes the characterizing parameters of four commercial cameras. To achieve high frame rates, an appropriate interface between the camera and the computer is required, and a fast PC is recommended to handle the data stream. Modern cameras utilize USB 2.0 or special PCI interface cards.

4.3.2 General Camera Aspects

A sensitive low-light camera is composed of the CCD-chip, a cooling system, appropriate electronics circuitry for camera control and clocking the data out

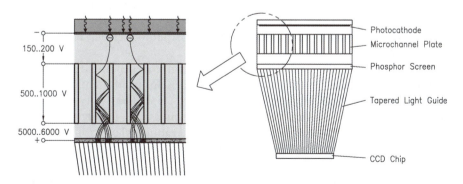

Figure 4.6 Layout of intensified camera showing photocathode, microchannel plate electron multiplier, phosphor screen, tapered light guide and CCD chip.

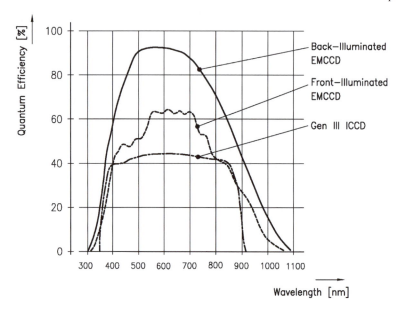

Figure 4.7   Quantum efficiency for different camera types with front illuminated chip, with back illuminated chip and for a third-generation intensifier system.

of the chip, and a computer interface for transferring the data to a computer for further processing. Several aspects have to be considered when selecting a camera for low light imaging instrumentation. The following sub-sections describe important camera parameters.

4.3.2.1   Spectral Response and Quantum Efficiency (QE)

The spectral response is given either by the silicon bulk material and its coating or by the selected photocathode material of the intensifier systems. Figure 4.7 shows that the spectral response covers the dominant wavelength range of CL imaging between 400 and 700 nm.

The QE is defined as the average amount of photoelectrons released per incident photons.

Figure 4.7 shows QE as a function of wavelength. Depending on the architecture and coating of the chip, the peak-QE is 60% for front-illuminated chips and up to 90% for back thinned back-illuminated chips. The two types of chips differ because in front-illuminated chips light must pass through the semi-transparent gate electrodes prior to charge generation, whereas in back illuminated chips the light hits the thinned bulk of silicon on top of the gate electrode directly.

For intensifier systems the QE of the photocathode material can reach a maximum of 40%. Sensitivity of an imaging system is, to a large extent, determined by the QE.

4.3.2.2 Spatial Resolution

The spatial resolution is governed by both the density of pixels on the chip area and the optical quality of imaging. The resolution is limited by the pixel size while the optical spatial resolution is determined by the modulation transfer function (MTF) of both chip and lens. The MTF is a measure of how many lines per mm can be resolved by an optical system. If the lens does not limit the resolution, then the size of about three pixels is the smallest image detail that can be resolved. Pixel sizes range from 8 μm^2 to about 26 μm^2. Small pixels yield a lower dark current and associated noise. It must be emphasized, however, that small pixels have a lower capacity of collecting electrons and thereby reduce the dynamic range of a camera.

4.3.2.3 Binning

Some low-light applications require a higher signal without a noise increase. This can be accomplished by binning, a process where charges of adjacent pixels are integrated. Since the integration of pixels is performed prior to read-out, the signal-to-noise ratio and the sensitivity improve. The situation of a 2×2 binning is sketched in Figure 4.4, item 8, where the charges of four adjacent pixels are clocked into the same output register before reading out the summed charges. The binning area is software selectable. Obviously, binning of pixels reduces the spatial resolution.

4.3.2.4 Cooling

The amount of thermally released electrons in bulk silicon is temperature dependent and is halved with every 6–7 °C temperature decrease. Therefore, for sensitive signal recovery the thermal dark noise must be reduced by cooling. Modern 4 stage thermo-electrical coolers reach an almost 100 °C temperature difference relative to ambient, resulting in dark noise levels of 0.0001 electrons per pixel per s.[9] To avoid condensation and loss of sensitivity, cooling has to be performed in a high vacuum.

In EMCCD chips the gain of impact ionization is inversely proportional to the temperature. Consequently, the temperature must be stabilized precisely to maintain a constant multiplication gain.[10] It is not advisable to cool below − 100 °C, since many sensors then no longer operate properly.

4.3.2.5 Dynamic Range

The maximum amount of charges a pixel can accept is named the full well capacity (FWC). If more light hits a pixel, charges begin to fill neighboring pixels, a process called blooming. The pixel is saturated and the image becomes blurred. The FWC increases with the size of the pixels and ranges from 30 000 electrons for a $8 \times 8 \mu m$ square pixel to 500 000 electrons for a $26 \times 26 \mu m$ square pixel. The dynamic range is limited by the dark current and noise contributions at the lower end of the range, and by the FWC at the upper end of the range. Dynamic range is normally defined as a ratio of FWC to camera noise. Larger pixels show a higher dynamic range although their dark noise is increased.

The dynamic range is transformed into grey levels of the image by means of the charge to voltage amplifier, ADC and also by the imaging software. Theoretically, a 16 bit ADC is able to represent 65536 gray levels, which are reduced by noise and offset signals. Almost all cameras apply 14–16 bit AD converters; some of them are software selectable.

4.3.3   Transfer Optics

The performance of a CL imager system is determined not only by the camera but also by the lens. It has to image the object on the chip, must collect light with high efficiency and must achieve various aspect ratios from small details to whole animals. Modern lenses have relative apertures up to 0.85 and are coated in the visible range and reach more than 90% transmission. Since the collection efficiency of CL and fluorescence emission is dependent on the solid angle that the aperture subtends from an object point, the object-to-lens distance should be as short as possible. Therefore, the lens must be able to image objects at small working distances with large viewing angles. Since off-axis objects are subject to an intensity decrease of the image, a so-called flat-field correction is performed by software. The focal length of the lens is dependent on optical and mechanical constraints and ranges from about 15 to 50 mm. Some lenses have a variable iris diaphragm to control the intensity and depth of focus.

To cover a broad range of magnifications, in some instruments a lens with constant focal length is moved precisely along the optical axis to adjust for the size and height of a sample.[1] Other instruments move the sample table or use zoom lenses.[11]

4.3.4   System Sensitivity

Low-level light imaging requires the most sensitive camera and lens systems available. Faint signals down to single photon level must be detected and all sources of noise must be reduced as far as possible. In the following, the most important parameters for high system sensitivity are described, starting at the object.

Given the radiance of a CL object, the photon flux to the camera is determined by the lens. The shorter the working distance and the larger the lens aperture, the more photons per second arrive at the camera chip. Certainly, the flux per pixel is dependent on precise focusing of the lens.

Next, the QE of the chip determines the number of electrons generated by the incident photons. A higher QE at the wavelength of interest directly increases the sensitivity.

An important parameter is the pixel size. A larger pixel could collect more photoelectrons of an image without significantly raising the noise level. The reasoning is identical to the method of binning.

Aside from optical and structural chip properties the system sensitivity is limited by noise of different origins.

First, there is photon shot noise, which is due to the statistical nature of the impinging photon flux. If N photoelectrons are released, the related Poisson-noise is $\sqrt{N}$, which is the predominant noise for large photon fluxes, where other noise sources are negligible.

The other noise components also exist in the dark. The second type of noise is dark shot noise. It is dependent on the chip temperature, as stated in Section 4.3.2.4 (Cooling), and on the size of the pixel. For example, a $24 \times 24\,\mu m^2$ pixel has a dark current of 8×10^{-4} electrons per s at $-80\,°C$.[12] For long exposure times T, the noise increased with $\sqrt{T}$. Therefore, cooling is an essential requirement for long integration times.

Independent of the exposure time, a third noise component exists: read-out noise. It is the sum of noise contributions from the processes of charge conversion into voltage, its amplification, the analog to digital conversion and the resetting of these elements. It is the limiting noise component in conventional CCD chips, but not in EMCCD chips and intensified systems as explained above. Read-out noise increases with the read-out clocking frequency, since the read-out amplifier requires a larger bandwidth for amplifying higher frequencies. For example, slow scan cameras operated at 50 kHz generate a read-out noise of about 2.5 electrons.

All noise components add up quadratically to yield the total noise squared, according to the laws of error propagation. A signal-to-noise-ratio of 3 is an accepted definition of pixel sensitivity.

In summary, a CL imager can detect the lowest signals, provided it is equipped with a large aperture lens at short working distances, a CCD chip with highest QE, the largest acceptable pixels, low temperature cooling, and performs at the lowest noise levels.

4.3.5 Sample Illumination

In most instruments, illumination is provided by white LED sources surrounded by diffuse plastic balls. A photo of the sample is then taken, and the software produces an overlay of the photo and the low-level CL image.

4.3.6 Sample Holder

The sample to be investigated is fixed on a holder, which is mounted on a pull-out base plate. In some instruments the holder can be rotated and the height of the sample can be adjusted. For many applications the sample holder has to be heated and temperature stabilized.

4.3.7 Accessories for Dedicated Applications

Modern imagers can be equipped with several accessories for plant, animal and *in vitro* imaging. Important accessories are:

1 filter option
2 microplate imaging
3 blot and gel detection
4 temperature and humidity control
5 daylight simulation
6 multiport flange
7 animal beds.

4.3.7.1 Filter Option

Most modern CL imagers can be equipped with a fluorescence option, consisting of a light source that complies with the required spectral range, and appropriate excitation and emission filters. Interference filters with high transmission and blocking of unwanted light are necessary. The excitation filter restricts the spectrum of the light source to the absorption peak of the fluorescent dye, and the function of the emission filter is to transmit the bandwidth of the dye's emission and block the excitation light. The position of the filter in a slider or wheel can be in front of or behind the lens. Both positions are realized in commercial instruments. Filters are also necessary in bioluminescence resonance energy transfer (BRET) assays.[13] Since CCD-cameras are sensitive above 700 nm, the phosphorescence of chlorophyll-containing food or plastic materials must be blocked by additional IR-cut-off filters.

4.3.7.2 Microplate Imaging

A CL imager can also be used to image microplates consisting of an array of 96 up to 1536 wells. A novel class of instruments has been designed as high-throughput screening imagers (HTS imagers) with a typical processing time of less than 1 min for 1536 well plates or about 100 000 well tests per hour.

In some instruments the camera measures the plate from above. In this case the camera is equipped with a telecentric lens, or with a supplemental Fresnel lens in front of the microplate. These lenses allow an unrestricted view even to the bottom of each well and thus support the measurement of lowest CL signals. Nevertheless, a flat-field correction is necessary to compensate for intensity losses for off-axis wells. As for luminometers, white opaque plates are normally used. Injection of reagents is not possible while the camera is running a measurement. For this reason, a measurement from above is restricted to glow luminescence.

If the camera views the microplate from below, the top-side of the plate is freely accessible for reagent injection and dispensing. This allows the detection of flash luminescence. A transparent plate with opaque side walls is required.

Some instruments apply a tapered fiber optic, which directly connects the bottom of the plate and the camera.[14]

4.3.7.3 Blot and Gel Detection

Electrophoretic separation of proteins followed by a transfer step to a (nitrocellulose) membrane (Western blot) and subsequent detection steps is an indispensable method in protein research as well as in diagnostics (*e.g.*, BSE). The detection procedure usually starts with the addition of a specific antibody (primary antibody) to the protein of interest followed by the addition of a rather general antibody (secondary antibody which is labeled with HRP) against the primary antibody. After the addition of the substrate luminol, a chemiluminescent signal is produced at the protein locations. Traditionally, these signals have been detected by exposing the treated membrane to X-ray films. Modern chemiluminescence imagers offer the advantage of digital documentation of the experiment, variable definition of exposure times during the measurement and a larger dynamic range of >4 orders of magnitude (versus 2 of films).

A so-called transilluminator accessory is used for gel documentation. The gel is illuminated from below with appropriate excitation light to match the spectral properties of the dyes in the gel.

4.3.7.4 Temperature and Humidity Control

Temperature control is often required in animal and in plant research. For example, nude mice lose a considerable amount of body temperature within a few minutes, and plant metabolism is temperature dependent. Enzymatic reactions are also temperature dependent, and, furthermore, the firefly luciferase shows a spectral redshift with increasing temperature.[15] Consequently, temperature control is available in most imaging systems.

Control of humidity is advantageous for plant applications in the light-tight cabinet. Some instruments are equipped with a combined temperature–humidity unit, which is incorporated in the instrument.

4.3.7.5 Daylight Simulation

In plant research, there is a need to define environmental sunlight conditions to find out stress parameters. For example, LED panels, installed in the cabinet, simulate the sunlight from dawn to sunset with varying intensity and spectral composition.

4.3.7.6 Multiport Flange

A multiport flange allows access of external components to the interior of the cabinet. It offers, for example, light-tight tubing lead-through for gases or liquids, thereby enabling anesthesia control, special atmospheres for anaerobe bacteria, supplying oxygen-enriched water for fish, or automatic watering of plants. An optional second camera can be mounted to view samples from an orthogonal perspective. Especially in plant research this camera allows us to observe leaves that might be hidden in a top view.

4.3.7.7   Animal Beds

If animals are imaged in a sequential order in different types of instruments, special animal beds are provided, ensuring a stable position of the animal during transfer from one instrument to the other. Animal beds are to be heated and temperature stabilized for anesthetized small animals like mice.

4.3.8   *In Vivo* Imaging

The measurement of light-producing reactions in animals and plants, following their intensity and position over time, provides important additional information to *in vitro* experiments.[16] *In vivo* imaging has been applied successfully in research on time-dependent gene expression (*e.g.*, circadian rhythms), tracking of bacterial infections and proliferation and metastasis of tumors in animal models, *etc.*[17] Imaging low level bioluminescence emitted from small animals, in particular mice, has become a major application of imagers and will be treated more extensively in Chapter 12 in this book.

4.3.8.1   Real Time Imaging

In traditional *in vivo* imaging with slow-scan CCD cameras animals have to be anaesthetized. This may cause undesirable effects on animal physiology, including altered drug response, or may change the LD_{50} dose of bacteria during infection studies.[8] Calcium flux in animals or plants is typically a fast transitory effect with a time scale of a few seconds.[7] Both applications require the acquisition of sequences of images with very short individual exposure times and high frame rates. ICCD or EMCCD camera systems are suitable devices.

4.3.8.2   3D Imaging

Straightforward planar luminescence imaging leads to qualitative results only. Because light emitted from cells is partially absorbed and scattered when passing through body material, both the intensity and the 3D location of the source cannot be determined quantitatively without further information.

The most ambitious goal is three-dimensional quantitative bioluminescence imaging or bioluminescence tomography (BLT). This is made possible by spectral analysis of the emitted light, based on the following facts.

Below 600 nm, light is strongly attenuated in animal tissue, due to absorbers like hemoglobin or melanin. This results in a spectral shift to longer wavelengths (redshift) of the light leaving the body relative to the spectrum at the source, increasing with tissue depth.[11] Based on knowledge of the spectral intensity distribution, which is measured with discrete optical filters, the depth of the source can be calculated.[18] For this reason, state-of-the-art 3D imagers use spectral information.

Wang *et al.* have developed a prototype BLT system for simultaneous acquisition of multiview and multispectral data, using an arrangement of four mirrors to image four aspects of a mouse onto the CCD.[19]

A modern BLT instrument is available from Caliper Life Science (Ivis 200® series). The system has single-view 3D tomography capability, based on spectral and topographical data, and theoretical models of light scattering and absorption in tissue. For this purpose the Ivis® imager uses six interchangeable filters.

In addition, the surface topography of a mouse is created from an image of a structured light pattern that is projected onto the animal by means of a scanning laser galvanometer. This topographic information is used in combination with a wavelength dependent scattering model in tissue. Together with the spectral data, a 3D image is reconstructed with an improved spatial resolution of 1 to 3 mm.

Tomographic evaluation on the basis of BLI alone, even with spectral analysis, still leaves room for improvement. A combination of BLI with other imaging technologies can lead to further improved results, or new information. For example, Allard *et al.* have combined high-resolution magnetic resonance imaging (MRI) with multispectral multiview bioluminescence images, and improved the positional and intensity accuracy of BLT.[20] MRI and BLI images were acquired simultaneously in the same session.

4.3.9 Multimodal Imaging

Other modalities have been adapted to small-animal imaging as well and are listed briefly without explanation: ultrasonic imaging (USI), X-ray computerized tomography (CT), magnetic resonance imaging (MRI), positron emission tomography (PET), single-photon emission computed tomography (SPECT) and 3D-fluorescence or fluorescence mediated tomography (FMT). Each modality has specific advantages and provides unique information.[21,22]

A superposition of different modalities is a promising strategy to collect further information. For example, a combination of BLI and PET imaging in a single instrument leads to a precise localization of the signal source.[23,24] The combination of MRI and BLI has already been mentioned.

A new multimodal instrument is the Caliper Lumina XR®, which combines fluorescence, bioluminescence and X-ray technologies. The X-ray image is obtained in seconds and can be overlaid with bioluminescence, fluorescence and photographic images.

4.3.10 Software

The software of an imaging system consists of two functions: instrument control and image processing. The controller software defines all instrument, accessory and camera parameters such as sample temperature and the exposure time, field of view, focus adjustment, filter setting and storage of raw data.

The image processing software performs the overlay of a black and white photograph with the low-level-light image for identification purposes and converts the measured intensities into pseudo-colors for better visualization. Furthermore, it controls the image brightness, contrast and transparency, subtracts background images, applies flat-field corrections and sets a region of interest manually or automatically. Additional functions are available like line plots for checking background intensities, surface plots for visualizing true 16-bit images, look-up table setups, annotation possibilities, multicolor overlay for spectral unmixing and different arithmetic functions for image manipulations.

Most software packages and file formats for imagers are proprietary, because special parameters like the working distance, sample height and pixel defects have to be taken into account in the calculation of signal intensities. This data would be lost if stored in common file formats like TIF. However, most imager software allows exporting of raw data to standard formats for further processing.

An alternative is to export into DICOM-format, where all information is stored, even the pixel- or voxel-size and the instrument classification.

Software to overlay the individual images derived from different modalities is available today (*e.g.*, VINCI, Pmod).

4.3.11   Plant Imager

An example for a complete system is Berthold Technologies' NightSHADE imager, which is specifically designed for plant research (Figures 4.8 and 4.9).

The camera (Andor® model Luca R 604) is mounted in a fixed position on top of the light-tight cabinet. The camera is Peltier cooled at $-20\,^\circ$C, and equipped with an EMCCD chip of 1004×1002 pixels, $8 \times 8\,\mu\text{m}^2$ size. The pixels are read out by a constant clocking frequency of 13.5 MHz, digitized at 14 bits, and supplied to a PC *via* a USB 2.0 interface. The corresponding frame rate is 12.4 fps without binning. Read-out noise is less than 1 electron rms (root mean square) in EMCCD mode, and the dark current is typically 0.17 electrons per pixel per s. Full well capacity is 30 000 electrons for an active light sensitive pixel.

The focal length of the C-mount lens is 25 mm with a relative aperture of 0.95.

Daylight simulation is performed by an LED panel with illumination control. The panel is composed of about 700 LEDs covering wavelengths of 470, 660, 730 nm and white light. Light at these wavelengths regulates the metabolism of plants. The LEDs are programmable by software to represent, for example, a circadian plant rhythm.

An optional temperature and/or humidity control unit is provided in the cabinet. The sample holder can be rotated with an axis that is displaced relative

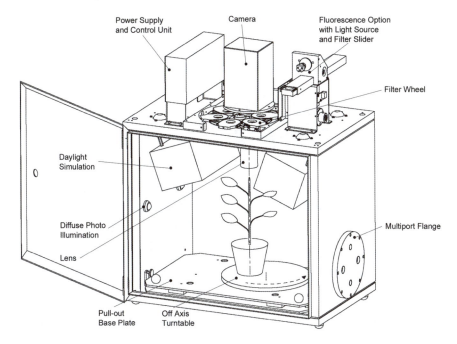

Figure 4.8 Plant Imaging System NightSHADE from Berthold Technologies. The cooled CCD camera is mounted on top of the light-tight cabinet (shown with door open, top cover removed), looking down. An excitation light source with a slider holding excitation filters allows fluorescence measurements (optional). A filter wheel in front of the camera holds emission filters for fluorescence or CL. An off-axis turntable can hold several specimens or plants, which can be positioned sequentially into the field of view of the camera. A multi-LED unit can simulate different sunlight spectra, while diffuse illumination can be switched on for photographic exposure. The cabinet can be temperature and humidity controlled.

to the optical axis of the camera lens, and which allows imaging multiple samples sequentially.

An optional second camera could be attached to the multiport flange, allowing one to image samples horizontally (Figure 4.9).

A major application is imaging of agar plates, which are kept in a vertical position. Plant roots are typically growing positively geotropic whereas shoots are following a negative geotropism. To enable unhindered root development and to allow full view of the whole seedling, the agar plates are kept in a vertical position with the seedling aligned along the vertical axis.[25]

The turntable can present up to six vertically oriented agar plates successively and in a repetitive fashion to the camera during one run. Of course, the side view camera is also useful to view plants horizontally, *e.g.* when lower leaves cannot be imaged from above.

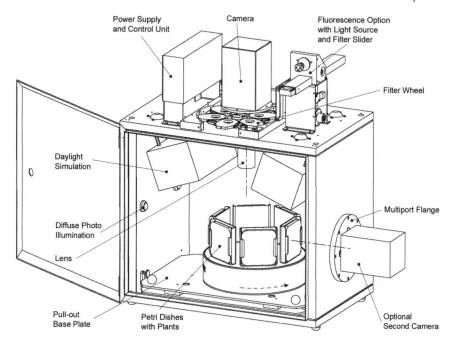

Figure 4.9 Plant Imaging System NightSHADE with side-on camera to capture sequential images from Petri dishes positioned vertically on a turntable.

References

1. F. Berthold, K. Herick and R. M. Siewe, in *Methods in Enzymology*, ed. M. M. Ziegler and T. O. Baldwin, Academic Press, San Diego, 2000, pp. 62–87.
2. M. Bidlingmaier, J. Suhr, A. Ernst, Z. Wu, A. Keller, C. J. Strasburger and A. Bergmann, *Clinial Chem.*, 2009, **55**, 445–453.
3. *A Practical Guide to Industrial Uses of ATP-Luminescence in Rapid Microbiology*, ed. P. E. Stanley, R. Smither and W. J. Simpson, Cara Technology Ltd, Lingfield, Surrey, UK, 1997.
4. C. D. MacKay, *ARA&A*, 1986, **24**, 255.
5. C. G. Coates, D. J. Denvir and E. K. Conroy, in *Bioluminescence & Chemiluminescence: Progress & Current Applications*, ed. P. E. Stanley and L. J. Kricka, World Scientific, Singapore, 2002, pp. 201–204.
6. Promet Consortium, *Prostate Cancer Molecular-oriented Detection and Treatment of Minimal Residual Disease*, Contract no: LSHC-CT-2006-018858, 2008.
7. K. L. Rogers, S. Picaud, E. Roncali, R. Boisgard, C. Colasante, J. Stinnakre, B. Tavitian and P. Brûlet, *PLoS ONE*, 2007, **2**(10), e974.
8. I. J. Hildebrandt, H. Su and W. A. Weber, *ILAR J.*, 2008, **49**(1), 17–26.

9. Andor Product Portfolio 2009 (http://www.andor.com/pdfs/downloads/product_portfolio.pdf).
10. Longevity in EMCCD and ICCD, Andor Technical Note, 2006.
11. B. W. Rice, M. D. Cable and M. B. Nelson, *J. Biomed. Opt.*, 2001, **6**(4), 432–440.
12. An introduction to Scientific Imaging Charge Coupled Devices, SITe® Corporation (http://astrosun2.astro.cornell.edu/academics/courses/astro310/SITe-CCD.pdf).
13. A. De, A. M. Loening and S. S. Gambhir, *Cancer Res.*, 2007, **67**(15), 7175–7183.
14. *US Pat.* 4922092; inventor John Rushbrooke *et al.*, 1990.
15. H. Zhao, T. C. Doyle, O. Coquoz, F. Kalish, B. W. Rice and C. H. Contag, *J. Biomed. Opt.*, 2005, **10**(4), 41210.
16. A. Maggi and P. Ciana, *Nat. Rev. Drug Discov.*, 2005, **4**(3), 249–255.
17. L. F. Greer 3rd and A. A. Szalay, *Luminescence*, 2002, **17**(1), 43–74.
18. G. Wang, H. Shen, K. Durairaj, X. Qian, and W. Cong, *Int. J. Biomed. Imag.*, 2006, article ID 58601.
19. G. Wang, W. Cong, K. Durairaj, X. Qian, H. Shen, P. Sinn, E. Hoffman, G. McLennan and M. Henry, *Opt. Express*, 2006, **14**(17), 7801–7809.
20. M. Allard, D. Côté and L. Davidson, *et al.*, *J. Biomed. Opt.*, 2007, **12**, 034018.
21. M. Lecchi, L. Ottobrini, C. Martelli, A. Del Sole and G. Lucignani, *Q. J. Nucl. Med. Mol. Imag.*, 2007, **51**(2), 111–126 (review).
22. M. L. Dustin, *Arthritis Res Ther.*, 2003, **5**(4), 165–171. Epub 2003 May 1 (review).
23. D. L. Prout, R. W. Silverman and A. Chatziioannou, *IEEE Trans. Nucl. Sci.*, 2004, **51**(3), 752–756.
24. G. Alexandrakis, F. R. Rannou and A. F. Chatziioannou, *Phys. Med. Biol.*, 2005, **50**, 4225–4241.
25. T. J. Mulkey, K. M. Kuzmanoff and M. L. Evans, *Planta*, 1981, **152**, 239–241.

Part 2
Analytical Applications of
Chemiluminescence and Bioluminescence

CHAPTER 5

"Classical" Applications of Chemiluminescence and Bioluminescence

MASSIMO GUARDIGLI,[a] ARNE LUNDIN[b] AND ALDO RODA[a]

[a] Department of Pharmacentuical Sciences, University of Bologna, Via Belmeloro 8, 40126, Bologna, Italy; [b] BioThema AB, Stationsvägen 17, 136 40 Handen, Sweden

5.1 Introduction

Chemiluminescence (CL) and bioluminescence (BL) are powerful detection techniques due to features that make them superior to other detection principles involving light, such as spectrophotometry and fluorometry. They offer the unique advantage that light is emitted by a specific reaction involving the analyte, thus avoiding interferences from light scattering and background emission due to sample matrix components. Thanks to the wide dynamic range, samples can be measured over several decades of concentration without dilution or modification of the analytical procedure. In addition, the onset of light emission usually takes place in seconds or minutes, thus rendering CL/BL techniques very rapid.

Especially in recent decades, the advantages of CL/BL detection have been exploited in various analytical techniques, ranging from simple assays for single analytes to whole-cell BL biosensors and "*in vivo*" imaging applications for the monitoring of physiological and pathological processes. This chapter reviews analytical applications of CL and BL in batch analytical formats (*e.g.*, tubes or

Chemiluminescence and Bioluminescence: Past, Present and Future
Edited by Aldo Roda
© Royal Society of Chemistry 2011
Published by the Royal Society of Chemistry, www.rsc.org

microtiter plates). The first part deals with applications of CL, from the detection of hydrogen peroxide in environmental samples to the selective and sensitive measurement of enzymes, enzyme substrates and enzyme inhibitors by means of coupled enzymatic reactions. Biologically-oriented applications such as the evaluation of reactive oxygen species production in cells are also described, as well as the CL techniques used for assessing the antioxidant activity of clinical and food samples. The second part of the chapter is devoted to analytical BL, especially on ATP-related assays based on the firefly luciferase BL reaction. The central role of ATP in all living cells makes it possible to detect bacterial or other living cells, which is useful, for example, for rapid microbiology and hygiene monitoring, and to follow a wide range of enzymatic reactions, thus allowing the development of diagnostic assays for enzymes and metabolites. Analytical applications of other luciferases, such as bacterial luciferase, are also briefly reported.

5.2 Luminol Chemiluminescence

The oxidation of luminol is probably the best-known liquid-phase CL reaction. This reaction is triggered by a range of catalysts, more or less specific for different oxidizing species. Peroxidase enzymes, particularly horseradish peroxidase, are considered the most efficient catalysts and possess a quite high specificity for hydrogen peroxide as the oxidizing agent. Several transition metal ions (including Fe^{2+}, Co^{2+}, Cu^{2+} and others) and their complexed forms, as well as electrochemical oxidation, can be also used to trigger the reaction. Moreover, the intensity and duration of the luminol CL emission can be greatly increased by adding in the substrate solution molecules (*e.g.*, 4-iodophenol) that work as enhancers.

5.2.1 Analytical Applications

The luminol CL reaction is widely employed in analytical chemistry.[1] For example, in flow injection analysis it allows the detection of compounds able to inhibit, enhance or catalyze the CL reaction. However, due to the lack of specificity of the reaction, interferences and cross reactivity problems could impede the analysis of complex samples, unless a preliminary separation step is performed before the measurement. The luminol CL reaction is also commonly used in bioanalysis (*e.g.*, in immunoassays and gene probe assays) for the determination of peroxidase-labeled biospecific probes. Regarding batch analysis, various applications of luminol CL have also been reported and some representative examples are listed below.

5.2.1.1 Detection of Hydrogen Peroxide

A CL one-shot sensor has been used for the determination of hydrogen peroxide in rainwater.[2] The sensor consisted of a hydroxyethyl cellulose matrix containing cobalt chloride and sodium lauryl sulfate casted on a microscope cover glass. The water sample, previously mixed with luminol and phosphate buffer, was applied on the membrane by a micropipette and the resulting CL signal was measured by a portable laboratory-built luminometer. The signal was proportional to the hydrogen peroxide concentration in the range 20–1600 ppb, with a detection limit of 9 ppb of hydrogen peroxide, which made the CL sensor suitable for the analysis of hydrogen peroxide in environmental samples. A highly sensitive CL method based on the Co^{2+}-catalyzed oxidation of luminol by hydrogen peroxide has been employed for its accurate determination in plant tissues.[3] Use of Co^{2+} as catalyst greatly enhanced the sensitivity of the CL reaction, and hydrogen peroxide could be detected at the nanomolar level. Plant extracts could thus be highly diluted before analysis to avoid the quenching effects of phenols and ascorbic acid, which are normally present at high concentrations in plant tissues, on the CL reaction. This analytical method represented a significant advance over previously reported procedures, which required sample pre-treatment steps to remove these quenchers before analysis. The luminol hydrogen peroxide-dependent CL also allowed the measurement of hydrogen peroxide levels in mother's milk at different times of postpartum period.[4] The maximum hydrogen peroxide levels (of the order of 20–30 μM) were found in the first week of the postpartum period, while significantly lower concentrations were observed at longer times. In addition, the stability of hydrogen peroxide levels upon storage at freezing point, at least for a period of one month, was assessed.

5.2.1.2 Detection of Species that Affect the CL Reaction

A sensitive assay for the detection of Co^{2+} based on luminol CL has been proposed recently.[5] Thanks to a pre-concentration step of cobalt ions on a chitosan membrane the method achieved a remarkably high sensitivity, with a detection limit of about $4 \, fg \, L^{-1}$ of Co^{2+}. The analytical performance of the assay was suitable for biomedical applications, such as the determination of the cobalt-containing vitamin B-12 in pharmaceuticals and biological samples, as well as for the detection of cobalt in soils, waters and other environmental matrices. A luminol-based CL method has been used to measure bivalent iron in seawater.[6] Analysis of bivalent iron in seawater is a complicated task due to the extremely low concentration and its tendency to undergo rapid oxidation before analysis. The reported CL assay allowed for sensitive and rapid analyses of seawater samples with minimal sample preparation and no pre-concentration, and proved suitable for field application with satisfactory analytical performance. Assays for determining phenol in water and in organic solvent mixtures based on the evaluation of the effects of phenol on the emission intensity of a *para*-iodophenol-enhanced HRP-catalyzed CL reaction have been developed.[7] Detection limits of a few ppm of phenol were achieved, suggesting that these methods could be applied for the analysis of phenol in environmental samples.

5.2.1.3   Coupling with Enzymatic Reactions

The specificity of luminol-based CL assays is greatly enhanced when detection of specific compounds is performed by employing enzymatic reactions. Coupling of hydrogen peroxide-generating enzymatic reactions involving the target analytes (usually enzyme substrates) with the luminol CL system has allowed development of CL assays for the determination of a wide variety of compounds in complex matrices.[8] In flow-based assays the enzyme(s) necessary for the obtainment of the CL signal are usually immobilized in flow reactors. An alternative approach is immobilization of the enzyme(s) in membrane- or polymer-based layers in contact with a fiber optic, which has lead to the development of sensitive and specific analytical systems.[9]

A CL system based on three simultaneous coupled enzymatic reactions involving acetylcholinesterase (AChE), choline oxidase (ChOx) and horseradish peroxidase (HRP) with luminol as the CL substrate [Reactions (5.1)–(5.3)] has been employed to assess the potency of acetylcholinesterase inhibitors:[10]

$$\text{acethylcholine} \xrightarrow{\text{AChE}} \text{choline} + \text{acetic acid} \tag{5.1}$$

$$\text{choline} \xrightarrow{\text{ChOx}} \text{betaine} + H_2O_2 \tag{5.2}$$

$$2H_2O_2 + \text{luminol} \xrightarrow{\text{HRP}} \text{3-aminophthalic acid} + N_2 + 2H_2O + \text{light} \tag{5.3}$$

The analytical procedure, based on measurement of the kinetics of the CL emission, was very rapid and suitable for the high-throughput screening of acetylcholinesterase inhibitors. The assay can be performed either in 96- or 384-well microtiter plates using conventional microplate luminometers (in the latter case, up to 30 compounds could be assayed in a single analytical session), although for 384-well microtiter plates a luminograph was preferable. A CL assay of lipase activity using a synthetic substrate as pro-enhancer for the luminol CL reaction has been described.[11] The assay employed the lauric acid ester of 2-(4-hydroxyphenyl)-4,5-diphenylimidazole as an enzyme substrate, which liberates 2-(4-hydroxyphenyl)-4,5-diphenylimidazole (an enhancer of the luminol CL reaction) upon enzymatic hydrolysis. The method was simple and rapid and different lipases could be determined with satisfactory detection limits. A CL assay of free fatty acids in three steps using acyl-CoA synthetase (ACS), reduced Coenzyme A (CoASH), inorganic pyrophosphatase (PPiase) and acyl-CoA oxidase (ACO) has been described [Reactions (5.4)–(5.6)].[12] The assay has a linear range of 0.05–5 nmol with different free fatty acids (C_{10}–C_{18}):

$$\text{Step 1: FFA} + \text{ATP} + \text{CoASH} \xrightarrow{\text{ACS}} \text{acyl-CoA} + \text{AMP} + \text{PPi} \tag{5.4}$$

This step is performed in the presence of PPiase to drive the reaction to the right:

$$\text{Step 2: removal of excess CoASH by Affi-Gel 501}$$
$$\text{Step 3: acyl-CoA} + O_2 \xrightarrow{\text{ACO}} 2,3\text{-}trans\text{-enoyl-CoA} + H_2O_2 \tag{5.5}$$

$$2H_2O_2 + \text{luminol} \xrightarrow{\text{HRP}} 3\text{-aminophthalic acid} + N_2 + 2H_2O + \text{light} \tag{5.6}$$

A similar assay employed mutarotase and glucose oxidase (GOD) to measure the glucose concentration in a two-step procedure [Reactions (5.7)–(5.9)].[13] Both steps could be automatically performed in a 1251 Luminometer in one run, and the assay gave a linear response between 0.01 and 1 nmol glucose:

$$\text{Step 1: } \alpha\text{-D-glucose} \xrightarrow{\text{mutarotase}} \beta\text{-D-glucose} \tag{5.7}$$

$$\beta\text{-D-glucose} + O_2 + H_2O \xrightarrow{\text{GOD}} \text{D-gluconic acid} + H_2O_2 \tag{5.8}$$

$$\text{Step 2: } 2H_2O_2 + \text{luminol}$$
$$\xrightarrow{\text{HRP}} 3\text{-aminophthalic acid} + N_2 + 2H_2O + \text{light} \tag{5.9}$$

5.2.1.4 Other Applications

Luminol has been employed as a reducing cyclooxygenase (COX) co-substrate to perform the direct CL measurement of cyclooxygenase activity [Reactions (5.10)–(5.12)].[14] The CL assay presented significant advantages, such as simplicity and rapidity, over other methods used for evaluating cyclooxygenase activity (*e.g.*, measurement of oxygen consumption by oxygen-sensitive electrodes or detection of prostaglandin products by radioactive or immunometric techniques). A method for screening cyclooxygenase inhibitors in the 96-well and 384-well microtiter plate formats based on this principle was developed and used for the rapid identification of new inhibitors selective for the cyclooxygenase COX-2 isoform:

$$\text{arachidonic acid} \xrightarrow{\text{COX, oxygen}} \text{prostaglandin G}_2 \tag{5.10}$$

$$\text{prostaglandin G}_2 + \text{luminol} \xrightarrow{\text{COX}} \text{prostaglandin H}_2 + \text{luminol radical} \tag{5.11}$$

$$\text{luminol radical} \xrightarrow{\text{oxygen}} 3\text{-aminophthalic acid} + N_2 + \text{light} \tag{5.12}$$

A further analytical application of luminol CL was the detection of viable microorganisms. Addition of quinone to a sample containing viable microorganisms results in the production of reactive oxygen species, which could be detected and quantified by a luminol-based CL assay employing a molybdenum complex as catalyst.[15] When the assay was performed in 96-well microplate plates, detection limits of several thousand colony-forming units per milliliter were achieved. Single-cell detection was also possible after 4-h enrichment by cultivation, making this CL assay useful for the rapid detection of viable bacteria and yeasts in food analysis.

5.3   Peroxyoxalate Chemiluminescence

The peroxyoxalate chemiluminescent reaction (POCL) involves oxidation of an aryl oxalate ester by hydrogen peroxide in the presence of a fluorophore that acts as an energy acceptor. Therefore, this reaction can be used for the detection of either fluorophores or hydrogen peroxide, as well as for the determination of analytes following their conversion into hydrogen peroxide.

5.3.1   Analytical Applications

The main analytical application of POCL is the detection of native fluorescent and fluorescent derivatized compounds in flow injection analysis, HPLC and capillary electrophoresis.[16] In most cases, the fluorophore and the other reactants necessary for the development of the CL signal are added to a flow stream containing the analyte(s). However, since the fluorescent energy acceptor is not degraded during reaction, immobilized fluorescent acceptors can be also used to reduce the consumption of fluorophore in the analysis. A flow cell for POCL has been developed using polycyclic aromatic hydrocarbons covalently immobilized onto functionalized polymer and glass beads, which were packed in a cell mounted adjacent to a photomultiplier tube.[17] The best results were obtained using 3-aminoperylene and 3-aminofluoranthene fluorophores immobilized on porous methacrylate beads, and large bead surface areas and high degrees of surface functionalization were found to increase the CL signal.

5.3.1.1   Detection of Hydrogen Peroxide

An innovative CL contrast agent, termed peroxalate micelles, has been proposed recently for the detection of hydrogen peroxide.[18] The peroxalate micelles consisted of amphiphilic peroxalate-based copolymers, a fluorescent dye (rubrene) and a poly(ethylene glycol) coating to avoid macrophage phagocytosis. The limit of detection (about 50 nM of hydrogen peroxide) was within the range of the physiological concentrations of this species. Thanks to its high sensitivity and biocompatibility, the peroxalate micelles have physical/chemical properties suitable for *in vivo* imaging of hydrogen peroxide. Sensors based on the peroxyoxalates TCPO [bis(2,4,6-trichlorophenyl)oxalate] and DNPO [bis(2,4-dinitrophenyl)oxalate] have been developed for the direct determination of hydrogen peroxide in washing powders containing sodium perborate and percarbonate.[19] Comparison of the analytical performance of the sensors with that of the standard iodometric method for determination of hydrogen peroxide in per-salts showed that they could be applied successfully for the analysis of hydrogen peroxide in the bleaching component of washing powder. Peroxyoxalate CL has been also used for the evaluation of the peroxide value in olive oil.[20] The assay is based on the reaction of TCPO with hydrogen peroxide or organic peroxides in the presence of Mn^{2+} as catalyst and 9,10-dimethylanthracene as fluorophore. The procedure allowed measurement of the peroxide value with good accuracy and precision using a simple manual measurement, thus avoiding the use of complicated and time-consuming analytical procedures.

5.3.1.2   Coupling with Enzyme Reactions

The POCL reaction can be used for the detection of enzymes that produce hydrogen peroxide, such as oxidases, and for the quantification of their substrates. Among the various CL reactions suitable for the analysis of hydrogen peroxide, POCL has the advantage that the reaction can be carried out at pH 7, which is close to the optimal pH for many enzymes. Again, analytical applications mostly pertain to flow-based assays and are not reported here. A simple and sensitive analytical procedure for the evaluation of the activity of rasburicase (a recombinant urate oxidase) has been described.[21] The assay, based on the CL detection of the hydrogen peroxide produced by rasburicase in the presence of its substrate (uric acid) by means of the peroxyoxalate TCPO [Reactions (5.13) and (5.14)], was performed in 96-well microtiter plates. It was used to study the pharmacokinetics of rasburicase after a single-dose administration for the treatment of hyperuricemia in chronic kidney disease patients:

$$\text{uric acid} \xrightarrow{\text{rasburicase, oxygen}} H_2O_2 + \text{5-hydroxyuric acid} \qquad (5.13)$$

$$H_2O_2 + \text{TCPO} \xrightarrow{\text{fluorophore}} \text{reaction products} + \text{light} \qquad (5.14)$$

5.4   Other CL Systems

Analytical applications have also been reported concerning other, less common CL systems. A homogeneous enzymatic CL assay for the determination of free choline has been described.[22] The assay was based on the enzymatic reaction catalyzed by choline oxidase that, in the presence of its substrate choline, produced hydrogen peroxide, which was then detected through its reaction with the CL probe acridinium-9-carboxamide. The analysis was performed in 96-well microtiter plates and required a short time and a minimal sample volume, being suitable for the determination of free choline both in human plasma and whole blood. The CL reaction between hypochlorite and fluorescein has been exploited for the analysis of hypochlorite in different types of water.[23] The assay employed a hypochlorite-sensitive reusable test strip made of anionic cellulose paper containing sodium fluorescinate. Measurement of the CL signal in a luminometer upon addition of 1 mL of sample permitted the quantitative determination of hypochlorite with a detection limit of $0.4\,\mathrm{mg\,L^{-1}}$ and a linear calibration range from 2.0 to $50.0\,\mathrm{mg\,L^{-1}}$.

5.5   Detection of Free Radicals by Chemiluminescence

Chemiluminescence is one of the most useful methods for the detection of radical production in biological systems, as extensively reviewed recently.[24,25] During their normal function, cells continuously produce radicals such as the superoxide anion-radical ($^{\cdot}OO^{-}$) and nitrogen monoxide ($^{\cdot}NO$). These highly reactive species quickly transform into other aggressive compounds, such as hydrogen peroxide (HOOH), hydroxyl radical ($^{\cdot}OH$), peroxyl radicals (ROO$^{\cdot}$), alkoxyl radicals (RO$^{\cdot}$), hydroperoxyl radical (HOO$^{\cdot}$), hypochlorite (ClO^{-}), singlet oxygen ($^{1}O_{2}$), peroxynitrite (ONOO^{-}), *etc.*, which are involved in lipid peroxidation and other processes leading to cell damage (Figure 5.1). These species are commonly referred as "reactive oxygen species" (ROS).[26] However, the general notion of ROS is quite vague because such molecules possess different properties and reactivities and carry out different functions in the cell. While some of the species listed above can undoubtedly be considered as ROS, hypochlorite could be called a "reactive chlorine species" and nitrogen monoxide and peroxynitrite, together with other radical and non-radical species of higher nitrogen oxides, should rather be defined as "reactive nitrogen species." The peroxyl and alkoxyl radicals involved in the chain lipid oxidation possess a further distinct behavior and therefore might be considered as "reactive lipid species."

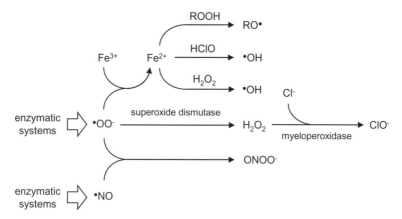

Figure 5.1   Schematic representation of the main processes responsible for the production of ROS (reactive oxygen species) in human and animal cells.

Most ROS are very reactive and cannot be isolated and purified. Hence, they are usually measured indirectly by quantifying stable molecular species originating from their reactions. However, this approach does not allow immediate determination of their nature and/or concentration. An exception is represented by electron paramagnetic resonance (EPR), which permits direct detection and identification of many radicals. Nevertheless, instrumentation for EPR is quite complicated and expensive and sensitivity is low and, thus, this technique is not particularly suited for the analysis of biological samples, which are often available in limited quantities and contain very low concentration of radicals.

Chemiluminescence methods are particularly useful for the detection of radical species, because they are very sensitive in detecting highly reactive radicals and can be used to quantify radical production in almost any kind of sample (solutions, cells, or even whole tissues and organs) in which luminescence can be recorded. Moreover, CL methods permit real-time monitoring of the formation of these species because the CL signal depends on the rate of production of radicals in the sample.

5.5.1 Ultraweak Chemiluminescence

Radical production in biological systems creates itself an ultraweak intrinsic CL, which has been studied to investigate the role of free radicals in cell functions. However, this emission is difficult to measure and is not specific, because many reactive species can produce ultraweak CL in biological samples. Therefore, analytical application of ultraweak CL is quite limited and in many cases enhanced CL (see below) is preferred.

5.5.1.1 Lipid Peroxidation Processes

Most of the experimental studies on intrinsic CL concern the lipid peroxidation process (see ref. 24). Investigations on the ultraweak CL of mitochondria, homogenates and cells suspensions showed that this process is the major responsible component of ultraweak CL emission and that there is a significant correlation between the intensity of intrinsic CL and the accumulation of lipid peroxidation products. Moreover, these studies demonstrated that the Fe^{2+} ions play a key role in activation of lipid peroxidation. Ultraweak CL been also associated with other processes in live cells, *e.g.*, reactions involving oxygen radicals (probably as the result of the production of singlet oxygen), NO biosynthesis and interaction of proteins with peroxynitrite (perhaps due to the generation of excited molecules during tryptophan oxidation). However, the intensity of such CL emissions is usually much lower than that of the CL deriving from chain lipid oxidation.

Intrinsic CL was observed in activated phagocytes, being associated with the production of superoxide radical anion and other ROS during phagocytosis.[27] It was used, for example, to assess activation of lipid peroxidation of blood plasma low-density lipoproteins by initiated neutrophils.[28] This event is of great importance for the production of oxidized forms of plasma lipoproteins, which are thought to be involved in the development of atherosclerosis. However, at present ROS production in phagocytes is mainly studied in the presence of chemical CL enhancers such as luminol, due to the much more intense emission (up to several orders of magnitude higher) that makes the measurement easier and permits use of smaller samples. Nevertheless, it should be pointed out that intrinsic CL depends on the rate of lipid peroxidation in the system, while the CL signal obtained with chemical CL enhancers is related to the concentrations of species such as hypochlorite, hydrogen peroxide, and hydroxyl and superoxide radicals. Thus, information obtained by intrinsic and enhanced CL is not identical.

5.5.1.2   Evaluation of Oxidative Status of Skin

An interesting recent application of ultraweak CL was the non-invasive monitoring of the oxidative status of skin.[29] Physical or chemical environmental stressors (exposure to ozone, ultraviolet irradiation or even cigarette smoke) generate ROS in deeper (living) skin layers, which cause an ultraweak CL emission that is recordable *in vivo* with sensitive photomultiplier systems. In this paper, the measurement of ultraweak CL following ultraviolet A (UVA) irradiation was used to assess variations in the oxidative status of the skin. For validation purposes, the influence on the CL signal of parameters such as skin thickness, humidity, temperature, pH and atmosphere composition was studied, and then the CL emission was measured in the presence of topically applied antioxidants (*e.g.*, vitamin C). A dose-dependent correlation between the CL signal and the antioxidants was observed, suggesting that this approach could represent a reliable method for the *in vivo* evaluation of topical antioxidants' potency.

5.5.2   Enhanced Chemiluminescence

To increase intensity and specificity of the CL signal due to the production of free radicals, CL enhancers (either chemical or physical, according to their mechanism of action) are used. Physical CL enhancers act as energy acceptors: the energy of an excited product of a reaction involving radicals is transferred to the enhancer, which then emits. For example, addition of a suitable coumarin derivative to a system undergoing a lipid peroxidation reaction causes a 1500-fold increase of the CL emission.[30] Chemical CL enhancers react with ROS or other free radicals to form light-emitting excited-state products. Among these enhancers, the best known are lucigenin, luminol and isoluminol (Figure 5.2), which differ in reactivity with radicals and also in intra/extracellular distribution.

5.5.2.1   Lucigenin

Lucigenin was the first CL enhancer used to study ROS production in neutrophils. It is well established that lucigenin is specific for the superoxide anion-

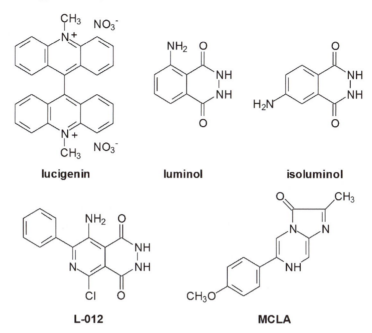

Figure 5.2 Structures of the most common chemical CL enhancers used for the detection of ROS.

radical and that it does not react with other ROS, such as hydrogen peroxide, nitric oxide and peroxynitrite.[31] The measurement of lucigenin-enhanced CL is therefore considered the most sensitive and specific test for assessing production of the superoxide anion-radical in biological systems. However, use of lucigenin presents some disadvantages. First, since lucigenin is not able to enter within cells it allows the quantification of only extracellular superoxide anion-radical.[32] In addition, this CL enhancer has the lowest intensity in CL response in comparison to other enhancers such as luminol and isoluminol.

5.5.2.2 Luminol

Luminol is the most used CL enhancer since its introduction by Allen and Loose in 1976.[33] It produces much stronger CL signals than lucigenin, and thanks to its lipophilicity and small molecular size it can diffuse across biological membranes. Therefore, both intra- and extracellular reactive species can be measured.[34] To detect only intracellular reactive species, membrane-impermeable enzymes such as superoxide dismutase and catalase should be added to remove secreted extracellular ROS. Luminol is a CL enhancer less specific than lucigenin, though the exact nature of the species that react with luminol and their relative importance are still under debate. It has been proved that the superoxide radial-anion causes the luminol CL emission, but involvement of hydroxyl radical and nitric oxide and/or nitric oxide donors (through the production of peroxynitrite and other metabolites) have been proposed. At least in some cases, the discrepancy between the findings reported by different authors could be related to the different types of activator used, which may lead to generation of reactive species through distinct chemical pathways. A limitation of luminol and isoluminol is the dependence of the CL signal on pH. The quantum yield of luminol-enhanced CL increases with increasing pH, because some of the species that participate in the CL reaction are stable only in alkaline medium, and the optimum pH (8.3–9.0 and higher) is quite high in comparison to physiological standards.[35]

5.5.2.3   Isoluminol

Isoluminol differs from luminol in respect to the position of the amino group in the phthalate ring of the molecule, which makes the molecule less lipophilic and thus less permeable to biological membranes. Therefore, this CL enhancer only allows detection of extracellular reactive species, especially of the superoxide radical-anion.[36] In general, the intensity of the isoluminol-enhanced CL is lower than that of the luminol-enhanced one, presumably because isoluminol does not react with intracellular reactive species.

5.5.2.4   Other Chemical CL Enhancers

In recent years, other chemical CL enhancers have been proposed for the detection of ROS. The L-012 derivative (Figure 5.2) is a highly sensitive luminol-like CL enhancer able to react with superoxide anion-radical, hydrogen peroxide and/or their metabolites.[37] It has been used to detect ROS generated not only by isolated neutrophils but also by whole blood samples, affording CL signals 100-times stronger than those obtained with luminol. The CL enhancer MCLA is highly specific for extracellular superoxide radical-anion. Even though this enhancer showed some disadvantages (including a strong background signal due to its auto-oxidation) a recent comparative study of different probes for detection of ROS showed that this probe provides the highest sensitivity.[38] Pholasin is a bioluminescent photoprotein from *Pholas dactylus*, a marine bivalve mollusc.[39] Pholasin emission is produced from the reaction with various extracellular ROS (superoxide radical-anion, singlet oxygen, hydroxyl radical, peroxynitrite), but the exact reaction mechanism is still unclear. The intensity of the CL signal depends on several factors (concentration of pholasin, concentrations and nature of oxidants, buffer, other components of the cell culture, pH) and in general is one or two orders of magnitude higher than that of luminol.

5.5.3   Study of ROS Production

Chemical CL enhancers are widely used to assess the capacity of neutrophils to produce ROS and other reactive species, which reflects the functional state of the phagocytic link. In fact, the interaction of phagocytizing cells with a foreign material triggers a complex cascade of physiological and metabolic processes, which ends with the formation of superoxide radicals and the biosynthesis of numerous physiologically active compounds.[40] Chemiluminescence-based detection of ROS using CL enhancers has been applied in the study of the function of the immune system, metabolic disorders, ischemic processes and oncological and inflammation diseases, as well as of many other diseases whose pathogenesis is related to oxidative stress. This term indicates an imbalance between pro-oxidants and antioxidants within an organism, leading to accumulation of damaged DNA bases, protein oxidation and lipid peroxidation products, as well as to decreased levels of antioxidants and increased susceptibility to the action of free radicals. Even though the term is still somewhat vague and it is not clear when the balance between pro-oxidants and antioxidants transforms into an imbalance, it has been postulated that oxidative stress plays an important role in the pathogenesis of many diseases. It is also thought to be involved in aging, although the fundamental mechanisms of this process (manifested at genetic, molecular, cellular, organ and system levels) are not clearly understood.

5.5.3.1 Ischemic Damage

The cell damage that takes place on reperfusion of ischemic tissue is mediated by many factors, including an increased free radical production after reestablishing the blood flow due to the reduced activity of the antioxidant system. Such a phenomenon is particularly dangerous for cells intensively consuming oxygen, such as cardiomyocytes and neurocytes. To investigate the cellular signaling pathways that cause cerebral vascular injury in response to hypoxia/reoxygenation the production of ROS has been evaluated in cerebral resistance arteries.[41] Hypoxia/reoxygenation determined a significant increase the production of the superoxide anion-radical detected by a lucigenin-based CL assay. Selective inhibition studies demonstrated that superoxide mediates cerebral endothelial dysfunction after hypoxia/reoxygenation mainly *via* activation of NADPH oxidase. Free radical production has been studied in the cytoplasmic fraction of normal and ischemized rat myocardium.[42] The intensity of the Fe^{2+}-induced CL measured in the cytoplasm of cardiomyocytes from ischemized myocardium suggested a significant increase in the free radical production, as also indicated by the high concentration of lipid peroxidation products.

5.5.3.2 Endothelium Dysfunction

The vascular endothelium performs many metabolic and signaling functions, in particular the regulation of vasodilatation and vasoconstriction processes. Its location in the interface between blood and tissues makes it vulnerable to attack by free radicals, which cause cell damage leading to pathologies such as angiopathy, atherosclerosis, *etc.* The relationship between diabetes-induced hyperglycemia, ROS production and endothelium-mediated arterial function was investigated in an animal diabetes model.[43] Measurement of ROS formation in blood and the aorta by luminol-enhanced CL showed a significant increase of ROS production, suggesting that hyperglycemia-induced ROS production plays an important role in mediating endothelial dysfunction in experimental diabetes. This finding was also supported by the observation that treatment with antioxidants restored the function of the endothelium.

5.5.3.3   Nervous System Diseases

Reactive oxygen species and other free radicals are involved in the pathogenesis of central nervous system diseases, such as Alzheimer's disease and brain damage caused by epileptic seizures, even though it is not clear whether they are a cause or a consequence of these pathological processes. The capacity of a fragment of the amyloid beta protein to induce production of ROS in human neutrophil granulocytes has been investigated by different analytical techniques, including a luminol-based CL assay.[44] The protein fragment stimulated formation of ROS in a concentration and time dependent manner, and use of selective inhibitors demonstrated the production of hypochlorite, superoxide anion-radical and hydroxyl radical, which have a cytotoxic potential, presumably involving NADPH oxidase.

5.5.3.4   Rheumatoid Arthritis

Rheumatoid arthritis is a systemic chronic inflammatory disease, in which the stationary neutrophil activity determines ROS production and consequent tissue damage. Therefore, evaluation of ROS production represents an important tool in assessing onset and development of the disease, as well as in investigating the therapeutic effectiveness of treatments. The effect of glucomannan (a natural polysaccharide) on the production of ROS has been tested by means of a luminol/isoluminol-enhanced CL assay.[45] Glucomannan significantly decreased extracellular ROS generation both in isolated human neutrophils and in the whole blood of animal models of arthritis. Moreover, CL of the joint of arthritis animal models was also reduced in comparison with untreated animals. These findings demonstrated the antioxidant effect of glucomannan, presumably due to its free radical scavenger activity and to interaction with different receptors and/or modulation of signaling pathways. A luminol-enhanced CL assay has been used to compare oxidative activation of blood and synovial fluid neutrophils from patients with rheumatoid and other types of arthritis with blood donor neutrophils.[46] Synovial fluid neutrophils from patients with rheumatoid arthritis, but not with other types of arthritis, showed high baseline intracellular ROS production, indicating that they are in an activated state. This finding indicated that synovial neutrophils are engaged in the processing of endocytosing material, which could have implications for the pathogenesis of rheumatoid arthritis.

5.5.3.5 Oncologic Diseases

Oncologic diseases are supposed to be the result of gene dysfunctions; thus all the factors causing DNA damage are potential carcinogens. Reactive oxygen species, such as the hydroxyl radical, can damage directly mitochondrial and nuclear DNA and, consequently, possess a mutagenic and carcinogenic action. Chemiluminescence methods have been used to estimate the effect of chemotherapy on the proliferation and activation of tumor cells.[47] The CL signal due to the production of ROS by gastrointestinal carcinoma cell lines closely correlated with the dynamics of cell cycle, thus reflecting oxidation metabolism and activation of proliferation of tumor cells. Assessment of the effect of drugs (*i.e.*, mitomycin C) on metabolism and proliferation of tumor cells was also possible, thus suggesting that the CL assay could be used for the screening of chemotherapeutic drugs.

5.5.3.6 Response of the Immune System to Pathogens

Measurement of ROS production can be used to evaluate the general state of the immune system and to study the response of the immune system's cells to specific pathogens. To investigate the cause of liver damage in leptospiral and borrelial infections, the production of ROS and the expression of inducible nitric oxide synthase (iNOS) have been evaluated in rat isolated Kupffer cells stimulated by *Leptospra interrogans* and *Borrelia burgdorferi*.[48] A luminol-based CL assay demonstrated that both pathogens induced a rapid ROS production, while iNOS overexpression lasted for several hours after infection. These findings suggested that liver damage could be initially mediated by oxygen radicals, and then maintained (at least in part) by nitric oxide.

5.6 Evaluation of Antioxidant Activity

The total antioxidant activity of a given sample can be determined using different approaches. First, the separate concentrations of all of the molecules recognized as antioxidants could be measured, thus obtaining a detailed description of the antioxidant properties of the sample. This procedure is time-consuming, expensive and technically demanding, and may fail if some antioxidants are not determined, or even known. In addition, cooperative interactions between antioxidants are not taken into account. Alternatively, the capacity of the sample under investigation to oppose a known oxidative stress could be measured using a suitable analytical method.[49] This approach avoids most of the problems reported above and represents a "physiological" measurement, which gives a quantitative evaluation of the functional aspects of antioxidant activity rather than simple antioxidant concentrations.

However, it should be pointed out that the result depends on the nature of the applied oxidative stress and on the index used to quantify the oxidative damage of the system. Thus, interpretation of total antioxidant activity assays is not straightforward and sometimes it is not clear which assay offers the most relevant information. In addition, correlation between the results obtained with the two approaches is difficult and often the total antioxidant activity does not correspond to the sum of the concentrations of single antioxidants. Nevertheless, it is important to establish a model that translates individual antioxidant concentrations into total antioxidant activity to enable interrogation of epidemiological databases that contain information only about individual antioxidants. A study performed on human serum samples showed that the stronger predictor of total antioxidant activity measured by luminol-enhanced CL is uric acid, followed by vitamins A, C and E.[50]

5.6.1 Chemiluminescence Methods

Chemiluminescence-based assays have been extensively used to evaluate the total antioxidant capacity. Compounds with antioxidative activity are able to quench the emission of CL systems because they act as scavengers for the radicals that, in most cases, drive the CL reaction. Their action provides either a decrease of the CL intensity or a delay (lag phase) in the onset of the CL emission, both of which are related to the concentration and activity of the antioxidants in the sample. To perform a quantitative evaluation, these effects are compared to those observed in a similar system in the presence of a known amount of antioxidant. Therefore, the total antioxidant capacity of the samples under investigation is reported in equivalents of standard antioxidant, *e.g.*, ascorbic acid or Trolox® (a water-soluble analogue of vitamin E). Thanks to the simple instrumentation required and the high sensitivity of the CL measurement, CL-based methods are particularly suited for screening assays in high-density analytical formats (*e.g.*, 96- and 384-well microtiter plates).

5.6.1.1 TRAP Assay

The total peroxyl radical-trapping antioxidant parameter (TRAP) assay is probably the most used assay for the evaluation of total antioxidant activity. The method is based on the constant production of water-soluble peroxyl radicals by the thermal decomposition of a radical source, such as the 2,2'-azo-bis(amidinopropane) dihydrochloride (ABAP), in the presence of linoleic acid. The peroxyl radicals initiate lipid peroxidation, unless a sample containing antioxidants is added to the system. In this case the onset of lipid peroxidation is delayed until the antioxidants have been consumed. In the original TRAP assay, lipid peroxidation was followed by measuring the oxygen consumption using an oxygen electrode.[51] However, the assay time was quite long and reproducibility was affected by the scarce stability of the oxygen electrode. Therefore, the TRAP assay has been modified by introducing a CL reporting system based on luminol, which is oxidized by peroxyl radicals with the generation of light.[52] Moreover, use of lipid-soluble radical sources such as the 2,2'-azo-bis(2,4-dimethylvaleronitrile) (AMVN) allowed measurement of antioxidants in the lipid phase.[53] A modified TRAP assay based on the quenching of luminol-enhanced CL and employing 2,2′-azo-bis(2-amidinopropane) dihydrochloride (AAPH) as the free radical source has also been proposed.[54] In such an assay, quantification of the antioxidant capacity by evaluating the area under the CL kinetic profile was proposed to permit analysis of samples that do not present a lag phase, which is a limitation of conventional TRAP methods. The assay has been validated by comparison with standard antioxidants, proving satisfactory in terms of linearity, intra- and inter-assay precision, robustness and limits of detection and quantitation.

5.6.1.2 ECL Assay

The enhanced chemiluminescence (ECL) antioxidant assay is another CL assay widely used for the measurement of the total antioxidant activity.[55] It is based on the luminol/oxidant/enhancer CL system that, in the presence of a catalyst such as the horseradish peroxidase enzyme, displays a CL emission mediated by the continuous production of enhancer radicals. Addition of antioxidant compounds interrupts light output because these compounds act as scavengers for such radicals. Again, the extent and the duration of the lag phase are proportional to the antioxidant concentration and activity (Figure 5.3).

5.6.2 Clinical Applications

Most of the clinical investigations have been performed on blood or serum, although to study the role of oxidative stress in specific pathologies the total antioxidant activity has also been measured in other biological fluids.

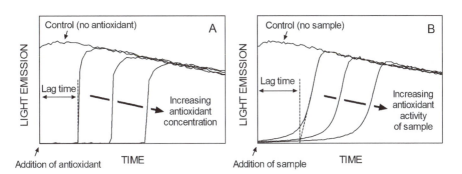

Figure 5.3 Determination of total antioxidant activity by means of an ECL assay: (A) and (B) show the CL kinetics profiles obtained for standard antioxidant solutions and for biological samples (*e.g.*, serum samples) with different total antioxidant activity, respectively. The lag times (*i.e.*, the parameter commonly used to estimate the antioxidant activity) for the standard antioxidant and the biological sample with the lowest antioxidant activities are also shown. Note that biological samples, which contain a mixture of different antioxidant species, present a more gradual onset of the CL emission than standard antioxidants.

5.6.2.1   Antioxidant Activity of Blood

Markers of oxidative stress have been measured in blood of patients affected by age-related neurological diseases (Alzheimer's, vascular dementia, Parkinson's disease) with different analytical techniques, including a TRAP CL assay for plasma antioxidant capacity.[56] Systemic oxidative stress was found statistically associated with all diseases, suggesting an imbalance between pro-oxidants and antioxidant defenses. Moreover, alterations observed in the levels of different markers indicated that different free radical metabolic pathways were involved. Hemodialysis is a common therapeutic strategy for patients with end stage renal failure, but neutrophil activation during the hemodialytic process causes release of ROS into the bloodstream, thus determining oxidative damage. Investigation by CL analysis in uremic patients undergoing hemodialysis showed a significant decrease in the plasma antioxidant activity, perhaps due to the loss of dialyzable antioxidants, such as uremic solutes, with small molecular weights.[57] Supplement of antioxidants capable of regaining antioxidant defense in plasma was thus proposed to prevent oxidative damage induced by hemodialysis. The effect of physical exercise on the oxidative status of trained and untrained healthy subjects has been investigated by measuring blood levels of lipid peroxides and enzymes involved in the control of oxidative stress and plasma total antioxidant capacity.[58] Trained subjects showed upon exercise a statistically significant increase in total antioxidant capacity and most of the enzyme levels, while lipid peroxidation decreased. Reduction of lipid peroxidation was attributed to the higher activity of enzymatic antioxidants.

5.6.2.2   Antioxidant Activity of other Biological Fluids

The total antioxidant activity of the aqueous humor of patients affected by two types of glaucoma, *i.e.*, glaucoma associated with exfoliation syndrome (XFG) and primary open-angle glaucoma (POAG), has been compared with that of patients with cataract.[59] The antioxidant activity was significantly lower in XFG than in POAG, and values for glaucomas were lower than those measured in patients affected by cataract. In addition, the levels of superoxide dismutase and glutathione peroxidase were increased, suggesting that the antioxidant status of the aqueous humor may play a role in the pathophysiology of glaucomas. Saliva constitutes the first line of defense against free radical-mediated oxidative stress promoted by foods. During gingival inflammation, alterations in saliva composition could have some role in controlling and/or modulating oxidative damages in the oral cavity. In recent years, the evaluation of antioxidant capacity of saliva has attracted increasing interest, as demonstrated by numerous studies concerning the nature and characteristics of oxidants, pro-oxidants and antioxidants in saliva and the development of several analytical methods for assaying its antioxidant activity.[60] Nevertheless, systematic studies of total antioxidant activity of saliva (even in healthy populations) are still lacking. Oxidative stress is considered an important factor in male infertility because it may impair the physiological function of spermatozoa at the molecular level. The measurement of ROS in semen samples from fertile donors and infertile patients using a luminol-based CL method allowed establishing a cutoff value of ROS level with a high sensitivity and specificity to differentiate infertile from fertile men.[61] This test could thus be used in the diagnosis of infertility, and it also may help clinicians treat patients with seminal oxidative stress.

5.6.3 Antioxidant Activity of Foods and Food Derivatives

The increasing recent interest in nutraceuticals and functional foods has led researchers to investigate the antioxidant potential of several foods, food supplements, fruits and plant derivatives and extracts. This interest is motivated by the assumption that an adequate intake of natural antioxidants could boost the natural body's defense mechanisms against free radicals. Chemiluminescence-based analytical methods used for the measurement of the total antioxidant activity of biological samples could, in principle, be used also for analyzing foods and food derivatives. Nevertheless, much effort has been devoted to the improvement of CL-based methods for food analysis to obtain more reliable results and increase analytical throughput.

5.6.3.1 Analytical Techniques

A sensitive and simple procedure for the evaluation of the antioxidant activity using the CL reaction of lucigenin with hydrogen peroxide has been reported recently.[62] To test this procedure, the antioxidant activity of 21 known hydrophilic and hydrophobic compounds was measured and compared with reference data. The method was applied to the estimation of total antioxidant activity of olive oils, proving suitable for their direct analysis without the need of preliminary extraction of antioxidant compounds in aqueous media. A high-throughput CL platform for the automatic and rapid evaluation of the total antioxidant activity has been developed based on the inhibition of the luminol CL initiated by 1,1-dipheny-2-picrylhydrazyl (DPPH) or hydrogen peroxide.[63] The platform, characterized by high analytical throughput (complete analysis in a 96-well microtiter plate format could be performed in about 10 min, including sampling and detection), was used to measure the antioxidant activity of pure compounds and plant extracts.

5.6.3.2 Antioxidant Activity of Fruits

Chemiluminescence-based assays have been used to evaluate the antioxidant activity of various foods, fruits and food additives. The antioxidant potential and reactivity of peach extracts (peels and flesh) have been measured using a luminol-enhanced CL assay.[64] All the extracts showed a concentration-dependent free radical scavenging activity that was significantly higher than that of the major antioxidant compound (chlorogenic acid) alone, which has been attributed to a synergistic or additive effect of other antioxidants present in the extracts. Different analytical methods, including an assay based on luminol-enhanced CL, were used to evaluate the radical scavenging activity of different extracts, fractions and residues of navel sweet orange (*Citrus sinensis*) peel.[65] Comparison between the results obtained using the different assays allowed the assessment of the contribution of various compounds (mainly flavonoids and other phenolic compounds) to the antioxidant activity. An ethyl acetate extract showed the highest activity and its use as an antioxidant in food and medicinal preparations was proposed.

5.6.3.3   Antioxidant Activity of Beverages

To study the winemaking process, the total antioxidant capacity of different wines was determined in different steps of winemaking by a CL assay.[66] The results showed that the antioxidant capacity significantly decreased (about by 25%) immediately after opening the bottle (thus immediate analysis or a correct sample storage is necessary) and that the levels of total phenolics and the antioxidant activity were affected by grape composition and winemaking technology. The antioxidant activity suffered the greatest decrease (30–50%) after the clarification procedure (perhaps due to the fining agents used and/or to oxygen contact), then remained constant in the subsequent steps. The polyphenolic content, antioxidant activity and phenolic profile of tea (Chinese green tea, black tea and Greek mountain tea) and herbal infusions have been studied.[67] Measurement of the total antioxidant activity by means of a $Co^{2+}/$ EDTA-induced luminol CL assay showed that the Chinese green tea extract possessed the highest antioxidant activity, followed by black tea and herbal extracts. The scavenging ability against specific radicals has also been measured.[68] A CL assay for evaluating peroxynitrite-scavenging capacity has been developed and used to measure the antioxidant ability of red wine samples to decrease peroxynitrite-initiated CL, which was then related to the phenolic content. Owing to its rapidity and reliability, the assay was proposed for the large-scale screening of aqueous food extracts.

5.7   Firefly Luciferase Reaction

Firefly luciferase catalyses the following reaction:

$$ATP + \text{D-luciferin} + O_2 \overset{\text{luciferase}}{\longrightarrow} AMP + PPi + \text{oxyluciferin} \\ + CO_2 + \text{light} \tag{5.15}$$

The fundamental work on firefly luciferase from *Photinus pyralis* was carried out by McElroy and DeLuca.[69] In two recent papers the structure and function of firefly luciferase have been reviewed in detail.[70,71] The following two paragraphs are a very short summary of some facts that may be of relevance for a discussion on analytical applications.

Firefly luciferase is found in many beetles, including fireflies, click beetles and glow-worms. Luciferase has been cloned from several species, *e.g.*, the American firefly *Photinus pyralis*, the Japanese fireflies *Luciola cruciata* and *Luciola lateralis*, the East-European firefly *Luciola mingrelica*, the click beetle *Pyrophorus plagiophthalamus*, the larval click-beetle *Pyrearinus termitilluminans* and the glow-worm *Lampyris turkestanicus*.[72–79] Genes have been modified to result in increased thermostability, changed emission spectrum, increased catalytic activity and fused gene products.[80–84] The 3D structure has been resolved.[85] The reaction mechanism is fairly well understood although some points remain to be clarified.[70] The quantum yield of the luciferase reaction has until recently been assumed to be close to unity but somewhat lower numbers are now discussed.[70] The energy to create the photon comes from the oxidative decarboxylation of D-luciferin, and ATP participates only to activate D-luciferin. The products PPi and oxyluciferin are inhibitors of the reaction.[86,87] Under normal analytical conditions in the presence of a low concentration of PPi they are, however, not accumulated in concentrations affecting the rate of the reaction, *i.e.*, the light intensity. Degradation products of D-luciferin like L-luciferin and dehydroluciferin and similar molecules are strongly inhibitory. Maximum light emission can therefore only be obtained with a very pure preparation of D-luciferin. The ATP site is highly specific for ATP. Among naturally occurring analogues only dATP gives some light, but in living cells dATP normally occurs in much lower concentrations than ATP.[88]

The molecular weight of firefly luciferase is *ca.* 60 kDa.[70] K_m values differ somewhat between species but are 1.3×10^{-4} M ATP and 1.4×10^{-4} M D-luciferin in thermostable recombinant luciferase.[89] The pH optimum is 7.8.[70] For analytical purposes the useful pH range is 6–8.[90] D-Luciferin is more stable at the low end of the interval. At high pH D-luciferin is subject to racemization to L-luciferin, which is a potent inhibitor of the luciferase reaction. D-Luciferin is also autoxidized to dehydroluciferin, which is another potent inhibitor. Under normal analytical conditions the stability of D-luciferin is not a problem, but it causes a limited shelf-life once the freeze-dried reagents are reconstituted. Reconstituted reagents containing D-luciferin can be stored for months at $-80\,^{\circ}$C but not at $-20\,^{\circ}$C. Wild-type luciferases are unstable above $25\,^{\circ}$C and are completely inactivated within 5 min at $50\,^{\circ}$C.[89] At the same temperature the thermostable mutant Lcr-T2171 retains 75% of the initial activity even after 10 min.[89] With thermostable luciferase it is possible to add stabilizers to obtain a half-life of reconstituted ATP reagents[i] at room temperature of more than a month (Lundin, unpublished). Combining three point mutations resulted in a 12.5-fold increased activity of *Photinus pyralis* luciferase, allowing the detection of 1 amol ATP.[83] With this luciferase a single bacterial cell could be detected.[91] Furthermore, *Salmonella enteritidis* could be identified in an immunochromatographical lateral flow test.[92]

---

[i] In the following an ATP reagent is defined as a reagent used for the determination of ATP concentrations, *i.e.*, a reagent containing luciferin, luciferase, magnesium ions and stabilizers. Similarly a luciferase kit is defined as a kit used to determine luciferase activity, *i.e.*, a kit containing ATP, D-luciferin and stabilizers and activators.

5.8 Kinetics of Light Emission

Following a 25 ms lag the light emission from the firefly luciferase reaction reaches a maximum intensity within a second.[93] Thereafter, it may decay or stay essentially stable. Decay is obtained if luciferase is inactivated or if a substantial percentage of the substrates ATP or D-luciferin is consumed per minute in the luciferase reaction.[94] Inactivation can be seen if luciferase is bound to surfaces or if a substance binding to luciferase inactivates the enzyme. Inactivation can be counteracted by stabilizing substances so that a stable light emission is obtained (half-life around 10 h).[95] Decay as a result of consumption of substrates requires a high luciferase concentration. It has been shown that under these conditions the peak light emission as well as the decay rate is proportional to the luciferase concentration (Figure 5.4).

With a low luciferase concentration the light emission is essentially stable and proportional to the ATP concentration at $ATP \ll K_m$, *i.e.* $ATP < 1 \mu mol$ L^{-1}. Under these conditions it is possible to continuously monitor ATP-forming or ATP-degrading reactions simply by having luciferase and D-luciferin present in the reaction mixture.[90,94–107] Such stable light emitting ATP reagents are ideal for monitoring kinetic or end-point assays of enzymes and metabolites.[90,94] They are also very useful for measuring ATP in cell cultures as even a single mammalian cell contains enough ATP to be detected (10–100 fmol ATP). For detection of low numbers of bacteria (1 amol per cell) higher luciferase levels are required. A series of three types of reagents has been developed:[94]

1 stable light ATP reagent with a decay rate of $0.5\% \, min^{-1}$ and a detection limit of 1000 amol ATP;
2 slow decay ATP reagent with a decay rate of $10\% \, min^{-1}$ and a detection limit of 10 amol ATP;
3 flash ATP reagent with a decay rate of $235\% \, min^{-1}$ (90% of the total light is emitted during the first minute) and a detection limit of 1 amol ATP or less.

With decay reagents the ATP depletion follows first-order kinetics when [D-luciferin] $\gg$ [ATP] and [ATP] $\ll K_m$ of luciferase, *i.e.*, normal analytical conditions.[94] If the first-order rate constant is, for example, $0.10 \, min^{-1}$, the ATP concentration and therefore the light intensity is 90% after 1 min, 82% after 2 min and 74% after 3 min compared to the initial ATP concentration ($[ATP_t] = [ATP_0]e^{-0.10t}$, where t is time in minutes). The rate constant depends on type and preparation of luciferase and reaction conditions. A rate constant of $0.10 \, min^{-1}$ may require 0.1–0.3 $\mu mol \, L^{-1}$ luciferase (depending on type and preparation) in a 50 mmol L^{-1} Tris-acetate buffer, pH 7.75, with optimal concentration of D-luciferin (around 0.7 mmol L^{-1}).

ATP reagents type 1 and 2 can be added manually to the sample as the decay rate is fairly slow. ATP reagent type 3 requires luminometers with automatic dispensers to measure the light signal accurately. Working with ATP reagent

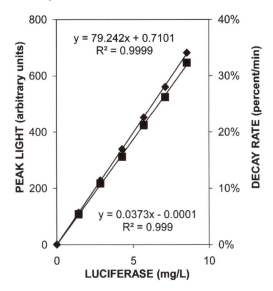

Figure 5.4 Kinetic effect of luciferase concentration on light emission. Peak light (♦) and decay rate (■) are proportional to luciferase activity. The decay rate is caused by ATP degradation in the luciferase reaction and is calculated by Equation (5.25). The graph shows that stable light can only be achieved at low luciferase levels, resulting in low levels of light.

type 3 also requires pipette tips and auxiliary reagents that are completely ATP free. Type 2 reagent is actually the most convenient to work with. If it is contaminated by the user, it will break down the ATP contamination by one order of magnitude every 23 min at room temperature. After 46 min only 1% of the contamination remains and it is likely that the reagent can be used.

5.9 Degradation of Adenine Nucleotides in Luminescent Reagents

Adenine nucleotides frequently contaminate raw materials for preparing reagents, extractants and buffers used in assays based on firefly luciferase. Microorganisms growing in the reagents may also be a problem. It is therefore essential to control the quality of all raw materials. ATP reagents with high levels of luciferase usually break down their own ATP background to AMP during the preparation. If such a reagent is used for determination of ATP/ADP/AMP the background will return in the assay of AMP. This problem can be counteracted by adding a low level of adenosine phosphate deaminase to the reagent.[108] This will convert the adenosine phosphates into the corresponding inosine phosphates, which do not give light in the luciferase reaction.

5.10 Reducing Contamination Problems in Assays of ATP

Performing ATP assays at the femtomol level is usually not a problem in a normal laboratory environment. One should avoid touching pipette tips and the interior of cuvettes or microplate wells with hands not covered with disposable powder-free gloves. In the author's laboratory assays at the attomol ATP level are frequently performed. It is then necessary to used ATP-free pipette tips mounted in racks. Such tips are available from several manufacturers. The lid of the rack should be on whenever possible to avoid contamination from air. ATP-free cuvettes or tubes are also available. They must come individually packed or at least only a small number of cuvettes per bag (ideally the number of cuvettes to be used in one experiment). Taking out cuvettes from a plastic bag with 1000 cuvettes will by necessity lead to ATP contamination sooner or later of some of the cuvettes even if gloves are used. In the assay of attomol ATP levels it is good practice to incubate the cuvette with the ATP reagent (type 2 or 3; *cf.* Section 5.8) for some time to allow luciferase to degrade contaminating ATP from the interior walls of the cuvette. Before adding the sample one should measure the blank. If the blank is higher than the blank without cuvette (*i.e.*, the instrument blank) one can either discard the cuvette or continue the incubation until the blank has disappeared.

Plastic or glass vials for preparing reagents, extractants or buffers must be ATP free. It is very difficult to remove ATP contamination by cleaning. Disposable vials should be used as often as is possible. ATP is destroyed at the high temperatures used during manufacture of both plastic and glass vials. The vials will remain ATP free provided they are packaged in a clean environment in sealed plastic bags.

Raw materials for extractants and buffers must not be contacted with anything that may be ATP contaminated, *e.g.*, spatulas. If spatulas cannot be avoided, they should be thoroughly cleaned ending with hot tap water, ATP-free water and ethanol.

5.11 Internal ATP Standard Technique

Luminometers detect only a small percentage of the photons emitted (*cf.* Chapter 4). The percentage depends on the geometry of the measuring chamber, the use of mirrors, what kind of light detector is used and if the photons are counted or measured in an analogue way. Even luminometers of the same model from the same manufacturer give different signals from the same amount of emitted light.

The light intensity is a measure of the rate of the luciferase reaction. This is affected by (a) the luciferase concentration, (b) inhibitors or enhancers of the reaction in the sample or in the reagent, (c) pH, (d) the nature and concentration of the buffer, (e) how long and under what conditions the reagent has been stored, as both luciferase and D-luciferin are slowly inactivated in solution, and (f) temperature (the temperature optimum is around 25 °C). Solely measuring the light intensity expressing the data in relative light units (RLU) is therefore a very unreliable estimate of ATP. Furthermore, it gives very little guidance to someone trying to do a similar experiment using another luminometer, other reagents or other conditions.

Preparing a standard curve for each single experiment does not solve the problems associated with variations in sample matrices or sample temperatures. The problems can, however, be solved by running each sample in two wells – one without and one with added ATP standard. Both procedures mean extra consumption of reagents and time.

The simplest and most reliable solution to all the problems mentioned above is the internal ATP standard technique. As illustrated in Figure 5.5 the internal ATP standard technique can be used both with stable light emitting and with flash type reagents. With stable light reagents one simply measures the light intensity from the sample before (I_{smp}) and after ($I_{smp+std}$) adding a known amount of ATP standard. This allows us to calculate the amount of ATP as:

$$\text{ATP in sample (mol)} = I_{smp}/(I_{smp+std} - I_{smp}) \times \text{ATP in standard (mol)} \tag{5.16}$$

With flash reagents three measurements are required:

1 immediately after adding the ATP reagent (I_{smp1}),
2 immediately before adding the ATP standard (I_{smp2}),
3 immediately after adding the ATP standard ($I_{smp2+std}$).

The amount of ATP in the sample is calculated as:

$$\text{ATP in sample (mol)} = I_{smp1}/(I_{smp2+std} - I_{smp2}) \times \text{ATP in standard (mol)} \tag{5.17}$$

Formula (5.17) only applies if the reason for the decay of the light is that ATP is consumed in the luciferase reaction, *i.e.*, luciferase is not inactivated in a time-dependent manner. Unfortunately, not all microplate luminometers can be programmed to perform the measurement of I_{smp2}, addition of ATP standard, mixing and immediate measurement of $I_{smp2+std}$. In these cases it is possible to perform two measurements after the addition of ATP standard, calculating the first-order decay rate constant of the light. This constant can then be used to calculate I_{smp2} by extrapolating from I_{smp1} (the same percentage decay per minute should apply before and after the addition of ATP standard).

Calculating the amount of ATP using an internal standard as described above results in a reliable and accurate measure of ATP, which can be compared between, for example, different laboratories and different conditions, provided:

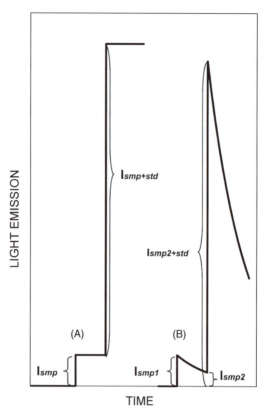

Figure 5.5 Calibration of ATP assays by the internal standard technique with stable light emitting reagents (A) and with flash reagents (B). The addition of the reagent to the sample results in the light emission I_{smp} in (A) and I_{smp1} in (B). The further addition of ATP standard results in the light emission $I_{smp+std}$ in (A) and $I_{smp2+std}$ in (B). With flash reagents it is also necessary to measure I_{smp2}. The amount of ATP is calculated by Equation (5.16) for stable light reagents and by (5.17) for flash reagents.

1 After the addition of the ATP standard the total ATP concentration is still in the linear range of the assay, *i.e.*, ATP $< 10^{-6}\,\text{mol}\,\text{L}^{-1}$.
2 The light measurement is performed within the linear range of the luminometer.
3 The volume of added ATP standard is small compared to the volume of the reaction mixture before the addition. This is because a high volume dilutes the reaction mixture, changing the reaction conditions. Manually as well as with most liquid dispensers it is difficult to add less than $10\,\mu\text{L}$ ATP standard accurately. It is therefore recommended to use $10\,\mu\text{L}$ for reaction mixtures in the interval 100–$1000\,\mu\text{L}$. If the sample matrix is not highly inhibitory or varies greatly between samples, the dilution effect can often be mathematically compensated for.

4 The amount of ATP in the added ATP standard is at least $10\times$ the amount of ATP in the reaction mixture. The additional light from the ATP standard is otherwise too small to be measured accurately. Furthermore, the additional light is strongly affected by the dilution effect when adding the standard. There is no need to use different standard concentrations for different samples as the light emission is linearly related to the ATP concentration and the measurement is performed within the linear range of the luminometer.

5 The ATP standard is reliable and accurate. Liquid-stable ATP standards are available on the market. These standards can be traced back to a reference ATP solution, which is validated by the SP Technical Research Institute of Sweden.[109]

5.12 ATP Biomass Testing

In the present context biomass means living cells, prokaryotic as well as eukaryotic, and includes the intracellular space, the cell membrane and, if present, the cell wall. Cells may be separated from each other or be linked to tissues or entire animals or plants. Outside the cell is the extracellular space.

5.12.1 ATP per Cell

ATP is the energy currency in the metabolism in all living cells. Numerous enzymes form or degrade ATP. Therefore, ATP has to be maintained at a carefully regulated intracellular level, which is similar in all cells. Consequently, small cells like bacteria contain little ATP (about 1 amol per cell) and big cells like most mammalian cells contain more ATP (typically 10 000–100 000 amol per cell).[103,110]

5.12.2 Pretreatment of Samples

The ATP pool in living cells has a turnover time of a few seconds. Changing physiological conditions may therefore affect the intracellular level in seconds. On the other hand, the cell tries to maintain the intracellular ATP concentration within a narrow interval as severe damage otherwise may occur. When sampling a culture it is good laboratory practice to avoid changing the conditions more than necessary and to stop metabolism with the extractant as rapidly as possible.

An important pretreatment is to remove extracellular ATP, which may be of levels similar to, or higher than, the intracellular level in many biological samples. This can be carried out by adding one or several ATP-degrading enzymes.[108] Usually, apyrase is employed but a combination of apyrase and adenosine phosphate deaminase has been reported to be a more efficient ATP eliminating reagent.[108] After an incubation time a strong extractant (*cf.* below) is added and inactivates the ATP-degrading enzyme at the same time as it extracts the intracellular ATP. It is important that the extractant inactivates the enzyme irreversibly. If it only inhibits the enzyme activity as, for example with EDTA, the activity will reappear when the extract is mixed with the ATP reagent, which contains magnesium ions.

If the sample is a mixture of mammalian and bacterial cells, *e.g.*, urine, the pretreatment described above can include a detergent like Triton X-100 to lyse the mammalian cells.[111-117] In this situation only bacterial ATP survives the pretreatment. Non-bacterial ATP is often reduced by three orders of magnitude.

Concentrating bacteria by filtration can sometimes be a way to detect low numbers of bacteria. Using a special filter holder it is possible to run water samples through a 0.2 µm filter with positive pressure so that the filter never becomes dry. The special filter holder has a free space above the filter, allowing extractant and ATP reagent to be added on top of the filter. The entire filter holder is then placed in a special luminometer and the light signal is read before and after adding the internal ATP standard. The method was used on boats participating in the Volvo Ocean Race 2008–2009 for measuring biomass in the oceans. This study has not yet been published.

Bacteria can be concentrated as a narrow band with equilibrium centrifugation in a density gradient. This has been carried out to detect bacteria in blood samples.[118] Collecting mammalian or bacterial cells by normal centrifugation does not work as the cells in the pellet have limited access to oxygen and nutrients.[103,119]

5.12.3   Extraction of ATP from Cells

To be accessible for luciferase the intracellular ATP must be released or extracted from the cell. If the cell is only surrounded by a cell membrane like mammalian cells, this can be achieved by a mild neutral detergent like Triton X-100. Cells surrounded not only by a membrane but also a strong cell wall like bacterial or plant cells require stronger releasing reagents like cationic or anionic detergents or strong acids like trichloroacetic acid or perchloric acid. Mycobacteria that are surrounded by a waxy cell wall require strong cationic detergents that are heated close to 100 °C.[120]

When using mild detergents one must remember that ATP-degrading enzymes will survive the extraction of ATP. Unless EDTA is included in the extractant this will result in unstable extracts. Even when doing so the ATP degrading activity may reappear when adding the ATP reagent containing a surplus of magnesium ions required for the luciferase reaction. This effect may cause severe problems if, for example, the ATP reagent is added to microplates outside the luminometer and allowed to stand in a stacker before the light is measured. Strong extractants should therefore be used whenever possible.

The problem with strong extractants is that they inhibit or inactivate luciferase. Inhibition can be compensated for by the internal ATP standard technique providing correct ATP results even if the inhibition is, for example, 50%. A time-dependent inactivation of luciferase means that the luciferase activity is higher when measuring before than after the addition of the internal ATP standard. Consequently, the denominator in Formulas (5.16) and (5.17) becomes too low, resulting in a too high calculated ATP value. This problem can, however, be avoided by having a compound in the ATP reagent that forms a complex with cationic or anionic detergents.[121] This obviates the inactivation of luciferase, and the internal ATP standard technique can be used.

Increased resistance to benzalkonium chloride (BAC), a frequently used strong extractant, has been achieved by random mutagenesis of the *Luciola lateralis* luciferase gene.[122] BAC interferes with the luciferase reaction by both inhibition and inactivation. Inhibition decreases the light emission, while inactivation causes the light to decay. With the best mutant, somewhat higher BAC concentrations as compared to wild-type *Photinus pyralis* luciferase could be used. A concentration of 0.06% BAC resulted in 50% inactivation after 10 min. The recombinant luciferase with increased resistance to BAC was used to develop an extraction cocktail consisting of 0.2% BAC in Tricine buffer, pH 12.0.[123] The extraction efficiency was compared to 10% trichloroacetic acid (TCA) recommended as the reference extraction method.[94,103] The percentage efficiency was determined in 24 Gram-negative bacteria, 13 Gram-positive bacteria and 17 yeast strains. The extraction efficiency compared to 10% TCA was 68–98% in Gram-negatives, 89–108% in Gram-positives and 77–151% in yeasts. The 151% efficiency was obtained with *Trichosporon ovoides*, which is a fungus-like yeast strain. The second highest yield was 109% in the yeast *Yarrowia lipolytica*, which can live under extreme conditions. ATP content in Gram-negatives was 0.4–2.7 amol CFU^{-1}, in Gram-positives 0.4–16.7 amol CFU^{-1} and in yeast 71.4–5460 amol CFU^{-1}. Having all these values collected in a single paper is a great advantage for those who measure CFU with ATP technology; especially since the BAC extraction method was validated by comparison with the TCA extraction method. One must, however, always consider the risk of interference with the extraction method. Extracts prepared from dextrose peptone broth were, for example, not as stable as those prepared from nutrient broth and trypto-soy broth. Extraction efficiency may also vary with physiological conditions and composition of the medium in which cells are suspended. A high content of proteins or lipids will decrease the extraction efficiency. An extraction method should always have an "over-kill" capacity to make sure that the ATP metabolism is stopped immediately at the addition of the extractant and that the extracts are stable. Using a much higher concentration of BAC and a more BAC resistant mutant of luciferase would be more reliable.

When correlating CFU with ATP one must consider that one CFU is one bacterial cell only in model experiments with pure cultures of some strains. Bacterial cells clump together and to other cells or particles and one CFU may consist of hundreds of cells.

5.12.4   Assay of ATP in Eukaryotes

5.12.4.1   Mastitis

The somatic cell count (SCC) in cow's milk has been routinely estimated by the assay of total ATP (intra- plus extracellular) at the National Veterinary Institute in Sweden for many years using a sampling device called Mastistrip.[124–126] Milk samples from all four teats are collected separately on the device and transported on paper disks to the institute, where one disk is used to assay ATP and the other disk is used for cultivating bacteria. The correlation between SCC is better for total ATP than for intracellular ATP.[127]

5.12.4.2   Cell Proliferation and Cytotoxicity

Enumerating cells in cell cultures is often carried out by ATP assay. It is a much more rapid, convenient and reliable measure than, for example, counting cells in the microscope.[103] The ATP assay has been compared to an MTT assay.[128] The ATP assay turned out to be more sensitive and more reproducible.

Monitoring cell death is the reverse of cell proliferation measurements. As long as the cell is alive it tries to maintain its normal ATP level. When cells die the ATP level goes down to zero, but with different time-courses depending on how they die. In apoptosis the cells have to maintain or even increase their cytosolic ATP level as apoptosis is an energy-requiring process.[129] The energy comes from glycolysis and to a minor extent from mitochondrial ATP synthesis. The high ATP level is maintained for up to six hours. In necrosis the cells do not need to maintain a high ATP level. How fast the ATP level goes down depends on the mode of action of the toxic agent and its concentration. Agents interfering with the energy metabolism are likely to cause a rapid decrease of ATP, while agents interfering with, for example, DNA synthesis are likely to give a less rapid decrease. Monitoring the time-course of changes in ATP, therefore, gives valuable information on the mechanism of the toxic agent. The time-course can be measured by transfecting the cell with the *luc* gene and by measuring the light in the presence of D-luciferin or by extracting the ATP and performing assays on the extracts.[129] The first continuous way of measuring gives more information on the time-course but may be exposed to analytical interference as the reaction conditions for the luciferase reaction may change during the measuring time. The second measurement requires frequent sampling to give detailed information of the time-course but should be used as a control for the first measurement.

5.12.4.3   Tumor Chemosensitivity Assay

Isolating tumor cells from patients and growing them in microplates in the presence of various cytostatic agents has been used by several groups to find out which is the most efficient drug towards the particular cancer of the patient. The technique has been shown for leukemic cells as well as solid tumours.[130,131]

Testing is commercially available in several countries. Advantages include: (1) a higher percentage of successful treatments, (2) avoiding treatments that have no effect on the cancer but still cause substantial suffering for the patient and (3) the possibility of choosing drugs causing less suffering or drugs that are less expensive.

5.12.5 Assay of ATP in Prokaryotes

5.12.5.1 Bacteriuria Testing

Urinary tract infection is a very common disease, particularly affecting women. Testing for bacteriuria is the most frequent clinical test, with several hundred million tests performed worldwide. ATP tests for bacteriuria have been evaluated in several studies.[111–117] The test requires non-bacterial ATP to be degraded in a short incubation with a detergent lysing somatic cells and an enzyme degrading extracellular ATP and ATP released from somatic cells. Bacterial ATP is extracted with a strong extractant that also kills the ATP-degrading enzyme. More than 1 million tests per year are performed in Spanish hospitals to avoid culturing of urines containing no bacteria. It has been shown that in a doctor's office situation the test has a diagnostic efficiency of 94%.[113] Avoiding prescribing antibiotics to patients not having urinary tract infection reduces the spread of resistance factors to antibiotics.

5.12.5.2 BCG Vaccine

A method for rapid estimation of viable units in lyophilized BCG vaccine has been developed and validated.[132,133] Lyophilized BCG vials were rehydrated in growth medium and incubated for 24 h before the assay to resuscitate the cells.

5.12.5.3 Susceptibility Testing and Assay of Antibiotic Levels

Exposing isolated bacterial strains to various antibiotics and measuring the bacterial ATP level after 1–2 h allows rapid susceptibility testing.[134,135] Antibiotic levels in clinical or food samples can be rapidly determined using a bacterial strain with known susceptibility to a certain antibiotic and ideally multiresistant to other antibiotics. The bacteria exposed to the sample containing the unknown level of antibiotics may: (1) stop growth, read as a decreased increase of the ATP level compared to a blank not containing antibiotics; (2) kill the cells, read as a decreased ATP level compared to the blank; and (3) lyse the cells, read as increased extracellular ATP levels.[136,137] Discrepancies between the ATP methods and standard cultivation methods can be explained by cells converting into filamentous forms being false resistant with ATP. Addition of 0.5% 2-amino-2-methyl-1,3-propanediol (AMPD) lysed the filamentous cells, and provided the correct classification with the ATP method.[135]

5.12.5.4 Food and Beverage Control

Rapid microbial control using the ATP method is widely used in food and beverage control. As most foods and beverages contain ATP even without bacterial growth the background is often substantial. This ATP can be reduced by a pre-incubation with an ATP-degrading enzyme. In aseptically packaged beverages like UHT-milk it is still necessary to allow the bacteria to grow at 28–32 °C for at least 48 h before performing the test. This allows release of the products from the factory store-room several days earlier, which is an important advantage. The testing can be completely robotized and many factories run millions of tests per year for UHT-milk as well as various fruit juices and other drinks.

5.12.5.5 Control of Personal Care Products

These products are tested in a similar way as food and beverages. In general the product needs to be diluted in a medium sustaining growth prior to the incubation at elevated temperature.

5.12.5.6 Environmental Testing

ATP biomass determination in oceans was one of the first applications of ATP technology.[138,139] The technology has also been used in groundwater testing and in sediment.[140,141] Recently, a new filtration technology has been developed (Lundin, unpublished). A filter holder was supplied with a 0.2 µm filter (diameter 37 mm). Bacteria from more than 1 L of tap water could be collected on the filter. Extractant was added into the filter holder and thereafter ATP reagent. The light was measured by inserting the entire filter holder into a luminometer. Finally, the assay was calibrated by the internal ATP standard technique. With slow decay reagents it should be possible to detect ten bacterial cells on the filter. As large volumes of water can be filtered we are now approaching a sterility control of water. The filter technique was used to measure ATP in the oceans on boats participating in the Volvo Ocean Race 2008–2009 (to be published).

5.13   Measurement of ATP/ADP/AMP

5.13.1   Energy Charge Concept

The measurement of all three adenine nucleotides is of interest as it allows us to calculate the energy charge (EC):

$$EC = (ATP + 0.5ADP)/(ATP + ADP + AMP) \qquad (5.18)$$

Energy charge is a concept introduced by Atkinson and reflects the energy status of the cell.[142] In a healthy cell the EC is above 0.9. If EC falls below 0.5

the cell is likely to die. A high EC means that all cells are healthy. A low EC is obtained if all cells are dying or if a few cells are healthy and the rest are dying. Measuring EC is therefore often of interest.

5.13.2 Measuring Energy Charge

The first method for measuring all three adenine nucleotides was based on the following reactions to convert ADP and AMP into ATP:[143]

$$\text{ADP} + \text{phosphoenolpyruvate} \xrightarrow{\text{pyruvate kinase}} \text{ATP} + \text{pyruvate} \qquad (5.19)$$

$$\text{AMP} + \text{ATP} \xrightarrow{\text{adenylate kinase}} 2\text{ADP} \qquad (5.20)$$

As both AMP and ATP concentrations are low, Reaction (5.20) becomes very slow and requires a lot of time. This problem has been solved by utilizing the fact that the ATP site of adenylate kinase accepts CTP just as well as ATP while CTP is not accepted by luciferase.[103] In this way the end-point of the reaction is reached at the same time regardless of the AMP concentration, if $\text{CTP} \gg \text{AMP}$ [Reaction (5.22)]:

$$\text{ADP} + \text{phosphoenolpyruvate} \xrightarrow{\text{pyruvate kinase}} \text{ATP} + \text{pyruvate} \qquad (5.21)$$

$$\text{AMP} + \text{CTP} \xrightarrow{\text{adenylate kinase}} \text{ADP} + \text{CDP} \qquad (5.22)$$

An alternative method is to use the enzyme pyruvate orthophosphate dikinase (PPDK) and pyruvate kinase:[144,145]

$$\text{ADP} + \text{phosphoenolpyruvate} \xrightarrow{\text{pyruvate kinase}} \text{ATP} + \text{pyruvate} \qquad (5.23)$$

$$\text{AMP} + \text{phosphoenolpyruvate} + \text{PPi} \xrightarrow{\text{PPDK}} \text{ATP} + \text{pyruvate} + \text{Pi} \qquad (5.24)$$

5.14 Hygiene Control

Hazard analysis critical control point (HACCP) is a technique for assuring that food does not expose the user to hazards. One of the highlights in the history of the HACCP system was in 1993 when the Codes Guidelines for the Application of HACCP system were adopted by the FAO/WHO Codex Alimentarius Commission. According to HACCP one should identify critical control points, *i.e.*, points, steps or procedures in a food manufacturing process at which control can be applied and, as a result, a food safety hazard can be prevented, eliminated or reduced to an acceptable level. For such points one should establish critical limits, *i.e.*, the maximum or minimum value to which a physical, biological or chemical hazard must be controlled at a critical control point to prevent, eliminate or reduce to an acceptable level. A critical control point must be monitored to be able to perform corrective actions. Conventional culture techniques are much too slow to fit into HACCP.

In the food industry, at the end of the day all food contact surfaces are cleaned and sometimes even disinfected. To assure that this is properly performed samples are taken with swabs or contact plates and cultivated. Results are only available after several days. Consequently, production has to start the following morning without knowing the results. If cleaning is regarded as a critical control point, results must be available before restarting production. Collecting samples from food contact surfaces with swabs and measuring the ATP content has become widespread. Essentially, all foods and beverages contain ATP and after cleaning there should be very little ATP left. ATP is used as an indicator of food residues in the same way as measurement of harmless bacteria is often measured as an indicator organism. ATP is a powerful indicator that biological dirt has not been properly removed. The ATP hygiene control can not only be used in the food industry but also in, for example, hospitals, nursery homes or even in normal offices as an objective measure of cleanliness. It can also be used for other products such as beverages, cosmetics and pharmaceuticals.

As ATP is degraded to ADP and AMP during cooking of food an even more sensitive measure of cleanliness is to use PPDK to measure ATP + AMP using Reaction (5.24).[145] In cooked food there is often more than 100× more AMP than ATP. If the ATP degradation continues in the food residues on food contact surfaces the assay of ATP + AMP is also more reliable as the timing of sampling is not important.

The PPDK enzyme can also be used to measure AMP in RNA after degradation with nuclease P1.[145] This will increase the sensitivity even further.

5.15 Enumeration of Bacterial Cells on Filter without Cultivation

Suspensions of bacterial cells were collected on a filter, which was thereafter sprayed with extractant and finally with an ATP reagent containing PPDK (*cf.* Section 5.13.2). Each luminescent spot on the filter corresponds to a bacterial CFU and the number of spots correlated with number of CFU. The luminescent spots were counted with a CCD camera.[146]

5.16 Coliform Test

In selective media coliforms can be detected by their high content of β-galactosidase. Using D-luciferin-*O*-β-galactopyranoside, it was possible to identify coliforms after culturing for 7 h rather than 24 h with conventional methods.[147]

5.17 Assays of Enzymes and Metabolites

All enzymes and metabolites participating in reactions that produce or degrade ATP can be assayed by ATP reagents. If the reaction is carried out in the presence of the ATP reagent we call it ATP monitoring as we can monitor continuously the ATP concentration by measuring the intensity of the emitted light. ATP monitoring provides more reliable data than stopping the reaction and then adding the ATP reagent. In particular, one may detect a strange kinetic behavior of an enzyme or that the endpoint has not been reached in an assay of a metabolite. The wide linear range of the ATP assay means that initial rate conditions apply for much longer than for, for example, spectrophotometric assays.

5.17.1 ATP Monitoring Concept

The ATP monitoring concept is based on adding luciferase and D-luciferin to a reaction mixture to continuously follow ATP formation or degradation.[90,94–107] The technique requires the following:

1 the percentage ATP degradation per minute in the luciferase reaction is negligible or can be compensated for;
2 luciferase does not suffer from increasing inhibition or inactivation during the measurement time;
3 reaction conditions are essentially unchanged during the measurement time and only the ATP concentration is changing to a measurable degree;
4 the ATP concentration is well below K_m for luciferase, resulting in a linear relation between ATP and light intensity.

Only stable light ATP reagents are used to fulfill the first requirement. Kinetic as well as end-point assays have been developed. Both enzymes and metabolites may be assayed kinetically for ATP-producing as well as ATP-degrading reactions. Metabolites may also be measured by end-point assays. For ATP-producing enzymes the assays are set up to produce less than micromolar levels of ATP and to consume only a small fraction of the substrates. Thus, initial rate conditions apply much longer than for less sensitive measurement technologies such as, for example, spectrophotometry or fluorometry. An advantage is that you do not need a stopping reagent. The assays are calibrated with an internal ATP standard.

For ATP-degrading reactions, kinetic assays are preferably set up as first-order reactions. Thus the ATP degradation rate must not be affected by the consumption of the second substrate. The logarithm of the light emission *versus* time is then a straight line and the rate constant can be measured as:

$$k = [\ln(I_1) - \ln(I_2)]/(t_2 - t_1) \qquad (5.25)$$

The rate constant can be expressed, for example, as percent decay per minute. The rate constant is proportional to the enzyme or the metabolite level. ATP can be added as a starting reagent (no need for an internal ATP standard at the end of the assay). Examples of different assays are given below.

5.17.2   Monitoring ATP Forming Enzyme Reactions

The first commercially available clinical kits based on luminescence were optimized kits for measuring total creatine kinase (CK) activity and creatine kinase B-subunit (CK-B) activity in serum from patients with suspected myocardial infarction.[100] In these assays the rate of ATP formation in the CK reaction is measured. Both kits fulfilled all the requirements for clinical use and the diagnosis could be made much earlier than with spectrophotometric methods. Furthermore, in those patients that did not fulfill the diagnosis of acute myocardial infarction but showed a small increase of CK-B the risk of obtaining a big infarction was considerably elevated (not published). A kit was also developed for measuring total CK in whole blood dried on paper disks for screening for Duchenne muscular dystrophy (DMD).[102] The testing was performed on 25–45-day-old boys to prevent the birth of a second child in the family by early diagnosis. DMD boys die very young. This assay has been used by Dr. Günter Sceuerbrandt for several years for screening for DMD at CK-Test Laboratorium, Breitnau, Germany. A technically interesting feature of the assay is that the high ATP content from the blood cells in the dried spot was efficiently degraded by the adenylate kinase from the cells by performing a preincubation in the presence of $1 \, \text{mmol} \, \text{L}^{-1}$ AMP. Thereafter the ADP and creatine phosphate were added and the CK activity could be accurately measured without interference from an ATP blank.[102]

ATP formation rate in the pyruvate kinase reaction has been used for measuring both pyruvate kinase and ADP.[90,148] Adenylate kinase can also be measured in this way.[149]

The endpoint assay of ATP, ADP and AMP has been described in Section 5.13.2.[90,103] In this assay it is possible to continuously monitor the light emission to assure that the endpoint is reached. The assay of all three adenine nucleotides can be determined in a single cuvette to reduce reagent consumption.[103]

5.17.3   Monitoring Oxidative Phosphorylation and Photophosphorylation

The first to describe a continuous measurement of oxidative phosphorylation by mitochondria was Lemasters and Hackenbrock.[150,151] Conditions were, however, far from those described in Section 5.17.1 for ATP monitoring. This led to complicated calculation of results. ATP monitoring of photophosphorylation of ADP into ATP in chromatophores under the conditions given in Section 5.17.1 was first described in 1977.[96,98,148,149,152–154] This was later extended to ATP monitoring of photophosphorylation to form PPi in chromatophores.[101] Oxidative phosphorylation in mitochondria isolated by needle biopsy is now a well-established assay for detecting mitochondrial diseases in clinical routine at the Karolinska University Hospital at Huddinge.[106,155–157] The assay is also used by at least eight other research groups, including the Mayo Clinic, as described by Dr Kevin Short.[158]

5.17.4 Monitoring ATP Degrading Enzyme Reactions

5.17.4.1 Glycerol

A kinetic assay of glycerol measuring the rate of degradation of ATP in the glycerol kinase reaction has been described.[104,105] The assay is set up as a first-order reaction. The paper also describes how results can be plotted as v/V_{max} *versus* $S/(K_m + S)$, *i.e.*, the two sides of the Michaelis–Menten equation. The result is a straight line even above the K_m value. The assay can also be used for measuring glycerol kinase and has been in routine use for over 20 years at the Karolinska Institute.

5.17.4.2 Urea

A kinetic assay of urea based on the ATP-hydrolyzing urease reaction (EC 3.5.1.45) has been developed according to the same principle as the glycerol assay above.[159] The assay was optimized with multivariate analysis and has a detection limit of 100 pmol.

5.17.4.3 Protein Kinase

A kit for the measurement of protein kinase or ATPase measurements has been developed based on similar principles to the glycerol assay.[95] The assay has been miniaturized and can be performed in 96-well or 384-well microplates for high-throughput screening (HTS) of drug libraries of potential inhibitors. The composition of the reagent has been optimized to result in a half-life of the light emission in the absence of kinase, *i.e.*, the blank, of around 10 h. The kinase reaction gives a more rapid degradation of ATP and consequently the light emission decays more rapidly. The light emission can be read twice or as many times as required to obtain an accurate result. Light readings from samples containing kinase are normalized by dividing with corresponding readings from the blank at the same times. The protein kinase reaction is set up to result in first-order kinetics, *i.e.*, an ATP concentration well below the K_m of the kinase and well below the concentration of the second substrate, *e.g.*, the peptide. First-order kinetics is easily confirmed by checking that the logarithm of the light readings *versus* time is a straight line. As compared to using high ATP concentrations the percentage ATP degradation per minute is higher and independent of the ATP concentration. This makes the assay more sensitive, allowing the use of low kinase concentrations in HTS, which is important as many kinases are very costly to prepare. The first-order rate constant of the kinase reaction is calculated by Equation (5.25). This first-order rate constant is linearly related to kinase concentration in an interval covering more than three orders of magnitude. It is therefore easy to decide a kinase activity that is suitable for HTS of compound libraries. In HTS the light is read at two points in time. It is also easy to optimize the second substrate and to measure IC_{50} of found inhibitors.

A major advantage of the assay is that variations in ATP concentration or luciferase inhibition from drugs cancel out in the calculations and do not affect the assay result. As can be seen in Figure 5.6(A), a kinase inhibitor not acting on luciferase results in a less rapid decay of the light, *i.e.*, a lower slope of the line representing the light emission. Figure 5.6(B) shows the effect of a luciferase inhibitor not acting on kinase. Kinase is not inhibited as the slope of the line is the same as the control. An inhibitor of both kinase and luciferase will give a line crossing the control line, as in Figure 5.6(C). The lines have been calculated based on 50% inhibition of kinase and 50% inhibition of luciferase. Measuring the slope, *i.e.*, the first-order rate constant, gives the same degree of kinase inhibition regardless of luciferase inhibition. If the light is measured only once, as with other luciferase-based assays of kinase, a kinase inhibitor is only identified, if the light emission at the measuring time is higher from the sample than from the control. This happens neither in Figure 5.6(B) nor if the measurement is performed too early in Figure 5.6(C).

The assay uses ATP well below K_m for kinase and well below the peptide substrate as it follows first-order reaction conditions. Therefore, the assay detects all competitive and non-competitive inhibitors. With previous ATP depletion assays based on luciferase as much as 2–3% of the drugs inhibited luciferase and were wrongly classified. A Z' value as high as 0.96 shows that the assay is highly reproducible. The assay is highly suited to HTS as the assay can be performed in 25 µL or even less. High kinase activities can be measured in minutes, and for low kinase activities the measurement can be extended to several hours. This gives a great flexibility depending on kinase cost and instrumentation available.

5.17.4.4   Pyrosequencing

The assay of inorganic pyrophosphate (PPi) by ATP monitoring was described by Nyrén and Lundin:[101]

$$PPi + \text{adenosine } 5'\text{-phosphosulphate} \xrightarrow{\text{ATP sulfurylase}} ATP + SO_4^{2-} \qquad (5.26)$$

This assay was later used for DNA sequencing.[160,161] A primed ss-DNA template is subjected to stepwise addition of deoxynucleotide, one at a time. If the nucleotide is complementary to the template, it is incorporated by a polymerase and PPi is formed. As PPi reacts in the sulfurylase reaction, ATP is formed and

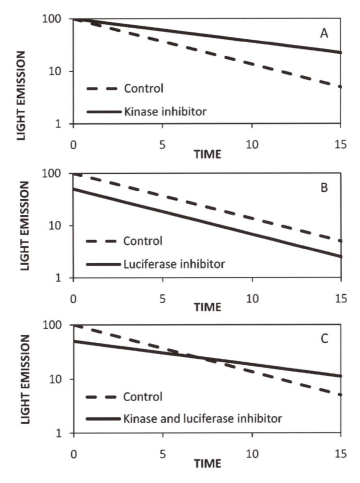

Figure 5.6   Light emission in the presence of a kinase inhibitor (A), a luciferase inhibitor (B) and an inhibitor of both kinase and luciferase (C) compared to a control with no inhibitor. Previously available bioluminescent methods measured the light at one point in time, which only gives correct results in the situation shown in (A). The new method described ensures correct results in all circumstances as the light is measured at a minimum of two points in time.

light is emitted. When the reaction is performed on a solid phase, the immobilized primer/template is washed and the cycle is repeated with a new nucleotide.[160] When performed in liquid phase, the reaction is carried out in the presence of a nucleotide-degrading enzyme, apyrase, degrading both formed ATP and unused deoxynucleotides. Therefore, a peak of light is obtained whenever a complementary deoxynucleotide is added. If there is more than one deoxynucleotide the size of the peak is bigger. Several companies have been involved in the commercialization of pyrosequencing.

5.18   Reporter Gene Assays

The *luc* gene coding for firefly luciferase has become a very important reporter gene. The reasons for this include that the gene is absent in all non-luminescent species, luciferase is easily detected at extremely low levels, a stable light emission suitable for HTS and a rapid mix-and-measure procedure. Reporter gene assays are treated in more detail elsewhere in this book.

5.19   Bacterial Luciferase

5.19.1   Enzymatic Assays

Bacterial luciferase catalyses the reaction between a long-chain aldehyde (RCHO), $FMNH_2$ and oxygen:

$$FMNH_2 + RCHO + O_2 \xrightarrow{\text{luciferase}} FMN + RCOOH + H_2O + \text{light} \qquad (5.27)$$

When used in assays of enzymes and metabolites it is often used together with NAD(P)H:FMN oxidoreductase:

$$NAD(P)H + FMN + H^+ \xrightarrow{\text{oxidoreductase}} NAD(P)^+ + FMNH_2 \qquad (5.28)$$

In various dehydrogenase reactions NAD(P)H can be formed:

$$SH_2 + NAD(P)^+ \xrightarrow{\text{dehydrogenase}} S + NAD(P)H + H^+ \qquad (5.29)$$

Reactions (5.27)–(5.29) can be used for assays of a large number of metabolites, such as $NAD^+/NADH$, $NADP^+/NADPH$, ethanol, malate, oxaloacetate and the corresponding dehydrogenases.[162–166] These assays can also be performed with immobilized enzymes.[167,168] Enzymes have been immobilized on, for example, Sepharose and nylon tubes. Flow-assisted analysis is described in detail in Chapter 6.

5.19.2   Cell-based Tests

A recent review describes numerous applications of natural luminescent bacteria, including determination of bacterial metabolic activity in food products; assays for lipase, phospholipase and esterase; long-chain fatty acids; lipopolysaccharide; antilipogenic compounds; oil oxidation; monoamine oxidase activity and its inhibitors; antibiotic activity; serum bactericidal activity; phagocytosis; cyclic adenosine 3′,5′-monophosphate; oxygen transmission rates through plastic films; and early indication of marine fish spoilage and several measurements of water quality.[169] Several of these tests have been commercialized. Chapters 14 and 15 are devoted to biosensors.

5.20 Conclusions

Chemiluminescent and bioluminescent reactions are used in analytical and bioanalytical chemistry for the direct detection and quantification of analytes, for monitoring several groups of enzymes (oxidases, kinases, ATPases, dehydrogenases, *etc.*) and for measuring their substrates and inhibitors, for evaluating the production of reactive oxygen species in living cells and for assessing antioxidant activity. Furthermore, the number of microbial and mammalian cells can be estimated, DNA can be sequenced and gene expression can be followed. The authors of this chapter have been involved in the development of CL and BL assays since the 1970s and 1980s. In those days CL/BL assays were often not recognized by the scientific community, being only performed by a small number of highly specialized scientists. Moreover, there were few commercially available reagents and luminometers and those available were of poor quality. Nowadays, CL and BL are well-established detection techniques and there are a large number of companies manufacturing luminometers and reagents, competing with more and more advanced products. Applications of CL/BL assays include such widely different areas as molecular biology, HTS drug discovery, clinical diagnosis, food microbiological control and hygiene control and environmental and forensic analysis. Over the years the analytical performance has improved continuously and the detection limits have reduced. Today we have reagents that can reveal attomoles of ATP (*i.e.*, the amount of ATP in a single bacterial cell) and we are approaching single-molecule detection. One may say that CL and BL are just a read-out technology, but it is a read-out that combines sensitivity, simplicity and low cost with a wide range of applications in life sciences and industry. A bright future for CL and BL is envisioned, especially in combination with technologies from molecular biology and immunology.

Acknowledgements

A.L. thanks all his collaborators throughout almost 40 years for participating in the development of the assays mentioned in this chapter. Special thanks go to Dr. Satoshi Fukuda, Kikkoman, for sending most of the Japanese reprints, to Professor Natalia Ugarova, Lomonosov Moscow State University, for sending the Russian reprints and to Professor Shimon Ulitzur, The Technion Institute, for sending his excellent review on cell bioreporters. The authors also thank Dr. Robert Inglis, BioThema AB, for correcting the English.

References

1. C. A. Marquette and L. J. Blum, *Anal. Bioanal. Chem.*, 2006, **385**, 546.
2. A. Tahirovic, A. Copra, E. Omanovic-Miklicanin and K. Kalcher, *Talanta*, 2007, **72**, 1378.

3. F. J. Perez and S. Rubio, *Plant Growth Regul.*, 2006, **48**, 89.
4. E. A. A. Al-Kerwi, A. H. M. Al-Hashimi and A. M. Salman, *Asia Pac. J. Clin. Nutr.*, 2005, **14**, 428.
5. B. Khadro, B. D. Leca-Bouvier, F. Lagarde, F. Barbier, L. J. Blum, C. Martelet, L. Marcotte, M. Tabrizian and N. Jaffrezic-Renault, *Sensor Lett.*, 2009, **7**, 833.
6. S. P. Hansard and W. M. Landing, *Limnol. Oceanogr. Methods*, 2009, **7**, 222.
7. A. D. Ilyina, J. L. M. Hernandez, J. E. M. Benavides, B. H. L. Lujan, E. S. Bogatcheva, J. R. Garcia and J. R. Martinez, *Luminescence*, 2003, **18**, 31.
8. L. J. Kricka, J. C. Voyta and I. Bronstein, *Methods Enzymol.*, 2000, **305**, 370.
9. D. J. Monk and D. R. Walt, *Anal. Bioanal. Chem.*, 2004, **379**, 931.
10. A. Andreani, M. Granaiola, M. Guardigli, A. Leoni, A. Locatelli, R. Morigi, M. Rambaldi and A. Roda, *Eur. J. Med. Chem.*, 2005, **40**, 1331.
11. T. Ichibangase, Y. Ohba, N. Kishikawa, K. Nakashima and N. Kuroda, *Luminescence*, 2004, **19**, 259.
12. B. Näslund, K. Bernström, A. Lundin and P. Arner, *J. Biolumin. Chemilumin.*, 1989, **3**, 115.
13. B. Näslund, P. Arner, J. Bolinder, L. Hallander and A. Lundin, *Anal. Biochem.*, 1991, **192**, 237.
14. A. Andreani, M. Granaiola, A. Leoni, A. Locatelli, R. Morigi, M. Rambaldi, A. Roda, M. Guardigli, S. Traniello and S. Spisani, *Eur. J. Med. Chem.*, 2004, **39**, 785.
15. S. Yamashoji, A. Asakawa, S. Kawasaki and S. Kawamoto, *Anal. Biochem.*, 2004, **333**, 303.
16. M. Tsunoda and K. Imai, *Anal. Chim. Acta*, 2005, **541**, 13.
17. E. Ponten, M. Stigbrand and K. Irgum, *Anal. Chem.*, 1995, **67**, 4302.
18. D. W. Lee, V. R. Erigala, M. Dasari, J. H. Yu, R. M. Dickson and N. Murthy, *Int. J. Nanomed.*, 2008, **3**, 471.
19. E. Omanovic and K. Kalcher, *Int. J. Environ. Anal. Chem.*, 2005, **85**, 853.
20. V. Stepanyan, A. Arnous, C. Petrakis, P. Kefalas and A. Calokerinos, *Talanta*, 2005, **65**, 1056.
21. E. Sestigiani, M. Mandreoli, M. Guardigli, A. Roda, E. Ramazzotti, P. Boni and A. Santoro, *Nephron Clin. Pract.*, 2008, **108**, C265.
22. M. Adamczyk, R. J. Brashear, P. G. Mattingly and P. H. Tsatsos, *Anal. Chim. Acta*, 2006, **579**, 61.
23. J. B. Claver, M. C. V. Miron and L. F. Capitan-Vallvey, *Anal. Chim. Acta*, 2004, **522**, 267.
24. Yu. A. Vladimirov and E. V. Proskurnina, *Biochemistry (Moscow)*, 2009, **74**, 1545.
25. M. Freitas, J. L. F. C. Lima and E. Fernandes, *Anal. Chim. Acta*, 2009, **649**, 8.

26. B. Halliwell and J. M. C. Gutteridge, *Free Radicals in Biology and Medicine*, Oxford University Press, Oxford-New York, 1999.
27. R. L. Stjernholm, R. C. Allen, R. H. Steele, W. W. Waring and J. A. Harris, *Infect. Immun.*, 1973, **7**, 313.
28. Yu. A. Vladimirov, S. R. Ribarov, P. G. Bochev, L. C. Benov and G. I. Klebanov, *Gen. Physiol. Biophys.*, 1990, **9**, 45.
29. R. Hagens, F. Khabiri, V. Schreiner, H. Wenck, K.-P. Wittern, H.-J. Duchstein and W. Mei, *Skin Res. Technol.*, 2008, **14**, 112.
30. Y. A. Vladimirov, V. S. Sharov, E. S. Driomina, A. V. Reznitchenko and S. B. Gashev, *Free Radic. Biol. Med.*, 1995, **18**, 739.
31. O. Myhre, J. M. Andersen, H. Aarnes and F. Fonnum, *Biochem. Pharmacol.*, 2003, **65**, 1575.
32. S. Kopprasch, J. Pietzsch and J. Graessler, *Luminescence*, 2003, **18**, 268.
33. R. C. Allen and L. D. Loose, *Biochem. Biophys. Res. Commun.*, 1976, **69**, 245.
34. V. Jancinova, K. Drabikova, R. Nosal, L. Rackova, M. Majekova and D. Holomanova, *Redox Rep.*, 2006, **11**, 110.
35. A. Krasowska, D. Rosiak, K. Szkapiak, M. Oswiecimska, S. Witek and M. Lukaszewicz, *Cell. Mol. Biol. Lett.*, 2001, **6**, 71.
36. M. Pavelkova and L. Kubala, *Luminescence*, 2004, **19**, 37.
37. I. Imada, E. F. Sato, M. Miyamoto, Y. Ichimori, Y. Minamiyama, R. Konaka and M. Inoue, *Anal. Biochem.*, 1999, **271**, 53.
38. M. Rinaldi, P. Moroni, M. J. Paape and D. D. Bannerman, *Vet. Immunol. Immunopathol.*, 2007, **115**, 107.
39. E. J. Swindle, J. A. Hunt and J. W. Coleman, *J. Immunol.*, 2002, **169**, 5866.
40. B. M. Babior, *Curr. Opin. Hematol.*, 1995, **2**, 55.
41. H. Xie, P. E. Ray and B. L. Short, *Stroke*, 2005, **36**, 1047.
42. L. V. Medvedeva, T. N. Popova, V. G. Artyukhov, L. V. Matasova and M. A. A. P. de Carvalho, *Biochemistry (Moscow)*, 2002, **67**, 696.
43. J. Zurova-Nedelcevova, J. Navarova, K. Drabikova, V. Jancinova, M. Petrikova, I. Bernatova, V. Kristova, V. Snirc, V. Nosal'ova and R. Sotnikova, *Neuroendocrinol. Lett.*, 2006, **27**, 168.
44. J. M. Andersen, O. Myhre, H. Aarnes, T. A. Vestad and F. Fonnum, *Free Radic. Res.*, 2003, **37**, 269.
45. K. Drabikova, T. Perecko, R. Nosal, K. Bauerova, S. Ponist, D. Mihalova, G. Kogan and V. Jancinova, *Pharmacol. Res.*, 2009, **59**, 399.
46. J. Cedergren, T. Forslund, T. Sundqvist and T. Skogh, *J. Rheumatol.*, 2007, **34**, 2162.
47. C. Chen, F. K. Liu, X. P. Qi and J. S. Li, *World J. Gastroenterol.*, 2003, **9**, 242.
48. A. Marangoni, S. Accardo, R. Aldini, M. Guardigli, F. Cavrini, V. Sambri, M. Montagnani, A. Roda and R. Cevenini, *World J. Gastroenterol.*, 2006, **12**, 3077.
49. L. M. Magalhães, M. A. Segundo, S. Reis and J. L. F. C. Lima, *Anal. Chim. Acta*, 2008, **613**, 1.

50. S. R. Maxwell, T. Dietrich and I. L. C. Chapple, *Clin. Chim. Acta*, 2006, **372**, 188.
51. D. D. M. Wayner, G. W. Burton, K. U. Ingold and S. Locke, *FEBS Lett.*, 1985, **187**, 33.
52. T. Metsä-Ketelä and A. L. Kirkkola, *Free Radical Res. Commun.*, 1992, **16**, 215.
53. V. E. Kagan, M. Tsuchiya, E. Serbinova, L. Packer and H. Sies, *Biochem. Pharmacol.*, 1993, **45**, 393.
54. M. T. K. Dresch, S. B. Rossato, V. D. Kappel, R. Biegelmeyer, M. L. M. Hoff, P. Mayorga, J. A. S. Zuanazzi, A. T. Henriques and J. C. F. Moreira, *Anal. Biochem.*, 2009, **385**, 107.
55. T. P. Whitehead, G. H. G. Thorpe and S. R. J. Maxwell, *Anal. Chim. Acta*, 1992, **266**, 265.
56. J. A. Serra, R. O. Dominguez, E. R. Marschoff, E. M. Guareschi, A. L. Famulari and A. Boveris, *Neurochem. Res.*, 2009, **34**, 2122.
57. T. S. Chen, S. Y. Liou and Y. L. Chang, *Ren. Fail.*, 2008, **30**, 843.
58. C. D. Schneider, J. Barp, J. L. Ribeiro, A. Bello-Klein and A. R. Oliveira, *Can. J. Appl. Physiol.*, 2005, **30**, 723.
59. S. M. Ferreira, S. F. Lerner, R. Brunzini, P. A. Evelson and S. F. Llesuy, *Eye*, 2009, **23**, 1691.
60. M. Battino, M. S. Ferreiro, I. Gallardo, H. N. Newman and P. Bullon, *J. Clin. Periodontol.*, 2002, **29**, 189.
61. N. Desai, R. Sharma, K. Makker, E. Sabanegh and A. Agarwal, *Fertil. Steril.*, 2009, **92**, 1626.
62. D. Christodouleas, C. Fotakis, K. Papadopoulos, E. Yannakopoulou and A. C. Calokerinos, *Anal. Chim. Acta*, 2009, **652**, 295.
63. H. Yao, B. Wu, Y. Cheng and H. Qu, *Food Chem.*, 2009, **115**, 380.
64. S. B. Rossato, C. Haas, M. D. C. B. Raseira, J. C. F. Moreira and J. A. S. Zuanazzi, *J. Med. Food*, 2009, **12**, 1119.
65. M. A. Anagnostopoulou, P. Kefalas, V. P. Papageorgiou, A. N. Assimopoulou and D. Boskou, *Food Chem.*, 2006, **94**, 19.
66. S. Girotti, F. Fini, L. Bolelli, L. Savini, E. Sartini and G. Arfelli, *Luminescence*, 2006, **21**, 233.
67. A. K. Atoui, A. Mansouri, G. Boskou and P. Kefalas, *Food Chem.*, 2005, **89**, 27.
68. S. Alvarez, T. Zaobornyj, L. Actis-Goretta, C. G. Fraga and A. Boveris, *Ann. NY Acad. Sci.*, 2002, **957**, 271.
69. M. DeLuca and W. D. McElroy, *Methods Enzymol.*, 1978, **57**, 3.
70. H. Fraga, *Photochem. Photobiol. Sci.*, 2008, **7**, 146.
71. N. N. Ugarova, in *Chemical and Biological Kinetics. New Horizons. Volume II. Biological Kinetics*, ed. E. B. Burlakova and S. D. Varfolomeev, Koninklijke Brill NV, Leiden, The Netherlands, 2005, p. 205.
72. J. R. De Wet, K. V. Wood, D. R. Helinsky and M. DeLuca, *Proc. Natl. Acad. Sci. USA*, 1985, **82**, 7870.
73. H. Tatsumi, T. Masuda and E. Nakano, *Agric. Biol. Chem.*, 1988, **52**, 1123.

74. K. V. Wood, Y. A. Lam, H. H. Seliger and W. D. McElroy, *Science*, 1989, **244**, 700.
75. J. H. Devine, G. D. Kutuzova, V. A. Green, N. N. Ugarova and T. O. Baldwin, *Biochim. Biophys. Acta*, 1993, **1173**, 121.
76. H. Tatsumi, T. Masuda, N. Kajiyama and E. Nakano, *J. Biolumin. Chemilumin.*, 1989, **3**, 75.
77. H. Tatsumi, N. Kajiyama and E. Nakano, *Biochim. Biophys. Acta*, 1992, **1131**, 161.
78. V. R. Viviani, A. C. R. Silva, G. L. O. Perez, R. V. Santelli, E. J. Bechara and F. C. Reinach, *Photochem. Photobiol.*, 1999, **70**, 254.
79. B. S. Alipour, S. Hosseinkhani, M. Nikkhah, H. Naderi-Manesh, M. J. Chaichi and S. K. Osaloo, *Biochem. Biophys. Res. Commun.*, 2004, **325**, 215.
80. N. Kajiyama and E. Nakano, *Biochemistry*, 1993, **32**, 13795.
81. N. Kajiyama and E. Nakano, *Biosci. Biotechnol. Biochem.*, 1994, **58**, 1170.
82. N. Kajiyama and E. Nakano, *Protein Eng.*, 1991, **4**, 691.
83. H. Fujii, K. Noda, Y. Asami, A. Kuruda, M. Sakata and A. Tokida, *Anal. Biochem.*, 2007, **366**, 131.
84. H. Tatsumi, S. Fukuda, M. Kikuchi and Y. Koyama, *Anal. Biochem.*, 1996, **243**, 176.
85. E. Conti, N. P. Franks and P. Brick, *Structure*, 1996, **4**, 287.
86. R. Fontes, D. Fernandes, F. Peralta, H. Fraga, I. Maio and J. C. Esteves da Silva, *FEBS J.*, 2008, **275**, 1500.
87. T. Goto, I. Kubota, N. Suzuki and Y. Kishi, in *Bioluminescence*, ed. M. J. Cormier, D. M. Hercules and J. Lee, Plenum Press, New York, 1973, p. 325.
88. R. T. Lee, J. L. Denburg and W. D. McElroy, *Arch. Biochem. Biophys.*, 1970, **141**, 38.
89. K. Hirokawa, N. Kajiyama and S. Murakami, *Biochim. Biophys. Acta*, 2002, **1597**, 271.
90. A. Lundin, A. Rickardsson and A. Thore, *Anal. Biochem.*, 1976, **75**, 611.
91. K. Noda, Y. Matsuno, H. Fujii, T. Kogure, M. Urata, Y. Asami and A. Kuroda, *Biotechnol. Lett.*, 2008, **30**, 1051.
92. M. Urata, R. Iwata, K. Noda, Y. Murakami and A. Kuroda, *Biotechnol. Lett.*, 2009, **31**, 737.
93. M. DeLuca and W. D. McElroy, *Biochemistry*, 1974, **13**, 921.
94. A. Lundin, *Methods Enzymol.*, 2000, **305**, 346.
95. A. Lundin and J. Eriksson, *Assay Drug Dev. Technol.*, 2008, **6**, 531.
96. A. Lundin, A. Thore and M. Baltscheffsky, *FEBS Lett.*, 1977, **79**, 73.
97. A. Lundin and I. Styrélius, *Clin. Chim. Acta*, 1978, **87**, 199.
98. A. Lundin and M. Baltscheffsky, *Methods Enzymol.*, 1978, **57**, 50.
99. A. Lundin, *Methods Enzymol.*, 1978, **57**, 56.
100. A. Lundin, B. Jäderlund and T. Lövgren, *Clin. Chem.*, 1982, **28**, 609.
101. P. Nyrén and A. Lundin, *Anal. Biochem.*, 1985, **151**, 504.
102. G. Scheuerbrandt, A. Lundin, T. Lövgren and W. Mortier, *Muscle Nerve*, 1986, **9**, 11.

103. A. Lundin, M. Hasenson, J. Persson and Å. Pousette, *Methods Enzymol.*, 1986, **133**, 27.
104. A. Lundin, P. Arner and J. Hellmér, *Anal. Biochem.*, 1989, **177**, 125.
105. J. Hellmér, P. Arner and A. Lundin, *Anal. Biochem.*, 1989, **177**, 132.
106. R. Wibom, A. Lundin and E. Hultman, *Scand. J. Clin. Invest.*, 1990, **50**, 143.
107. R. Wibom, K. Söderlund, A. Lundin and E. Hultman, *J. Biolum. Chemilum.*, 1991, **6**, 123.
108. T. Sakakibara, S. Murakami, N. Hattori, M. Nakajima and K. Imai, *Anal. Biochem.*, 1997, **250**, 157.
109. B. Lyvén, *Report P8 03765/Eng*, 2008, SP Technical Research Institute of Sweden.
110. N. Hattori, T. Sakakibara, N. Kajiyama, T. Igarashi, M. Maeda and S. Murakami, *Anal. Biochem.*, 2003, **319**, 287.
111. A. Thore, S. Ánséhn, A. Lundin and S. Bergman, *J. Clin. Microbiol.*, 1975, **1**, 1.
112. A. Thore, A. Lundin and S. Ånséhn, *J. Clin. Microbiol.*, 1983, **17**, 218.
113. H. Hallander, A. Kallner, A. Lundin and E. Österberg, *Acta Path. Microbiol. Immunol. Scand. Sect. B*, 1986, **94**, 39.
114. A. Lundin, H. Hallander, A. Kallner, U. Karnell Lundin and E. Österberg, *J. Biolumin. Chemilumin.*, 1989, **4**, 381.
115. B. Gästrin, R. Gustafsson and A. Lundin, *Scand. J. Infect. Dis.*, 1989, **21**, 409.
116. E. Österberg, H. Åberg, H. Hallander, A. Kallner and A. Lundin, *J. Infect. Dis.*, 1990, **161**, 942.
117. E. Österberg, H. Hallander, A. Kallner, A. Lundin and H. Åberg, *Eur. J. Clin. Microbiol.*, 1991, **10**, 70.
118. Ö. Molin, L. Nilsson and S. Ånséhn, *J. Clin. Microbiol.*, 1983, **18**, 521.
119. A. Lundin and A. Thore, *Appl. Microbiol.*, 1975, **30**, 713.
120. S. E. Hoffner, C. A. Jimenez-Misas and A. Lundin, in *Bioluminescence and Chemiluminescence*, ed. A. K. Campbell, L. J. Kricka and P. E. Stanley, John Wiley & Sons, Chichester, 1994, p. 442.
121. A. Lundin, J. Anson and P. Kau, in *Bioluminescence and Chemiluminescence*, ed. A. K. Campbell, L. J. Kricka and P. E. Stanley, John Wiley & Sons, Chichester, 1994, p. 399.
122. N. Hattori, N. Kajiyama, M. Maeda and S. Murakami, *Biosci. Biotechnol. Biochem.*, 2002, **66**, 2587.
123. N. Hattori, T. Sakakibara, N. Kajiyama, T. Igarashi, M. Maeda and S. Murakami, *Anal. Biochem.*, 2003, **319**, 287.
124. T. Olsson, K. Sandstedt, O. Holmberg and A. Thore, *Biotechnol. Appl. Biochem.*, 1986, **8**, 361.
125. U. Emanuelson, T. Olsson, O. Holmberg, M. Hageltom, T. Mattila, L. Nelson and G. Åström, *J. Dairy Sci.*, 1987, **70**, 880.
126. U. Emanuelson, T. Olsson, G. Mattila, G. Åström and O. Holmberg, *J. Dairy Res.*, 1988, **55**, 49.

127. V. Frundzhyan, I. Parkhomenko, L. Brovko and N. Ugarova, *J. Dairy Res.*, 2008, **75**, 279.
128. R. Petty, L. Sutherland, E. Hunter and I. Cree, *J. Biolumin. Chemilum.*, 1995, **10**, 29.
129. M. V. Zamaraeva, R. Z. Sabirov, E. Maeno, Y. Ando-Akatsuka, S. V. Bessonova and Y. Okada, *Cell Death Different.*, 2005, **12**, 1390.
130. A.-S. Rhedin, U. Tidefelt, K. Jönsson, A. Lundin and C. Paul, *Leukemia Res.*, 1993, **17**, 271.
131. S. Sharma, M. Neale, F. Di Nicolantonio, L. Knight, P. Whitehouse, S. Mercer, B. Higgins, A. Lamont, R. Osborne, A. Hindley, C. Kurbacher and I. Cree, *BMC Cancer*, 2003, **3**, 19.
132. S. E. Jensen, P. Hubrechts, B. M. Klein and K. R. Hasløv, *Biologicals*, 2008, **36**, 308.
133. M. M. Hoa, K. Markeya, P. Rigsbyb, S. E. Jensen, S. Gairola, M. Seki, L. R. Castello-Branco, Y. López-Vidal, I. Knezevicc and M. J. Corbela, *Vaccine*, 2008, **26**, 4754.
134. H. Höjer, L. Nilsson, S. Ånséhn and A. Thore, *Scand. J. Infect. Dis. (Suppl.)*, 1976, **9**, 58.
135. N. Hattori, M. Nakajima, K. O'Hara and T. Sawai, *Antimicrob. Agents Chemother.*, 1998, **42**, 1406.
136. L. Nilsson, H. Höjer, S. Ånséhn and A. Thore, *Scand. J. Infect. Dis.*, 1977, **9**, 232.
137. L. Nilsson, *Antimicrob. Agents Chemother.*, 1978, **14**, 812.
138. O. Holm-Hansen and C. R. Booth, *Limnol. Oceanogr.*, 1966, **11**, 510.
139. D. M. Karl, *Microbiol. Rev.*, 1980, **44**, 739.
140. H. C. S. Eydal and K. Pedersen, *J. Microbiol. Methods*, 2007, **70**, 363.
141. P. K. Egeberg, in *Proceedings of the Ocean Drilling Program, Scientific Results*, ed. C. K. Paull, R. Matsumoto, P. J. Wallace and W. P. Dillon, Ocean Drilling Program, College Station, TX, 2000, vol. 164, p. 393.
142. D. E. Atkinson, *Ann. Rev. Microbiol.*, 1969, **23**, 47.
143. A. Pradet, *Physiol. Vég.*, 1967, **5**, 209.
144. N. Eisaki, H. Tatsumi, S. Murakami and T. Horiuchi, *Biochim. Biophys. Acta*, 1999, **1431**, 363.
145. T. Sakakibara, S. Murakami, N. Eisaki, M. Nakajima and K. Imai, *Anal. Biochem.*, 1999, **268**, 94.
146. T. Sakakibara, S. Murakami and K. Imai, *Anal. Biochem.*, 2003, **312**, 48.
147. I. Masuda-Nishimura, S. Fukuda, A. Sano, K. Kasai and H. Tatsumi, *Lett. Appl. Microbiol.*, 2000, **30**, 130.
148. A. Lundin, A. Rickardsson and A. Thore, in *Proceedings of the 2nd Bi-Annual ATP Methodology Symposium*, ed. G. A. Borun, SAI Technology Company, San Diego, 1977, p. 205.
149. A. Lundin, U. Karnell Lundin and M. Baltscheffsky, *Acta Chem. Scand., B*, 1979, **33**, 608.
150. J. J. Lemasters and C. R. Hackenbrock, *Biochem. Biophys. Res. Commun.*, 1973, **55**, 1262.

151. J. J. Lemasters and C. R. Hackenbrock, *Methods Enzymol.*, 1978, **57**, 36.
152. A. Lundin, A. Thore and M. Baltscheffsky, *FEBS Lett.*, 1977, **79**, 73.
153. M. Baltscheffsky and A. Lundin, in *Cation Flux Across Membranes*, ed. Y. Mukohaka and L. Packer, Academic Press, New York, 1979, p. 209.
154. M. Baltscheffsky and A. Lundin, in *Cyclotrons to Cytochromes*, ed. N. Kaplan, Academic Press, New York, 1982, p. 347.
155. R. Wibom and E. Hultman, *Am. J. Physiol.*, 1990, **259**, E204.
156. R. Wibom, *Studies of Mitochondrial ATP-production Rate in Skeletal Muscle*, Thesis at the Karolinska Institute (ISBN 91-628-0358-1), Stockholm, 1991.
157. R. Wibom, L. Hagenfeldt and U. von Döbeln, *Anal. Biochem.*, 2002, **311**, 139.
158. K. R. Short, *Am. J. Physiol. Regul. Integr. Comp. Physiol.*, 2004, **287**, R243 (http://ajpregu.physiology.org/cgi/content/full/287/1/R243).
159. B. Näslund, L. Ståhle, A. Lundin, B. Anderstam, P. Arner and J. Bergström, *Clin. Chem.*, 1998, **44**, 1964.
160. M. Ronaghi, S. Karamohamed, B. Pettersson, M. Uhlén and P. Nyrén, *Anal. Biochem.*, 1996, **242**, 84.
161. M. Ronaghi, M. Uhlén and P. Nyrén, *Science*, 1998, **281**, 363.
162. P. E. Stanley, *Anal. Biochem.*, 1971, **39**, 441.
163. S. E. Brolin, E. Borglund, L. Tegner and G. Wettermark, *Anal. Biochem.*, 1971, **42**, 124.
164. S. E. Brolin, C. Berne and U. Isacsson, *Anal. Biochem.*, 1972, **50**, 50.
165. C. Berne, S. E. Brolin and A. Ågren, *Horm. Metab. Res.*, 1973, **5**, 141.
166. B. Anderstam, A. Gutierrez, A. Lundin and A. Alvestrand, *Scan. J. Clin. Lab. Invest.*, 1998, **58**, 89.
167. A. Roda, S. Girotti, S. Ghini and G. Carrea, *Methods Enzymol.*, 1988, **137**, 161.
168. K. Kurkijärvi, P. Turunen, T. Heinonen, O. Kolhinen, R. Raunio, A. Lundin and T. Lövgren, *Methods Enzymol.*, 1988, **137**, 171.
169. S. Ulitzur, in *Handbook of Biosensors and Biochips*, ed. R. S. Marks, D. Cullen, C. Lowe, H. Wetall and I. Karube, John Wiley & Sons, Ltd., Chichester, 2007, p. 143.

CHAPTER 6
Flow-assisted Analysis

ALDO RODA,[a, c] MARA MIRASOLI,[a, c] BARBARA
RODA[b, c] AND PIERLUIGI RESCHIGLIAN[b, c]

[a] Department of Pharmaceutical Sciences, University of Bologna, Via
Belmeloro 6, 40126, Bologna, Italy; [b] Department of Chemistry "G.
Ciamician", University of Bologna, Via Selmi 2, 40126, Bologna, Italy;
[c] National Institute of Biostructure and Biosystems, N.I.B.B., Interuniversity
Consortium, Rome, Italy

6.1 Introduction

Flow-assisted techniques were developed in response to the need for high
sample throughput and automation for quantitative detection of analytes in
real, complex matrices in many analytical fields. Among these techniques, flow
injection analysis (FIA) has emerged as an increasingly used laboratory tool in
chemical analysis, and its employment for online sample treatment and online
measurement in chemical process control is a growing trend.

FIA, which was first introduced by Rutzicka and Hansen in 1975, is a dispersion technique, in which a precise aliquot of sample is injected in a carrier
stream of reagents. The injected sample forms a zone that is transported into a
detector that continuously monitors the change in a sample property and yields
an analytical signal depending on the analyte concentration.[1,2]

The basic advantages of FIA with respect to batch measurements are high
reproducibility, large throughput and reduction of sources of contamination
between consecutive analyses.[3] The result of a determination in FIA depends on
the chemistry applied, reproducibility of operations and controlled sample dispersion connected to the dynamics of the occurring process. These factors,
together with involved online operations of pre-concentration or separation of the
analyte from the matrix, affect the selectivity of flow determinations. The controlled flow conditions of measurement compared to batch measurements may

Chemiluminescence and Bioluminescence: Past, Present and Future
Edited by Aldo Roda
© Royal Society of Chemistry 2011
Published by the Royal Society of Chemistry, www.rsc.org

additionally enhance the analytical signal due to convection, or provide additional kinetic discrimination of interferences. Through the appropriate configuration of a flow system, a multicomponent determination can also be realized.

FIA is then a well-established technique for rapid, automated, quantitative analysis by combining online chemical and physical sample treatment with a range of flow-through detection systems.[4] Among them, the use of chemiluminescence (CL), bioluminescence (BL) and electrogenerated chemiluminescence (ECL) represents a promising approach thanks to simple instrumentation, high sensitivity and low background signal. However, since FIA is not a separative technique, pre- or post-column treatments are frequently required, and thus its combination with separative techniques is a promising approach. Several examples have shown the use of flow injection techniques with chromatography, capillary electrophoresis and multisyringe chromatography.[5]

This chapter describes the basic principles, features and applications of flow-assisted techniques with CL-BL-ECL detection. Moreover, an overview of conceptually different flow-assisted techniques, referred to as field-flow fractionation (FFF), is also presented. Field-flow fractionation methods are separative techniques suitable for the fractionation of high molar weight analytes and, when coupled with CL detection, they have shown interesting features for the development of new assay formats, with advantages such as short analysis time, simple instrumental setup and sensitivity.

6.2 Fundamental and Analytical Setup

6.2.1 Flow Injection Analysis

The first successful flow measurement was achieved by segmenting a flowing stream with the introduction of air bubbles to assure homogeneity of samples and avoid the risk of crossover between consecutive analyses.[6] However, the use of bubbles increases analysis time due to bubble introduction/removal and decreases the reproducibility. The main milestone in the evolution of flow measurements occurred in the mid-1970s by Rutzicka and Hansen, who eliminated the stream segmentation by the implementation of a system with narrow internal diameter tubing and low flow rates. The proposed methodology took the name of flow injection analysis (FIA). The analytical features of FIA rely on a well-defined volume of injected sample, easy operation, low costs and high reproducibility. The fast response of FIA makes analytical information available at real-time, allowing also for the analysis of rapid transformation of target analytes.[7]

FIA is based on the measurement of a transient signal obtained after the injection of a defined volume of sample in a flowing carrier stream. A typical output of FIA analysis is represented by a peak, from which several parameters (height, area, width at fixed height or elapsed time between two points) can be used for quantitative determinations.

During transport to the detector, axial dispersion due to the continuous flow of stream and radial dispersion due to diffusion occur. Since FIA is performed under laminar flow conditions, a concentration gradient is developed, which determines a radial diffusion to equilibrate the different concentrations, thus allowing low sample dilution and a short time between injections.

The dispersion (D) can be defined by Equation (6.1):

$$D = \frac{c_0}{c} \tag{6.1}$$

where c_0 represents the sample concentration in the injection volume and c is the peak concentration at the detector.

The dispersion is affected by: (a) sample volume, since increased injection volumes reduce analyte dilution by carrier; (b) length of tube for the flow system, since short tubes reduce the time available for diffusion and consequently yield low dispersion; (c) flow rate; and (d) tube internal diameter and geometrical properties.[8] Since D represents the degree of sample dilution, its value affects the sensitivity of a FIA analysis. A calibration procedure to determine the value of D with a pure dye can be performed if necessary. Flow injection assays work with a controlled D value, which is selected according to the application and the detector employed. For convenience, sample dispersion has been defined as: (i) limited ($D = 1$–3), characterized by high sensitivity and high sample throughput; (ii) medium ($D = 3$–10); and (iii) large ($D > 10$), characterized by lower sensitivity and sample throughput and used for reactions needing time to equilibrate.[9] Flow injection assays with CL-BL-ECL detection are generally performed under low or medium dispersion.

A modern FIA system usually consists of a multichannel peristaltic pump, an injection valve, a coiled reactor and a detector (Figure 6.1). Additional components may include, for example, a flow through heater, a column for sample extraction and a filter for particulate removal.

Sample injection is performed by means of an injection valve, which allows for a rapid injection of a specified volume of analyte (generally in the range 5–200 μL) into the flow stream without flow perturbations.

Tubes with small diameters (<0.1–1 mm id) for carrier and reagent handling and reactors coil (<50 cm long), used to improve mixing, are employed. Generally, FIA provides high analytical throughput, usually ranging around 2–3 samples per minute.

6.2.2 Sequential Injection Analysis

Thanks to the advent of computer-controlled devices, at the beginning of the 1990s, an evolution of FIA based on a different approach and named sequential

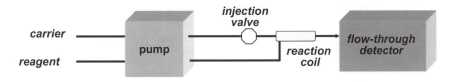

Figure 6.1 Instrumental setup of a FIA system.

injection analysis (SIA) was proposed.[10] While in a conventional flow injection technique a sample is injected into a flowing carrier and auxiliary reagents are merged with it on the way to the detector, in SIA the use of a multiport selector valve allows us to handle several flows with much higher versatility. Figure 6.2 shows a scheme of a conventional SIA system. In SIA, a well-defined volume of sample is aspirated through a port of the valve in a holding coil positioned between the valve and the propulsion system. Fixed volumes of one or more reagents are sequentially aspirated in the holding coil, forming a zone that is in contact with the previously aspirated sample. The stacked zones are then propelled through a reaction coil and the product is formed on the overlapped regions of the sample and reagent zones.[11]

Owing to the use of robust syringe pumps and computer control, the advantages of SIA over the more traditional FIA are low reagent consumption and production of far less waste. Sequential injection analysis is then the recommended approach for online and remote-site monitoring, where robustness and low reagent use is critical. One disadvantage of SIA is its tendency to run slower than FIA because of the periodic refill of the liquid and the stacking of the entire set of reagents in the holding coil.

6.2.3   Multisyringe Flow Injection Analysis and Multicommutation in Flow Analysis

A multisyringe flow injection analysis (MSFIA) was first described in 1999 as a novel multichannel technique combining the advantages of FIA and SIA.[12] By use of parallel moving syringes as liquid drivers, MSFIA overcomes the shortcomings of peristaltic pumping, such as pulsation and required recalibration. As in SIA, flow rates and propelled volumes are precisely known and defined by software-based remote control. The potential of MSFIA may lie in the simultaneous or multi-parameter determination and enhanced sensitivity due to confluent mixing of reagent and sample as in FIA. However, periodically refilling of the syringe involves a certain delay time, resulting in slightly lower sampling rates than in FIA.[13]

Multisyringe flow injection techniques have proven to be an effective tool for miniaturization and automatization. Figure 6.3 shows a scheme of MSFIA instrumental setup. The device is based on a multiple-channel piston pump containing up to four syringes, which are moved by means of a

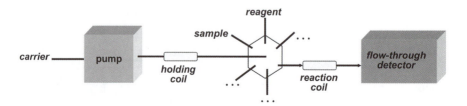

Figure 6.2   Instrumental setup of a SIA system.

(a)

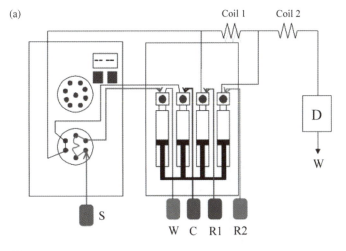

(b)

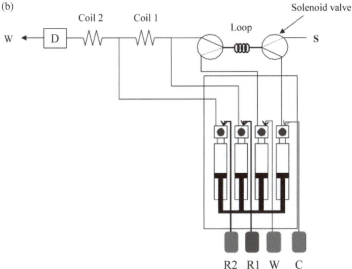

Figure 6.3 Example of instrumental configuration with MSFIA: (a) Multisyringe piston pump burette. S: sample, C: carrier, R1 and R2: reagents, D: detector; W: waste. (b) Volume-based multisyringe flow setup with solenoid commutation valves for injection. Reproduced from reference 13 with permission from Elsevier.

computer-controlled motor. Each head of the syringe is connected with a commutation valve, connecting the syringe either to the manifold lines or to the solution reservoir. Syringes of different volumes can be employed to create specific flow rates for each reagent.

A multicommutation approach was also applied to flow analysis. Three- or two-way valves for reagents/samples/carrier delivery operating at different time schedules can be placed in serial or in a centered position (Figure 6.4). Flow is

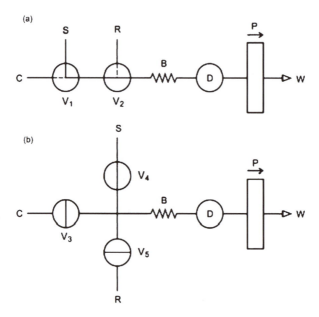

Figure 6.4 A design for multi-commuted flow systems with serial (a) or centered (b) valves. S = sample; C = carrier stream; R = reagent; B = reactor; D = detector; P = pump; W = waste; Vi = valves. Reproduced from reference 15 with permission from Elsevier.

aspirated through the channel by means of a syringe or peristaltic pump placed after the detection system. Different flow paths can be established by modifying the valves position and time course, with a resultant system that is simple, versatile and easy to control.

The use of different computers to manage all the operations allows us to improve performance.[14,15] The use of solenoid valves of small size allows the easy, complete automation of the process with low sample and reagent consumption and it permits the development of compact, integrated systems and of portable equipment for on-site analyses.

All present flow techniques can be combined (dual SIA systems, SIA-MSFIA systems) to achieve higher sampling throughput and high versatility prior to the analyte detection.

6.3 Flow Injection Analysis with Bioluminescence, Chemiluminescence and Electrogenerated Chemiluminescence Detection

6.3.1 Fundamentals

The "marriage" of flow-assisted techniques with CL-BL-ECL detection is a powerful means of exploiting the advantages of both systems. Indeed, the specific features of CL-BL-ECL detection, such as instrument simplicity due to the absence of a light excitation source, low background signals, high detectability and wide linear dynamic ranges, make them particularly attractive for automated flow techniques. On the other hand, analytical CL-BL-ECL measurements are very sensitive to various experimental factors; reproducibility of experimental conditions is crucial to reach high precision. High reproducibility is indeed offered by the ability to automate flow-assisted techniques to provide rapid, highly reproducible and well-controlled mixing of the reagents and, if necessary, online generation of unstable reagents.

The use of CL detection has also some shortcomings, such as the difficulty in handling reactions with complex kinetics and the limited selectivity of CL, because a class of similar compounds usually elicits a positive response, rather than a single analyte. The selectivity of the CL analysis, which is particularly important when real sample matrices are analyzed, can be improved by employing online analyte extraction procedures (*e.g.*, by means of solvent extraction in micellar aqueous–organic systems or by employing packed molecular imprinted polymer particles) or substrate-specific enzymes.

Automated flow techniques can be coupled with CL detection by employing either a direct or an indirect approach. In direct methods the CL signal is emitted due to the interaction between the analyte and the CL reagent (usually a redox reaction); in indirect methods the analyte acts as an inhibitor or sensitizer influencing the intensity of a CL reaction (typically the oxidation of luminol, lucigenin, lophine, sulfite or peroxyoxalates).[16] If the mixing of reagents takes place very close to the photodetector window, even fast CL reactions can be recorded. Since the timing of mixing is highly reproducible, it is unnecessary to measure the whole emission profile.

There are few flow-assisted techniques with BL detection,[17–19] and most often they are referred to coupled enzymatic reactions or immunoassays performed in flow manifolds, rather than FIA applications.[20,21] Such subjects are treated extensively in Chapters 5 and 8 and will not be reported here.

Electrochemiluminescence (ECL) is a chemiluminescent electron-transfer reaction in which the reactants are generated electrochemically at the surface of an electrode. When flow-assisted techniques are coupled to ECL, reagents are usually generated *in situ* and the ECL reaction takes place at the surface of an electrode as a result of an applied potential. This setup usually provides high detectability and the possibility of controlling the reaction using the applied potential, but it suffers from electrode fouling, a narrow linear range due to the small area of the electrode, and complicated design of the flow cell.

6.3.2 Instrumental Setup

Flow-assisted techniques are particularly suited to accurately monitoring transient light emission and to correlating it to the analyte concentration, due to the rapid and reproducible mixing of sample and reagents in close proximity to the detector. To ensure high detectability, crucial factors are mixing efficiency and geometry of the detection cell. Poor mixing causes a large consumption of reagents or a loss of sensitivity, while the size and shape of the detection cell regulate the duration and magnitude of the signal, as well as light collection efficiency.

Each CL reaction, which starts as soon as the analyte mixes with the reagents flow, presents a typical emission profile, reaching the maximum signal intensity after a few seconds to several minutes. Instrumental and experimental parameters (*e.g.*, volume of the detection cell, flow rates, length and diameter of the tubing between the mixing point and the detector flow cell) must be optimized so that the CL solution flows through the detection cell during the time of maximum light emission. If the kinetics of the reaction is very fast, mixing can be performed immediately before or even in front of the detector. On the other hand, in the case of slower kinetics, a delay coil, or even the stopped-flow technique, can be employed to allow the reaction to proceed extensively before entering the detection cell.

6.3.3 Detectors and Flow Cells

Chemiluminescence detection in a flow-assisted technique is typically performed by employing a flow-through transparent cell placed in such a way as to expose its maximum surface in close proximity to, or even in contact with, the detector window, which is usually a photomultiplier tube (PMT). Even if improvements in photodiodes have made them suitable for some applications, PMTs are still the detectors of choice for measuring extremely low levels of light.

As in any other luminescence detection, external light interference must be excluded from the detector, which is not always straightforward, since the tubes connected to the detection flow-through cell can act as optical fibers, leading light directly into the cell. This limits the number of reagent and/or carrier streams and the instrumental design.

Since Rule and Seitz first proposed the use of CL detection in FIA in 1979,[22] various configurations of flow-cells for CL/BL measurements have been proposed.

Most commonly, the flow cell is a flat spiral of glass or plastic tubing placed in front of the detector, in which solutions enter immediately after merging at a T- or Y-shaped junction. This configuration has, however, some drawbacks, since the fraction of solution in view of the detector is low and the restrictions inherent in the tube flexibility reduce the possible geometries. In addition, polymer tubing is most often translucent rather than transparent. The use of refractive surfaces, lenses and optical fibers to increase the light collection efficiency has also been adopted.[23]

Detection cells with spiral channels have been also created by etching or machining channels into polymer blocks or chips.[24-27]

Recent studies have shown that the design of channels that produce sudden changes in the direction of flow, such as serpentine channels, increases the signal detectability, especially in the case of fast reactions, mainly because of an enhanced mixing efficiency; this aspect will also be reported in Section 6.4 (miniaturized flow systems).

As an alternative, the fountain flow cell has been proposed in order to bring a high fraction of the solution in contact with a flat surface facing the photodetector. In this cell, the reacting mixture enters an open, shallow cylindrical space *via* a central inlet, directed perpendicularly to a flat optical surface and, after forming a thin, flat, radially symmetric flow path, it drains into a ring-shaped well around the edge, which contains the outlet hole.[28]

Various flow cell designs have been proposed in attempts to increase analytical performances and/or device simplicity and portability, some of which offer the advantage of great versatility in their applicability to various detection techniques,[29–31] such as chemiluminescence, spectrophotometry, laser-induced fluorescence and electrochemistry, by the use of flow cells combining fiber optics, reflecting surfaces, flow channels and membranes.[32,33] Figure 6.5 gives examples of flow cell designs for CL measurement.

Flow cells for ECL measurements also need to contain the working, counter and reference electrodes, placed in such a way as to ensure proper positioning into the flow stream. In addition, a potentiostat is required to control the cell potential during ECL measurements.

6.3.4 Applications

6.3.4.1 Pharmaceutical Analysis

The main features of FIA and related techniques, *i.e.*, high sample throughput, high precision, possibility of automation and low sample and reagents consumption, are particularly advantageous in pharmaceutical analysis, especially in quality control of pharmaceutical formulations or pharmacokinetics and pharmacodynamic studies that require a high number of analyses. Various FIA, SIA and MSFIA assays with CL detection have been developed for pharmaceutical analysis applications, as recently reviewed.[16,33] Employing this technique, analytes are quantified in drug formulations or biological fluids, typically with limits of detections in the nM or $ng\,mL^{-1}$ range and analytical throughputs of about 100 samples per hour. Table 6.1 reports selected applications published since 2007.

In a multicommutation approach, additional components such as mini-column, reactors, cartridges, *etc.* can be inserted into the analytical flow path. Applications to pharmaceutical analysis are shown using different strategies such as in-line anion exchange column or binary sampling injection consisting of inserting sequential aliquots of sample and reagents into the analytical path, managing small volumes with high precision. Hence, these systems would be especially appropriate for quality control in pharmaceutical companies.

Selected examples are reported in Table 6.1. The main handicap of multicommutated systems is the lack of commercially available instrumentation and software.

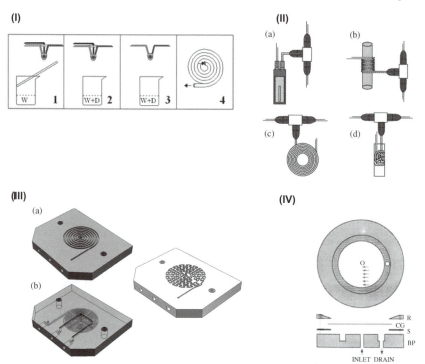

Figure 6.5 Examples of flow-cells. **(I)** Droplet detectors: (1–3) different configurations, (4) conventional spiral-shaped flow cell. Reprinted from reference 30 by permission of The Royal Society of Chemistry. **(II)** Assemblies of flow-cells: (a) quartz flow cell, (b) helix cell, (c) spiral cell and (d) bundle cell. Reproduced from reference 31 with permission from Elsevier. **(III)** Three-dimensional model of a spiral-configuration flow-cell: (a) top face, (b) bottom face and chip with serpentine channel configuration. Reprinted from reference 27 by permission of The Royal Society of Chemistry. **(IV)** Fountain cell: a circular Teflon base plate (BP) containing an inlet channel surrounded by a ring-shaped outlet well with a drain. The flow path is formed by a circular Teflon spacer (S), a cover glass (CG) with a ring (R). Reproduced from reference 28 with permission from American Chemical Society.

6.3.4.2 Applications to Biological Fluids

Flow injection techniques also find many applications in assays on biological fluids; the determination of amino acids, proteins or other analytes in biological fluids such as serum, urine or extracts from biological tissues has been performed based on inhibition or catalytic effects of analytes on a specific CL reaction. Examples of application for FIA, SIA and MSFIA have been reviewed.[34,35]

Even multicommutation systems have been used extensively for the analysis of animal and human biological fluids. Table 6.1 reports recent selected examples.

Table 6.1 Application to drug formulations and biological fluids analysis.

Analyte	Matrix	Detection system	LOD	Linear range	Remarks	Throughput (samples per h)	Ref.
Promazine hydrochloride	Human serum, drug formulations	$Ru(phen)_3^{2+}$ Ce(IV) in sulfuric acid medium	$0.012 \mu g\,mL^{-1}$	CL intensity was proportional to the concentration of the drug in solution over the ranges 0.020–0.32 and 0.32–32 $\mu g\,mL^{-1}$	FIA	100	36
Indomethacin	Pharmaceutical formulations and biological samples	$KMnO_4$ in polyphosphoric acid medium with 4% formaldehyde	$6.0 \times 10^{-10}\,g\,mL^{-1}$	1.0×10^{-9}– $1.0 \times 10^{-6}\,g\,mL^{-1}$	FIA	100	37
Retinol, derivatives and α-tocopherol	Pharmaceutical preparations	$KMnO_4$ formaldehyde in acidic medium	In the range 5.0×10^{-9}– $1.0 \times 10^{-7}\,mol\,L^{-1}$	5.0×10^{-8} to 2.5– $5.0 \times 10^{-6}\,mol\,L^{-1}$ for all-*trans*-retinol and all-*trans*-retinal Values for retinoic acid, retinyl acetate, retinyl palmitate and α-tocopherol are also reported.	FIA	100	38
Ciprofloxacin, enrofloxacin	Pharmaceutical formulations	$Ru(bpy)_3^{2+}$	Ciprofloxacin: $0.2\,\mu mol\,L^{-1}$, enrofloxacin: $0.04\,\mu mol\,L^{-1}$	Ciprofloxacin: 0.200–$100\,\mu mol\ L^{-1}$ enrofloxacin: 0.100– $25.0\,\mu mol\,L^{-1}$	FIA Other quinolone and fluoroquinolone antibiotics were investigated	120	39
Puerarin	Pharmaceutical formulations and human urine	Glyoxal $KMnO_4$ in a sulfuric acid medium	$3.0\,ng\,mL^{-1}$	$10.0\,ng\,mL^{-1}$– $7.0\,\mu g\,mL^{-1}$	FIA	120	40

Table 6.1 (*continued*)

Analyte	Matrix	Detection system	LOD	Linear range	Remarks	Throughput (samples per h)	Ref.
Clindamycin	Pharmaceutical preparations and human urine	Luminol/$K_3Fe(CN)_6$ in alkaline medium	$0.2\ ng\ mL^{-1}$	$0.7-1000\ ng\ mL^{-1}$	FIA	120	41
Antioxidants	Pharmaceutical preparations	Luminol/oxidant (either O_2^- or NO)	L-Ascorbic acid: $9.4\ \mu g\ mL^{-1}$ IC_{50} against NO, $47.5\ \mu g\ mL^{-1}$ IC_{50} against O_2^-. Values for other antioxidants are reported	L-Ascorbic acid: $3.5-52.8\ \mu g\ mL^{-1}$ against NO, $17.6-211.4\ \mu g\ mL^{-1}$ against O_2^-. Values for other antioxidants are reported	SIA and FIA Quasi-simultaneous determination of antioxidative activities against O_2^- and NO	30	42
Chlorpheniramine	Pharmaceutical preparations	$Ru(phen)_3^{2+}$ in sulfuric acid medium	$0.04\ \mu g\ mL^{-1}$	$0.1-10\ \mu g\ mL^{-1}$	SIA	N.r.[a]	43
Salicylic acid	Pharmaceutical preparations	Oxidation with $KMnO_4$	$0.3\ \mu g\ mL^{-1}$	$1-30\ \mu g\ mL^{-1}$	Multicommutation with anion exchange	60	44

Analyte	Matrix	Reaction	Detection limit	Linear range	Technique		Reference
Clomipramine	Pharmaceutical preparations	Oxidation of sulfite by Ce(IV)	0.7 µg mL⁻¹ for 10-s sampling time	2.5–10 mg L⁻¹ for 10-s sampling time	Multicommutation with binary reagent approach	19–32	45
H$_2$O$_2$	Pharmaceutical preparations	Luminol/H$_2$O$_2$ catalyzed by hexacyanoferrate(III)	1.8 µmol L⁻¹	2.2–210 µmol L⁻¹	Multicommutation	200	46
Propranolol hydrochloride	Pharmaceutical preparations	KMnO$_4$ in acidic medium	N.r.	20–150 mg L⁻¹	Multicommutation	27	47
5-Aminosalicylic acid	Pharmaceutical preparations	KMnO$_4$	300 ng mL⁻¹	1–20 µg mL⁻¹	Multicommutation	17	48
Glucose	Animal serum	Luminol/H$_2$O$_2$ in the presence of hexacyanoferrate(III)	12.0 mg L⁻¹	50–600 mg L⁻¹	Multicommutation with glucose oxidase immobilized on porous silica beads packed in a microcolumn	60	49
Karbutilate	Human urine	Photodegradation	10 µg L⁻¹	20 µg L⁻¹–20 mg L⁻¹	Multicommutation	17	50
Herbicide benfuserate		Photo-induced CL	0.1 µg L⁻¹	1 µg L⁻¹–4 mg L⁻¹	Multicommutation	22	51

[a]N.r.: not reported.

6.3.4.3   *Environmental Applications*

Flow injection methodologies with CL-ECL-BL detection have been applied to a wide range of analytes of environmental interest on different matrices such as water, wastewater or soil.[34,35,52]

A wide range of applications of multicommutation in flow analysis has been applied to environmental problems, *e.g.* to detect pesticides or metal, even for multicomponent analysis. Table 6.2 gives selected recent applications.

6.3.4.4   *Application to the Analysis of Food and Beverage*

In recent years the field of food and drink analysis has become an area of great interest. Analytical applications of flow injection techniques are summarized in many reviews, presenting examples of methods based on CL detection for the determination of analytes such as ascorbic acid, sulfite and carbohydrates in food and wine, and BL detection based on reactions involving ATP detection by means of luciferase for bacteria-contamination analysis.[34,52,60,61]

Multicommutation analysis employing columns packed with functionalized beads have been used for food analysis. Table 6.3 gives selected recent applications.

6.3.4.5   *Applications to Multicomponent Analysis*

Although flow injection analysis has been applied mainly to single-component analysis, its general features of low reagent consumption, great versatility and ability to accommodate different operations have encouraged the development of some multicomponent determinations in a single analysis with CL detection. Different strategies can be exploited to perform multicomponent analysis in flow injection techniques, ranging from the use of multichannel manifolds, the direct incorporation of solid-phase extraction or separation columns, the use of specific detectors placed in series, to data processing with multivariate chemometric tools.

An assay has been proposed that combines the advantages of the kinetic methods that use stopped-flow mixing technique with high precision and minimized interferences. Two examples involved the use of chemometric treatments. Table 6.4 reports details of analytical performances.

6.4   Miniaturized Flow Injection Analysis

6.4.1   Lab-on-valve and Lab-on-chip Systems

The miniaturization of flow-assisted techniques is recognized as one of the most important trends in analytical chemistry. Miniaturization of flow injection techniques has been pursued to add features, such as low reagent consumption and reduced analysis times, to their characteristic reliability, rapidity and robustness. This has allowed accommodation of various assays even for

Table 6.2 Environmental applications.

Analyte	Matrix	Detection system	LOD	Linear range	Remarks	Through-put (samples per h)[a]	Ref.
Hydrazine	Water	Reaction with either sodium dichloroisocyanurate (SDCC) or trichloroisocyanuric acid (TCCA) in alkaline medium	3×10^{-7} mol L^{-1}	5×10^{-7} – 6×10^{-5} mol L^{-1}	FIA	65	53
Fenitrothion	Water	Luminol/H$_2$O$_2$ in the presence of sodium chloride as enhancer	4×10^{-9} g mL^{-1}	1×10^{-8} – 2×10^{-6} g mL^{-1}	FIA	N.r.	54
Aldicarb	Water	Photodegradation	0.069 μg L^{-1}	2.2–100.0 μg L^{-1}	Multicommutation	17	55
Fluometuron	Water	Photodegradation	N.r.[a]	0.1–5 mg L^{-1}	Multicommutation	16	56
Diphenamid	Water	Photodegradation	1 μg L^{-1}	0.005–20 mg L^{-1}	Multicommutation	20	57
Acrolein	Soil	Photodegradation	0.1 μg L^{-1}	5–100 μg L^{-1}	Multicommutation	52	58
Cr/Co	Water	Luminol/H$_2$O$_2$	0.2 μg L^{-1}	0.5–2 μg L^{-1}	Multicommutation with stopped-flow manifolds for single or simultaneous analysis	N.r.	59

[a]N.r.: not reported.

Table 6.3 Applications to food and beverage.

Analyte	Matrix	Detection system	LOD	Linear range	Remarks	Throughput (samples per h)[a]	Ref.
Parathion	Rice	Luminol/H_2O_2	$0.008\,mg\,L^{-1}$	0.02–$1.0\,mg\,L^{-1}$	FIA	N.r.	62
Patulin	Apple juice	Luminol/H_2O_2	$0.01\,ng\,mL^{-1}$	0.05–$80\,ng\,mL^{-1}$	FIA	Complete analysis in 25 s	63
Ethanol	Wine	Peroxide developed after enzymatic reaction	0.3% (v/v)	2.5–25% (v/v)	Multicommutation with alcohol oxidase immobilized on beads	23	64
Lactic acid	Yoghurt	Luminol/H_2O_2 catalyzed by hexacyanoferrate(III) after enzymatic reaction with lactate	$1.2\,mg\,L^{-1}$	10–$125\,mg\,L^{-1}$	Multicommutation with lactate oxidase immobilized on beads	55	65

[a]N.r.: not reported.

Table 6.4 Multicomponent applications.

Analyte	Matrix	Detection system	LOD	Linear range	Remarks	Ref.
Citrate and pyruvate	Human urine	Ru(bpy)$_3^{3+}$	Citrate: 0.1 µg mL^{-1} Pyruvate: 0.3 ng mL^{-1}	Citrate: 0.38–38 µg mL^{-1} Pyruvate: 8.7–1300 ng mL^{-1}	Stopped-flow	66
Co^{2+} and Cu^{2+}	Water	Luminol/H$_2$O$_2$	Co^{2+}: 0.08 ng mL^{-1} Cu^{2+}: 6 ng mL^{-1}	Co^{2+}: 0.0002–0.4 µg mL^{-1} Cu^{2+}: 0.02–20 µg mL^{-1}	CL intensity was measured and recorded at different reaction times and the obtained data were processed by the chemometric approach of partial least-squares	67
Rifampicin (RFP) and isoniazid (INH)	Pharmaceutical preparation	Oxidation with alkaline N-bromosuccinimide	RFP: 0.005 µg mL^{-1} INH: 0.03 µg mL^{-1}	RFP: 0.01–10 µg mL^{-1} INH: 0.06–40 µg mL^{-1}	Different kinetic spectra of the analytes in their CL reaction with data processed chemometrically by use of an artificial neural network	68

limited-reagent assays, handling of high hazardous chemicals and critical waste products. Ultrasensitive CL-BL-ECL is a suitable detection technique for miniaturized flow assisted systems.

The sequential injection lab-on-valve (LOV) concept was the first example described recently for downscaling the level of fluidic manipulation.[69] In a LOV system different manifold components are integrated in a microconduit incorporated on top of a conventional multiposition valve. Usually, in a LOV system a high-precision syringe pump is employed to deliver the reagents and samples with a programmable flow to a central flow-through channel used to connect all the other flow-through ports. In one of the valve ports, a multi-purpose flow cell is permanently incorporated to allow different microfluidic manipulations such as sample dilution, sample and reagents mixing and incubation, and real-time detection by means of a side-on photomultiplier tube. The entire system is compact and software-controlled.[70,71] Figure 6.6 shows a typical LOV system.

The LOV system has been used to facilitate real time detection; sensitivity can be increased also for slow kinetics reactions by means of the stopped-flow technique.

The application of LOV for bead injection to perform online enrichment and separation of trace-level analytes has also been developed. In this case, the central part of the system incorporates microcolumns where sorbent beads are aspirated and trapped while flow passes freely through the column with high reproducibility for each run.

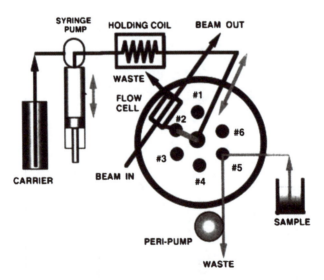

Figure 6.6   A typical LOV system. The central sample processing unit, integrated with a flow cell, is mounted on a top of a six-position valve. In the Figure, the flow through cell is working in absorbance mode. Reprinted from reference 69 by permission of The Royal Society of Chemistry.

Alternative approaches towards the miniaturization of flow systems are represented by micro-machined lab-on-chip (LOC) devices characterized by the design of a fixed architecture onto a single chip to allow predetermined chemistry, with the advantages of rapid separation of complex mixtures, portability, low reagent/solvent consumption and low cost.[72-74] Low sample volumes, ranging from nano- to picoliters are involved, thus enabling single-molecule detection techniques. The use of microcolumns entrapping biospecific reagents (*e.g.*, enzymes, molecular imprinted polymers) has also been exploited in order to enhance selectivity.

6.4.2 Miniaturized System with CL-ECL-BL Detection

6.4.2.1 LOV System

The use of CL detection in LOV systems was proposed with the advantages of improving the detection limit and reducing reagent and sample consumption with respect to SIA.[75] The light emission is usually monitored by placing in direct contact a side-on photomultiplier tube to the side of the multipurpose flow-cell.

6.4.2.2 LOC System

An important improvement in LOC systems for flow injection analysis results from the recent development of microvalves and the use of injection pumps with accurate time control. In fact, the commonly employed electro-osmotic flow to drive sample and reagents zones finds limited application with CL detection, because of the gaseous products involved in many CL reactions. Various schemes to couple microfluidic devices with CL detection have been proposed. Optimization of cell materials and geometry has been performed to maximize emission and detection of light while the reaction mixture passes through the cell. To significantly improve performances offered by conventional detection systems (such as the coiled-tube flow cell placed against a PMT, where reagent and sample solutions flow after merging into a junction) different micro-chip configurations offering high proximity between reaction area and detector have been studied to obtain high sensitivity. As an example, a three-sampling integrated structure was designed in a poly(methyl methacrylate) chip incorporating (a) different injection ports for reagents, solution and sample controlled by injection pumps, (b) a mixing area and (c) a 1.2-µL transparent reaction area from where light emission is detected by means of a PMT.

This micro-system for flow analysis has shown interesting results for different applications to clinical analysis, food control, pharmaceutical studies and *in vivo* determinations.[76] It was also demonstrated that the introduction of changes in the direction of flows by means of serpentine or sinusoidal configurations on a chip increased mixing efficiency; moreover, the use of an opaque chip allowed enhancement of the transmission of light to a PMT.[27]

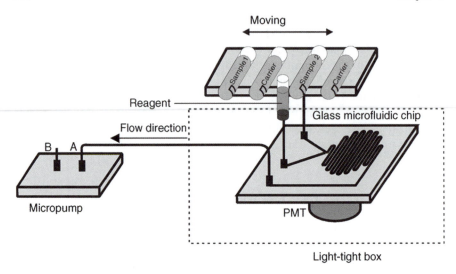

Figure 6.7 Microfluidic injection system based on micropump operating on eva-
poration and capillary effects and glass chip. Reproduced from reference
74 with permission from Elsevier.

Microfluidic devices equipped with a micro-scale detection window able to
detect rapid CL reaction, based on the principle of liquid core waveguide, have
also been proposed. The performance of a micropump operating on evapora-
tion and capillary effects, developed for microfluidic (lab-on-a-chip) systems,
has been studied by employing it as the fluid drive in a microfluidic flow
injection system, with CL detection of luminol. Figure 6.7 reports a scheme of
the microfluidic system.

The micropump has a simple structure, small dimensions, low fabrication
cost and stable and adjustable flow-rates during long working periods.[74]

Electrogenerated chemiluminescence (ECL) provides another effective
detection tool in miniaturized systems because of its stability and reproducibility. A microcell for ECL detection can be mounted close to a photomultiplier
tube (PMT), or bonded to a silicon photodiode, thus becoming an intrinsic
component of the device itself.[77] An integrated silicon ECL chip was constructed by Fiaccabrino *et al.* based on a device including two identical cells,
each with a gold interdigitated microelectrode array and a photodiode in a
single $5 \times 6 \, mm^2$ silicon chip.[78] Figure 6.8 gives a schematic view of the integrated probe.

A similar cell was constructed with silicon/SU-8, a photosensitive epoxy resin
characterized by low optical absorption in the near-UV and good mechanical
properties.[79] A poly(methyl methacrylate) (PMMA)-based micro flow ECL cell
has also been proposed to detect $Ru(bpy)_3^{2+}$ at a rather low concentration in a low
sample volume.[80] A glass miniaturized ECL flow-through cell has been developed
where flow can be continuously delivered under hydrostatic pressure, thus eliminating the need for pumps and resulting in a simplified experimental setup.[81]

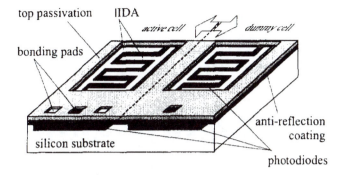

Figure 6.8 Schematic view of a miniaturized probe for electrochemiluminescence measurement which integrates electrode transducer and photodetector in a silicon chip. Reproduced from reference 78 with permission from Elsevier.

6.4.3 Applications

LOV and LOC systems for flow analysis have been coupled to CL and ECL detection for several applications, such as analysis of biological fluids, environmental samples and foodstuff.

6.4.3.1 LOV Systems

To obtain improved sensitivity and limit of detection, a LOV system equipped in one port with a bismuthate-immobilized microcolumn for the oxidation of KBr inserted through a different port has been developed. The bromine generated was used for the CL-reaction of analyte. The system was employed for the determination of tetracycline in milk. In this system a Z-type cell was employed, which is characterized by simple fabrication and high versatility, being compatible with various detection modes. The main limitation of the method resides in the necessity of an external and manually operated sample pretreatment procedure based on solid-phase extraction.

A fully automated, portable LOV instrument with CL detection for *in situ* real-time monitoring of iron concentration in sea water samples has also been developed.

Selected applications of LOV employing CL detection are reported in Table 6.5.

6.4.3.2 LOC Systems

Table 6.5 reports examples of the application of LOC systems with CL detection in many fields. To reach high detectability, integrated structures to improve mixing and efficiency of CL reaction were designed, different geometries were adopted and micro-scale detection systems were proposed.

Table 6.5 Application of miniaturized flow injection analysis.

Analyte	Matrix	Detection system	LOD	Linear range	Remarks	Throughput (sample per h)[a]	Ref.
Tetracycline	Milk	Bromine *in situ* generated/H_2O_2	$2.0\ \mu g\ L^{-1}$	$6.0–10000\ \mu g\ L^{-1}$	LOV: immobilized microcolumn	120	82
Iron	Seawater	Luminol oxidation	$21\ pmol\ L^{-1}$	$21–2000\ pmol\ L^{-1}$	LOV	60-s sample load time	83
Uric acid	Human serum urine	Luminol/$Fe(CN)_6^{4-}$	$0.5\ mg\ L^{-1}$	$0.8–30\ mg\ L^{-1}$	LOC: three-sampling integrated structure in poly(methyl methacrylate)	N.r.	73
Benzoyl peroxide	Flour	Luminol oxidation	$4 \times 10^{-7}\ g\ mL^{-1}$	$8 \times 10^{-7}–1 \times 10^{-4}\ g\ mL^{-1}$	LOC: three-sampling integrated structure in poly(methyl methacrylate)	N.r.	76
Chromium(III)	Rabbit blood – *in vivo* determination	EDTA-luminol/H_2O_2	N.r.[a]	N.r.[a]	LOC: three-sampling integrated structure in poly(methyl methacrylate); flow rate 5 μL min^{-1}; dyalisate sample was 400 nL	N.r.	84
Terbutaline	Human serum	Luminol/$Fe(CN)_6^{4-}$	$4.0\ ng\ mL^{-1}$	$8.0–100\ ng\ mL^{-1}$	LOC online enrichment with molecularly imprinted polymer	10	85
1-Aminopyrene	Acetonitrile	1,1′-Oxalyldiimidazole derivate CL reaction	0.05 fmole injection	0.4–2.3 fmole injection	LOC: liquid core waveguide	N.r.	86

Analyte	Sample/matrix	Reagent	LOD	Linear range	System	Remarks	Ref.
Morphine	Acidic aqueous polyphosphate solutions	$KMnO_4$	N.r.	N.r.	LOC: sinusoidal/serpentine geometry Concentrations in the range 1×10^{-10}–1×10^{-5} mol L^{-1} were tested	N.r.	27
Luminol	Carbonate buffer	$Fe(CN)_6^{4-}/H_2O_2$	N.r.	0.2–1.4 mmol L^{-1}	LOC evaporation and capillary effect as pumping system	Samples injected for 10 s at 110 s intervals	74
Ru(byp)$_3^{2-}$ Codeine	Phosphate buffer Pharmaceutical preparations	Tripropylamine Ruthenium complexes	0.5 µmol L^{-1} 100 µmol L^{-1}	0.5–50 µmol L^{-1} 100 µmol L^{-1}– 2 mmol L^{-1}	LOC-ECL LOC-ECL	N.r. N.r.	78 79
Glucose	Phosphate and carbonate buffer	Luminol/H_2O_2	50 µmol L^{-1}	50–500 µmol L^{-1}	LOC-ECL with immobilized enzyme	N.r.	79
Oxalate, tripropylamine, proline	Phosphate buffer	Ru(bpy)$_3^{2+}$	Oxalate: 4.4×10^{-8} mol L^{-1} Tripropylamine: 1.5×10^{-8} mol L^{-1} Proline: 3.1×10^{-8} mol L^{-1}	N.r.	LOC-ECL-glass chip	N.r.	81
DNA strands	Phosphate buffer	Ru(bpy)$_3^{2+}$	40 fmol LOD corresponds to 6×10^8 molecules – 30 thermal PCR cycles	N.r.	LOC-ECL-glass chip	N.r.	87

^{a}N.r.: not reported.

Integrated silicon ECL chips were studied for the optimization of detection using ruthenium complexes, even with the immobilization of biospecific reagents, to improve selectivity.

Example applications of ECL detection to LOC systems are reported in Table 6.5.

6.5   Field-flow Fractionation

6.5.1   Principle and Instrumental Setup

Among flow-assisted techniques, field-flow fractionation (FFF) has emerged recently. In FFF, separation is performed by exploiting an external field perpendicularly applied to a mobile phase flow. Field-flow fractionation is ideally suited to separating macromolecules, nano-sized and micro-dispersed particles of any origin, and, among samples of biological origin, proteins, protein complexes, microorganisms and cells.[88,89] A FFF fractionation is performed in an empty capillary channel where a sample zone is injected and sample components are driven by the mobile phase toward the channel outlet, connected to a flow-through detector. The main feature of FFF lies in the absence of a stationary phase, which results in a "soft" fractionation mechanism, offering several advantages, such as the possibility of maintaining analytes in their native form after fractionation, reduction of sample carry-over, high biocompatibility and easy application even to complex matrices.

The mobile phase flow generates a laminar flow within the capillary channel. Sample components differing in molar mass, size and/or other physical properties are driven by the applied field into different velocity regions within the parabolic flow profile of the mobile phase across the channel. Analytes are then carried downstream through the channel at different speeds, and reach the outlet after different retention times. Fractionation mechanisms in FFF are summarized in Figure 6.9.

Various fields (such as gravitational and multi-gravitational, hydrodynamic, electrical, thermal or magnetic field) can be applied to structure the separation, giving rise to different FFF variants characterized by particular channel design and potential applications.

The use of FFF has interesting perspectives in many fields, ranging from environmental, food analysis and biotechnology, for biophysical sample characterization and fractionation.[90] However, the integration of FFF with other orthogonal highly sensitive analytical methods and the design of new analytical formats for FFF-based assays soon showed appealing possibilities, with the aim of substantially amplifying the analytical information. Coupling with multi-angle laser scattering (MALS) or luminescence detection, flow cytometry (FC) and soft-impact mass spectrometry have proven to be promising in many applications, from the biophysical characterization of bio-nanoparticles and of very/ultra-large proteins and protein aggregates under native conditions to pre-MS sample separation in proteomics, from the

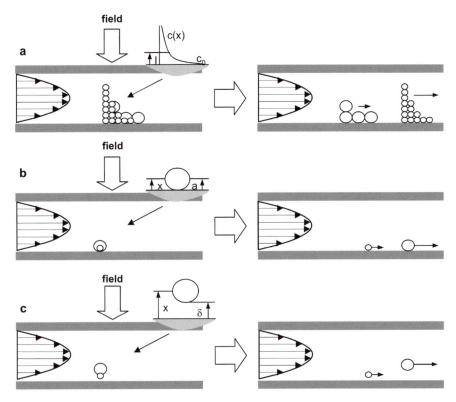

Figure 6.9 Fractionation principles of FFF. (a) Normal mode for macromolecules and submicrometer particles. The field creates a concentration gradient that causes sample diffusion away from the wall. The lower the molar mass or size of the sample component, the greater the component cloud elevation, the deeper the cloud penetration into the faster streamlines of the parabolic flow profile and the shorter the time required by the component to exit. (b) Steric and (c) hyperlayer mode, for micron-sized particles with negligible diffusion. Particles are driven by the field directly to the accumulation wall. Particles of a given size form a thin layer of a given thickness, hugging the wall. Retention in steric FFF then depends only on particle size. During elution, however, the micron-sized particles make very little contact with the wall (steric mode), as a consequence bigger particles elute earlier than smaller particles. Instead, they are driven from the wall by a distance that is greater than their diameter because of the opposed by mobile phase flow-induced lift forces (hyperlayer mode). Retention in hyperlayer mode is still reversed with respect to particle size and it depends on the various physical features of the particles, which will have a varying influence on the intensity of the flow-induced lift forces.

development of multi-analyte, flow-assisted immunoassays in dispersed phase to tag-less cell sorting methods.[91,92]

A typical FFF system, as reported in Figure 6.10, consists of a reserve of carrier flow, the composition of which is chosen on the basis of separative needs

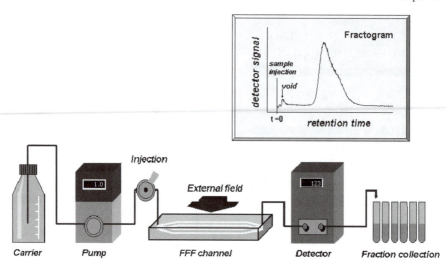

Figure 6.10   A typical instrumental setup for FFF and analytical output (fractogram: detector signal vs. retention time). Reproduced from reference 90 with permission from Elsevier.

and sample features; examples of carrier flow composition are buffers, cell culture media and saline solutions. The flow is then delivered into the capillary channel by a pump, which can be a simple peristaltic pump, a syringe pump or an HPLC pump; an injection valve or autosampler can be used to introduce narrow zone of sample into the flow stream. Sample components are then driven toward a flow-through detector connected to the channel outlet where a fraction collector can be, finally, placed. Specific components used to manage the applied field can be added to the above-described general instrumental setup. The FFF output consists of a fractogram, which is a graph recording the signal generated by the eluted samples over time.

Among FFF sub-techniques, gravitational field-flow fractionation (GrFFF) and flow field-flow fractionation (FlFFF) are the variants that have been coupled with CL detection up to now.

GrFFF, the humblest FFF variant, employs the Earth's gravity field for separation. The capillary fractionation channel, which has a rectangular geometry, is cut from a thin plastic foil and sandwiched between two plastic walls. The channel can be laboratory-built; therefore, different materials or even bio-functionalized walls can be used to improve selectivity of the fractionation process. GrFFF is suited for the fractionation of micron-sized analytes and its main features are its simple instrumental setup, simple use, low cost, easy integration with other analytical modules and possibilities of miniaturization.[93–95]

Conventional FlFFF employs a rectangular flat channel, in which at least one wall must be permeable to enable a cross flow to generate the field used for the fractionation. The system may require an additional pump or additional valves to control the intensity of the applied field. Flow FFF is characterized by a more complex instrumental setup and the choice of materials is limited; however, the possibility to modify the field intensity provides FlFFF with a high versatility that enables high size-based selectivity of fractionation for nano- and micro-sized analytes.[96]

6.5.2 Field-flow Fractionation with Chemiluminescence Detection

In FFF, sample detection has mostly been performed by spectroscopic methods, among which UV/Visible (UV/Vis) spectroscopy still constitutes the workhorse. However, the main problem of UV/Vis methods in FFF lies in the complex dependence of the detector's response to sample features, in addition to its intrinsic low sensitivity and specificity. Therefore, to increase the analytical information deductible from an FFF experiment, FFF has been coupled to other detection systems, such as flow-through or off-line laser and quasi-elastic laser scattering or fluorescence detection.[97–99] However, for all the detection methods based on a spectroscopic response, interaction of the incident radiation with the sample is complicated because of the heterogeneous nature of the sample/dispersing medium system.

CL detection has been successfully applied to several flow-through analytical techniques, providing a significant gain in detection sensitivity and selectivity, reduced assay times and the possibility of system miniaturization and automatization. Although CL detection is highly specific, separation of the bio-analytes before detection may still be required, as in heterogeneous immunoassays. However, protein complexes, nucleic acids, enzyme-labeled bio-specific probes supported on nano- or micro-beads, viruses, bacteria and cells can hardly be separated by conventional, flow-assisted separation techniques. Field-flow fractionation can thus play the role of an ideal flow-assisted separation method for many CL bio-applications. For instance, immunology employing CL probes can take advantage of FFF to separate mixtures of biological macromolecules supported on beads at high resolving power, relatively low cost and short analysis time. Because of the highly specific response, FFF-CL can also be used in conjunction with non-specific UV/Vis turbidity for selective detection of CL-inducing analytes that are fractionated from CL-inactive species. All the above features of FFF and CL have strongly encouraged the development of coupled FFF-CL methods.

6.5.3 Pioneering Applications

A pioneering application of GrFFF with CL detection was proposed for the separation of silica sorbent particles of different size. The setup consisted of a mixing coil placed at the channel outlet, where luminol and hydrogen peroxide were injected by means of a peristaltic pump for the determination of iron on particles; the emitted light was monitored by a modified commercial liquid scintillation counter with a spiral flow cell. GrFFF-FIA-CL was demonstrated for size-based speciation with interesting perspectives in the field of environmental science for the determination of elements associated with particles.[100,101]

Subsequently, CL was applied to GrFFF to the online detection of the separated analytes by continuously detecting the steady-state CL emission generated by eluting analytes upon the addition of a suitable CL substrate directly into the mobile phase. The CL signal was detected online by employing either a flow-through luminometer or a CCD camera with ultralow detection limits. In particular, a luminometer adopting a photomultiplier tube technology characterized by high stability and low-noise electronics was custom modified to allow flow-though measurements, by including a coiled transparent Teflon tube within the sample vial originally designed for batch measurements. Different flow-cell volumes to fulfill the needs of the specific applications were obtained by modifying the tube internal diameter and length. In addition, owing to the transparent channel walls, imaging detection using a low-light luminograph equipped with a highly sensitive, back-illuminated, Peltier-cooled CCD camera was performed to continuously measure the CL emission from samples eluting within the transparent channel. In particular, the entire GrFFF system was placed in the dark chamber of the luminograph; elution was then tracked by taking consecutive images. Digital images were analyzed to obtain quantitative results. Peroxidase-immobilized micrometer beads and the CL cocktail luminol/H_2O_2/p-iodophenol added to the mobile phase were used as a model sample and CL system. The possibility of fully separating and quantifying free and bead-immobilized enzymes has been reported, as a step towards the development of multi-analyte, ultra-sensitive, micron-sized beads-based flow-assisted immunoassays.[102] Figure 6.11 shows an example of CL signals obtained with an online luminometer and with imaging detection.

The highly versatile FlFFF variant was also coupled with CL detection, employing an off-line configuration using micrometer-sized polystyrene beads coated with enzymes suitable for CL detection as model samples. The high size-based selectivity of FlFFF and specificity of CL detection allowed the simultaneous and highly sensitive detection of different enzymes linked to beads of different size, in a single run and with very short analysis time.[103]

6.5.4   Recent Trends

Based on the instrumental setup and configuration used to couple FFF and CL detection, different assays formats have been developed with the aim of simplifying complex matrices prior to analyte detection and to produce rapid and versatile analytical assays with improved analytical performance.

6.5.4.1   *GrFFF and FlFFF with CL Detection as New Assay Formats*

The GrFFF-CL system has been applied for a new non-competitive immunoassay for pathogenic bacteria detection. The method combines the high

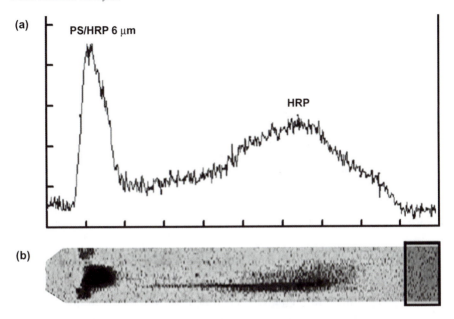

Figure 6.11 GrFFF-CL of a mixture of PS-bound HRP (PS/HRP) and free HRP (HRP): (a) GrFFF-CL fractogram; (b) GrFFF-CCD image. Reproduced from reference 102 with permission from Elsevier.

sensitivity of CL detection with the GrFFF capability for efficient separation of the bound bacterium–antibody complexes from the free antibodies. The immunoassay employs a HRP-labeled monoclonal antibody (MAb) produced against the target bacterium *Yersinia enterocolitica*. The immunological reaction takes place in the GrFFF channel and a CL substrate was post-column injected after fractionation. Figure 6.12 gives a schematic view of the assay.

The resulting CL fractogram consisted of two resolved peaks: a void peak and a retained peak corresponding to the free and bound MAb fractions, respectively. The area of these peaks was used to determine the calibration curve, which was linear between 10^8 and $10^{10}\,CFU\,L^{-1}$; the detection limit of the method was $10^9\,CFU\,L^{-1}$. Analytical validation of the method was performed with enriched human fecal samples.

This GrFFF-assisted, whole-cell, noncompetitive immunoassay offers many advantages with respect to the conventional solid-phase sandwich-type immunoassay format: only a single MAb is required for the detection of the analyte, costs and analysis time are significantly decreased and the method is suitable for automation. Moreover, multiplexed immunoassays can be developed for simultaneous analysis of bacterial mixtures, since pathogenic bacteria with different morphological properties can elute differently in the GrFFF channel.[104]

The GrFFF-CL system was also applied at the forefront of cell sorting with interesting perspectives. The GrFFF fractionation method is "soft" as no tags

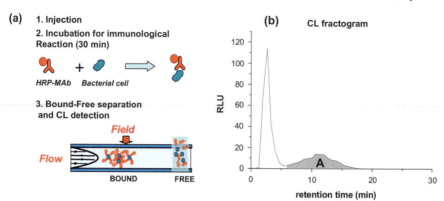

Figure 6.12 (a) schematic view of the GrFFF-CL immunometric method for pathogenic bacteria determination; (b) representative fractograms. A = peak area of the retained immuno-complex used for the calibration curve.

are needed and any flow composition can be used for cell isolation; therefore, intact cells can be sorted, even from complex matrices, and reused. Cells can exhibit spontaneous BL (*e.g.*, some marine bacteria) or BL as a consequence of manipulation, such as surface labeling or genetic modification.[105,106] In addition, cell components can be coupled with an appropriate substrate to generate CL. A study was carried out using red blood cells as sample model, which were detected by exploiting the CL luminol reaction catalyzed by red blood cell heme group; different instrumental setups by which to couple GrFFF and CL detection were tested. The best performance was obtained with the online coupling format, performed in post-column flow-injection mode. A detection limit of a few hundred injected cells was found.[107] The GrFFF-CL system with imaging detection was used also to study the possibility of exploiting GrFFF channels with a bio-functionalized wall to modulate the FFF fractionation process, with a view to combing biospecific analyte recognition systems with the "soft" fractionation process based on native morphological properties. This approach appears to be promising for the development of microfluidic integrated devices with high versatility able to perform multiplexed assays in a short time. In particular, a capture antibody was immobilized in the injection area of the accumulation wall, so that specific analytes were bound while non-bound analytes could be fractionated. Subsequently, bound analytes could be eluted with a mobile phase gradient with a solvent mixture able to break the antibody–antigen bond. Peroxidase-coated polystyrene particles were chosen as model sample to optimize experimental conditions; then the hybrid method was applied to a bacteria mixture.[108]

The FlFFF-CL system has been exploited for the development of a competitive immunoassay for chloramphenicol (CAF). Antibody-coated microspheres were employed as a solid phase, with CAF-HRP as a tracer. A mixture of CAF, tracer and antibody-coated microspheres was injected into a FlFFF

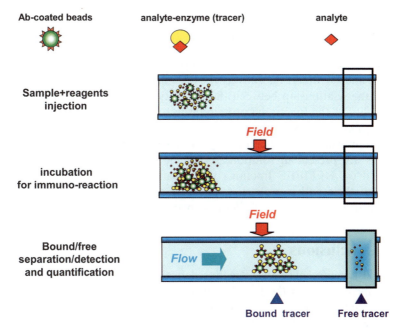

Figure 6.13 Schematic view of the CL-FFF-based mutianalyte immunoassay: reagents injection; relaxation/incubation; separation of the free tracer fraction and the various microsphere populations, with on-line CL measurement of enzyme activity and quantification calculated as the area of the unretained and retained peak respectively in CL fractograms.

channel, where the incubation step was performed. Free and bound tracer fractions were then separated during the FlFFF run and measured by online CL detection after a post-column flow injection of CL substrate. Figure 6.13 shows a schematic view of the method.

A dose–response curve for CAF was obtained by plotting the peak area of the retained peak in the CL fractogram against CAF concentration.

A linear range of 0.03–15 μmol L^{-1} (0.01–5 ppm) and a limit of detection of 0.3 μmol L^{-1} (0.1 ppm) were evaluated. The method is rapid; the immunological binding is preserved during fractionation and no run-to-run carry-over effects are present. Moreover, the use of differently-sized microbeads coated with antibodies specific for different analytes can be adopted for the development of a multi-analyte immunoassay.[109]

6.5.4.2 Miniaturization of FFF with CL Detection

Miniaturization of the FFF channel is still an open issue. A scale-down of channel volume was obtained by the use of a hollow fiber as separative channel, employing hydrodynamic flow as applied filed (HF FlFFF). Low channel volume (ranging from 100 to 30 μL) reduces sample dilution with an increase in detection sensitivity. The possibility of disposable usage of the HF channels eliminates the risk of run-to-run sample carryover. With these advantages, nanometer- or micrometer-sized particles of different origin have been separated with HF FlFFF.[110–112] A method combining HF FlFFF with matrix-assisted laser desorption-ionization time-of-flight mass spectrometry (MALDI-TOFMS) and CL detection to assay the enzymatic activity has been developed to study the relationship between the impurity profile and conformation and functionality of the functional protein urate oxidase. The results showed that the recombinant uricase samples obtained from different microorganisms have different impurities and different enzymatic activity and that different uricase oligomers are present in solution. A specific enzymatic CL assay has been performed on HFFlFFF fractions where samples are collected with maintenance of their native properties, allowing the specific assay results to be related to these properties with improvement in analytical output.[113]

6.6   Conclusions

CL-ECL-BL represents very attractive detection techniques for flow-assisted analysis mostly due to the low-detection limit, low background signals and relatively simple instrumentation. On the other hand, flow-assisted analysis add advantages such as reproducibility of reagent mixing and reaction conditions, even when the use of unstable reagents is necessary, and the potential kinetic discrimination of interferences. Moreover, they make possible the direct analysis of complex matrices with consequent improvement of assay robustness and analytical performances. By the use of flow injection analysis techniques in different formats or combined with specific reagents in order to improve selectivity, different applications have been developed with high precision, automation and versatility in many fields ranging from pharmaceutical, environmental and food analysis, even in a miniaturized format that allows for low reagent consumption, short analysis times and portability. Furthermore, the use of flow-assisted separative techniques, such as field-flow fractionation, has been implemented with CL detection as a new analytical format for several assays with advantages of short analysis time, simple instrumental setup and potential multiplexed analysis.

However, some drawbacks are still present, mostly due to the lack of standardization in detection instruments. The potential marriage of flow-assisted techniques and CL/BL/ECL will be further explored, aimed at the development of fully controlled systems able to maintain well-defined conditions both for detection and flow parameters, for multicomponent analysis with low reagent consumption and high throughput.

References

1. J. Rutzicka and E. Hansen, *Anal. Chim. Acta*, 1975, **78**, 145.
2. M. Valcarcel, M. D. Luque and De Castro, *Flow Injection Analysis: Principles and Applications*, Ellis Horwood, Chichester, 1985.

3. M. Gisin and C. Thommen, *TrAC*, 1989, **8**, 62.
4. G. Hanrahan, S. Ussher, M. Gledhill, E. P. Achterberg and P. J. Worsfold, *TrAC*, 2002, **21**, 23.
5. P. Chocholous, P. Solich and D. Satiínskyý, *Anal. Chim. Acta*, 2007, **600**, 129.
6. L. T. Skeegs, *Am. J. Clin. Pathol.*, 1957, **28**, 311.
7. J. Rutzicka and E. H. Hansen, *Flow Injection Analysis*, 2nd edn., Wiley-Interscience, New York, 1988.
8. J. T. Vanderslice, K. K. Stewart, A. G. Rosenfeld and D. J. Higgs, *Talanta*, 1981, **28**(1), 11.
9. B. Rocks and C. Riley, *Clin. Chem.*, 1982, **28**(3), 409.
10. J. Rutzicka and G. D. Marshall, *Anal. Chim. Acta*, 1990, **237**, 329.
11. A. Economou, *TrAC*, 2005, **24**, 416.
12. V. Cerdà, J. M. Estela, R. Forteza, A. Cladera, E. Becerra, P. Altimira and P. Sitjar, *Talanta*, 1999, **50**, 695.
13. M. Mirò, V. Cerdà and J. M. Estela, *TrAC*, 2002, **21**, 199.
14. M. A. Segundo and A. O. S. S. Rangel, *J. Flow Inject. Anal.*, 2002, **19**(1), 3.
15. M. A. Feres, P. R. Fortes, E. A. G. Zagatto, J. L. M. Santos and J. L. F. C. Lima, *Anal. Chim. Acta*, 2008, **618**, 1.
16. K. Mervatová, M. Polášek and J. M. Calatayud, *J. Pharm. Biomed. Anal.*, 2007, **45**, 367.
17. G. Gamborg and E. H. Hansen, *Anal. Chim. Acta*, 1994, **285**, 321.
18. S. M. Gautier, P. E. Michel and L. J. Blum, *Anal. Lett.*, 1994, **27**, 2055.
19. S. D. Clerc, R. A. Jewsbury, M. G. Mortimer and J. Zeng, *Anal. Chim. Acta*, 1997, **339**, 225.
20. S. Girotti, M. Muratori, F. Fini, E. N. Ferri, G. Carrea, M. Koran and P. Rauch, *Eur. Food Res. Technol.*, 2000, **210**, 216.
21. J. A. Ho and M.-R. Huang, *Anal. Chem.*, 2005, **77**, 3431.
22. G. Rule and W. R. Seitz, *Clin. Chem.*, 1979, **25**, 1635.
23. P. K. Dasgupta, Z. Genfa, J. Li, C. B. Boring, S. Jambunathan and R. Al-Horr, *Anal. Chem.*, 1999, **71**, 1400.
24. L. M. Magalhaes, M. A. Segundo, S. Reis, J. L. F. C. Lima, J. M. Estela and V. Cerdà, *Anal. Chem.*, 2007, **79**, 3933.
25. J. M. Terry, J. L. Adcock, D. C. Olson, D. K. Wolcott, C. Schwanger, L. A. Hill, N. W. Barnett and P. S. Francis, *Anal. Chem.*, 2008, **80**, 9817.
26. Y.-X. Guan, Z.-R. Xu, J. Dai and Z.-L. Fang, *Talanta*, 2006, **68**, 1384.
27. S. Mohr, J. M. Terry, J. L. Adcock, P. R. Fielden, N. J. Goddard, N. W. Barnett, D. K. Wolcottc and P. S. Francis, *Analyst*, 2009, **134**, 2233.
28. K. M. Scudder, C. H. Pollema and J. Ruzicka, *Anal. Chem.*, 1992, **64**, 2657.
29. N. Ibáñez-García, M. Puyol, C. M. Azevedo, C. S. Martínez-Cisneros, F. Villuendas, M. R. Gongora-Rubio, A. C. Seabra and J. Alonso, *Anal. Chem.*, 2008, **80**, 5320.
30. Y. Wen, H. Yuan, J. Mao, D. Xiao and M. M. F. Choi, *Analyst*, 2009, **134**, 354.

31. P. Campıns-Falcó, L. A. Tortajada-Genaro and F. Bosch-Reig, *Talanta*, 2001, **55**, 403.
32. J. L. Pérez Pavón, E. Rodriguez Gonzalo, G. D. Christian and J. Ruzicka, *Anal. Chem.*, 1992, **64**, 923.
33. M. Miro, J. M. Estela and V. Cerda, *Anal. Chim. Acta*, 2005, **541**, 55.
34. P. Fletcher, K. N. Andrew, A. C. Calokerinos, S. Forbes and P. J. Worsfold, *Luminescence*, 2001, **16**, 1.
35. C. E. Lenehan, N. W. Barnett and S. W. Lewis, *Analyst*, 2002, **127**, 997.
36. B. Rezaei and A. Mokhtari, *Luminescence*, 2009, **24**, 183.
37. Z. F. Fu and L. Wang, *Chem. Anal-Warsaw*, 2009, **54**, 163.
38. A. Waseem, L. Rishi, M. Yaqoob and A. Nabi, *Anal. Sci.*, 2009, **25**, 407.
39. M. S. Burkhead, H. Y. Wang, M. Fallet and E. M. Gross, *Anal. Chim. Acta*, 2008, **613**, 152.
40. G. L. Wei, C. L. Wei, G. C. Dang, H. C. Yao and H. Li, *Anal. Lett.*, 2007, **40**, 2179.
41. G. L. Wei, G. Dang and H. Li, *Luminescence*, 2007, **22**, 534.
42. A. Miyamoto, K. Nakamura, N. Kishikawa, Y. Ohba, K. Nakashima and N. Kuroda, *Anal. Bioanal. Chem.*, 2007, **388**, 1809.
43. F. E. O. Suliman, M. M. Al-Hinai, S. M. Z. Al-Kindy and S. B. Salama, *Luminescence*, 2009, **24**, 2.
44. E. J. Llorent-Martinez, P. Ortega-Barrales and A. Molina-Diaz, *Anal. Chim. Acta*, 2006, **580**, 149.
45. K. L. Marques, J. L. M. Santos and J. L. F. C. Lima, *Anal. Chim. Acta*, 2004, **518**, 31.
46. O. D. Leite, O. Fatibello-Filho, H. J. Vieira, F. R. P. Rocha and N. S. M. Cury, *Anal. Lett.*, 2007, **40**, 3148.
47. K. L. Marques, J. L. M. Santos and J. L. F. C. Lima, *J. Pharm. Biomed. Anal.*, 2005, **39**, 886.
48. E. J. Llorent-Martínez, P. Ortega-Barrales, M. L. Fernández de Córdova and A. Ruiz-Medina, *Anal. Bioanal. Chem.*, 2009, **394**, 845.
49. C. K. Pires, P. B. Martelli, B. F. Reis, J. L. F. C. Lima and M. L. M. F. S. Saraiva, *J. Automated Methods Management Chem.*, 2003, **25**, 109.
50. C. M. P. G. Amorima, J. R. Albert-Garcia, M. C. B. S. Montenegro, A. N. Araujo and J. Martınez Calatayud, *J. Pharm. Biomed. Anal.*, 2007, **43**, 421.
51. J. R. Albert-García and J. M. Calatayud, *Talanta*, 2008, **75**(3), 717.
52. J. M. Estela and V. Cerda, *Talanta*, 2005, **66**, 307.
53. A. Safavi and M. A. Karimi, *Talanta*, 2002, **58**(4), 785.
54. L. Aifang, W. Chengmou and L. Jiuru, *Anal. Lett.*, 2007, **40**, 2737.
55. M. Palomeque, J. A. G. Bautista, M. C. Icardo, J. V. G. Mateo and J. M. Calatayud, *Anal. Chim. Acta*, 2004, **512**, 149.
56. F. Sá, D. L. Malo and J. M. Calatayud, *Anal. Lett.*, 2007, **40**, 2872.
57. A. Czescik, D. Lopez Malo, M. J. Duart, L. L. Zamorac, G. M. A. Fos and J. M. Calatayud, *Talanta*, 2007, **73**, 718.
58. T. G. Climent, J. R. Albert-Garcı and J. M. Calatayud, *Anal. Lett.*, 2007, **40**, 629.

59. L. A. Tortajada-Genaro, P. Campıns-Falcó and F. Bosch-Reig, *Anal. Chim. Acta*, 2003, **488**, 243.
60. C. Ruiz-Capillas and F. Jimenez-Colmenero, *Food Chem.*, 2009, **112**, 487.
61. R. Perez-Olmos, J. C. Soto, N. Zarate, A. N. Araujo, J. L. F. C. Lima and M. L. M. F. S. Saraiva, *Food Chem.*, 2005, **90**(3), 471.
62. X. Liu, J. Du and J. Lu, *Luminescence*, 2003, **18**, 245.
63. H. Liu, X. Gao, L. Niu, X. Li and Z. Song, *J. Sci. Food Agric.*, 2008, **88**, 2744.
64. E. N. Fernandes and B. F. Reis, *J. AOAC Int.*, 2004, **87**, 920.
65. P. B. Martelli, B. F. Reis, A. N. Araujo, M. Conceição and B. S. M. Montenegro, *Talanta*, 2001, **54**, 879.
66. T. Perez-Ruiz, C. Martinez-Lozano, V. Tomas and J. Fenolo, *Anal. Chim. Acta*, 2003, **485**, 63.
67. B. Li, D. Wang, J. Lv and Z. Zhang, *Talanta*, 2006, **69**, 160.
68. B. Li, Y. He, J. Lv and Z. Zhang, *Anal. Bioanal. Chem.*, 2005, **383**, 817.
69. J. Ruzicka, *Analyst*, 2000, **125**, 1053.
70. J. Wang and E. H. Hansen, *TrAC*, 2003, **22**(4), 225.
71. X.-W. Chen and J.-H. Wang, *Anal. Chim. Acta*, 2007, **602**, 173.
72. M. Mirò and E. H. Hansen, *Anal. Chim. Acta*, 2007, **600**, 46.
73. D. He, Z. Zhang, Y. Huang, Y. Hu, H. Zhou and D. Chen, *Luminescence*, 2005, **20**, 271.
74. Y.-X. Guan, Z.-Run Xua, J. Dai and Z.-L. Fang, *Talanta*, 2006, **68**, 1384.
75. J. H. Wang, Y. Xu and M. Yang, *Anal. Chem.*, 2006, **78**, 5900.
76. Z. Zhang, D. He, W. Liu and Y. Li, *Luminescence*, 2005, **20**, 377.
77. J. Yan, X. Yang and E. Wang, *Anal. Bioanal. Chem.*, 2005, **381**, 48.
78. G. C. Fiaccabrino, N. F. de Rooij and M. Koudelka-Hep, *Anal. Chim. Acta*, 1998, **359**, 263.
79. E. L'Hostis, P. E. Michel, G. C. Fiaccabrino, D. J. Strike, N. F. de Rooij and M. Koudelka-Hep, *Sens. Actuators, B*, 2000, **64**, 156.
80. A. Arora, A. J. de Mello and A. Manz, *Anal. Commun.*, 1997, **34**, 393.
81. J. F. Liu, J. L. Yan, X. R. Yang and E. K. Wang, *Anal. Chem.*, 2003, **75**, 3637.
82. M. Yang, Y. Xu and J. H. Wang, *Anal. Chem.*, 2006, **78**, 5900.
83. A. R. Bowie, E. P. Achterberg, S. Ussher and P. J. Worsfold, *J. Automated Methods Management Chem.*, 2005, **2**, 37.
84. B. Li, ZJ. Zhang and Y. Jin, *Anal. Chim. Acta*, 2001, **432**, 95.
85. D. He, Z. Zhang, H. G. Zhou and Y. Huang, *Talanta*, 2006, **69**, 1215.
86. Y.-T. Kim, S. O. Ko and J. H. Lee, *Talanta*, 2009, **78**, 998.
87. Y. T. Hsueh, S. D. Collins and R. L. Smith, *Sens. Actuators, B*, 1998, **49**, 1.
88. J. C. Giddings, *Science*, 1993, **260**, 1456.
89. M. E. Schimpf, K. Caldwell and J. C. Giddings, *Field-Flow Fractionation Handbook*, Wiley-Interscience, New York, 2000.
90. P. Reschiglian, A. Zattoni, B. Roda, E. Michelini and A. Roda, *TRENDS Biotechnol.*, 2005, **23**(9), 475.
91. B. Roda, A. Zattoni, P. Reschiglian, M. H. Moon, M. Mirasoli, E. Michelini and A. Roda, *Anal. Chim. Acta*, 2009, **635**(2), 132.

92. B. Roda, G. Lanzoni, F. Alviano, A. Zattoni, R. Costa, A. Di Carlo, C. Marchionni, M. Franchina, F. Ricci, PL. Tazzari, P. Pagliaro, SZ. Scalinci, L. Bonsi, P. Reschiglian and G. P. Bagnara, *Stem Cell Rev.*, 2009, **5**(4), 420.

93. B. Roda, P. Reschiglian, A. Zattoni, P. L. Tazzari, M. Buzzi, F. Ricci and A. Bontandini, *Anal. Bioanal. Chem.*, 2008, **392**, 137.

94. A. Bernard, C. Bories, P. M. Loiseau and P. J. P. Cardot, *J. Chromatogr. B*, 1995, **664**, 444.

95. B. Roda, P. Reschiglian, F. Alviano, G. Lanzoni, G. P. Bagnara, F. Ricci, M. Buzzi, P. L. Tazzari, P. Pagliaro, E. Michelini and A. Roda, *J. Chromatogr. A*, 2009, **1216**(52), 9081.

96. J. R. Silveira, G. J. Raymond, A. G. Hughson, R. E. Race, V. L. Sim, S. F. Hayes and B. Caughey, *Nature*, 2005, **437**, 257.

97. K. D. Caldwell and J. Li, *J. Colloid Interface Sci.*, 1989, **132**, 256.

98. H. Thielking, D. Roessner and W.-M. Kulicke, *Anal. Chem.*, 1995, **67**, 3229.

99. M. Andersson, B. Wittgren and K.-G. Wahlund, *Anal. Chem.*, 2001, **73**, 4852.

100. R. Chantiwas, J. Jakmunee, R. Beckett, I. Mckelvie and K. Grudpan, *Anal. Sci.*, 2001, **17**, i423.

101. R. Chantiwas, R. Beckett, J. Jakmunee, I. D. McKelvie and K. Grudpan, *Talanta*, 2002, **58**, 1375.

102. D. Melucci, M. Guardigli, B. Roda, A. Zattoni, P. Reschiglian and A. Roda, *Talanta*, 2003, **60**, 303.

103. P. Reschiglian, A. Zattoni, D. Melucci, B. Roda, M. Guardigli and A. Roda, *J. Sep. Sci.*, 2003, **26**, 1417.

104. M. Magliulo, B. Roda, A. Zattoni, E. Michelini, M. Luciani, R. Lelli, P. Reschiglian and A. Roda, *Clin. Chem.*, 2006, **52**, 2151.

105. S. Daunert, G. Barrett, J. S. Feliciano, R. S. Shetty, S. Shrestha and W. Smith-Spencer, *Chem. Rev.*, 2000, **100**, 2705.

106. A. Keane, P. Phoenix, S. Ghoshal and P. C. K. Lau, *J. Microbiol. Methods*, 2002, **49**, 103.

107. D. Melucci, B. Roda, A. Zattoni, S. Casolari, P. Reschiglian and A. Roda, *J. Chromatogr. A*, 2004, **1056**, 229.

108. B. Roda, S. Casolari, P. Reschiglian, M. Mirasoli, P. Simoni and A. Roda, *Anal. Bioanal. Chem.*, 2009, **394**(4), 953.

109. A. Roda, M. Mirasoli, D. Melucci and P. Reschiglian, *Clin. Chem.*, 2005, **10**, 1993.

110. B.-R. Min, S. J. Kim, K.-H-. Ahn and M. H. Moon, *J. Chromatogr. A*, 2002, **950**, 175.

111. P. Reschiglian, B. Roda, A. Zattoni, B.-R. Min and M. H. Moon, *J. Sep. Sci.*, 2002, **25**, 490.

112. P. Reschiglian, A. Zattoni, B. Roda, L. Cinque, D. Melucci, B.-R. Min and M. H. Moon, *J. Chromatogr. A*, 2003, **985**, 519.

113. A. Roda, D. Parisi, M. Guardigli, A. Zattoni and P. Reschiglian, *Anal. Chem.*, 2006, **78**, 1085.

CHAPTER 7

Analytical Applications of Chemiluminescence in Chromatography and Capillary Electrophoresis

ANA M. GARCÍA-CAMPAÑA, LAURA GÁMIZ-GRACIA, JOSÉ F. HUERTAS-PÉREZ AND FRANCISCO J. LARA

Department of Analytical Chemistry, Faculty of Sciences, Campus Fuentenueva, University of Granada, E-18071 Granada, Spain

7.1 Introduction

Chemiluminescence (CL) is considered a well-established spectrometric branch of analytical chemistry.[1,2] CL measurements are strongly modified by experimental factors, including temperature, pH, ionic strength, solvent and solution composition, in the case of liquid phase, and gas composition, temperature, and pressure in the case of gas-phase measurements. Because of the dependence of the reaction rate on the concentration, CL techniques may be satisfactorily used for quantitative analysis of a wide variety of species that can participate in the CL process, such as CL precursors, reagents, species that affect the rate or efficiency of the CL reaction such as activators, catalysts or inhibitors, species that are not directly involved in the reaction but that can react with other reagents to generate a product that participates in the CL reaction or species that can be derivatized with some CL precursors or fluorophores.

Some limitations should be considered in CL analysis, such as the control of factors that affect the CL emission, the lack of selectivity because a CL reagent is not limited to just one analyte, and, finally, like other mass flow detection approaches, since CL emission is not constant but varies with time (light flash

Chemiluminescence and Bioluminescence: Past, Present and Future
Edited by Aldo Roda
© Royal Society of Chemistry 2011
Published by the Royal Society of Chemistry, www.rsc.org

composed of a signal increase after reagent mixing, passing through a maximum, then declining to the baseline), and this emission *versus* time profile can
widely vary in different CL systems, care must be taken so as to detect the signal
in the flowing stream at strictly defined periods. In recent years, several reviews
have been published about the use of CL as an analytical technique.[3–7]

Owing to its simplicity, low cost and high sensitivity, CL-based detection has
become a useful tool in gas chromatography (GC),[8] liquid chromatography
(LC)[9] and capillary electrophoresis (CE),[10] making this technique an interesting
research field in clinical,[11,12] biomedical,[13] environmental and food analysis.[14]
Minimal instrumentation is required and because no external light source is
needed the optical system is quite simple. Hence, strong background light levels
are excluded, reducing the background signal, the effects of stray light and the
instability of the light source, leading to improved detection limits. In this
chapter we discuss the main couplings of CL as detection mode in separation
techniques, such as GC-CL, HPLC-CL and CE-CL. General characteristics,
advantages and drawbacks, different CL reactions applied and important
applications in different fields are included.

7.2 Gas Chromatography Coupled to Chemiluminescence Detection

Gas-phase CL consists of a chemical reaction forming an excited-state product
that then undergoes one or more relaxation processes to attain its ground state.
The production of CL emission in the UV–visible spectral region requires
highly exothermic reactions such as atomic or radical recombinations, or
reaction of reduced species such as hydrogen atoms, olefins, and several sulfur
and phosphorous compounds with strong oxidants such as ozone, fluorine and
chlorine dioxide. In gas-phase CL detection, radiative emission is usually
competitive with non-radiative processes, and both the quantum yield of the
reaction and the emission spectrum vary with physical conditions such as bath
gas composition, temperature and pressure. CL detection in the gas phase has
been reviewed previously, and a broad spectrum of CL reactions has been
studied and reported.[8,15,16]

CL detectors in GC are valuable because they have greater selectivity and
sensitivity for compounds with particular atoms, compared to the flame ionization detector and the thermal conductivity detector. Commercially available
CL detectors for GC include the flame photometric detector (FPD), or the
nitrogen and sulfur chemiluminescence detectors (NCD and SCD, respectively), based on the reaction with O_3.

7.2.1 Flame Photometric Detector (FPD)

The most commonly used and commercially available gas-phase CL detector is
the FPD.[17] This detector has, mostly, been coupled to GC, being considered a
relatively robust and low-cost system. In FPD, the high temperature of a flame

promotes chemical reactions that form key reaction intermediates and may provide additional thermal excitation of the emitting species. FPD may be used to selectively detect compounds containing sulfur, nitrogen, phosphorous, boron, antimony, arsenic and even halogens under special reaction conditions, but commercial detectors are normally configured only for P and S detection. In a GC-FPD, the GC column extends to the sample inlet where it is mixed with oxygen (or air) and with hydrogen fuel prior to the burner head, and before entering the CL cell. Fuel-rich, hydrogen–oxygen flames are preferred, but the combustion mixture must be optimized for each analyte. The main limitation to sensitivity in FPD is the noise associated with the background signal, arising primarily from other flame emissions. Significant quenching by the presence of hydrocarbon solvents is another reported problem. To optimize the limit of detection (LOD), the PMT (photomultiplier tube) is positioned appropriately, and lenses are used to view a region of the flame where the signal-to-noise ratio is greatest, while filters are employed to reduce background contributions of flame emissions.

Many efforts have focused on solving some of the above-mentioned problems, and other devices that incorporate similar operating principles have been developed. One example is the dual flame FPD, which incorporates a dual flame burner that oxidizes carbon in the first flame, and measures sulfur CL in the second flame, obtaining a more uniform sulfur response and reducing quenching.[18,19] The reactive flow detector is a stable, several centimeters-long luminescent column easily formed by a hydrogen–air mixture ascending through a glass capillary toward a glow-sustaining flame on top, which combusts the excess hydrogen with auxiliary air.[20,21] Sometimes these efforts have focused on developing novel burners designs, for example, the so-called counter-current FPD (ccFPD), in which the fuel and oxidant are introduced as opposing gas streams from burners directly opposite (usually above and below) one another, producing a highly stable flame.[22] This system is easy to miniaturize and has been employed for developing an improved μFPD, making it possible to adapt it to a micro-analytical and portable format.[23]

The most interesting and well-established improvement has been the pulse flame photometric detector (PFPD), which employs pulsed flame and time-resolved emission detection with gated electronics.[24,25] In this design, the burner is constructed to generate a non-continuous flame, which is reignited at a frequency of about 1–10 Hz. This periodic interruption allows the acquisition of the time-dependent emissions from the various excited-state species present in the detector. As each species presents different fluorescence lifetimes of the order of milliseconds, they can be differentiated in the time domain between flame pulses. The added time domain information is used to enhance element-specific detection and to increase the selectivity over hydrocarbons, improving substantially the overall performance of the FPD. The improvements include a sensitivity enhancement of one or two orders of magnitude, about an order of magnitude of increased selectivity and reduced quenching effects. Figure 7.1 shows a scheme of a PFPD detector.[25]

In summary, a combustible gas mixture of hydrogen and air (3) (Figure 7.1) is continuously fed into the small pulsed flame chamber (6) together with the

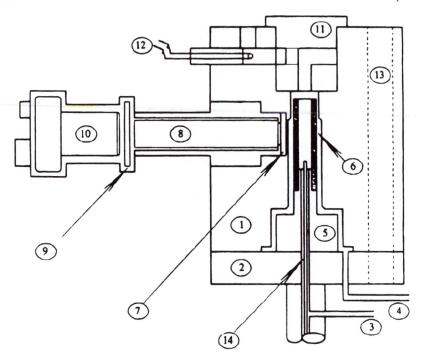

Figure 7.1 Schematic diagram of the PFPD design. (1) PFPD body, (2) GC-heated
detector base, (3) central hydrogen-rich H_2/air mixture tube leading to the
combustor, (4) outer bypass H_2/air mixture tube, (5) combustor holder, (6)
quartz combustor tube, (7) sapphire window, (8) light guide, (9) colored
glass filter, (10) PMT, (11) spiral igniter light shield, (12) heated wire
igniter, (13) assembly guiding rod in a guiding hole and (14) column.
(Reproduced with permission from ref. 25.)

sample molecules that are eluted in the usual way from the GC column (14).
The combustible gas mixture also flows separately (4) to a light-shielded,
continuously heated, wire igniter (12). The ignited flame is propagated back to
the gas source through the pulsed flame chamber (6), and is self terminated in a
few milliseconds, since the pulsed flame cannot propagate through the small
hole of the combustor holder (5) at the bottom of the pulsed flame chamber (6).
The continuous gas flow creates additional ignition after a few hundred milliseconds in a pulsed periodic fashion (~3 Hz). The emitted light is transferred
with a light pipe (8) through a broad band filter (9) and detected with a PMT
(10). Some analytical applications are described in the following sections.

7.2.1.1 Pesticide Analysis

The most important application of FPD and PFPD in the field of pesticide
analysis is its coupling with GC for the determination of organophosphorous
compounds (OPPs). Its suitability for the analysis of OPPs in all kind of food

samples and beverages has been demonstrated in numerous publications and has been reviewed recently.[26] For instance, Di Muccio *et al.* have developed a method for the determination of 45 OPP residues in vegetable oils by GC-FPD.[27] The method performs an on-column extraction and clean-up of OPP residues in a single step by a three-cartridge system. A solution of 1 g oil in *n*-hexane is loaded into an Extrelut-NT3 cartridge (large-pore diatomaceous material), and the OPP residues eluted in acetonitrile are cleaned-up by passing through a silica and a C18 cartridge connected on-line to the Extrelut-NT-3 cartridge. In addition, the determination of 24 OPPs in vegetables has been reported using liquid–liquid extraction (LLE) without further clean-up.[28] The reliability of results obtained by GC-PFPD was assessed by analyzing 20 samples of different fresh fruits and vegetable matrices (green beans, cucumbers, peppers, tomatoes, melons, eggplants, watermelons and zucchini), all of them with a high water content.

Although less frequently than for OPPs, FPD has also been used as a specific detector for sulfur-containing pesticides in food, environmental and biological samples.[26] Thus, the fungicide probenazole has been determined in soil, rice plant and paddy water by GC-FPD.[29] Also, several *N,N*-dimethyldithiocarbamate (thiram, ziram) and ethylenebis(dithiocarbamate)s (maneb, zineb, mancozeb) have been analyzed in fruits and vegetables by GC-FPD,[30] reaching LODs and limits of quantification (LOQs) in the range of 0.005–$0.1\,\text{mg kg}^{-1}$, with recoveries higher than 80% (RSD $<$ 20%).

Several fast GC systems with direct resistive heating as a principle have become commercially available, in which a capillary column is inserted into a resistively heated metal tube (resistive heating-gas chromatography, RH-GC) or enclosed in resistively heated toroid-formed assembly (low thermal mass chromatography, LTM-GC). More recently, RH-GC has been coupled with FPD and PFPD for the rapid, sensitive and selective analysis of pesticides; for example, the screening method proposed by Patel *et al.* for 37 OPPs.[31] Reporting limits were $0.01\,\text{mg kg}^{-1}$ for all pesticides in peach and grapes, and for 36 of them in sweet pepper, and mean recoveries ranged from 70 to 116% with an RSD $<$ 20%. The same authors coupled RH-GC-FPD with programmable temperature vaporization (PTV) for the sensitive analysis of 20 OPPs in apple, pear and orange juice samples, carrying out the separation in less than 6 min.[32]

In recent years, most advances and research efforts concerning the analysis of pesticides have focused on the development of improved sample preparation, to remove interfering compounds and to achieve sufficient sensitivity using more environmental friendly, economical and miniaturized processes. One example, among others, is the method proposed by Yu *et al.* for the determination of 11 OPPs in honey, orange juice and pakchoi (*e.g.*, Chinese vegetable) using solid-phase micro-extraction (SPME).[33] The LODs varied from 0.003 to $1\,\text{ng g}^{-1}$ for OPPs in food samples, and recoveries obtained for samples spiked at $20\,\mu\text{g L}^{-1}$ were from 74.4 to 105.2%. Liquid phase micro-extraction (LPME) has been used in conjunction with GC-FPD for the development of a rapid and sensitive analytical method for OPPs in different types of natural water samples.[34] Recently introduced, dispersive solid-phase extraction (DSPE) was used in a method for the determination of 17 OPPs in spinach,[35] reaching LOQs in the range $10-20\,\mu g\,kg^{-1}$. Other sample treatment techniques introduced more recently for the analysis of pesticides, such as dispersive liquid–liquid micro-extraction (DLLME), single drop micro-extraction (SDME), stir-bar sorptive extraction (SBSE), pressurized liquid extraction (PLE), *etc.*, have also been used for the selective and sensitive determination of OPPs by GC-FPD/ PFPD.[26] FPD and PFPD detectors in GC have been used simultaneously with other selective detector systems such as electron capture detection (ECD), halogen specific detector (XSD) or mass spectrometry (MS), allowing the development of multi-residue analytical methods, in which several families of pesticides are analyzed at the same time, taking advantage of their intrinsic characteristics and the selectivity of each detector. For example, in a recent method, the determination of 108 OPPs in dried ground ginseng root has been proposed.[36] Pesticides were extracted from the sample using acetonitrile–water saturated with salts, followed by DSME clean-up, and analyzed by GC-MS and GC-FPD in phosphorus mode. Quantification was achieved from 0.050 to $5.0\,\mu g\,g^{-1}$ ($R^2 > 0.99$) for most of the pesticides using both detectors. A fast method based on LTM-GC-MS/PFPD has been proposed for the analysis of pesticides in complex samples.[37] The method was evaluated for the separation of 82 pesticide mixtures, including 27 OPPs, and applied to the analysis of pesticides at the $pg\,mL^{-1}$ levels in a brewed green tea sample with SBSE.

7.2.1.2   Organotin Compounds

Organotin compounds (OTCs) are widely used as marine antifouling on ships and channel facilities, in the plastic industry as catalyst and stabilizers, and in agriculture as insecticides, fungicides and biocides. Their use, especially in the marine industry, has been restricted or even forbidden due to their toxic and non-specific action.[38] GC analysis with a selective detection technique after alkylation with Grignard reagents (RMgX) or, more preferred lately, hydride formation with $NaBH_4$ or ethylation with $NaBEt_4$ shows generally excellent performance. For example, Grignard reagents have recently been used as derivatization agents for optimizing a, method for the determination of monobutyltin (MBT), dibutyltin (DBT), tributyltin (TBT) and triphenyltin (TPhT) in human breast milk by GC-FPD.[39] Recoveries ranged from 80 to 105%, and LODs from 1.3 to $2.5\,ng\,mL^{-1}$. The method was used to monitor OTCs in 70 breast milk samples with the aim of correlating these concentrations with the consumption of fish by the mothers. Recently, a method based on DLLME and GC-FPD has been proposed for the speciation of butyl and phenyltin compounds in water samples with high enrichment factors (825–1036); LODs in the low-ppt could be reached.[40] Recoveries from river and seawater spiked at 10.0 and $100\,ng\,L^{-1}$ (as Sn) were in the range 82.5–104.7%. In addition, a method for organotin speciation in mussel and oyster tissue as well as in marine sediments has been developed using ultrasonic probe extraction (UPE) for the quantitative leaching of MBT, DBT, TBT and tripropyltin (TPrT) in a very short time.[41] The extracts are then cleaned up by SPE using a molecularly imprinted polymer (MISPE) especially designed for TBT. LODs and LOQs of 3 and $10 \mu g \, kg^{-1}$, respectively, were obtained for all the OTCs.

A PFPD detector has been used in conjunction with GC separation for more sensitive and selective determination of OTCs, decreasing the LODs 25–50-fold compared to those obtained by FPD.[42] In this sense, Bravo *et al.* have studied and identified the interferences that occurred when analyzing OTCs in harbor surface sediments in order to optimize a new method based on head space (HS)-SPME after ethylation with NaBEt$_4$ for the analysis of eight OTCs in real harbor sediment samples.[43] LODs in the range $0.003–0.73 \, ng(Sn) \, L^{-1}$ were obtained, which are lower than those obtained by other methods previously reported for the simultaneous determination of butyl, phenyl and octyltin compounds.

7.2.1.3 Volatile Sulfur Compounds

Another important application of GC-FPD/PFPD is found in the analysis of volatile sulfur compounds (VSCs), which constitute a significant source of atmospheric pollution and are responsible for the taste and aroma (or off-flavors and odors) of food and beverages (such as wine, beer, *etc.*). VSCs have been determined most commonly through GC, with FPD being traditionally the most frequently applied detector due to the advantages previously noted. Also in this field, most research carried out in the last decade has focused on improving sample treatment and/or instrumentation. For example, Xiao *et al.* have compared three different sample treatment methods based on direct-SDME, HS-SDME and HS-SPME for the determination of dimethyl sulfide (DMS), diethyl sulfide (DES), dimethyl disulfide (DMDS), dipropyl disulfide (DPrDS) and dipropyl trisulfide (DPrTS) in beer, using GC-FPD, and obtaining the highest enrichment factor with HS-SDME.[44] In another recent study, carried out by Mochalski *et al.*, different pre-concentration processes, such SPME and sorbent trapping, a SPE with thermal desorption (TD), were compared, using FPD or MS detection.[45] Excellent LODs (at the ppt level) were obtained with the FPD detector when sorbent trapping was used as enrichment method; therefore, it could successfully replace the MS. HS-SPME followed by GC-PFPD has been also used for the determination of VSCs in wine.[46] The method implies a previous dilution of the samples to avoid matrix effects and short linear ranges due to saturation of the used fiber (carboxen-polydimethylsiloxane). After optimization, the linear dynamic ranges cover the normal ranges of occurrence of these analytes in wine, with recoveries ranging from 89–119% and LOQs $\leq 1.7 \mu g \, L^{-1}$. The method was applied successfully to the analysis of 34 Spanish wines.

Von Hobe *et al.* have developed a portable and automated analyzer for *in situ* analysis of VSCs in atmospheric samples.[47] The equipment consists of a dual sampling system, an electrically cooled cryotrap (small newly designed gas chromatograph) and a FPD detector. The equipment presents LODs around

20 ng L^{-1} when typical sampling routing conditions are applied, and could even be enhanced further by increasing the trapping time or using PFPD instead of FPD. A GC-PFPD dual-mode analytical system has been developed for the quantification of S-compounds over a wide range of concentrations in different environmental samples, such as high-(source-affected) and low-concentration (ambient air) samples, equipped with both a loop injection (LI) system for high concentrated samples (above 10 ppb) and a Peltier cooling (PC), a cryogenic pre-concentration technique, *e.g.* electrically cooled focusing trap with thermal desorption (TD) system, for low concentration mode (below 10 ppb).[48]

7.2.2 Nitrogen and Sulfur Selective Chemiluminescence Detectors (NCD and SCD)

Nitrogen and sulfur chemiluminescence detectors (NCD and SCD) are similar in terms of operating principles and both of them follow the same general detection scheme: a prior high-temperature pyrolytic reaction to convert all or nearly all nitrogen/sulfur (N/S) compounds into their respective chemilumi-nescent species, and a subsequent reaction of these products with ozone for the chemiluminescence detection. The reaction involved on these detectors as well as the principles, main characteristics and applications have been thoroughly described.[8,15,16,49]

Owing to the simplicity of the universal N/S CL detection processes, NCD and SCD present some inherent interesting characteristics for analytical che-mists, one of the most important being the *equimolarity*. Because all com-pounds are converted into common intermediate CL species, the detector yields equal responses to equal amounts of analytes on a molar basis, regardless of the molecular structure of the analytes. Therefore, quantitation of the N/S content of each analyte present in the sample (even unknown analytes) is possible without necessity of a calibration for each analyte. They are highly selective detectors, since the sample matrix and other non-N/S components are trans-formed into non-CL reagents (mostly CO_2 and H_2O) in the first step. On the other hand, NCD and SCD allow the *sensitive* determination of N/S containing compounds over a wide linear range because in CL methods the light emission is collected in a dark background, and because the CL reaction in the second step is of first order when the ozone is in excess.

7.2.2.1 Nitrogen Chemiluminescence Detectors (NCD) and Thermal Energy Analyzer (TEA)

NCDs were developed to detect all nitrogen-containing organics.[50] This is accomplished by the oxidative conversion, with or without catalyst, at around 800–1000 °C of any nitrogen-containing compound into nitric oxide (NO), which is the nitrogen CL species that then reacts with O_3 to produce a CL emission. The emission range is 600–3000 nm and it is centered on 1200 nm. Since at this spectral region little interferences are found, and because any carbon and hydrogen is oxidized to CO_2 and H_2O in the first oxidation reaction, the presence of any N-compound can be detected very selectively. The entire reaction mechanism can be illustrated as:

$$R-N + R-H + O_2 \rightarrow NO + CO_2 + H_2O \tag{7.1}$$

$$NO + O_3 \rightarrow NO_2^* \rightarrow NO_2^* + h\nu \text{ (near-IR)} \tag{7.2}$$

LODs have been demonstrated to be in the picogram range for organic analytes that contain at least one nitrogen atom.[51] The response is not always equimolar to compounds containing more than one nitrogen atom,[52] and, for example, the ratio of responses to *N*-nitrosodimethylamine and pyridine is not 3 : 1, as expected from their nitrogen molar ratio, but 2 : 1.[51]

A variation of the universal NCD is the so-called thermal energy analyzer (TEA), which provides a highly sensitive and selective response to *N*-nitrosamines (R–N–N=O). The N–NO bond is the weakest in the molecule and can be selectively broken under moderate conditions (at lower pyrolyzer temperature), which is easily achieved by changing the operating conditions of NCD without any hardware modification.[53] Although the simplest design uses only a heated tube as the pyrolysis chamber, these detectors have more often made use of catalysts, such as WO_3 and $W_{20}O_{58}$, to detect a wider variety of compounds at lower temperatures.[51,53]

7.2.2.2 Sulfur Chemiluminescence Detectors (SCD)

In SCD, introduced by Benner and Stedman,[54] the CL species are not simply produced by oxidative combustion – a high temperature reduction with H_2 is necessary after the oxidation to obtain the species that produce the CL emission when they react with O_3. This reductive pyrolysis may be attained either using a fuel-rich H_2O_2 flame or with H_2 as a separate reducing reagent following the oxidative combustion. It has been widely accepted that the signal is derived from the reaction of SO with O_3, with SO_2^* being the intermediate emitting species.[55] As several doubts still remain, the reduction products that react with O_3 to produce the CL emission will be referred as S-CL species, as in the following scheme:

$$R-S + R-H + O_2 \rightarrow SO_2 + CO_2 + H_2O \tag{7.3}$$

$$SO_2 + H \rightarrow S-CL \text{ species} \tag{7.4}$$

$$S-CL \text{ species} + O_3 \rightarrow SO_2^* \rightarrow SO_2 + h\nu \text{ (near-IR)} \tag{7.5}$$

The SCD is linear over four or more orders of magnitude and exhibits a nearly equimolar response to all S-compounds.[54,56] As in the case of the NCD, selectivity values of 10^7 for the flameless version and of 10^8 for the flame version have been routinely reported since the potential interferences, such as olefins, are combusted to form non-CL species such as CO_2 and H_2O.[15,56] LODs of

$25\,\mathrm{fg(S)\,s^{-1}}$ have been reported.[56] In comparison to the FPD, the SCD is more sensitive and exhibits a linear response.

7.2.2.3 Petrochemical Industry

Since their introduction, NCD and more largely SCD coupled with GC have found most applications in the petroleum industry, where S/N-compounds are undesirable in process and product streams, among other reasons because low concentrations of these compounds strongly poison the activity of expensive desulfuration catalysts and they are precursor of nitrogen and sulfur oxides (NO_x and SO_x) which are released after combustion. Many applications of GC-NCD/SCD have been developed, most of them in the late 1980s and early 1990s, for monitoring and checking the efficiency of the desulfuration and denitrogenation processes, and for the speciation of N/S-containing compounds in all types of petroleum fractions (from natural gas and light refinery streams to crude oil).[8]

There has been in the last ten years a growing interest in the use of comprehensive 2D GC (GC $\times$ GC) coupled to NCD/SCD, which has become in a powerful technique for improving characterization and identification of S-compounds due to higher resolution power and enhanced sensitivity. The potential of GC $\times$ GC-SCD for the separation of individual S-compounds in petroleum products streams has been demonstrated by Blomberg *et al.*[57] In this respect, an analytical method for the determination of S-containing compounds and their groups in different diesel oils by GC $\times$ GC-SCD was developed,[58] showing that the distribution of S-compounds in diesel oils from different process units are different. The speciation of these compounds in crude oil fractions was also achieved by the same authors, who were able to detect a total of 3620 compounds, including 1722 thiols/thioethers/disulfides/one-ring thiophenes, 953 benzothiophenes, 704 dibenzothiophenes and 241 benzonaphthothiophenes, in one injection.[59] This technique has also been compared with standard methods for speciation of S-containing compounds in middle distillates and more accurate and detailed results for benzothiophenes and dibenzothiophenes were obtained by GC $\times$ GC-SCD.[60] Wang *et al.* have developed a method for the speciation of S-containing compounds in diesel, and demonstrated the usefulness of this technique for the characterization of different diesels and to reflect desulfurization process efficiency.[61] They also demonstrated the capability of GC $\times$ GC-NCD for the speciation of N-containing compound classes in diesel.[62] Figure 7.2(A) illustrates a traditional GC-NCD chromatogram of a typical diesel sample with approximately 40 wppm total nitrogen. Because of the low concentration of individual nitrogen compounds, the NCD detector can barely detect them. Figure 7.2(B) demonstrates a GC $\times$ GC-NCD chromatogram of the same sample. Improvements between the two cases are seen both in sensitivity and in resolution. However, at higher sensitivity, there are large numbers of co-eluting/overlapping peaks in the chromatogram. To take full advantage of the two-dimensional character of the separation, the data in Figure 7.2(B) need to be processed. Figure 7.2(C)

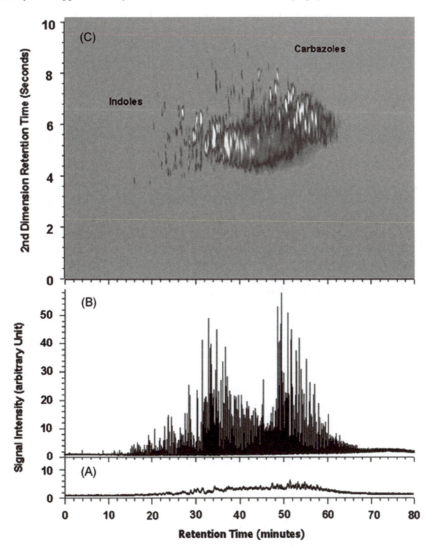

Figure 7.2 (A) A traditional GC-NCD chromatogram of a typical diesel sample with 40 wppm total nitrogen. (B) A GC × GC-NCD chromatogram of a typical diesel sample with 40 wppm total nitrogen. (C) A GC × GC-NCD chromatogram processed to illustrate the two-dimensional separation of a typical diesel sample with 40 wppm total nitrogen. (Reproduced with permission from ref. 62.)

displays the two-dimensional chromatogram. Major peaks have been identified by matching retention times with pure compounds or peaks were identified by alternative techniques. For example, materials corresponding to peaks were isolated by LC and identified by MS. Among the identified compounds, two major types are present, basic (*i.e.*, pyridines and quinolines) and neutral (*i.e.*, indoles and carbazoles) nitrogen compounds. Figure 7.2(C) demonstrates two well-separated bands of nitrogen-containing compounds as well as a broader range of basic nitrogen-containing compounds. The upper band contains carbazoles and the lower band contains indoles. With this resolution, each subgroup as well as the class can be integrated independently and quantified.

Adam *et al.* have optimized the separation and identification of N-compounds in middle distillates by GC × GC-NCD, and applied the method to a wide range of diesels feedstocks and demonstrated that the detailed speciation of N-compounds in different diesel and coal samples is possible with no use of MS.[63,64]

7.2.2.4   N-Nitrosamine Compounds

NCD and TEA have been largely used for the analysis of nitrosamines, many of which are known carcinogens. Although a classical application of this detector is found in the tobacco industry, these compounds can be found in such diverse matrices as foods, cosmetics and environmental samples of soil and water. Grebel *et al.* have developed a method for the analysis of seven *N*-nitrosamines by SPME in groundwater.[65] When three detectors, NCD, FPD and MS, were compared after the GC separation, NCD provided most reliable results in terms of both selectivity and reproducibility. A sensitive method for the determination of *N*-nitrosodimethylamine (NDMA) at ppt level in groundwater by means of a TEA detector has been described by Tomkins.[66] More recently, Andrade *et al.* have proposed a method based on HS-SPME with GC-TEA for the analysis of volatile nitrosamines, NDMA, *N*-nitrosodiethylamine (NDEA), *N*-nitrosopiperidine (NDEA) and *N*-nitrosopyrrolidine (NPYR) in sausages.[67] LODs and LOQs were 3 and $10\,\mu g\,kg^{-1}$ for the crude meat, and recoveries ranged from 105 to 110% with RSD ($n = 6$) lower than 6%, except for *N*-nitrosopyrrolidine (12%).

7.2.2.5   Food, Biological and other Analysis

SCD and NCD have also been widely applied to the determination of flavors for the characterization of beverages and food. In conjunction with GC, SCD with other detectors such as FID, FPD and MS has been used for analysis of VSCs for the sensitive determination of these compounds in wines, beers, molasses flavor isolates, cooked milk,[8] *etc.*

Some applications have been also developed in the pharmaceutical field. One example is the improved GC-SCD method developed by Okamoto *et al.* for the analysis of S-1452, an S-containing antiasthma drug, and its nine metabolites in human plasma, reaching LOQs in the range 0.5–$1.0\,ng\,mL^{-1}$ in this complex matrix with no interferences from biological components present.[68] Furne *et al.* have optimized a simple technique to measure sulfide in fecal homogenates or any other liquid milieu, which involves acidification followed by measurement of H_2S by GC-SCD. The use of a SCD facilitated these measurements, since it specifically and sensitively responds to S-containing compounds.[69]

A TEA detector can also be used to quantify nitroaromatics, a class of compounds that includes many explosives and various reactive intermediates used in the chemical industry. Several of them are known carcinogens and are found as environmental contaminants. Application of GC-TEA to the analysis of explosives containing N-compounds has been reviewed by Jiménez *et al.*, illustrating the important role that GC-TEA technique plays in this area.[70] Nitroaromatic compounds have been repeatedly identified in organic aerosol particles, formed from the reaction of polycyclic aromatic hydrocarbons with atmospheric nitric acid at the particle surface.[16]

7.3 Liquid Chromatography Coupled to Chemiluminescence Detection

A wide variety of analytical methods has been developed in different fields since the first development of CL-based techniques as a means of detection for HPLC in the 1980s. The application of CL as detection system in HPLC requires the generation of CL emission by a post-column reaction of the analytes in the column eluent with the reagent(s). Therefore, a high efficiency of the CL reaction is required. In addition, this assembly requires additional pump(s) to deliver post-column CL reagent or the incorporation of devices for a rapid mixing of the column eluate with the reagent solution(s) to obtain a stable baseline. Considering that the CL reactions used in HPLC are very rapid and are accompanied by intense CL, it is necessary to use a reaction coil between the mixing device and the detector, in which the length and diameter must be adjusted to optimize the CL reaction time. This will allow the measurement of CL intensity at the time that the maximum CL emission is observed, by using the maximum S/N ratio as a criterion. However, the most important problem of HPLC-CL coupling is the compatibility between the chemical environments required for an efficient chromatographic separation (mainly the composition of the mobile phase) and those required for an efficient and sensitive CL emission (reaction temperature, pH, solvent, nature of the CL precursor and coexisting compounds such as catalyst and enhancer affecting the CL reaction yield). Thus, careful optimization of all the variables involved in the HPLC-CL coupling is important. Concerning the design of a HPLC-CL system, the detection cell is usually a spiral coil with volumes ranging 60 to 120 µL, and placed just in front of a photomultiplier tube. The flow rates of the mobile phase and post-column reagents must be optimized to prevent alteration of the flow rate, while obtaining a good mixing efficiency and CL reaction yield. In addition, pumps capable of delivering pulse-less flows are required for mobile phase and CL reagents to obtain reproducible detections.[71]

Different configurations have been considered to achieve a sensitive and selective determination of analytes, related to the good resolution of the HPLC separation, adequate efficiency of the CL system, stability of the reagents or the compatibility of the CL reaction conditions and the requirement for the HPLC separation, most of them in a post-column manner.[71] Figure 7.3 shows these

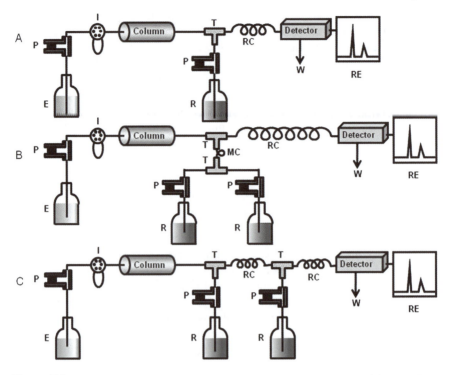

Figure 7.3 Most common CL detection systems in combination with HPLC. P, pump; I, injector; T, mixing tee; RC, reaction coil; MC, mixing coil; RE, recorder; E, eluent; R, reagent; W, waste.

different assemblies. Figure 7.3(B) and (C) shows the most common approaches, where the reagent solutions are pumped after the separation column and first combined and mixed with the eluent or sequentially added to the eluent. Applications of HPLC-CL coupling in several fields, such as clinical, environmental and food analysis[9] or the syntheses of analytical reagents for CL detection for HPLC-CL analysis of drugs of abuse,[72] have been published recently. Although different reagents have been proposed for HPLC-CL systems, the most commonly used will be discussed here along with some selected applications.

7.3.1 Peroxyoxalate Reaction

The peroxyoxalate (PO) reaction involves the H_2O_2 oxidation of an aryl oxalate ester in the presence of a fluorophore, following a chemically initiated electron exchange luminescence mechanism *via* a high-energy intermediate 1,2-dioxetanedione.[73] The high sensitivity, versatility and the relative absence of interferences make this system very useful.[74] The most widely used oxalate ester is bis-(2,4,6-trichlorophenyl) oxalate (TCPO) followed by bis-(2,4-dinitrophenyl) oxalate (DNPO) or bis[4-nitro-2-(3,6,9-trioxadecyloxycarbonyl)phenyl] oxalate (TDPO). The main disadvantage of this system resides in the insolubility of the above-mentioned compounds in water and their instability towards hydrolysis, which requires the use of organic solvents such as acetonitrile, dioxane, *tert*-butanol and ethyl acetate. This reaction can be used to determine a great number of species such as hydrogen peroxide, compounds that are highly fluorescent (*e.g.* polycyclic aromatic hydrocarbons) or compounds that do not exhibit native fluorescence but can be chemically derivatized using dansyl chloride, *ortho*-phthalaldehyde (OPA) or fluorescamine, among others.[75] However, a problem with PO-CL when it is applied to post-column detection in HPLC is the relatively long duration of the light-emitting reaction, resulting in a light emission profile that lasts for more than 60 s, and can produce extra-column band broadening. For this reason, the use of a catalyst to speed up the reaction while maintaining reasonable high quantum efficiency is mandatory. This enables a "chemical narrowing" of analytes separated in HPLC and allows the effective cell volume to become dependent on the speed of mixing of eluate and reagent, rather than on the total detection cell volume. Imidazole has been proved to be an efficient catalyst for the PO-CL reaction[76,77] due to its ability to destabilize the PO by forming a less stable intermediate, which is thus more susceptible to nucleophilic attack by hydrogen peroxide.[78] In the HPLC-PO-CL systems the configuration shown in Figure 7.3(B) is usually employed, where two reagents solutions (POs and peroxide) are mixed on-line, and the mixture is then added to the HPLC effluent. Analytical applications of PO-CL system, including its coupling with HPLC, have been covered in several reviews.[74,75] Selected applications noted in this section are included in Table 7.1.

7.3.1.1 Clinical and Pharmaceutical Analysis

Considering clinical applications of the HPLC-PO-CL coupling, the determination of catecholamines (CAs) is of great interest. As their concentration in biological tissues is very low, very sensitive methods are required. Different approaches have been proposed for their determination in different samples, using SPE for analyte extraction and different derivatization reagents before their post-column CL reaction using POs and hydrogen peroxide.[79] Imai's group has extensively studied the determination of CAs by HPLC-PO-CL, proposing different automated methods for the determination of CAs in rat plasma[80,81] and studying different aspects of the role of CAs and their derivatives.[82–86]

TCPO has been the PO of choice in the determination of phenols in urine samples by an integrated derivatization-HPLC-PO-CL detection unit[87] and 5-hydroxyindoles in human platelet-poor plasma.[88] However, other POs have also been used in different HPLC-PO-CL methods, such as bis(2,4,5-trichloro-6-carbopentoxyphenyl)oxalate (CPPO) in the determination of stimulants in human hair sample,[89] TDPO in the determination of penbutolol and its metabolite 4-hydroxypenbutolol in rat plasma,[90] DNPO for the detection of estradiol in plasma samples,[91] DNPO for the indirect determination of

Table 7.1 Selected applications of HPLC-CL using POs.

Analyte	CL system/HPLC characteristics[a]	LOD	Applications	Ref.
Catecholamines (nor-epinephrine, epinephrine and dopamine)	TDPO–H$_2$O$_2$– imidazole. TSK gel ODS-120T (5 μm, 250 × 4.6 mm i.d.), Tosoh Co.; imidazole buffer (120 mM, pH 5.8)–methanol–acetonitrile (13 : 4 : 18, v/v/v), 0.8 mL min⁻¹, 24–25 °C	40–120 amol	Human blood plasma	79
Nitrocatecholamines	TDPO–H$_2$O$_2$–imidazole. Unison UK-C$_{18}$ (250 × 4.6 mm i.d.), Imtakt; potassium acetate buffer (75 mM, pH 3.2)–potassium dihydrogen phosphate buffer (50 mM, pH 3.2)–acetonitrile containing sodium 1-heptanesulfonate (5 mM) (86.5 : 4.5 : 9, v/v/v), 0.5 mL min⁻¹.	75–150 fmol	Rat brain	84
Phenol, 4-methylphenol	TCPO–H$_2$O$_2$. C$_{18}$ Nova-Pack (4 μm, 250 × 4.6 mm i.d.), Waters; water–acetonitrile (30 : 70, v/v), 1 mL min⁻¹	0.3–3.1 μg L⁻¹	Human urine	87
5-Hydroxyindoles	TCPO–H$_2$O$_2$–imidazole. Wakosil-II 5C$_{18}$ RS (5 μm, 150 × 4.6 mm i.d.) Wako Pure Chemicals; acetonitrile–Tris–HCl buffer (50 mM, pH 6.09 (2 : 3, v/v), 0.8 mL min⁻¹, 20–23 °C	0.5–1.2 fmol in 100 μL	Platelet-poor plasma	88
3,4-Methylenedioxyme-tham-phetamine, 3,4-methylene-dioxyamphetamine, methamphetamine, amphetamine	CPPO–H$_2$O$_2$–imidazole. Capcellpak C$_{18}$ UG120 (5 μm, 250 × 1.5 mm i.d.), Shiseido; imidazole (20 mM)/HNO$_3$ buffer (pH 7.0)–acetonitrile–THF (51.5 : 45 : 3.5, v/v), 0.1 mL min⁻¹	0.02–0.16 ng mg⁻¹	Hair	89

Analyte	Conditions	Detection limit	Sample	Ref.
Penbutolol, 4-hydroxypenbutolol	TDPO–H_2O_2–imidazole. Nucleosil ODS 100-5C_{18}, (5 μm, 150 × 4.6 mm i.d.), GL 18 Science Inc.; imidazole buffer (0.01 M, pH 7.0)–acetonitrile (80 : 20, v/v), 0.9 mL min^{-1}	9.9 and 15 fmol for penbutolol and 4-hydroxypenbutolol	Rat plasma	90
Estradiol	DNPO–H_2O_2–imidazole. CapcelPak ODS (250 × 4.6 mm i.d.), Shiseido; acetonitrile–water (85 : 15, v/v), 0.5 mL min^{-1}, 40 °C	44 fmol	Rat plasma	91
Artemisinin	DNPO–H_2O_2–imidazole. Discovery HS C_{18} (5 μm, 250 × 4.6 mm i.d.), Supelco; imidazole–HNO_3 buffer (20 mmol L^{-1}, pH 8.50) containing 70% acetonitrile, 0.5 mL min^{-1}	0.062 μmol L^{-1}	Human serum	92
Vitamin K homologues (phylloquinone (PK), menaquinone-4 (MK-4) and menaquinone-7 (MK-7)	TDPO–H_2O_2–imidazole. Develosil ODS UG-5 (50 × 1.5 mm i.d.), Nomura Chemicals; imidazole–HNO_3 buffer (600 mM, pH 9.0)–acetonitrile (5 : 95, v/v), 0.25 mL min^{-1}	32–85 fmol	Human plasma	93
Aliphatic amines and polyamines	TCPO–H_2O_2–imidazole. LichroCart RP 18 (5 μm 4 × 125 mm), Merck; acetonitrile–imidazole solution (1 mM, pH 7.0) (50 : 50, v/v) in the gradient elution mode, 1.5 mL min^{-1}	0.5–0.9 μg L^{-1}	Tap, irrigation ditch, lake, residual and marine water	94
Nitropolycyclic aromatic hydrocarbons	TCPO–H_2O_2. Cosmosil C_{18} (5 μm, 4.6 × 250 mm), Nacalai Tesque; acetonitrile–10 mM imidazole–$HClO_4$ buffer (pH 7.6), 1.0 mL min^{-1}, 40 °C	1×10^{-11}–40×10^{-11} M	Airborne particulates	96
N-Methylcarbamates (carbaryl, carbofuran and propoxur)	TCPO–H_2O_2. Nova-Pack C_{18} (4 μm, 250 × 4.6 mm i.d.), Waters; acetonitrile–water-methanol (55 : 37 : 8, v/v), 1 mL min^{-1}	2–3 μg L^{-1}	Fruit juices (apple, pineapple and grapefruit)	99

Table 7.1 (*continued*)

Analyte	CL system/HPLC characteristics[a]	LOD	Applications	Ref.
Organic peroxides (benzoyl peroxide, *tert*-butyl hydroperoxide, *tert*-butyl perbenzoate, cumene hydroperoxide), hydrogen peroxide	TCPO–H_2O_2–imidazole. Chemcosorb 5-ODS-UH (5 μm, 150 × 4.6 mm i.d.), Chemco; imidazole–HNO_3 buffer (20 mM, pH 7.5)–acetonitrile (40 : 60, v/v), 0.5 mL min^{-1}	1.1–31.3 μM	Benzoyl peroxide in wheat flour	102
Sulfonamides	TCPO–H_2O_2–imidazole. Luna (5 μm, 150 × 4.6 mm i.d.), Phenomenex; tri-sodium citrate 5 1/2 hydrate–citric acid mono-hydrate buffer (10 mM, pH 5.2)–acetonitrile–THF (65 : 22 : 13, v/v/v), 1.0 mL min^{-1}	6.2–13.6 μg L^{-1}	Bovine raw milk	103
Bisphenol A	TDPO–H_2O_2. Daisopak-SP-120-5-ODS-BP (5 μm, 250 × 4.6 mm i.d.), Daiso; imidazole–HNO_3 buffer (40.0 mM, pH 7.0)–acetonitrile (17 : 83 v/v), 1 mL min^{-1}	0.38 μg L^{-1}	Hot water in contact with commercially available baby bottle	104

[a]TDPO: bis [(2-(3,6,9-trioxadecanyloxycarbonyl)-4-nitrophenyl] oxalate; TCPO: bis-(2,4,6-trichlorophenyl) oxalate; CPPO: bis(2,4,5-t:ichloro-6-carbopentox-yphenyl) oxalate; DNPO: bis-(2,4-dinitrophenyl) oxalate.

artemisinin in human serum[92] and TDPO for the indirect determination of vitamins in human plasma,[93] both of them previously irradiated under UV light to produce hydrogen peroxide.

7.3.1.2 Environmental and Food Analysis

Some applications concerning environmental analysis using HPLC-PO-CL can also be found in the literature. Thus, amines[94] and ammonium[95] have been determined in different water samples by using TCPO and dansyl chloride on solid sorbents as derivative reagent. The reaction product between TCPO and H_2O_2 is 2,4,6-trichlorophenol, which can produce a quenching effect on CL. To avoid this product, both reagents were prepared separately and mixed during the reaction procedure, affording an increase in the sensitivity of $\times 5$ when the TCPO reagent was included in the system before the H_2O_2 reagent and an acetonitrile–tetrahydrofuran mixture was used as TCPO solvent. In addition, an HPLC-PO-CL system with on-line cleanup, reducer and concentrator columns was developed for determining nitropolycyclic aromatic hydrocarbons (NPAHs) in airborne particulates[96,97] and mutagenic nitrobenzanthrone isomers.[98]

In the field of food analysis, the TCPO-CL reaction has been proposed for the determination of some pesticides by HPLC after their hydrolysis[99,100] in different food samples, biogenic polyamines in olives[101] and organic peroxides in spiked wheat flour previously converted into hydrogen peroxide by on-line UV irradiation.[102] Instead of TCPO, TDPO was chosen for the analysis of sulfonamides in raw milk, previously derivatized with fluorescamine; detection limits in the low $\mu g\,L^{-1}$ were obtained, which are below the maximum residue limits established by the European legislation.[103] Figure 7.4 shows a chromatogram of a spiked raw milk sample. The use of TDPO provided higher sensitivities and stabilities, avoiding precipitation problems that eventually occur with other oxalates. TDPO was also used in the determination of Bisphenol A in hot water in contact with commercially available baby bottles, previously derivatized with a fluorescent labeling reagent [4-(4,5-diphenyl-1*H*-imidazol-2-yl)benzoyl chloride].[104]

7.3.2 Tris(2,2′-bipyridine)ruthenium(II) Reaction

Tris(2,2′-bipyridine)ruthenium(II) [$Ru(bpy)_3^{2+}$] is another CL reagent used for post-column CL reaction in HPLC. $Ru(bpy)_3^{2+}$ is the stable species in solution and the reactive species, $Ru(bpy)_3^{3+}$, can be generated from $Ru(bpy)_3^{2+}$ by oxidation. $Ru(bpy)_3^{3+}$ can react with analytes containing tertiary, secondary and primary alkyl amine groups to form the excited state [$Ru(bpy)_3^{2+}$]*, which will decay to the ground state emitting orange light at 610 nm. Three main methods have been reported to obtain $Ru(bpy)_3^{3+}$ from $Ru(bpy)_3^{2+}$: chemical oxidation, photochemical oxidation [where a photoreactor is included in the system for the on-line oxidation of $Ru(bpy)_3^{2+}$ to $Ru(bpy)_3^{3+}$] and electrogenerated chemiluminescence (ECL), where the production of light following the oxidation of $Ru(bpy)_3^{2+}$ to $Ru(bpy)_3^{3+}$ is carried out on an electrode surface. Among them,

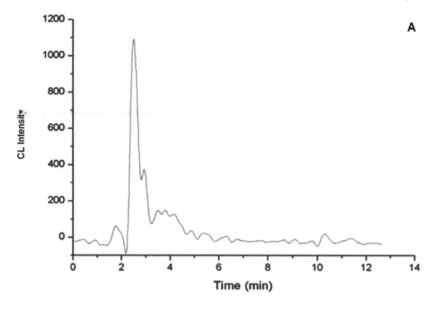

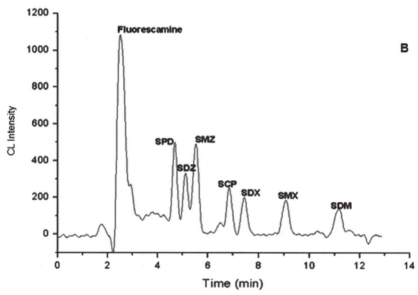

Figure 7.4 Chromatogram of a blank sample (A) and a bovine raw milk sample spiked with $20\,\mu g\,L^{-1}$ of each sulfonamide (B). SPD, sulfapyridine; SDZ, sulfadiazine; SMZ, sulfamethazine; SCP, sulfachloropyridazine; SDX, sulfadoxine; SMX, sulfamethoxazole; and SDM, sulfadimethoxine. (Reproduced with permission from ref. 103.)

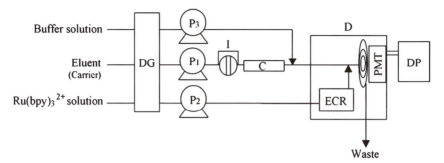

Figure 7.5 Schematic diagram of an HPLC-ECL detection system. P, pump; DG, degasser; I, injector; C, column; D, detector; ECR, electrochemical reactor; PMT, photomultiplier tube; DP, data processor. (Reproduced with permission from ref. 121.)

ECL is the most widely used for quantitative analysis; however, it has certain limitations, including loss of sensitivity due to dilution of the reagent and analytes, together with electrode fouling, the need to recondition the electrode surface to retain sensitivity and additional instrumentation, such as a potentiostat. On the other hand, the external electrochemical generation is simpler and has the advantages of reproducibility, simplicity and cost performance. Some contributions have summarized the analytical usefulness of $Ru(bpy)_3^{2+}$, including proposed devices for HPLC-CL couplings.[105–110] Figure 7.5 shows a typical HPLC system coupled to ECL.

Different factors affect the coupling of HPLC with $Ru(bpy)_3^{2+}$ CL, especially the necessity of a medium compatible not only with the CL reaction but also with the chromatographic separation.[111] Other aspects of this CL reaction should also be taken into account, such as the pH, flow rate and the configuration of the flow cell, especially in the case of ECL. For example, it is well known that hydroxide ions will react with $Ru(bpy)_3^{3+}$ to generate light; therefore, acidic conditions are often required. Thus, if the pH of the separation buffer and the pH of the detection solution are different, a post-column pH adjustment is used.[105] Some selected applications of this system are noted here and included in Table 7.2.

7.3.2.1 Clinical and Pharmaceutical Analysis

The ECL system using $Ru(bpy)_3^{2+}$ has become a powerful tool for the determination of compounds containing tertiary amine groups due to its high sensitivity and selectivity, or compounds that can be converted into derivatives suitable for ECL detection by tagging appropriate molecules with an analyte. In this sense, 2-(2-aminoethyl)-1-methylpyrrolidine, *N*-(3-aminopropyl) pyrrolidine (NAPP)[112] and 3-isobutyl-9,10-dimethoxy-1,3,4,6,7,11b-hexahydro-2*H*-pyrido-[2,1-*a*]isoquinolin-2-ylamine (IDHPIA)[113] have proven to be sensitive derivatization reagents for the determination of carboxylic acids. They were tested in a

Table 7.2 Selected applications of HPLC-CL using the tris-(2,2'-bipyridyl)ruthenium(II) system.

Analyte	CL system/HPLC characteristics	LOD	Applications	Ref.
Carboxylic acids (fatty acids and phenylbutyric acid)	ECL Ru(bpy)$_2^{2+}$. Inertsil C$_8$ (150 × 4.6 mm i.d.) GL Sciences Inc.; 52% acetonitrile containing 0.05% TFA, 1.0 mL min^{-1}, 23 ± 2 °C	0.6 fmol (phenylbutyric acid), 0.5 fmol (myristic acid)	Human serum	113
β-Blockers	ECL Ru(bpy)$_2^{2+}$. XTerra RP18 (5 μm, 150 × 4.6 mm i.d.), Waters; 15% acetonitrile in borate buffer (10 mM, pH 9.0), 1.5 mL min^{-1} 25 ± 2 °C	0.5 μM (atenolol), 0.08 μM (metoprolol)	Human urine	115
Oxycodone, noroxycodone	ECL Ru(bpy)$_2^{2+}$. Develosil ODS HG-5 (5 μm, 150 × 4.6 mm i.d.), Nomura Chemical; phosphate buffer (0.1 M, pH 2.6) containing sodium dodecyl sulfate 5 mM (pH 2.6)–methanol (53 : 47, v/v), 1.0 mL min^{-1}, 50 °C	0.5 ng mL^{-1} (oxycodone), 1 ng mL^{-1} (noroxycodone) (as LOQ)	Human plasma	116
Tricyclic antidepressants	ECL Ru(bpy)$_2^{2+}$. YMC-Pack TMS (5 μm, 150 × 4.6 mm i.d.), YMC; sodium phosphate buffer (50 mM, pH 7.0)–acetonitrile (55 : 45, v/v), 1.0 mL min^{-1}	12–105 fmol (20 μL)	Human plasma	118
Hydroxyproline	ECL Ru(bpy)$_2^{2+}$. Develosil C30 UG-5 (250 × 4.6 mm i.d., 5 μm), Nomura Chemicals; acetate buffer (100 mM, pH 5) containing sodium 1-octanesulfonate (6 mM) and copper(II) sulfate (1.2 mM), 0.5 mL min^{-1}	30 nmol mL^{-1} (as LOQ)	Rat urine	119
Amino acids	ECL Ru(bpy)$_2^{2+}$. L-column ODS (150 × 4.6 mm i.d.), Kagakuhin; acetonitrile–KH$_2$PO$_4$ (10 mM) (15 : 85, v/v), 1.0 mL min^{-1}, 50 °C	0.04–8.0 pmol	Human plasma	120

Analyte	Method	Detection limit	Sample	Ref.
Pipecolic acid	ECL Ru(bpy)$_2^{2+}$. Chromolith Performance RP-18e column (100 × 4.6 mm i.d.), Merck (× 2); acetic acid solution (20 mM, pH 3.5) containing sodium octanesulfonate (15 mM), 0.7 mL min^{-1}	24 fmol	Human serum, cow's milk, beer, apple juice	121
Erythromycin	Sol–gel-immobilized Ru(bpy)$_3^{2+}$ ECL sensor. XTerra RP18 (5 μm, 150 × 4.6 mm i.d.), Waters; 30% acetonitrile in phosphate buffer (0.05 M, pH 7.0), 1.0 mL min^{-1}, (25 ± 2 °C)	1.0 μM	Human urine	122
Positron emission tomography, radiopharmaceuticals	KMnO$_4$ for chemical generation of Ru(bpy)$_3^{3+}$ (1) Xterra RP18 (150 × 3.9 mm i.d.), Waters; sodium phosphate buffer (100 mM, pH 7.2) (2) Xterra RP8 (150 × 3.9 mm i.d.), Waters; CH$_3$CN–ammonium acetate (100 mM) (30 : 70, v/v) (3) Xterra RP8 (150 × 3.9 mm i.d.), Waters; CH$_3$CN–sodium phosphate buffer (100 mM, pH 2.6) (30 : 70, v/v), 1.0 mL min^{-1}	0.6–6.5 ng mL^{-1}	Pharmaceutical samples	124
Papaver somniferum alkaloids	Solid lead dioxide for chemical generation of Ru(bpy)$_2^{3+}$. ChromolithTM SpeedROD RP-18e (50 × 4.6 mm i.d.), Merck; acetonitrile (0.1% TFA)–water (0.1% TFA), 3 mL min^{-1}, gradient elution program	1×10^{-10} M (codeine), 1×10^{-9} M (thebaine)	Industrial alkaloid extracts	125
Psilocin, psilocybin	KMnO$_4$ for chemical generation of Ru(bpy)$_2^{3+}$. Synergi Max-RP C$_{12}$ (4 μm, 150 × 4.6 mm i.d.), Phenomenex; methanol–10 mM ammonium formate (pH 3.5) (95 : 5, v/v), 0.5 mL min^{-1}	1.2×10^{-8} mol L^{-1} (Psilocin) 3.5×10^{-9} mol L^{-1} (Psilocybin)	Hallucinogenic mushrooms	127

Table 7.2 (continued)

Analyte	CL system/HPLC characteristics	LOD	Applications	Ref.
N-Methylcarbamates (bendiocarb, carbaryl, promecarb and propoxur)	Photo-oxidation of Ru(bpy)$_2^{2+}$. Ultrasphere C$_{18}$ (5 μm, 4.6 × 45 mm), Beckman; acetonitrile–water–imidazole, gradient elution mode, 1.2 mL min^{-1}	3.9–36.7 ng L^{-1}	Mineral, tap, ground and irrigation ditch water, pear and apple	130
Catechins [(–)-epicatechin, (–)-epigallocatechin gallate]	ECL Ru(bpy)$_2^{2+}$; on-line photochemical derivatization. Chromolith-Performance RP-18e (5 μm, 100 × 4.6 mm i.d.), Merck; 20 mM phosphoric acid–methanol (80 : 20, v/v), 0.5 mL min^{-1}	1.2, 0.8 pmol	Tea	133
Quinolones	Ce(SO$_4$)$_2$–HNO$_3$ for chemical generation of Ru(bpy)$_2^{3+}$. Zorbax Eclipse XDB-C$_8$ (5 μm, 150 × 4.6 mm i.d.), Agilent Technologies; acetonitrile–methanol–ammonium acetate buffer (containing 7.5 × 10^{-4} M tetra-butylammonium bromide, 0.8% (v/v) trie-thylamine, 1.0 × 10^{-4} M ammonium acetate, pH 3.65) (3 : 15 : 82, v/v/v), 1.0 mL min^{-1}, 30 °C	0.36–10.4 ng mL^{-1}	Prawn	135

reversed-phase HPLC-CL with ECL using $Ru(bpy)_3^{2+}$ added to HPLC effluent with a spiral flow cell. The methods were applied to the determination of free fatty acids in human plasma. A similar system was used for the determination of erythromycin A and erythromycin B, and its metabolite decladinosyl erythromycin A in rat plasma and urine.[114] Other reported applications of the ECL reaction are the determination of β-blockers in human urine,[115] oxycodone and its metabolite noroxycodone in human plasma[116] and in dog plasma and urine,[117] tricyclic antidepressants in human plasma based on the detection of aliphatic secondary or tertiary amino moieties,[118] hydroxyproline in urine,[119] several amino acids in plasma after derivatization with divinyl sulfone[120] and pipecolic acid (a cyclic amino acid that acts as a biomedical marker for human peroxisomal disorders) in different samples, including human serum.[121]

Another possible use of ECL is the immobilization of $Ru(bpy)_3^{2+}$ on the surface of an electrode. This approach has been used for constructing a sol–gel-immobilized $Ru(bpy)_3^{2+}$ ECL sensor used as detector in HPLC for the determination of erythromycin in urine samples,[122] and erythromycin in spiked urine samples and phenothiazine derivatives.[123]

The FLB457 preparation (a useful radioligand for measuring the dopamine D_2 receptor density) in pharmaceutical samples[124] is an example of the generation of $Ru(bpy)_3^{3+}$ from $Ru(bpy)_3^{2+}$ by means of chemical oxidation. In addition, the use of monolithic columns has been proposed in a dual system, which offers the possibility of using two different CL reagents: $Ru(bpy)_3^{2+}$ in an acidic solution with solid lead dioxide as oxidant, or acidic potassium permanganate (depending on the analytes to be determined) in a rapid and highly sensitive method to monitor the extraction process of opiate alkaloids from *Papaver somniferum*.[125] Following the same strategy, a hybrid FIA/HPLC system using also monolithic columns and acidic potassium permanganate or $Ru(bpy)_3^{2+}$ as CL reagents, depending on the target analyte,[126] and a dual CL reagent for the determination of psilocin and psilocybin in hallucinogenic mushrooms,[127] where the CL reagent consisted on a mixture of acidic potassium permanganate and $Ru(bpy)_3^{2+}$, have also been proposed. In the latter case both reagents were used together, as potassium permanganate can oxidize $Ru(bpy)_3^{2+}$ to $Ru(bpy)_3^{3+}$; thus, the former acted as both the oxidant for the latter and as a CL reagent in its own right.

Concerning the generation of $Ru(bpy)_3^{3+}$ from $Ru(bpy)_3^{2+}$ by photo-oxidation, an example is the determination of amiodarone and desethylamiodarone in serum and pharmaceutical formulations, based on the post-column photolysis of the analytes into photoproducts that are active in the $Ru(bpy)_3^{2+}$ system. The method utilizes two photoreactors, one for the on-line generation of $Ru(bpy)_3^{3+}$ and the other for the photodegradation of the analytes.[128]

7.3.2.2 Environmental and Food Analysis

Although applications of this system in environmental analysis are not very common, some applications of this reaction by the generation of $Ru(bpy)_3^{3+}$ from $Ru(bpy)_3^{2+}$ by photo-oxidation can be found in the literature, such as the determination of nitrosamines in water without derivatization by means of an automatic and sensitive SPE-HPLC-CL method.[129] The system consists of two photoreactors, one for the on-line generation of $Ru(bpy)_3^{3+}$ and the other for the photodegradation of the analytes to the corresponding amines. A similar assembly has been proposed for the determination for N-methylcarbamate residues[130] and for the determination of aminopolycarboxylic acids,[131] both in water samples. For aminopolycarboxylic acids, only one photoreactor was required, as the analytes did not need a previous photo-oxidation.

Some of the applications of ECL reaction in food analysis are the determination of pipecolic acid in apple juice, cow's milk and beer,[121] domoic acid (a neurotoxic amino acid associated with shellfish poisoning, which contains a secondary amine moiety) in blue mussel,[132] catechins in tea based on their on-line photochemical derivatization,[133] and α-solanine and α-chaconine (tri-saccharide glycosides with a common tertiary amine, which act as reducing agents in the $Ru(bpy)_3^{3+}$ CL reaction) in potatoes.[134] The determination of N-methylcarbamates pesticide in pear and apple samples[130] and EDTA in canned foods[131] by photo-oxidation of $Ru(bpy)_3^{2+}$, and quinolones in prawn by means of chemical oxidation using the $Ce(IV)–Ru(bpy)_3^{2+}$ system,[135] have also been reported.

7.3.3 Luminol and Derivative Reaction

The best known example in direct CL reactions is the oxidation of luminol (3-aminophthalylhydrazide) in alkaline medium, to produce the excited 3-aminophthalate anion, which emits light when relaxing to the ground state. Several oxidants such as permanganate, periodate, hexacyanoferrate(III) and hydrogen peroxide are commonly used. Luminol-type compounds can be used as derivatization reagents, allowing the analytes to be detected at very low levels.[136,137] The luminol reaction, when hyphenated with HPLC, provides a high efficiency in separation and low limits of detection (LODs) inherent to CL. Some selected applications are included in Table 7.3.

7.3.3.1 Clinical and Pharmaceutical Analysis

One of the applications of HPLC-CL with post-column luminol-type reaction in clinical analysis is the determination of endogenous lipid hydroperoxides (LOOHs), which react with cytochrome *c*, resulting in the formation of lipid alkoxyl radical, which subsequently reacts with luminol to generate CL.[138–140]CAs have also been determined by means of this reaction, as they produce an enhancement effect on the CL emission of the luminol system using potassium ferricyanide as oxidant;[141] CAs also strongly inhibit the CL of the luminol–I_2 system and this effect has been used in a HPLC-CL method for the quantification of norepinephrine, epinephrine and dopamine in urine samples.[142] These analytes have also been determined using gold nanoparticles, as it was found that monoamine neurotransmitters could inhibit the CL from

Table 7.3 Selected applications of HPLC-CL using the luminol reaction.

Analyte	CL system/HPLC characteristics	LOD	Applications	Ref.
Phosphatidylcholine hydroperoxides	Luminol–cytochrome *c*. Mightysil RP-18 GP (5 μm, 250 × 4.6 mm i.d.), Kanto Chemical Co.; methanol–isopropanol (20 : 1, v/v), $0.5\,mL\,min^{-1}$	1.0 pmol	Human plasma	140
Catecholamines (epinephrine, noradrenaline, dopamine)	Luminol–$K_3[Fe(CN)]_6$. C-18 (150 × 4.6 mm i.d.); potassium hydrogen phthalate (0.01 M, pH 4.5)–methanol (92 : 8, v/v), $1\,mL\,min^{-1}$	8.0×10^{-10}– $1.0 \times 10^{-9}\,mg\,L^{-1}$	Human serum	141
Catecholamines (nor-epinephrine, epinephrine, dopamine, L-dopa)	Luminol–iodine. Lichrosorb LC-8 (5 μm, 150 × 4.6 mm i.d.), Supelco; potassium dihydrogen phosphate (50 mM)–methanol ($100\,mL\,L^{-1}$)–acetonitrile ($200\,mL\,L^{-1}$)– sodium dodecyl sulfate ($500\,mg\,L^{-1}$)– EDTA ($250\,mg\,L^{-1}$), pH 3.5, $1.0\,mL\,min^{-1}$	0.26–$0.73\,\mu g\,L^{-1}$	Human urine	142
Levodopa	Luminol–$K_3[Fe(CN)]_6$. Microsorb C_{18} silica (5 μm, 150 × 4.6 mm i.d.), Rainin; potassium hydrogen phthalate (0.01 M, pH 4.5)–methanol (92 : 8 v/v), $1.0\,mL\,min^{-1}$	$3 \times 10^{-9}\,g\,mL^{-1}$	Rabbits blood	144
Amikacin	Luminol–H_2O_2–Cu^{2+}. Phenomenex C_{18} Synergi RP 80A (4 μm, 250 × 4.6 mm i.d.), Waters; potassium hydrogen phthalate (10^{-2} M, pH 3.35)–acetonitrile (90 : 10, v/v), $1.0\,mL\,min^{-1}$	$50\,\mu g\,L^{-1}$	Human plasma and urine	146
Oxacillin	Luminol–H_2O_2–Co^{2+}. GFFII-S5-80 (5 μm, 150 × 4.6 mm i.d.), Regis Technologies Inc.; potassium phosphate buffer (67 mM, pH 7.4, ionic strength 0.17), $0.3\,mL\,min^{-1}$, 20 °C	0.12 μM	Human serum	147

Table 7.3 (continued)

Analyte	CL system/HPLC characteristics	LOD	Applications	Ref.
Primary and secondary amines	Luminol–H_2O_2– $K_3[Fe(CN)]_6$. Cosmosil 5C – MS (5 μm, 250 × 4.6 mm i.d.), Nacalai Tesque; sodium phosphate buffer (50 mM, pH 7.0)–acetonitrile (55 : 45, v/v), 22 ± 4 °C, 1.0 mL min^{-1}	Between 210 and 560 amol	Human plasma	154
Organophosphorous insecticides (diclorvos, isocarbophos, methylparation)	Luminol–H_2O_2–CTMAB. Hypersil ODS (5 μm, 4.6 × 100 mm i.d.), Thermo Electron Corporation; methanol–water (60 : 40, v/v), 1 mL min^{-1}	0.03, 0.05, 0.1 μg mL^{-1}	Chervil leaves, cucumber peels and leaves from trees	155
N-Methylcarbamates (carbofuran, carbaryl and methiocarb)	Luminol–$KMnO_4$. C_{18} Luna (5 μm, 4.6 × 250 mm), Phenomenex; acetonitrile–water (50 : 50, v/v), 1 mL min^{-1}	38.6, 6.4, 58.3 ng L^{-1}	Mineral, river and ground waters, and cucumber	156
Aminoglycoside antibiotics	Luminol–H_2O_2–Cu(ii). Hamilton PRP-X200 (4.1 × 250 mm), Phenomenex; sodium acetate (5.0 × 10^{-3} M), sodium chloride (0.65 mol L^{-1}), (pH 6), 1 mL min^{-1}	0.7–10 μg L^{-1}	Tap, river and wastewater	158
Polyamines (spermine, spermidine, putrescine)	Luminol–H_2O_2–Cu^{2+}. Symmetry C_{18} (5 μm, 150 × 3.9 mm i.d.), Waters; methanol–water (25 : 75, v/v), 1 mL min^{-1}	2 pmol	Apple leaves, strawberry	160
β$_2$-Agonists (terbutaline, salbutamol and clenbuterol)	Luminol–$[Cu(HIO_6)_2]^{5-}$ on-line electro-generated. Nucleosil C_{18} (5 μm, 250 × 4.6 mm i.d.), Macherey-Nagel; acetonitrile–ammonium acetate (20 mM, pH 4.0) (90 : 10, v/v), 1 mL min^{-1}	0.007–0.01 ng g^{-1}	Pig liver	162

Analyte	Conditions	LOD	Sample	Ref.
Sudan dyes	Luminol–BrO^- on-line electrogenerated. Nucleosil RP-C_{18} (5 μm, 250 × 4.6 mm i.d.), Macherey-Nagel; methanol–0.2% aqueous formic acid (90 : 10, v/v), 1.0 mL min^{-1}, 35 °C	4–8 μg kg^{-1}	Chilli tomato sauce, hot chilli pepper	164
trans-Resveratrol	Luminol–$K_3[Fe(CN)]_6$. Zorbax Eclipse XDB-C_8 (5 μm, 150 × 4.6 mm i.d.), Agilent Technologies; methanol–water (35 : 65, v/v), 1 mL min^{-1}, 25 °C	0.166 μg L^{-1}	Red wine	165
Corticosteroids	Luminol–$K_3[Fe(CN)]_6$- $K_2[Fe(CN)]_6$. Luna (5 μm, 150 × 4.6 mm i.d.), Phenomenex; acetonitrile–water (35 : 65, v/v), 0.8 mL min^{-1}	0.13–8.40 μg L^{-1}	Bovine liver	168
Citric, lactic, malic, oxalic and tartaric acids	Luminol–Fe^{3+}–UO_2^{2+}. Ultrasphere C_{18} (5 μm, 250 × 4.6 mm i.d.), Beckman; 0.005 M H_2SO_4, 0.3 mL min^{-1}	50–540 fmol	Milk, wine, beer, fruit juices, soft drinks	169
Co^{2+}, Fe^{2+}, Mn^{2+}	Luminol–H_2O_2–Co^{2+}, Fe^{2+} or Mn^{2+}. ION-PAC CS5A, Dionex; 140 mM oxalic acid, 200 mM NaCl, 200 mM HCl–200 mM NaCl, 200 mM HCl–320 mM $LiClO_4$–water (with gradient composition), 1 mL min^{-1}	0.24, 0.50, 375 nM	Co^{2+} in commercial Cu chelates for animal feeding	170

the luminol–AgNO$_3$–Au colloid system. Based on the inhibited CL, a method for simultaneous determination of monoamine neurotransmitters and their metabolites in a mouse brain microdialysate was proposed.[143] The concentration of levodopa, a neurotransmitter used for treatment of neural disorders, in blood has also been on-line determined based on the luminol–potassium ferricyanide reaction.[144]

Other examples of this system are the determination of busulfan (a drug used in the treatment of leukemia) in serum,[145] amikacin in body fluids, based on the inhibitory effect of the aminoglycoside on the CL reaction between luminol and hydrogen peroxide catalyzed by Cu(II),[146] or the determination of unbound oxacillin[147] and its interaction with cefoperazone[148] in human serum albumin solutions, based on high-performance frontal analysis (HPFA), which uses restricted-access type HPLC columns, by means of the luminol–H$_2$O$_2$–Co^{2+} system. A hyphenated technique, HPLC-UV/nano-TiO$_2$-CL, has been established for the determination of selenocystine and selenomethionine in rabbit serum and garlic, based on the fact that UV decomposed products of selenocystine or selenomethionine can inhibit the CL produced by the luminol reaction, which takes place after the UV/nano-TiO$_2$ photocatalytic reaction device.[149] Figure 7.6 shows a scheme of this system.

An original application has been the determination of phenolic acids in *Danshen*[150,151] and flavonoids in *Ginkgo biloba*[152] by means of HPLC-CL with the luminol system to establish the activity-integrated fingerprints of these herbal medicines for quality control, according to the antioxidant activity as radical scavengers of the selected analytes.

In addition, some luminol-type reagents have been proposed for HPLC-CL in clinical applications, such as 6-aminomethylphthalhydrazide (6-AMP) for the quantification of 5-hydroxyindole-3-acetic acid (the most abundant end-product of serotonin metabolism) in human urine[153] and 4-(6,7-dihydro-5,8-dioxothiazolo[4,5-g]phthalazin-2-yl)benzoic acid *N*-hydroxysuccinimide ester,

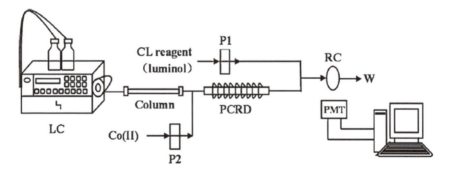

Figure 7.6   Schematic diagram of the LC-UV/Nano-TiO$_2$-CL system. P, peristaltic pump; PCRD, photocatalysis reaction device; RC, reaction cell; PMT, photomultiplier tube; and W, waste. (Reproduced with permission from ref. 149.)

for the determination of primary and secondary amines in the presence of triethylamine to give highly CL derivatives, which produce emission by reaction with H_2O_2 in the presence of potassium hexacyanoferrate(III).[154] This last method has been applied for the determination of amantadine in human plasma.

7.3.3.2 Environmental and Food Analysis

Pesticides, such as organophosphorus insecticides in leaves, have been determined by a reaction using cetyltrimethylammonium bromide (CTMAB) as a sensitizer[155] and N-methylcarbamates (NMCs) in water,[156] based on the enhancing effect of these analytes on the CL emission generated by the oxidation of luminol with potassium permanganate. A method for the simultaneous determination of N,N-dimethylaniline and phenol in water samples using the luminol–$K_3Fe(CN)_6$ system in alkaline medium has also been proposed,[157] as well as a method based on the inhibited luminol CL emission for the direct analysis of aminoglycoside antibiotics (AGs) in water samples[158] and a HPLC-CL system using the luminol analogue L-012 (8-amino-5-chloro-7-phenylpyrido[3,4-d]pyridazine-1,4-(2H,3H)dione) for the determination of quinones in airborne particulates.[159]

In food analysis, the determination of polyamines in apple leaves and strawberry fruit has been proposed, taking advantage of the catalytic effect of unsaturated complexes of Cu(II) with a polyamine on the luminol–H_2O_2 CL reaction.[160] Concerning the use of Cu as a catalyst of the luminol reaction, trivalent copper has been considered to be an uncommon oxidation state, unstable in aqueous solution. However, the complex of trivalent copper and periodate, $[Cu(HIO_6)_2]^{5-}$, could be electrogenerated on-line near the surface of a platinum electrode, and the reaction between $[Cu(HIO_6)_2]^{5-}$ and luminol in alkaline medium was accompanied with a weak emission of light. Glucocorticoids, β_2-agonist and some sugars exhibit strong enhancing effects on this CL intensity, and a HPLC-CL system has been proposed for the analysis of glucocorticoids[161] and β_2-agonist (terbutaline, salbutamol and clenbuterol) in spiked pig liver,[162] glucose and fructose in grape samples and lactose in milk samples.[163] A similar system for the determination of Sudan dyes in hot chilli pepper, based on the on-line electrogeneration of unstable reagents (in this case, BrO^-) in flow CL systems has also been proposed.[164]

Other examples of the use of the luminol reaction are the determination of *trans*-resveratrol (a stilbene produced by plants in response to fungal infection or abiotic stresses) in wine samples, as it enhances strongly the CL emission from the luminol–ferricyanide system,[165] dithiocarbamates in cucumber and apple[166] and corticosteroids in bovine liver[167] and bovine urine[168] by reaction with luminol in the presence of hexacyanoferrate(III) in alkaline solution, NMCs in cucumber by increasing the CL emission of the oxidation of luminol with potassium permanganate in alkaline medium[156] or the coupling of photochemical-CL reaction for the determination of organic acids in different samples[169] (wines, beer, milk, fruit and soft drinks), when they are irradiated with visible or UV light in the presence of Fe(III) or UO_2^{2+}. The influence of different experimental parameters on the luminol–H_2O_2 CL reaction used as a detection technique for Co(II), Fe(II) and Mn(II) has also been studied by means of ion chromatography, and the cobalt amount was determined in four commercial copper chelates used for animal feeding.[170]

7.3.4 Other Oxidation Reactions

Strong oxidants, such as MnO_4^- (in acidic or alkaline medium), ClO^-, Ce(IV), H_2O_2, IO_4^-, Br_2 or N-bromosuccinimide, have been tested under different chemical conditions to produce a CL emission from different analytes. Selected applications are included in Table 7.4.

7.3.4.1 Potassium Permanganate

In the clinical and pharmaceutical analysis some applications can be considered. For example, an assembly using monolithic columns has been used to characterize the CL reaction of several adrenergic amines with potassium permanganate in a sodium polyphosphate solution as CL reagent, including the determination of synephrine in dietary supplements.[171] A hybrid FIA/HPLC system using a monolithic column for the determination of phenolic alkaloids using potassium permanganate[126] and biogenic amines in urine samples[172] based on the CL emission generated from the permanganate oxidation has also been proposed.

Based on the sensitizing effect of formaldehyde on the CL reaction of different analytes with acidic potassium permanganate, different HPLC-CL detection methods for the determination of propylthiouracil and methylthiouracil in human serum[173] and arbutin and L-ascorbic acid in cosmetics[174] have been proposed. In addition, ibuprofen, naproxen and fenbufen have been determined in spiked plasma samples by HPLC-CL based on the CL emission produced by their reaction with $KMnO_4$ and Na_2SO_3 in sulfuric acid medium.[175]

Most of the compounds containing a phenolic and/or amine moiety show a CL response in acidic potassium permanganate system. Taking advantage of this fact, a simple HPLC-CL method for the determination of polyhydroxybenzenes based on their reaction with acidic potassium permanganate and the sensitizing effect of formic acid on the CL system has been proposed for their determination in river water.[176]

Tetracyclines enhance the CL of the potassium permanganate–sodium sulfite–β-cyclodextrin system. Based on this, a method has been developed for the determination of tetracycline residues in honey by coupling HPLC with this CL reaction,[177] suggesting that the excited state sulfur dioxide and the excited state inclusion complex of β-cyclodextrin and anhydro-derivatives of tetracyclines might be the emitters in this system.

7.3.4.2 Cerium(IV)

The determination of three flavonols in phytopharmaceuticals by reversed-phase HPLC-CL detection has been developed, based on the CL enhancement by flavonols of the Ce(IV)–Rhodamine 6G system in sulfuric acid medium.[178] The CL enhancement by phenolic compounds of the Ce (IV)-Tween 20 system in a sulfuric acid medium was the basis of a HPLC-CL method for the determination of salicylic acid and resorcinol in dermatitis clear tincture.[179]

A HPLC-CL method based on the great enhancement produced by products obtained from the irradiation of tetracyclines on the CL emission from Rhodamine B oxidation by Ce(IV) in sulfuric medium has been proposed for the determination of these compounds in surface waters.[180] After separation of tetracyclines, they were photochemically derivatizated and the resulting products were mixed with Rhodamine B and subsequently with Ce(IV) in H_2SO_4 inside the flow cell of the CL detector.

In food analysis, the CL reaction of Ce(IV) with different compounds has been used in different HPLC-CL methods. Thus, it was found that the reaction between Ce(IV) and Rhodamine 6G in strong sulfuric acid medium underwent weak CL, which could be greatly enhanced by different phenolic compounds. Phenolic hydroxyls were the main active groups for the generation of CL, and the magnitude of the intensity was related to the type and position of substituent in the benzene ring. A possible mechanism for this reaction was proposed, as well as a HPLC-CL method for the simultaneous detection of 20 phenolic compounds, and the quantification of six of them in wine samples.[181] Based on a similar effect, a HPLC-CL method for the determination of parabens in food samples (orange juice, soy sauce, pickle, strawberry jam, vinegar and cola soda) and cosmetic products was proposed.[182] The CL enhancement by phenolic compounds of the Ce(IV)–Tween 20 system in a sulfuric acid medium is the basis of a HPLC-CL method for the determination of *p*-hydroxybenzoic acid in apple juices.[179] The column effluent was mixed with Tween 20 solution at a mixing tee and then combined with Ce(IV) solution.

7.3.4.3 Other Oxidants

Other oxidants have been used for direct CL post-column detection in HPLC, as H_2O_2 in an alkaline borate solution containing acetonitrile to oxidize the CL derivatives of cefaclor (a β-lactam antibiotic) with the fluorescent reagent, 4-(2′-cyanoisoindolyl)phenylisothiocyanate (CIPIC), allowing its determination in spiked serum samples[183] or electrogenerated Mn(III) to determinate indomethacin in pharmaceutical preparations and human urine.[184]

Soluble Mn(IV) has also been used as oxidant, being less selective than either acidic potassium permanganate or tris(2,2′-bipyridyl)ruthenium(III).[185] Thus, freshly precipitated Mn(IV) dioxide can be dissolved in orthophosphoric acid to produce a soluble colloidal Mn(IV) species. High CL intensity was observed from the reaction of this Mn(IV) reagent with a range of analytes, and has been used for the determination of six opiate alkaloids.

Table 7.4 Selected applications of HPLC-CL using direct oxidations.

Analyte	CL system/HPLC characteristics	LOD	Applications	Ref.
Adrenergic amines	$KMnO_4$ – polyphosphates. Chromolith SpeedROD RP-18 monolithic column, Merck; aqueous solution of trifluoroacetic acid (pH 2.5), $1\,mL\,min^{-1}$	1×10^{-9}–$1 \times 10^{-8}\,M$	Synephrine in dietary supplements containing *C. aurantium* extracts	171
Propylthiouracil, methylthiouracil	$KMnO_4$–formaldehyde. Nucleosil C_{18} (5 μm, 250×4.6 mm i.d.), Macherey-Nagel; methanol–water (40 : 60, v/v), $1.0\,mL\,min^{-1}$	$0.03\,\mu g\,mL^{-1}$ (propylthiouracil), $0.03\,\mu g\,mL^{-1}$ (methylthiouracil)	Human serum	173
Non-steroidal anti-inflammatory drugs (ibuprofen, naproxen and fenbufen)	$KMnO_4$–Na_2SO_3. Zorbax C_{18} (5 μm, 150×4.6 mm i.d.), Agilent; acetonitrile–water (0.5% phosphoric acid v/v, pH 2.2), $0.8\,mL\,min^{-1}$, gradient elution mode	0.5, 0.05, $0.5\,ng\,mL^{-1}$	Human plasma	175
Polyhydroxybenzenes (catechol, resorcinol, hydroquinone and 1,2,4-benzenetriol)	$KMnO_4$–formic acid. Cosil C_{18} (5 μm, 4.6×150 mm). Supelco; methanol–phosphoric acid (0.5%) (30 : 70, v/v), $0.6\,mL\,min^{-1}$	3.2–$5.2 \times 10^{-3}\,mg\,L^{-1}$	River water	176
Tetracyclines (oxytetracycline, tetracycline, metacycline)	$KMnO_4$–Na_2SO_3–β-cyclodextrin. Zorbax Eclipse XDB-C_{18} (5 μm, 150×2.1 mm i.d.), Agilent Technologies; acetonitrile–0.001 M aqueous phosphoric acid, $0.5\,mL\,min^{-1}$, $25\,^{\circ}C$	0.9–$5.0\,ng\,mL^{-1}$	Honey	177
Phenolic compounds	Ce(iv)–Tween 20. Zorbax Eclipse XDB-C_8 (5 μm, 150×4.6 mm i.d.), Agilent Technologies; methanol– aqueous 1.5% acetic acid (either isocratic (35 : 65, v/v) or elution mode), $25\,^{\circ}C$, $1\,mL\,min^{-1}$	1.40–$5.02\,ng\,mL^{-1}$	Dermatitis clear tincture and apple drinks	179

Analyte	Conditions	LOD/Range	Matrix	Ref.
Tetracyclines (tetra-, oxy-, chlorotetracycline and demeclocycline, doxycycline and meclocycline)	Ce(IV)–Rhodamine B. Aquasil C$_{18}$ (5 μm, 4.6 × 150 mm), Thermo Electron Corporation; acetonitrile–phosphate buffer (0.1 M, pH 2.6), gradient elution mode, 1 mL min^{-1}	0.05–0.1 μg mL^{-1}	Surface water	180
Phenolic compounds	Ce(IV)–Rhodamine 6G Zorbax Eclipse XDB-C$_8$ (5 μm, 150 × 4.6 mm i.d.), Agilent Technologies; methanol–1% aqueous AcOH (1 : 99, v/v, pH 2.75), gradient elution mode, 1.0 mL min^{-1}, 25 °C	1.5–82.1 ng mL^{-1}	Red wine	181
Cefaclor	H$_2$O$_2$–acetonitrile TSKgel ODS-80TM RP (5 μm, 150 × 4.6 mm i.d.), Tosoh.; water–acetonitrile–triethylamine (0.01 M, pH 8.5) (55 : 35 : 10, v/v/v), 0.6 mL min^{-1}	1 pmol	Human serum	183
Indomethacin	Mn(III) Nucleosil RP-C$_{18}$ column (5 μm, 250 × 4.6 mm i.d.), Macherey-Nagel; methanol–water–acetic acid (67 : 33 : 0.1, v/v), 1.0 mL min^{-1}, 20 °C	8 ng mL^{-1}	Human urine, tablets	184
Pyrethroid insecticides	K$_3$[Fe(CN)]$_6$–ACN Gemini C$_{18}$ (5 μm, 150 × 4.6 mm i.d.), Phenomenex; acetonitrile–water, gradient elution mode, 1.4–1.5 mL min^{-1}	0.009–0.049 μg mL^{-1}	Tomato	186

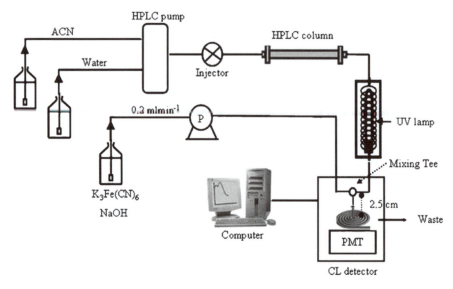

Figure 7.7   Schematic diagram of the HPLC-CL system used in the determination of benzoylureas. CL, chemiluminescence; PMT, photomultiplier tube. (Reproduced with permission from ref. 187.)

Also, $K_3[Fe(CN)_6]$ has been used as oxidant in food quality control, in the determination of nine pyrethroids in tomato[186] and five benzoylurea insecticides in cucumber.[187] The CL emission was generated by post-column irradiation of the pesticides with UV light and subsequent oxidation of the irradiated pesticides with $K_3[Fe(CN)_6]$ and NaOH, whose CL signal increased with the percentage of acetonitrile in the reaction medium. Figure 7.7 shows the proposed HPLC-CL system. As can be seen, in this case the configuration is not the usual, with two different channels for the CL reagents, as only one channel is required for introduction of the oxidant.

7.4   Capillary Electrophoresis Coupled to Chemiluminescence Detection

CE presents, as main advantages, low reagent and sample consumption, high separation efficiency and reduced analysis time. However, due to the ultra-small sample volumes introduced in the system and because of the small internal diameter of the capillaries, poor detection limits are encountered, limiting its usefulness. Owing to the advantages of CL detection in terms of sensitivity and considering its potential when combined with the high separation ability offered by CE, research in this area significantly increased from the 1990s onwards. In the last decade, some reviews have appeared related to the coupling of CE-CL.[10,188–189,190,191,192]

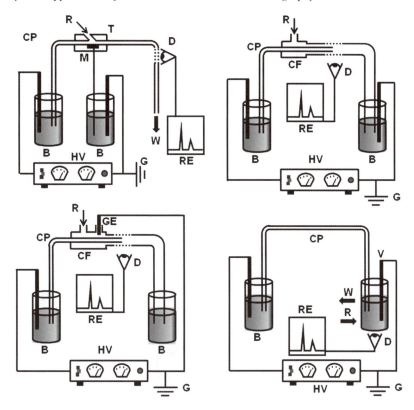

Figure 7.8 Schematic devices using CE-CL with different interfaces. CP, capillary separation; HV, high voltage supply; G, grounding; B, buffer reservoir; R, CL reagent; D, detector; M, semipermeable membrane; T, four-way connector; W, waste; CF, coaxial flow connector; GE, grounding electrode; V, reaction and grounding vial; RE, recorder.

Despite its great potential, this coupling is not widespread, mainly because there are no commercially available instruments, except in the case of CE coupled with ECL. Basically, a CE-CL instrument is a common CE instrument with some modifications to incorporate a channel with the CL reagents to carry out the CL reaction once the analytes have been separated in the electrophoretic capillary. Most of the efforts to improve actual CE-CL instruments are focused on the design of the detection mode, *i.e.*, the CE–CL interface. In this sense, Figure 7.8 shows different interfaces used to introduce this channel accomplished with the CL requirements: on-column, off-column and end-column interfaces, depending on the way the analytes and the CL reagents are mixed.[188] Various CL reagents have been used in CE-CL, mainly luminol and some derivatives such as isoluminol isothiocyanate (ILITC) or *N*-(4-aminobutyl)-*N*-ethylisoluminol (ABEI), ruthenium complexes, peroxyoxalates or strong oxidants or reductants, under different chemical conditions.

Table 7.5 Selected applications of CE-CL using POs.[a]

Analyte	CL system/CE buffer	Detection mode	LOD	Applications	Ref.
Amino-PAHs and compounds tagged with chemilumophores	TCPO–H_2O_2 catalyzed by PMP and triazole non-aqueous CE	On-column	0.939–2060 nM	–	193
3-Amino-fluoranthene, 1-aminopyrene, 1-aminoanthracene, dansyl hydrazine	TCPO–H_2O_2 CEC (90 : 10 acetonitrile–H_2O, 1 mM tris, 10 mM PMP, 5 mM H_2O_2, 20 mM triazole)	End-column	53.7–863 nM	–	194
Bovine serum albumin, lysozyme, cytochrome c	TDPO or TDPO–H_2O_2 derivatized with FR; phosphate buffer (pH 3.0)	On-column flow-type detector	6×10^{-7} M lysozyme (TCPO) 6×10^{-8} M lysozyme (TDPO)	–	196
α-Amino acids, proteins, and phenolic compounds	TDPO–H_2O_2–dansyl chloride/FITC Tris-borate buffer (pH 7.0) Tris-borate buffer (pH 8.4)–0.1% CMC–1 mM EDTA phosphate buffer (pH 8.0)/50% acetonitrile/2 mM SDS	Multicapillary device for simultaneous operation; end-column detection cell	–	–	197
Phenolic compounds	TDPO–H_2O_2 derivatization with dansyl chloride; phosphate buffer (pH 8.0)–50% acetonitrile–2 mM SDS	Comparison of two batch-type CL detection cells End-column	1–5×10^{-7} M	Surface and reused waters	198

Dansyl-lysine, dansyl-glycine	TDPO–H$_2$O$_2$ Tris-borate buffer (pH 7.0)	Microchip	1×10^{-5} M dansyl-lysine	—	199
Dansyl-amino acids	TDPO–H$_2$O$_2$ CE– Tris-borate buffer (pH 7.0); MECK–borate buffer (pH 8.3)/150 mM SDS	Batch-type and flow-type detection cells	0.43 fmol (batch-type) 1.3 fmol (flow-type) for dansyl-Trp	—	201
DNA and proteins (lysozyme, cytochrome C, RNase A)	TCPO–H$_2$O$_2$ Derivatization with FR Tris-glycine buffer (pH 8.4)/0.5% CMC/5 mM EDTA/2.5% dextran sulfate for DNA Tris-borate buffer (pH 8.4)–0.5% CMC–5 mM EDTA for proteins	Batch-type and flow-type detection cells	6×10^{-7} M lysozyme	—	202
Human serum albumin	TCPO–H$_2$O$_2$–dyestuff (Eosin Y-containing liposome) carbonate buffer (pH 9.0)	On-column coaxial flow	1×10^{-6} M	Human serum	245

[a]CEC: capillary electrochromatography; CMC: carboxylmethylcellulose; FITC: fluorescein isothiocyanate; FR: fluorescamine; PMP: 1,2,2,6,6-pentamethylpiperidine; SDS: sodium dodecyl sulfate; TCPO: bis-(2,4,6-trichlorophenyl)oxalate; TDPO: bis [(2-(3,6,9-trioxadecanyloxycarbonyl)-4-nitrophenyl)]oxalate.

7.4.1   Peroxyoxalate Reaction

As mentioned before, the main disadvantage of this system resides in the insolubility of the POs (peroxyoxalates) in water and their instability towards hydrolysis, which requires the use of organic solvents, which is not adequate for CE separation. In addition, the high separation voltage can affect the stability of the POs, being also the most commonly used TCPO, DNPO and TDPO. Although impressive detection limits and satisfactory selectivity were obtained in the first contributions, applications to real samples were not reported until recent years (Table 7.5).

The efficiency of the POs in the CE-CL is limited by the relatively slow kinetics of the reaction, so the use of a catalyst such as imidazole is considered. Kuyper *et al.*[193] have presented a method for coupling an ultrafast PO-CL reaction to CE based on the use of the ancillary base 1,2,2,6,6-penta-methylpiperidine (PMP) to assist the nucleophilic reagent 1,2,4-triazole (triazole). Under certain conditions, this PO-CL reaction exhibits a half-life of less than 2 s with light yields that are twice that of the imidazole-catalyzed PO-CL reaction. Post-separation electrokinetic delivery of TCPO was accomplished using a micro-tee. Electrokinetic addition of TCPO allowed for precise control of the ratio of TCPO to the CL reagents (PMP and triazole), spiked into the running buffer. This method for CL reagent delivery avoided the problems and costs associated with using pressure or mechanical pumps and proved its usefulness for environmental analysis of amino-PAHs and of compounds tagged with dansyl hydrazine. To extend the range of analytes, the same group reported a new detector cell for use with an ultrafast co-catalyzed PO-CL reaction coupled to capillary electrochromatography (CEC).[194] The method avoids problems associated with tedious reactor configurations and expensive syringe pumps. In relation to their previous paper,[193] a significant improvement in separation efficiency employing electrokinetic reagent delivery was carried out when using this CL configuration. The use of separation buffers containing small percentages of water yields excellent CEC separations, while CL reagent degradation is minimized.

Improvements on previous CE-CL devices using the PO system, applying untreated fused-silica capillaries[195] have been carried out by the introduction of a polyacrylamide-coated capillary and the attachment of a flow line of an aqueous buffer solution and a mixing filter to provide the conductivity required for CE.[196] The stable and constant electric current in the system allowed the separation and detection of a mixture of proteins labeled with fluorescamine (FR). Although more expensive, TDPO was used in this case, providing better CL performance than TCPO. Considering that the CE-CL detection system, mainly carried out end-column, possesses a useful "micro-space area" for reaction/detection at the tip of the capillary outlet, Tsukagoshi *et al.*[197] have proposed to take advantage of this area for the simultaneous operation of multiple separation modes in the CE-CL detection system using multiple capillaries and different migration buffers. Analysis of α-AAs, proteins and phenolic compounds were simultaneously performed using three capillaries, including usual, polymer-containing and SDS-containing migration buffers for separation, previously described in individual papers for each one of the three types of compounds, respectively.[198,201,202] The eluted samples from the capillaries, which were inserted into the CL detection cell, were mixed with CL reagent at the tips of the capillaries to generate visible light.

In addition, an on-line quartz microchip CE-CL device has been used, applying the PO reaction for the detection of dansyl-AAs, using TDPO as CL reagent.[199] The advantages were rapid separation times, small and accurate sample injection, simplification and miniaturization. Also, by using microchip CE-CL, chiral recognition was demonstrated, using hydroxypropyl-β-cyclo-dextrin as an effective enantiomer selector for chiral dansyl-AAs.[200] Glass and PTFE tubes as detection cells have been put in small light-tight boxes to achieve miniaturization of batch- and flow-type CL detectors for CE.[201] More sensitivity was obtained for dansyl-tryptophan using the glass-type detection cell but the flow-type detector was more suitable to apply MEKC mode to a mixture of dansyl-AAs, because the fresh CL reagent solution is continuously delivered to a tip of capillary, so that the migration buffer solution including SDS is easily removed from the reaction area. Both types of detection cells have been compared, using a polymer solution as the separation medium for the analysis of biopolymers, such as ATP, DNA and proteins, using fluorescein as labeling reagent.[202] Another glass electrophoresis microchip integrating a flow-type CL detection cell was developed and evaluated by Su *et al.*[203] In this case, the chip pattern is a double-T-type electrophoretic sample injection and separation combined with a Y-type CL detector. The double-T geometry allows for high-efficiency sample injection and geometric definition of sample plug size (Figure 7.9). The Y branch was used as CL reagent channel, and the CL reagent was delivered by a laboratory-made micropump. Using TCPO-H_2O_2, dansyl-phenylalanine and -sarcosine were successfully separated and detected within 250 s. Detection limits in the low μM range were obtained due to the vigorous dilution of sample with CL reagent and timely removal of the waste solution from reaction area.

7.4.2   Tris(2,2′-bipyridine)ruthenium(ɪɪ) Reaction

In this CL system, for which derivatization is not required for many classes of compounds, the reagent is regenerated and can be recycled. ECL employing $Ru(bpy)_3^{2+}$ offers an attractive detection scheme for CE because of the solubility and stability of the reagents in aqueous media and the high efficiency over a wide pH range, making it compatible with most buffer systems used (Table 7.6).

Most of the first contributions in this area suffered from poor separation efficiency, low sensitivity and complicated instrumentation. In previous designs, an electric field decoupler was used to eliminate the influence of high voltage (HV) field on the ECL detector.[204] However, its use brings peak broadening and involves experimental skill for its manufacturing, while the capillary is easily blocked and also makes the alignment of capillary to electrode more difficult.

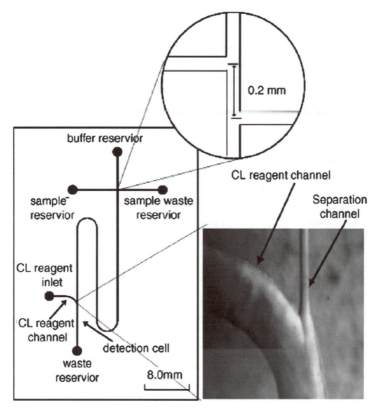

Figure 7.9   Schematic diagram of a microchip. (Reproduced with permission from ref. 203.)

Relevant improvements have been carried out on the design of the ECL end-column detection cell with the aim of applying CE-ECL to the analysis of bio-chemicals and pharmaceuticals in real samples. For example, the use of a narrow capillary and low conductivity buffer to eliminate the influence of the HV field on the electrochemical detector was proposed[205] (Figure 7.10). Using this approach, drugs containing tertiary amine groups have also been monitored in biological fluids and pharmaceutical formulations.[205–206,207] Although the employment of low-conductivity CE buffer and a narrow capillary effectively decreased the influence of the CE voltage on the ECL detection, it also limits the application of this technique. Consequently, the same group explored the use of 75 μm i.d. capillary and high-conductivity CE buffers to investigate the influence of HV field on the ECL detection without the use of an electric field decoupler.[208] The hydrodynamic cyclic voltammogram and corresponding ECL of $Ru(bpy)_3^{2+}$ under the HV field showed that the electrophoretic current did not annihilate the ECL of $Ru(bpy)_3^{2+}$, but made the ECL potential shift more positively and demonstrated that the capillary to electrode distance is crucial to achieving high sensitivity. Based on this, a new end-column ECL detector without the need for

Table 7.6 Selected applications of CE-CL using the tris-(2,2′-bipyridyl)ruthenium(II) system.[a]

Analyte	CL system/CE buffer	Detection mode	LOD	Applications	Ref.
Tramadol, lidocaine	ECL Ru(bpy)$_3^{2+}$ phosphate buffer (pH 9.0)	End-column ECL	6.0×10^{-8} M, 4.5×10^{-8} M	Human urine	205
Polyamines	ECL Ru(bpy)$_3^{2+}$ phosphate buffer (pH 2.0)	End-column ECL	1.9×10^{-9} M putrescine and cadaverine 7.6×10^{-9} M spermidine and spermine	Human urine	206
Benzhexol hydrochloride	ECL Ru(bpy)$_3^{2+}$ phosphate buffer (pH 8.0)	End-column ECL	6.7×10^{-9} M	Pharmaceutical formulations	207
Tripropylamine, Lidocaine	ECL Ru(bpy)$_3^{2+}$ phosphate buffer (pH 9.5)	End-column ECL	5.0×10^{-11} M, 2.0×10^{-8} M	Human urine	208
Ethambutol, methoxyphenamine	ECL Ru(bpy)$_3^{2+}$/ITO electrode phosphate buffer (pH 10.0)	End-column ECL	1.0 ng mL^{-1}, 0.9 ng mL^{-1}	Human plasma	215
Tripropylamine, proline, oxalate	ECL Ru(bpy)$_3^{2+}$ phosphate buffer (pH 9.0)	Miniaturized chip-type cell suitable for FI and CE; end-column ECL	1.5×10^{-8} M, 3.1×10^{-8} M, 4.4×10^{-8} M	—	216
Tertiary amines, lidocaine	ECL Ru(bpy)$_3^{2+}$ borate buffer (pH 8.0)	Flow cell (PDMSAO)-ITO electrode; end-column ECL	20–32 nM	Human urine	218
α-Ketocarboxylic acids	Ce(IV) [for chemical generation of Ru(bpy)$_3^{3+}$] phosphate buffer (pH 9.5)–0.7 mM CTAB (reverse EOF)	Off-column coaxial flow	1.3×10^{-9}– 3.7×10^{-8} M	Honey	246
Trimethylamine	ECL Ru(bpy)$_3^{2+}$ borate buffer (pH 9.2)	End-column ECL	8.0×10^{-9} M	Fish	247
Enrofloxacin, ciprofloxacin	ECL Ru(bpy)$_3^{2+}$ phosphate buffer (pH 8.5)	End-column ECL	10 ng mL^{-1}, 15 ng mL^{-1}	Milk	248

[a]CTAB: cetyltrimethylammonium bromide; ITO: indium/tin oxide; PDMSAO: poly(dimethoxysilane)–Al$_2$O$_3$.

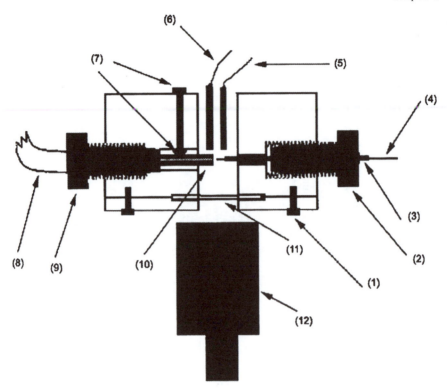

Figure 7.10 Schematic diagram of an ECL cell coupled with a separation capillary by end-column mode. (1) Stainless steel holding screw, (2) PVC capillary holder, (3) stainless steel tube, (4) separation capillary, (5) reference electrode, (6) counter electrode, (7) nylon screws for alignment, (8) working electrode cable, (9) PVC electrode holder, (10) working electrode, (11) optic glass window and (12) PMT. (Reproduced with permission from ref. 205.)

the electric field decoupler was proposed. The cell body was made of poly(methyl methacrylate) and the end of the separation capillary was inserted into a stainless steel tube that served as the ground electrode of electrophoresis. CE current flowed through the ground electrode of electrophoresis without a fracture on the column to separate CE current. In addition, by using a large Pt disk as working electrode, the amount of oxidized $Ru(bpy)_2^{3+}$ is enough to ascertain sufficient $Ru(bpy)_3^{3+}$ is generated at the end of the capillary. The lower layer of the cell was made of a piece of UV white optic glass, through which the photons were captured by the PMT. The reference electrode and counter electrode were inserted into the phosphate buffer containing $Ru(bpy)_3^{2+}$ in the cell above both the capillary and the working electrode to avoid blocking the flow of photons into the PMT. This device has been used to monitor lidocaine in urine.[208]

More recently, a commercial computer-controlled CE-ECL system has been developed, integrating CE HV power together with the electrochemical and

ECL system, and including data acquisition and data treatment.[209] Among the different applications carried out using this commercial device, relevant applications in clinical analysis have been developed.[210,211]

CE-solid ECL detection system is another option, by using a large inner diameter capillary without decoupler.[212] Although the solid ECL system gives high detection limits, it is cost-effective and regenerative. It is prepared by immobilizing $Ru(bpy)_3^{2+}$ in poly-(p-styrene-sulfonate)-silica-poly(vinyl alcohol) grafted 4-vinylpyridine copolymer films. Fabrication of this solid ECL detector only consumed 0.1 μL of 10 mM $Ru(bpy)_3^{2+}$, which was less than the hundreds of μL of $Ru(bpy)_3^{2+}$ in the general solution CE-ECL system. Sol–gel derived ZrO_2–Nafion composite films had also proven to be an effective matrix for the immobilization of $Ru(bpy)_3^{2+}$ at an electrode surface.[213]

Indium/tin oxide (ITO) has also been used as electrode to construct a cost-effective end-column ECL detector for CE.[214,215] The ITO electrode simplifies the alignment between the separation capillary and the working electrode. However, a need for machine work for fabrication of the working electrode, large void volumes at the detection cell and inconvenience in cell assembly are its limitations. Other drawbacks include spreading of epoxy glue in the working electrode and leakage of the solution from the contact area between the ITO electrode and the pipette vial.

Related to miniaturized systems, some instrumental developments have been addressed to the coupling of this CL detection with CE separations on a chip. In this sense, an instrument with miniaturized ECL detection in flow systems with a detection cell as an affiliated part of the instrument has been proposed.[216] The most important features of this chip-type ECL detection cell are its suitability for CE measurements and the continuous $Ru(bpy)_3^{2+}$ solution flow formed by hydrostatic pressure flowing through the capillary–electrode interface without the use of a syringe pump, thus simplifying the ECL instrument design. The CE effluent comes directly into contact with the surface of the working electrode in CE measurement mode, so the dead volume of the cell is greatly decreased. $Ru(bpy)_3^{3+}$ is generated *in situ* at the surface of the working electrode by electrochemical oxidation; however, the flow of effluent from the electrophoresis capillary over the electrode may reduce the concentration of $Ru(bpy)_3^{3+}$, thereby reducing the efficiency of the light-producing reaction. Consequently, an optimum distance between the separation capillary outlet and the working electrode should be taken into consideration to obtain higher separation efficiency and ECL signals.

The ITO electrode can be also integrated into an ECL detector for a microchip CE device.[217] The system uses an ITO-coated glass slide as the chip substrate with a photolithographically fabricated ITO electrode located at the end of the separation channel. The separation and injection channels were contained in a poly(dimethylsiloxane) (PDMS) layer, which was reversibly bound to the ITO electrode plate. With this construction, the photon-capturing efficiency was enhanced. To improve sensitivity, an approach was developed based on a pH junction and field amplification for determining tertiary amines by CE-ECL.[218] To simplify the connection between the separation capillary and ECL detector and to increase light collection efficiency, a flow cell of poly(dimethoxysilane)-Al_2O_3 (PDMSAO) was fabricated and used.

In the microchip area, Ding *et al.* have proposed a simpler and universal wall-jet configuration combined with an end-column ECL detection system in which both detection modes, *i.e.*, pre-column mode and post-column mode, were applied.[219] A glassy carbon disc electrode used as a working electrode and aligned with the outlet of separation channel electrochemically oxidizes the $Ru(bpy)_3^{2+}$ to the active $Ru(bpy)_3^{3+}$ form, which then reacts with analytes at the cathodic cell and produces light. The device was evaluated using local anesthetic containing tertiary amines.

In some previous studies, $Ru(bpy)_3^{2+}$ was usually immobilized on various working electrodes. These conventional modified electrodes work at room temperature and must be ground to obtain a fresh surface for reproducibility. In contrast, using an electrically heated carbon paste electrode (CPE)[220] presents several advantages: (i) with moderate heating, the peak width of analyte was decreased and the shape of peak was obviously improved, which is favorable for the separation in CE; (ii) the surface fouling effects could be minimized by moderate heating, which leads to satisfactory reproducibility and (iii) the application of a heated CPE in CE-ECL provided a higher sensitivity, wider linear range and lower detection limit.

7.4.3   Luminol and Derivatives

In this case, two interfaces (shown in Figure 7.8) have been mainly used: end-column (or batch-type) and on-column (or flow-type). The end-column mode simply immerses the outlet end of a separation capillary into an outlet reservoir containing certain CL reagents and electrolyte buffer. This mode has been applied for the detection of luminol or compounds labeled with isoluminol isothiocyanate[221] and its advantage is its simplicity and sensitivity but it can be problematic in reproducibility and separation efficiency. After several injections the outlet vial containing the CL reagent can be diluted, with the electrophoretic buffer needing to be renewed. In addition, possible contamination in the outlet reservoir could occur. The on-column detection mode consists of an electrophoretic capillary, a reagent tube and a reaction capillary connected by a three- or four-way joint.[222] This design has proven to be effective in many applications but it is more complex because the reagent flow affects the sensitivity and the possible mismatch between EOF and reagent flow can result in a turbulent mixing decreasing the peak efficiency; therefore, it needs to be optimized. Selected applications are included in Table 7.7.

As oxidant, H_2O_2 has been widely used, in presence of a catalyst, such as a metal ion [mainly Co(II) or Cu(II)] or horseradish peroxidase (HRP). One of the problems of using a metal ion is the incompatibility with the alkaline pH conditions of the CL emission. At basic pH, the metal ions precipitate as hydroxides, avoiding the catalytic effect and producing other problems such as clogging of the capillary. Different strategies must be used to avoid this, such as the utilization of an acidic electrophoretic buffer containing the metal ion and luminol. In this case, the alkaline CL reagent determines the pH of the reaction zone, as its flow is much higher than that of the electrophoretic capillary.[223] A different approach that avoids the precipitation is the use of a complexing agent. This strategy has been adopted to determine biomolecules, such as α-AAs, peptides and proteins.[224] Potassium sodium tartrate was used as masking agent to avoid the formation of $Cu(OH)_2$ and allowed the interaction of Cu(II) with biomolecules because it was found that Cu(II) was more catalytically active when it interacted with biomolecules to form Cu(II)-biomolecule complexes (Figure 7.11). In some cases a metal ion contained in a macromolecule is used to determine the macromolecule. This is the case for the determination of hemoglobin, which reacts with H_2O_2 and leads to fluorescent products and Fe(III). Zhou *et al.* have recently applied CE-CL to monitor the course of hemoglobin reacting with H_2O_2 and they found an intermediate that intensely enhances the CL emission of luminol–H_2O_2.[225] Using H_2O_2 as oxidant, luminol-based systems in CE can be a powerful tool for checking the contamination of waters by metal ions due to their enhancement effect. In this manner, Cr(III) and Cr(VI) (CrO_4^{2-}) have been determined in tap, surface, well, river and waste water samples, allowing speciation.[226,227] This method was based on in-capillary reduction, *i.e.*, Cr(VI) can be reduced by acidic sodium hydrogen sulfite to form Cr(III) while the sample is running through the capillary. Other metal ions were masked with EDTA, taking advantage of slow kinetics of Cr(III)-EDTA complex formation; therefore, Cr(III) can be determined selectively.

HRP has been widely used as catalyst in the luminol reaction but because the catalytic activity is quite weak it is conventional to find an appropriate enhancer, which strongly increases the sensitivity, such as *p*-iodophenol or *p*-iodophenylboronic acid. In this sense, a CE-immunoassay (CEIA) method to detect up to zeptomoles of bone morphogenic protein-2 (BMP-2) in rat vascular smooth muscle cell was developed.[228] HRP was linked to BMP-2 with a noncompetitive format, catalyzing the luminol/H_2O_2/*p*-iodophenol reaction. This method allows the baseline separation of HRP complexes from free HRP in less than 3 min, and has been applied successfully to arteriosclerosis pathology research.

Although H_2O_2 is the preferred reagent for macromolecules, $K_3[Fe(CN)_6]$ has some advantages for its utilization in CE: (i) it is both oxidant and catalyst at the same time and, hence, the CE-CL configuration is simpler, (ii) there are no problems with disruptions of current due to bubble formation occurring with H_2O_2, (iii) it is soluble in alkaline media and, therefore, (iv) it is suitable for the luminol reaction (pH 10–11).

The reaction of luminol with $K_3[Fe(CN)_6]$ produces an intense CL emission and the analytes can participate by enhancing or inhibiting the emission. Some analytes such as rutin or chlorogenic acid present both effects depending on the experimental conditions. For example, Jiang *et al.* have developed a method to determine these analytes in cigarettes by indirect CL;[229] however, better sensitivity was achieved using the direct method.[230] Phenothiazines have also been

Table 7.7 Selected applications of CE-CL using the luminol reaction.[a]

Analyte	CL system/CE buffer	Detection mode	LOD	Applications	Ref.
Hemoglobin	Luminol–H_2O_2 phosphate buffer (pH 10.0)	Off-column coaxial flow	1×10^{-9} M	Lysate of human red blood cell	225
Cr(III), Cr(VI)	Luminol–H_2O_2 acetate buffer (pH 4.7)	On-column coaxial flow	6×10^{-13} M, 8×10^{-12} M	Water	226
Cr(III), Cr(VI)	Luminol–H_2O_2 acetate buffer (pH 4.7)	On-column coaxial flow	6×10^{-13} M, 8×10^{-12} M	Water	227
Bone morphogenic protein-2	CEIA, HRP-*p*-iodophenol–luminol–H_2O_2 phosphate buffer (pH 6.5)	Off-column coaxial flow	6.2 pM	Rat vascular smooth muscle	228
Rutin, chlorogenic acid	Luminol–$K_3[Fe(CN)_6]$ borate buffer (pH 8.5)	Off-column coaxial flow	$0.22\ \mu g\ mL^{-1}$, $0.50\ \mu g\ mL^{-1}$	Pharmaceutical formulations and cigarettes	230
Phenothiazines	Luminol–$K_3[Fe(CN)_6]$ borate buffer (pH 8.5)	On-column coaxial flow	Injection by gravity: $80\ ng\ mL^{-1}$ promazine $334\ ng\ mL^{-1}$ promethazine; electrokinetic injection: $1\ ng\ mL^{-1}$ promazine	Pharmaceutical formulation and human urine	231
Co(II), Cu(II)	Luminol–$K_3[Fe(CN)_6]$ acetate buffer (pH 4.8)	On-column coaxial flow	7.5×10^{-11} M, 7.5×10^{-9} M	Electroplating wastewater	232
Polyphenols	Luminol–$K_3[Fe(CN)_6]$ phosphate buffer (pH 7.8)	Off-column coaxial flow	1.0×10^{-10} M catechol 1.0×10^{-9} M hydroquinone 3.0×10^{-11} M pyrogallol	River water	233

Analyte	Reagent / conditions	Detection mode	Concentration / LOD	Sample	Ref.
Biogenic amines	Analytes labeled with ABEI–H_2O_2–$K_3[Fe(CN)_6]$ borate buffer (pH 9.3)	Off-column coaxial flow	3.5×10^{-8} M diaminopropane 3.5×10^{-8} M putrescine 3.9×10^{-8} M cadaverine 1.2×10^{-7} M diaminohexane	Lake water	234
Amino acids	Luminol–BrO^- borate buffer (pH 9.2)	On-column coaxial flow	0.35×10^{-6}– 7.2×10^{-6} M	Pharmaceutical formulation	235
Amino acids	Luminol–BrO^- borate buffer (pH 9.4)	Off-column coaxial flow	1×10^{-7} M glutamic acid 1.3×10^{-7} M aspartic acid	Rat brain tissue and monkey plasma	236
Folic acid	Luminol–BrO^- borate buffer (pH 9.4)	Off-column coaxial flow	2.0×10^{-8} M	Pharmaceutical tablets, apple juices, human urine	237
Antioxidants	Cu(II)–Luminol–H_2O_2 carbonate buffer (pH 10.8)	Microchip	0.1 mM catechin 0.1 mM nitro-blue tetrazolium 0.05 mM superoxide dismutase	Catechin in commercial green tea beverages	238
Human serum albumin and immunosuppressive acidic protein	ILITC–H_2O_2–microperoxidase phosphate buffer (pH 7.3) 4 µM microperoxidase	Microchip	1.0×10^{-7} M HAS 1.0×10^{-7} M IAP	Human serum	240

[a]ABEI: *N*-(4-aminobutyl)-*N*-ethylisoluminol; CEIA: CE-immunoassay; HAS: human serum albumin; HRP: horseradish peroxidase; IAP: immunosuppressive acidic protein; ILITC: isoluminol isothiocyanate.

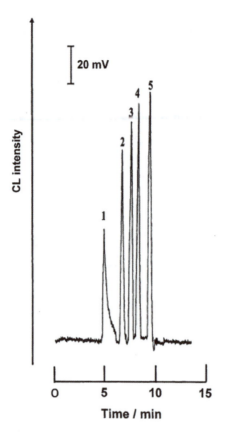

Figure 7.11 Electropherogram of a mixture of glycine and peptides: (1) glycine, (2) glycylglycine, (3) glycylglycylglycine, (4) glycylglycylglycylglycine and (5) glycylglycylglycylglycylglycine. Conditions: fused-silica capillary, 50 cm long and 50 μm i.d.; applied voltage, 12 kV; reagent, 10 mM phosphate buffer (pH 10.8) containing 5.0×10^{-5} M luminol, 5.0×10^{-6} M Cu(II) and 5.0×10^{-5} M potassium sodium tartrate in the inlet reservoir and 10 mM phosphate buffer (pH 10.8) containing 0.4 M hydrogen peroxide in the outlet reservoir; sample, 1.0×10^{-5} M. (Reproduced with permission from ref. 224.)

determined using the direct CL approach, allowing the separation of promethazine and promazine hydrochloride in less than 4 min.[231] The method was characterized using two different injection modes: by gravity, to determine promethazine as an impurity in thiazinamium methyl-sulfate ampoules, and electrokinetic injection to determine promazine in human urine. No interferences co-migrating with analytes were found.

$K_3[Fe(CN)_6]$ is the preferred oxidant for environmental applications and it has been used to determine Co(II) and Cu(II) in electroplating wastewater.[232] It is also known that polyphenols can enhance the CL emission of luminol with $K_3[Fe(CN)_6]$ as well, so they can be determined by CE-CL with detection limits

1–5-orders lower than that of other methods.[233] Also, micellar electrokinetic chromatography (MECK) has been used to determine biogenic amines with this system in lake water samples with prior labeling of the analytes with *N*-(4-aminobutyl)-*N*-ethylisoluminol.[234]

BrO$^-$ has also been used as an oxidant, mainly in clinical applications. For example, Yang *et al.*[235] determined seven underivatized AAs based on their inhibitory effect in the CL reaction between luminol and BrO$^-$ and the method was satisfactorily applied to AAs injections, achieving recoveries higher than 94%. In addition, Zhao *et al.* have developed a method to determine excitatory AAs (glutamic acid and aspartic acid) in rat brain tissue and monkey plasma.[236] In this case, amino acids enhanced the CL emission of luminol and BrO$^-$. Lower limits of detection were achieved with the direct approach. Folic acid (FA) has also been determined in apple juice, pharmaceutical tablets and human urine by its enhancement of the CL reaction between luminol and BrO$^-$ in alkaline aqueous solution.[237] No SPE extraction was needed due to the high selectivity.

Recently the suppression of CL from luminol in the presence of antioxidants has been for the first time introduced into a microchip CE-CL as the principle for antioxidant detection.[238] Luminol–H_2O_2 in the presence of Cu(II) was used in this system, where active oxygen species, such as superoxide radical anions, are generated. Nitroblue tetrazolium, superoxide dismutase and catechin were analyzed as model analytes of antioxidants. Negative peaks from baseline formed by the CL reaction were observed based on the reaction between active oxygen (superoxide radical anion) and antioxidants (analytes). The analytes were separated and detected within *ca.* 2 min.

The mixing of the reagent flow and the electrophoretic buffer using the luminol system is simpler in a microchip since crossing of different channels can be easily achieved.[239] Tsukagoshi's group have developed a micro total analysis system (μTAS) in which the CL reaction of isoluminol isothiocyanate (ILITC)–microperoxidase–H_2O_2 was used to determine human serum albumin or immunosuppressive acidic protein as a cancer marker in human serum.[240] The three processes (immune reaction, electrophoresis and CL) were compactly integrated onto the microchip to give the μ-TAS.

7.4.4 Direct Oxidations

In recent years, there has been a development of new CL reactions by testing the analyte with a wide range of strong oxidants, such as MnO_4^- (in acidic and alkaline medium), ClO$^-$, Ce(IV), H_2O_2, IO_4^-, Br_2 and *N*-bromosuccinimide, and reductants, under different chemical conditions. Lee and Whang[241] reported in 1997 the first methodology where aqueous acidic potassium permanganate was used to elicit CL by reaction with serotonin, dopamine, norepinephrine and catechol. This approach necessitated the fabrication of a cellulose acetate-coated porous polymer joint in the capillary (near the detector) to isolate the high voltage to prevent migration of the permanganate anion towards the anode. Using this device the EOF acted as a pump to propel the analytes along the length of capillary after the joint. Consequently, the effluent from the capillary merged with a flowing stream of reagent, generating the CL emission. Based on this, Barnett *et al.*[242] proposed a simple, robust CE-CL detection system for the determination of morphine, oripavine and pseudo-morphine, based upon the reaction of these analytes with acidic potassium permanganate in the presence of sodium polyphosphate. The reagent solution was contained in a quartz detection cell that also held both the capillary and the anode. The resultant CL was monitored directly using a PMT mounted flush against the base of the detection cell. To ensure that no migration of the permanganate anion occurred, the anode was placed at the detector end whilst the EOF was reversed by the addition of hexadimethrine bromide to the electrolyte. Satisfactory sensitivity (LODs: 2.5×10^{-7} M, 2.5×10^{-7} M and 5×10^{-7} M) and low relative standard deviation (RSD) for migration time and peak height were obtained.

Using also permanganate as the CL system, a microchip was designed on the principle of flow-injection CL-CE, including three main channels, five reservoirs and a detection cell.[243] As model samples, dopamine and catechol were sensitively and rapidly separated and detected, with LODs of 20 and $10\,\mu$M, respectively. The samples were electrokinetically injected into the double-T cross section, separated in the separation channel and then oxidized by the CL reagent delivered by a laboratory-made micropump to produce light in the detection cell. Recently, $K_3[Fe(CN)_6]$ in alkaline aqueous solution has been proposed for the first time in CE-CL for the simultaneous determination of vitamin B1 and vitamin B2.[244] After being separated by CE, the analytes were determined by direct CL emission on reaction with $K_3[Fe(CN)_6]$, with migration times shorter than 5 min and LODs of 0.01 and $0.03\,\mu$g mL^{-1}, respectively.

Acknowledgements

The Spanish Ministry of Science and Innovation (Project Ref. DEP2006-56207-C03-02) supported this work.

References

1. *Chemiluminescence in Analytical Chemistry*, ed. A. M. García-Campaña and W. R. G. Baeyens, Marcel Dekker, New York, 2001.
2. Y. Su, H. Chen, Z. Wang and Y. Lv, *Appl. Spectrosc. Rev.*, 2007, **42**, 139.
3. F. Li, C. Zhang, X. Guo and W. Feng, *Biomed. Chromatogr.*, 2003, **17**, 96.
4. Y. Ohba, N. Kuroda and K. Nakashima, *Anal. Chim. Acta*, 2002, **465**, 101.
5. A. M. García-Campaña, W. R. G. Baeyens, L. Cuadros-Rodríguez, F. Alés-Barrero, J. M. Bosque-Sendra and L. Gámiz-Gracia, *Curr. Org. Chem.*, 2002, **6**, 2001.
6. N. W. Barnett and P. S. Francis, Chemiluminescence: liquid-phase chemiluminescence, in *Encyclopedia of Analytical Science*, 2nd edn, ed. P. J.

Worsfold, A. Townshend and C. F. Poole, Elsevier, Oxford, 2005, pp. 511–520.
7. A. M. García-Campaña and F. J. Lara, *Anal. Bioanal. Chem.*, 2007, **387**, 165.
8. X. Yan, *J. Sep. Sci.*, 2006, **29**, 1931.
9. L. Gámiz Gracia, A. M. García Campaña, J. F. Huertas-Pérez and F. J. Lara, *Anal. Chim. Acta*, 2009, **640**, 7.
10. A. M. García Campaña, F. J. Lara, L. Gámiz Gracia and J. F. Huertas-Pérez, *Trends Anal. Chem.*, 2009, **28**, 973.
11. C. Dodeigne, L. Thunus and R. Lejeune, *Talanta*, 2000, **51**, 415.
12. L. J. Kricka, *Anal. Chim. Acta*, 2003, **500**, 279.
13. A. Roda, M. Guardigli, P. Pasini and M. Mirasoli, *Anal. Bioanal. Chem.*, 2003, **377**, 826.
14. L. Gámiz-Gracia, A. M. García-Campaña, J. J. Soto-Chinchilla, J. F. Huertas Pérez and A. González Casado, *Trends Anal. Chem.*, 2005, **24**, 927.
15. X. Yan, *J. Chromatogr. A*, 1999, **842**, 267.
16. J. E. Boulter and J. W. Birks, in *Chemiluminescence in Analytical Chemistry*, ed. A. M. García-Campaña and W. R. G. Baeyens, Marcel Dekker, Inc., New York, 2001, p. 349.
17. S. S. Brody and J. E. Chaney, *J. Gas Chromatogr.*, 1966, **4**, 42.
18. P. L. Patterson, *Anal. Chem.*, 1978, **50**, 339.
19. P. L. Patterson, *Anal. Chem.*, 1978, **50**, 345.
20. K. B. Thurbide and W. A. Aue, *J. Chromatogr. A*, 1994, **684**, 259.
21. K. B. Thurbide and W. A. Aue, *J. Chromatogr. A*, 1999, **858**, 245.
22. K. B. Thurbide, B. W. Coke and W. A. Aue, *J. Chromatogr. A*, 2004, **1029**, 193.
23. K. B. Thurbide and T. C. Hayward, *Anal. Chim. Acta*, 2004, **519**, 121.
24. S. Cheskis, E. Atar and A. Amirav, *Anal. Chem.*, 1993, **65**, 539.
25. A. Amira and H. Jing, *Anal. Chem.*, 1995, **67**, 3305.
26. L. Gámiz-Gracia, J. F. Huertas-Pérez, J. J. Soto-Chinchilla and A. M. García-Campaña, in *Handbook of Pesticides. Methods of Pesticide Residue Analysis*, ed. and L. M. L. Nollet, H. S. Rathore, CRC Press, Taylor & Francis Group, Boca Raton, 2010, p. 303.
27. A. Di Muccio, A. M. Cicero, A. Ausili and S. Di Muccio, *Methods Biotechnol.*, 2006, **19**, 263.
28. I. Martínez Salvador, A. Garrido Frenich, F. J. González Egea and J. L. Martínez Vidal, *Chromatographia*, 2006, **64**, 667.
29. X. Yi and Y. Lu, *Chemosphere*, 2006, **65**, 639.
30. E. Papadopoulou-Mourkidou, E. N. Papadakis and Z. Vryzas, *Methods Biotechnol.*, 2006, **19**, 319.
31. K. Patel, R. J. Fussell, R. Macarthur, D. M. Goodall and B. J. Keely, *J. Chromatogr. A*, 2004, **1046**, 225.
32. K. Patel, R. J. Fussell, D. M. Goodall and B. J. Keely, *J. Sep. Sci.*, 2006, **29**, 90.
33. J. Yu, C. Wu and J. Xing, *J. Chromatogr. A*, 2004, **1036**, 101.

34. M. R. Khalili-Zanjani, Y. Mamini, N. Yazdanfar and S. Shariati, *Anal. Chim. Acta*, 2008, **606**, 202.
35. L. Li, W. Li, J. Ge, Y. Wu, S. Jiang and F. Liu, *J. Sep. Sci.*, 2008, **31**, 3588.
36. J. Wong, M. Hennessy, D. Hayward, A. Krynitsky, I. Cassias and F. Schenck, *J. Agric. Food Chem.*, 2007, **55**, 1117.
37. K. Sasamoto, N. Ochiai and H. Kanda, *Talanta*, 2007, **72**, 1637.
38. M. A. Champ, *Sci. Total Environ.*, 2000, **258**, 21.
39. Y. Mino, F. Amano, T. Yoshioka and Y. Konishi, *J. Health Sci.*, 2008, **54**, 224.
40. A. P. Birjandi, A. Bidari, F. Rezaei, M. R. M. Hosseini and Y. Assadi, *J. Chromatogr. A*, 2008, **1193**, 19.
41. M. Gallego-Gallegos, M. Livia, R. Muñoz-Olivas and C. Cámara, *J. Chromatogr. A*, 2006, **1114**, 82.
42. C. Bancon-Montigny and G. Lespes, *J. Chromatogr. A*, 2000, **896**, 149.
43. M. Bravo, G. Lespes, I. De Gregori, H. Pinochet and M. P. Gautier, *Anal. Bioanal. Chem.*, 2005, **383**, 1082.
44. Q. Xiao, C. Yu, J. Xing and B. Hu, *J. Chromatogr. A*, 2006, **1125**, 133.
45. P. Mochalski, B. Wzorek, I. Sliwka and A. Amann, *J. Chromatogr. B*, 2009, **877**, 1856.
46. R. López, A. C. Lapeña, J. Cacho and V. Ferreira, *J Chromatogr.*, 2007, **1143**, 8.
47. M. Von Hobe, U. Kuhn, H. Van Diest, L. Sandoval-Soto, T. Kenntner, F. Helleis, S. Yonemura, M. O. Andreae and J. Kesselmeire, *Int. J. Environ. Anal. Chem.*, 2008, **88**, 303.
48. K.-H. Kim, *Int. J. Environ. Anal. Chem.*, 2006, **86**, 805.
49. X. Yan, *J. Chromatogr. A*, 2002, **976**, 3.
50. H. V. Drushel, *Anal. Chem.*, 1977, **49**, 932.
51. R. L. Shearer and R. E. Sievers, in *Chemiluminescence and Photochemical Reaction Detection in Chromatography*, ed. J. W. Birks, VCH publishers, New York, 1989, p. 71.
52. B. M. Jones and C. G. Daughton, *Anal. Chem.*, 1985, **57**, 2320.
53. D. H. Fine, D. Lieb and F. Rufeh, *J. Chromatogr.*, 1975, **107**, 351.
54. R. L. Benner and D. H. Stedman, *Anal. Chem.*, 1989, **61**, 1268.
55. P. L. Burrow and J. W. Birks, *Anal. Chem.*, 1997, **69**, 1299.
56. R. L. Shearer, *Anal. Chem.*, 1992, **64**, 2192.
57. J. Blomberg, T. Riemersma, M. van Zuijlen and M. H. Chaabani, *J. Chromatogr. A*, 2004, **1050**, 77.
58. R. Hua, Y. Li, W. Liu, J. Zheng, H. Wei, J. Wang, X. Lu, H. Kong and G. Xu, *J. Chromatogr. A*, 2003, **1019**, 101.
59. R. Hua, J. Wang, H. Kong, J. Liu, X. Lu and G. Xu, *J. Sep. Sci.*, 2004, **27**, 691.
60. R. Ruiz-Guerrero, C. Vendeuvre, D. Thiébaut, F. Bertoncini and D. Espinat, *J. Chromatogr. Sci.*, 2006, **44**, 566.
61. F. C.-Y. Wang, W. K. Robbins, F. P. Di Sanzo and F. C. McElroy, *J. Chromatogr. Sci.*, 2003, **41**, 519.

62. F. C.-Y. Wang, W. K. Robbins and M. A. Greaney, *J. Sep. Sci.*, 2004, **27**, 468.
63. F. Adam, F. Bertoncini, N. Brodush, E. Durand, D. Thiebaut, D. Espinat and M. C. Hennion, *J. Chromatogr. A*, 2007, **1148**, 55.
64. F. Adam, F. Bertoncini, C. Dartiguelongue, K. Marchand, D. Thiebaut and M. C. Hennion, *Fuel*, 2009, **88**, 938.
65. J. E. Grebel, C. C. Young and I. H. Suffet, *J. Chromatogr. A*, 2006, **1117**, 11.
66. B. A. Tomkins, W. H. Griest and C. E Higgins, *Anal. Chem.*, 1995, **67**, 4387.
67. R. Andrade, F. G. R. Reyes and S. Rath, *Food Chem.*, 2005, **91**, 173.
68. J. Okamoto, T. Matsubara, T. Kitagawa and T. Umeda, *Anal. Chem.*, 2000, **72**, 634.
69. J. K. Furne, J. Springfield, T. Koenig, F. Suarez and M. D. Levitt, *J. Chromatogr. B*, 2001, **754**, 503.
70. A. M. Jiménez and M. J. Navas, *J. Hazard. Mater.*, 2004, **106**, 1.
71. N. Kuroda, M. Kai and K. Nakashima, in *Chemiluminescence in Analytical Chemistry*, ed. A. M. García-Campaña and W. R. G. Baeyens, Marcel Dekker, New York, 2001, p. 393.
72. K. Nakashima, R. Ikeda and M. Wada, *Anal. Sci.*, 2009, **25**, 21.
73. R. Bos, N. W. Barnett, G. A. Dyson, K. F. Lim, R. A. Russell and S. P. Watson, *Anal. Chim. Acta*, 2004, **502**, 141.
74. M. Tsunoda and K. Imai, *Anal. Chim. Acta*, 2005, **541**, 13.
75. M. Stigbrand, T. Jonsson, E. Pontén, K. Irgum and R. Bos, in *Chemiluminescence in Analytical Chemistry*, ed. A. M. García-Campaña and W. R. G. Baeyens, Marcel Dekker, New York, 2001, p. 141.
76. M. Emteborg, E. Pontén and K. Irgum, *Anal. Chem.*, 1997, **69**, 2109.
77. T. Jonsson, M. Emteborg and K. Irgum, *Anal. Chim. Acta*, 1998, **361**, 205.
78. A. G. Hadd and J. W. Birks, *J. Org. Chem.*, 1996, **61**, 2657.
79. G. H. Ragab, H. Nohta and K. Zaitsu, *Anal. Chim. Acta*, 2000, **403**, 155.
80. K. Takezawa, M. Tsunoda, K. Murayama, T. Santa and K. Imai, *Analyst*, 2000, **125**, 293.
81. K. Takezawa, M. Tsunoda, N. Watanabe and K. Imai, *Anal. Chem.*, 2000, **72**, 4009.
82. M. Tsunoda, *Chromatography*, 2005, **26**, 95.
83. M. Tsunoda, M. Nagayama, T. Funatsu, S. Hosoda and K. Imai, *Clin. Chim. Acta*, 2006, **366**, 168.
84. M. Tsunoda, E. Uchino, K. Imai, K. Hayakawa and T. Funatsu, *J. Chromatogr. A*, 2007, **1164**, 162.
85. M. Tsunoda, E. Uchino, K. Imai and T. Funatsu, *Biomed. Chromatogr.*, 2008, **22**, 572.
86. M. Tsunoda, M. Yamagishi, K. Imai and T. Yanagisawa, *Anal. Bioanal. Chem.*, 2009, **394**, 947.
87. E. Orejuela and M. Silva, *Analyst*, 2002, **127**, 1433.

88. J. Ishida, M. Takada, N. Hitoshi, R. Iizuka and M. Yamaguchi, *J. Chromatography B*, 2000, **738**, 199.
89. S. Nakamura, M. Wada, B. L. Crabtree, P. M. Reeves, J. H. Montgomery, H. J. Byrd, S. Harada, N. Kuroda and K. Nakashima, *Anal. Bioanal. Chem.*, 2007, **387**, 1983.
90. K. Funato, T. Imai, K. Nakashima and M. Otagiri, *J. Chromatogr. B*, 2001, **757**, 229.
91. H. Yamada, Y. Kuwahara, Y. Takamatsu and T. Hayase, *Biomed. Chromatogr.*, 2000, **14**, 333.
92. A. Amponsaa-Karikari, N. Kishikawa, Y. Ohba, K. Nakashima and N. Kuroda, *Biomed. Chromatogr.*, 2006, **20**, 1157.
93. S. Ahmed, N. Kishikawa, K. Nakashima and N. Kuroda, *Anal. Chim. Acta*, 2007, **591**, 148.
94. S. Meseguer Lloret, C. Molins Legua, J. Verdú Andrés and P. Campíns Falcó, *J. Chromatogr. A*, 2004, **1035**, 75.
95. S. Meseguer Lloret, C. Molins Legua and P. Campíns Falcó, *Anal. Chim. Acta*, 2005, **536**, 127.
96. K. Hayakawa, K. Noji, N. Tang, A. Toriba, R. Kizu, S. Sakai and Y. Matsumoto, *Anal. Chim. Acta*, 2001, **445**, 205.
97. N. Tang, A. Toriba, R. Kizu and K. Hayakawa, *Anal. Sci.*, 2003, **19**, 249.
98. N. Tang, R. Taga, T. Hattori, K. Tamura, A. Toroba, R. Kizu and K. Hayakawa, *Anal. Sci.*, 2004, **20**, 119.
99. E. Orejuela and M. Silva, *J. Chromatog. A*, 2003, **1007**, 197.
100. E. Orejuela and M. Silva, *Anal. Lett.*, 2004, **37**, 2531.
101. M. Cobo and M. Silva, *Chromatographia*, 2000, **51**, 706.
102. M. Wada, K. Inoue, A. Ihara, N. Kishikawa, K. Nakashima and N. Kuroda, *J. Chromatogr A.*, 2003, **987**, 189.
103. J. J. Soto-Chinchilla, L. Gámiz-Gracia, A. M. García-Campaña, K. Imai and L. E. García-Ayuso, *J. Chromatogr. A*, 2005, **1095**, 60.
104. Y. Sun, M. Wada, O. Al-Dirbashi, N. Kuroda, H. Nakazawa and K. Nakashima, *J. Chromatogr. B*, 2000, **749**, 49.
105. B. A. Gorman, P. S. Francis and N. W. Barnett, *Analyst*, 2006, **131**, 616.
106. X. B. Yin, S. J. Dong and E. K. Wang, *Trends Anal. Chem.*, 2004, **23**, 432.
107. X. B. Yin and E. K. Wang, *Anal. Chim. Acta*, 2005, **533**, 113.
108. K. A. Fahnrich, M. Pravda and G. G. Guilbault, *Talanta*, 2001, **54**, 531.
109. A. J. Bard, *Electrogenerated Chemiluminescence*, Marcel Dekker, New York, 2004.
110. Y. Du and E. K. Wang, *J. Sep. Sci.*, 2007, **30**, 875.
111. A. W. Knight, in *Chemiluminescence in Analytical Chemistry*, ed. A. M. García-Campaña and W. R. G. Baeyens, Marcel Dekker, New York, 2001, p. 211.
112. H. Morita and M. Konishi, *Anal. Chem.*, 2002, **74**, 1584.
113. H. Morita and M. Konishi, *Anal. Chem.*, 2003, **75**, 940.

114. T. Hori, H. Hashimoto and M. Konishi, *Biomed. Chromatogr.*, 2006, **20**, 917.

115. Y. J. Park, D. W. Lee and W. Y. Lee, *Anal. Chim. Acta*, 2002, **471**, 51.

116. Y. Gemba, H. Hashimoto and M. Konishi, *J. Liquid Chromatogr.*, 2004, **27**, 1611.

117. Y. Gemba, M. Konishi, T. Sakata and Y. Okabayashi, *J. Liquid Chromatogr. Relat. Technol.*, 2004, **27**, 843.

118. H. Yoshida, K. Hidaka, J. Ishida, K. Yoshikuni, H. Nohta and M. Yamaguchi, *Anal. Chim. Acta*, 2000, **413**, 137.

119. T. Ikehara, N. Habu, I. Nishino and H. Kamimori, *Anal. Chim. Acta*, 2005, **536**, 129.

120. K. Uchikura, *Chem. Pharm. Bull.*, 2003, **51**, 1092.

121. H. Kodamatani, Y. Komatsu, S. Yamazaki and K. Saito, *J. Chromatogr. A*, 2007, **1140**, 88.

122. H. N. Choi, S. H. Cho and W. Y. Lee, *Anal. Chem.*, 2003, **75**, 4250.

123. H. N. Choi, S. H. Cho, Y. J. Park, D. W. Lee and W. Y. Lee, *Anal. Chim. Acta*, 2005, **541**, 49.

124. R. Nakao, K. Furutsuka, M. Yamaguchi and K. Suzuki, *Anal. Sci.*, 2007, **23**, 151.

125. J. W. Costin, S. W. Lewis, S. D. Purcell, L. R. Waddell, P. S. Francis and N. W. Barnett, *Anal. Chim. Acta*, 2007, **597**, 19.

126. J. L. Adcock, P. S. Francis, K. M. Agg, G. D. Marshall and N. W. Barnett, *Anal. Chim. Acta*, 2007, **600**, 136.

127. N. Anastos, S. W. Lewis, N. W. Barnett and D. N. Sims, *J. Forensic Sci.*, 2006, **51**, 45.

128. T. Pérez-Ruiz, C. Martínez-Lozano and M. D. García-Martínez, *Anal. Chim. Acta*, 2008, **623**, 89.

129. T. Pérez-Ruiz, C. Martínez-Lozano, V. Tomás and J. Martín, *J. Chromatogr. A*, 2005, **1077**, 49.

130. T. Pérez-Ruiz, C. Martínez-Lozano and M. D. García, *J. Chromatogr. A*, 2007, **1164**, 174.

131. T. Pérez-Ruiz, C. Martínez-Lozano and M. D. García, *J. Chromatogr. A*, 2007, **1169**, 151.

132. H. Kodamatani, K. Saito, N. Niina, S. Yamazaki, A. Muromatsu and I. Sakurada, *Anal. Sci.*, 2004, **20**, 1065.

133. H. Kodamatani, H. Shimizu, K. Saito, S. Yamazaki and Y. Tanaka, *J. Chromatogr. A*, 2006, **1102**, 200.

134. H. Kodamatani, K. Saito, N. Niina, S. Yamazaki and Y. Tanaka, *J. Chromatogr. A*, 2005, **1100**, 26.

135. G. H. Wan, H. Cui, Y. L. Pan, P. Zheng and L. J. Liu, *J. Chromatogr. B*, 2006, **843**, 1.

136. M. Yamaguchi, H. Yoshida and H. Nohta, *J. Chromatogr. A*, 2002, **950**, 1.

137. T. Fukushima, N. Usui, T. Santa and K. Imai, *J. Pharm. Biomed. Anal.*, 2003, **30**, 1655.

138. J. Adachi, R. Kudo, Y. Ueno, R. Hunter, R. Rajendram, E. Want and V. R. Preedy, *J. Nutr.*, 2001, **131**, 2916.
139. S. P. Hui, T. Murai, T. Yoshimura, H. Chiba, H. Nagasaka and T. Kurosawa, *Lipids*, 2005, **40**, 515.
140. S. P. Hui, H. Chiba, T. Sakurai, C. Asakawa, H. Nagasaka, T. Murai, H. Ide and T. Kurosawa, *J. Chromatogr. B*, 2007, **857**, 158.
141. F. N. Chen, Y. X. Zhang and Z. J. Zhang, *Chin. J. Chem.*, 2007, **25**, 942.
142. E. Nalewajko, A. Wiszowata and A. Kojło, *J. Pharm. Biomed. Anal.*, 2007, **43**, 1673.
143. N. Li, J. Guo, B. Liu, Y. Yu, H. Cui, L. Mao and Y. Lin, *Anal. Chim. Acta*, 2009, **645**, 48.
144. F. Chen, Z. Zhang, Y. Zhang and D. He, *Anal. Bioanal. Chem.*, 2005, **382**, 211.
145. M. Fukumoto, M. Saitoh and H. Kubo, *Anal. Sci.*, 2000, **16**, 97.
146. J. M. Serrano and M. Silva, *J. Chromatogr. B*, 2006, **843**, 20.
147. M. Qiao, X. Guo and F. Li, *J. Chromatogr. A*, 2002, **952**, 131.
148. F. Li, X. Guo, M. Quiao, Z. Xiong and D. Ziiou, *Sepu*, 2004, **22**, 349.
149. Y. Sua, H. Chen, Y. Gao, X. Li, X. Hou and Y. Lv, *J. Chromatogr. B*, 2008, **870**, 216.
150. Y. Chang, X. Ding, J. Qi, J. Cao, L. Kang, D. Zhu, B. Zhang and B. Yu, *J. Chromatogr. A*, 2008, **1208**, 76.
151. Y. Chang, D. Yan, L. Chen, X. Ding, J. Qi, L. Kang, B. Zhang and B. Yu, *Chem. Pharm. Bull.*, 2009, **57**, 586.
152. X. Ding, J. Qi, Y. Chang, L. Mu, D. Zhu and B. Yu, *J. Chromatogr. A*, 2009, **1216**, 2204.
153. T. Yakabe, J. Ishida, H. Yoshida, H. Nohta and M. Yamaguchi, *Anal. Sci.*, 2000, **16**, 545.
154. H. Yoshida, R. Nakao, T. Matsuo, H. Nohta and M. Yamaguchi, *J. Chromatogr. A*, 2001, **907**, 39.
155. G. Huang, J. Ouyang, W. R. G. Baeyens, Y. Yang and C. Tao, *Anal. Chim. Acta*, 2002, **474**, 21.
156. J. F. Huertas-Pérez and A. M. García-Campaña, *Anal. Chim. Acta*, 2008, **630**, 194.
157. C. Huang, G. Zhou, H. Peng and Z. Gao, *Anal. Sci.*, 2005, **21**, 565.
158. J. M. Serrano and M. Silva, *J. Chromatogr. A*, 2006, **1117**, 176.
159. S. Ahmeda, N. Kishikawa, K. Ohyama, T. Maki, H. Kurosaki, K. Nakashima and N. Kuroda, *J. Chromatogr. A*, 2009, **1216**, 3977.
160. Q. Wu, Y. Su, L. Yang, J. Li, J. Ma, C. Wang and Z. Li, *Microchim Acta*, 2007, **159**, 319.
161. Y. Zhang, Z. Zhang, Y. Song and Y. Wei, *J. Chromatogr. A*, 2007, **1154**, 260.
162. Y. Zhang, Z. Zhang, Y. Sun and Y. Wei, *J. Agric. Food Chem.*, 2007, **55**, 4949.
163. Y. Sun, Z. Zhang, Y. Zhang and Y. Wei, *Chromatographia*, 2008, **67**, 825.
164. Y. Zhang, Z. Zhang and Y. Sun, *J. Chromatogr. A*, 2006, **1129**, 34.

165. J. Zhou, H. Cui, G. Wan, H. Xu, Y. Pang and C. Duan, *Food Chem.*, 2004, **88**, 613.

166. H. Nakazawa, Y. Tsuda, K. Ito, Y. Yoshimura, H. Kubo and H. Homma, *J. Liquid Chromatogr. Relat. Tech.*, 2005, **27**, 705.

167. B. I. Vázquez, X. Feás, M. Lolo, C. A. Fente, C. M. Franco and A. Cepeda, *Luminescence*, 2005, **20**, 197.

168. Y. Iglesias, C. Fente, S. Mayo, B. Vázquez, C. Franco and A. Cepeda, *Analyst*, 2000, **125**, 2071.

169. T. Pérez-Ruiz, C. Martínez-Lozano, V. Tomás and J. Martín, *J. Chromatogr. A*, 2004, **1026**, 57.

170. D. Badocco, P. Pastore, G. Favaro and C. Maccà, *Talanta*, 2007, **72**, 249.

171. T. Slezak, P. S. Francis, N. Anastos and N. W. Barnett, *Anal. Chim. Acta*, 2007, **593**, 98.

172. J. L. Adcock, N. W. Barnett, J. W. Costin, P. S. Francis and S. W. Lewis, *Talanta*, 2005, **67**, 585.

173. Y. Wei, Z. J. Zhang, Y. T. Zhang and Y. H. Sun, *J. Chromatogr. B*, 2007, **854**, 239.

174. Y. Wei, Z. Zhang, Y. Zhang and Y. Sun, *Chromatographia*, 2007, **65**, 443.

175. X. Xiong, Q. Zhang, F. Xiong and Y. Tang, *Chromatographia*, 2008, **67**, 929.

176. S. L. Fan, L. K. Zhang and J. M. Lin, *Talanta*, 2006, **68**, 646.

177. G. H. Wan, H. Cui, H. S. Zheng, J. Zhoua, L. J. Liu and X. F. Yu, *J. Chromatogr. B*, 2005, **824**, 57.

178. Q. Zhang and H. Cui, *J. Sep. Sci.*, 2005, **28**, 1171.

179. H. Cui, J. Zhou, F. Xu, C. Z. Lai and G. H. Wan, *Anal. Chim. Acta*, 2004, **511**, 273.

180. R. Santiago Valverde, I. Sánchez Pérez, F. Franceschelli, M. Martínez Galera and M. D. Gil García, *J. Chromatogr. A*, 2007, **1167**, 85.

181. Q. Zhang, H. Cui, A. Myint, M. Lian and L. Liu, *J. Chromatogr. A*, 2005, **1095**, 94.

182. Q. Zhang, M. Lian, L. Liu and H. Cui, *Anal. Chim. Acta*, 2005, **537**, 31.

183. M. Kai, H. Kinoshita, K. Ohta, S. Hara, M. K. Lee and J. Lu, *J. Pharm. Biomed. Anal.*, 2003, **30**, 1765.

184. Y. Zhang, Z. Zhang, G. Qi, Y. Suna, Y. Wei and H. Ma, *Anal. Chim. Acta*, 2007, **582**, 229.

185. A. J. Brown, C. E. Lenehan, P. S. Francis, D. E. Dunstan and N. W. Barnett, *Talanta*, 2007, **71**, 1951.

186. M. Martínez Galera, M. D. Gil García and R. Santiago Valverde, *J. Chromatogr. A*, 2006, **1113**, 191.

187. M. D. Gil García, M. Martínez Galera and R. Santiago Valverde, *Anal. Bioanal. Chem.*, 2007, **387**, 1973.

188. X. J. Huang and Z. L. Fang, *Anal. Chim. Acta*, 2000, **414**, 1.

189. C. Kuyper and R. Milofsky, *Trends Anal. Chem.*, 2001, **20**, 232.
190. Y. M. Ming and J. K. Cheng, *J. Chromatogr. A*, 2002, **959**, 1.
191. A. M. García-Campaña, L. Gámiz-Gracia L, W. R. G. Baeyens and F. Alés-Barrero, *J. Chromatogr. B*, 2003, **793**, 49.
192. X. Huang and J. Ren, *Trends Anal. Chem.*, 2006, **25**, 155.
193. C. Kuyper, K. Denham, J. Dickson, J. Murray and R. Milofsky, *Chromatographia*, 2001, **53**, 173.
194. A. Carr, J. Dickson, M. Dickson and R. Milofsky, *Chromatographia*, 2002, **55**, 687.
195. K. Tsukagoshi, A. Tanaka, R. Nakajima and T. Hara, *Anal. Sci.*, 1996, **12**, 525.
196. K. Tsukagoshi, T. Kimura, T. Fuji, R. Nakajima and A. Arai, *Anal. Sci.*, 2001, **17**, 345.
197. K. Tsukagoshi, T. Tokunaga and R. Nakajima, *J. Chromatogr. A*, 2004, **1043**, 333.
198. K. Tsukagoshi, T. Kameda, M. Yamamoto and R. Nakajima, *J. Chromatogr. A*, 2002, **978**, 213.
199. M. Hashimoto, K. Tsukagoshi, R. Nakajima, K. Kondo and A. Arai, *J. Chromatogr. A*, 2000, **867**, 271.
200. B. F. Liu, M. Ozaki, N. Matsubara, Y. Utsumi, T. Hattori and S. Terabe, *Chromatography*, 2002, **23**, 5.
201. K. Tsukagoshi, Y. Obata and R. Nakajima, *J. Chromatogr. A*, 2002, **971**, 255.
202. K. Tsukagoshi, Y. Shikata, R. Nakajima, M. Murata and M. Maeda, *Anal. Sci.*, 2002, **18**, 1195.
203. R. Su, J. M. Lin, K. Uchiyama and M. Yamada, *Talanta*, 2004, **64**, 1024.
204. H. P. Hendrickson, P. Anderson, X. Wang, Z. Pittman and D. R. Bobbitt, *Microchem. J.*, 2000, **65**, 189.
205. W. Cao, J. Liu, H. Qiu, X. Yang and E. Wang, *Electroanalysis*, 2002, **14**, 1571.
206. J. Liu, X. Yang and E. Wang, *Electrophoresis*, 2003, **24**, 3131.
207. J. Yan, J. Liu, W. Cao, X. Sun, X. Yang and E. Wang, *Microchem. J.*, 2004, **76**, 11.
208. W. Cao, J. Liu, X. Yang and E. Wang, *Electrophoresis*, 2002, **23**, 3683.
209. X.-B. Yin, S. J. Dong and E. K. Wang, *Trends Anal. Chem.*, 2004, **23**, 432.
210. J. Yuan, T. Li, X.-B. Yin, L. Guo, X. Jiang, W. Jin, X. Yang and E. Wang, *Anal. Chem.*, 2006, **78**, 2934.
211. J. Yuan, H. Wei, W. Jin, X. Yang and E. Wang, *Electrophoresis*, 2006, **27**, 4047.
212. W. Cao, J. Jia, X. Yang, S. Dong and E. Wang, *Electrophoresis*, 2002, **23**, 3692.
213. S.-N. Ding, J.-J. Xu and H.-Y. Chen, *Electrophoresis*, 2005, **26**, 1737.
214. M.-T. Chiang, M.-C. Lu and C.-W. Whang, *Electrophoresis*, 2003, **24**, 3033.

215. Y.-C. Hsieh and C.-W. Whang, *J. Chromatogr. A*, 2006, **1122**, 279.
216. J. Liu, J. Yan, X. Yang and E. Wang, *Anal. Chem.*, 2003, **75**, 3637.
217. H. Qiu, J. Yan, X. Sun, J. Liu, W. Cao, X. Yang and E. Wang, *Anal. Chem.*, 2003, **75**, 5435.
218. M. Sreedhar, Y.-W. Lin, W.-L. Tseng and H.-T. Chang, *Electrophoresis*, 2005, **26**, 2984.
219. S.-N. Ding, J.-J. Xu and H.-Y. Chen, *Talanta*, 2006, **70**, 403.
220. Y. Chen, Z. Lin, J. Sun and G. Chen, *Electrophoresis*, 2007, **28**, 3250.
221. K. Tsukagoshi, T. Nakamura and R. Nakajima, *Anal. Chem.*, 2002, **74**, 4109.
222. K. Tsukagoshi, Y. Ouji and R. Nakajima, *Anal. Sci.*, 2001, **17**, 1003.
223. X. Huang and J. Ren, *J. Liq. Chromatogr. Relat. Technol.*, 2003, **26**, 355.
224. K. Tsukagoshi, K. Nakahama and R. Nakajima, *Anal. Chem.*, 2004, **76**, 4410.
225. S. L. Zhou, J. H. Wang, W. H. Huang, X. Lu and J. K. Cheng, *J. Chromatogr. B*, 2007, **850**, 343.
226. W.-P. Yang, Z.-J. Zhang and W. Deng, *Anal. Chim. Acta*, 2003, **485**, 169.
227. W.-P. Yang, Z.-J. Zhang and W. Deng, *J. Chromatogr. A*, 2003, **1014**, 203.
228. J. Wang, W. Huang, Y. Liu, J. Cheng and J. Yang, *Anal. Chem.*, 2004, **76**, 5393.
229. H.-L. Jiang, Y.-Z. He, H.-Z. Zhao and Y.-Y. Hu, *Anal. Chim. Acta*, 2004, **512**, 111.
230. S. Han, *Anal. Sci.*, 2005, **21**, 1371.
231. F. J. Lara, A. M. García-Campaña, L. Gámiz-Gracia, J. M. Bosque-Sendra and F. Alés-Barrero, *Electrophoresis*, 2006, **27**, 2348.
232. X. M. Guo, X. D. Xu, H. J. Zhang, Y. G. Hu and J. Zhang, *Chin. Chem. Lett.*, 2007, **18**, 1095.
233. E. B. Liu and J. K. Cheng, *Chromatographia*, 2005, **61**, 619.
234. Y. M. Liu and J. K. Cheng, *J. Chromatogr. A*, 2003, **1003**, 211.
235. W. Yang, Z. Zhang and W. Deng, *Talanta*, 2003, **59**, 951.
236. S. Zhao, C. Xie, X. Lu, Y. Song and Y.-M. Liu, *Electrophoresis*, 2005, **26**, 1745.
237. S. Zhao, H. Yuan, C. Xie and D. Xiao, *J. Chromatogr. A*, 2006, **1107**, 290.
238. K. Tsukagoshi, T. Saito and R. Nakajima, *Talanta*, 2008, **77**, 514.
239. K. Tsukagoshi, M. Hashimoto, T. Suzuki, R. Nakajima and A. Arai, *Anal. Sci.*, 2001, **17**, 1129.
240. K. Tsukagoshi, N. Jinno and R. Nakajima, *Anal. Chem.*, 2005, **77**, 1684.
241. Y.-T. Lee and C.-W. Whang, *J. Chromatogr. A*, 1997, **771**, 379.
242. N. W. Barnett, B. J. Hindson and S. W. Lewis, *Analyst*, 2000, **125**, 91.
243. R. Su, J.-M. Lin, F. Qu, Z. Chen, Y. Gao and M. Yamada, *Anal. Chim. Acta*, 2004, **508**, 11.
244. J. Bai, H. Xu, Q. Liu, J. Chen and L. Bai, *Microchim. Acta*, 2008, **160**, 165.

245. K. Tsukagoshi, H. Akasaka, Y. Okumura, R. Fukaya, M. Otsuka, K. Fujiwara, H. Umehara, R. Maeda and R. Nakajima, *Anal. Sci.*, 2000, **16**, 121.
246. X. Ji, Z. He and D. Pang, *Electrophoresis*, 2007, **28**, 3260.
247. M. Li and S. H. Lee, *Luminescence*, 2007, **22**, 588.
248. X. Zhou, D. Xing, D. Zhu, Y. Tang and L. Jia, *Talanta*, 2008, **75**, 1300.

CHAPTER 8

Chemiluminescence to Immunoassays

MICHAEL SEIDEL AND REINHARD NIESSNER

Chair for Analytical Chemistry & Institute of Hydrochemistry, TU München, Marchioninistraße 17, D-81377 München, Germany

8.1 Introduction

An immunoassay (IA) is a quantitative analytical technique that uses antibodies as specific recognition elements. Appropriate antibodies bind with high selectivity to their analytes. IA can be designed for a wide range of analytes with low limits of detection and wide calibration ranges. Antibodies are applied directly to the sample without prior extraction and purification steps, which speeds up the analysis time, reduces costs and permits fully automated platforms for analysis in difficult matrices. Therefore, IAs are widely applied in pharmaceutical analysis, toxicological analysis, bioanalysis, clinical chemistry, environmental and food analysis.[1–3] Ultrasensitive IAs have been developed, which strongly depend on the affinity of specific antibodies and on the sensitivity of the detection method. By combining the characteristics of antibody–analyte interactions with highly sensitive detection systems effective analytical platforms are feasible. Among detection methods, chemiluminescence (CL) is a very versatile and ultrasensitive method that is easy to adopt in automated platforms to obtain highly reproducible results and high throughput of samples.[4] Generally, chemiluminescence immunoassays (CLIA) are more sensitive than chromogenic ELISAs or fluorescence immunoassays and often exceed radioimmunoassays.[5,6] CL is the production of light by a chemical reaction. Since there is no need for sample illumination, problems with light scattering, unselective excitation and source instability are absent.[7] Therefore, the extreme sensitivity of chemiluminescence analysis is a result of the low background. The CL reaction has

Chemiluminescence and Bioluminescence: Past, Present and Future
Edited by Aldo Roda
© Royal Society of Chemistry 2011
Published by the Royal Society of Chemistry, www.rsc.org

additional advantages: minimal instrumentation is required, with a quite simple optical system because no external light source is needed and chemical light can be generated at each solid support.[8] CLIAs are applied in heterogeneous test formats, which means that unreacted reagents and matrix are separated from the immune complex at the solid phase. A large variety of solid supports have been applied on CLIA: multiwell plates, microparticles, affinity columns, capillaries, fibres, membranes or glass slides. CLIA is preferred for samples with complex matrices. It is mainly applied in clinical chemistry, environmental and food analysis due to its high sensitivity, wide calibration range and the ability for complete automation in immunoanalytical platforms.[9] The introduction of CL imaging by means of light detection with CCD or CMOS chips offers new perspectives on multiplexed analysis systems like CL microarrays, which often use horseradish peroxidase (HRP) as an enzymatic CL label.[10] Accurate localization of the target, down to micrometer scale, is required for CL microarrays. The light emission takes place close to the enzyme label and the produced excited species is characterized by a relatively short half-life.[11]

8.2   Principles of CLIAs

8.2.1   Immunorecognition of Elements

Each antibody in an IA reacts selectively by affinity reaction to its specific antigen. The antigen is the counterpart to the antibody and its chemical structure reacts to the paratope of the antibody with high affinity. The affinity reaction is an equilibrium reaction strongly depending on the concentration of analytes, if the antibody and the other reagents are present in constant concentrations. Antibodies bind to various important analytical candidates: small organic molecules like pesticides, hormones, pharmaceuticals or toxic small molecules as well as biopolymers (peptides, proteins, DNA) or the outer membrane of cells or organelles (viruses, bacteria, spores, fungi, protozoa, eukaryotic cells). Therefore, antibodies have been widely used for the rapid detection of analytes in biological, environmental or food samples. Monoclonal (mAb) and polyclonal (pAb) antibodies are available for immunoassays.[12] For routine analysis, defined antibody species are relevant that can be delivered over many years. Currently, this demand is met by monoclonal antibodies. In the past few years, alternative recognition elements have been developed with potential improved properties. Ease of preparation, low price and better physical and chemical stability are arguments for substitution of antibodies. In future, recombinant antibodies (rAbs) or even artificial recognition elements like aptamers could substitute them.[13]

8.2.2   Labels and Reagents for CLIAs

CL labels can be divided into chemical and enzyme-based labels. Chemical labels are consumed in the analytical reaction and have a defined sensitivity.

Luminol and luminol derivatives were first applied for CLIA. Under alkaline conditions, luminol is oxidized in the presence of a suitable catalyst to yield high CL signals. Transition metal cations can be used as catalysts; however, they were not compatible to immunoassays as low signal-to-noise ratio at high pH were obtained.[14] In contrast, acridinium compounds have no catalytic requirements and are remarkably resistant to microenvironmental interference.[15] The advantage of acridinium esters over luminol is that they have a higher quantum efficiency and they show no serious quenching effect when linked to proteins or haptens.[16] Aromatic acridinium esters generate light when they are oxidized by hydrogen peroxide. Abs or Ag-derived conjugates can be detected at high sensitivity. Exposure of an acridinium ester label to an alkaline hydrogen peroxide solution triggers a flash of light. The measurement process can be completed quickly, which increases the throughput of automated immunoassay analysers.[17] The most sensitive chemical label is the acridinium sulfonamide label, which is stabile, linkable to haptens and antibodies, and has a low CL lifetime.[18,19] Lai *et al.* have shown that acridinium esters could emit light by ultrasonic activation as an alternative strategy to peroxide triggering.[20]

Horseradish peroxidase (HRP) and alkaline phosphatase (ALP) are the main enzyme labels for CLIA and, nowadays, 80% of immunoassays for routine clinical analysis are based on the use of enzymatic labels.[21] Using enzymes as label in immunoassays affords a much more sensitive detection system in comparison to chemical labelling because of the great catalytic power resulting in generation of many product molecules from one enzyme molecule.[22] The enzyme label is coupled to primary detection antibodies, to tracers (enzymes coupled to an antigen), to a secondary antibody, which binds to the unlabelled primary antibody, or to streptavidin, which binds to biotinylated antibodies. For all enzyme labels it is important that the sensitivity of the CLIA partly depends on the enzyme activity. An enzymatic turnover rate depends on the temperature and the substrate concentration. Both parameters should be kept constant during the immune analysis. CLIA platforms have to be temperature controlled if the ambient temperature varies.

Enzyme labels require suitable CL substrates. HRP is frequently used as a label for antibodies or tracers. The chemiluminescence substrate luminol reacts with H_2O_2 in the presence of HRP by emitting light at 428 nm with a relatively low quantum yield of 1%.[23] The signal intensity per HRP molecule can be increased by using luminol analogues or enhancers like *p*-iodophenol (PIP), 4-(1-imidazolyl)phenol,[24] and other *p*-phenol derivatives,[25] *para*-phenylphenol and sodium tetraphenylborate as synergistic enhancer,[26] or $K_3Fe(CN)_6$ as electron mediator.[27]

Dioxetanes such as the stabilized adamantyl 1,2-dioxetane (3-(4-methoxyspiro [1,2-dioxetane-3,2′-tricyclo[3.3.1.13,7]decan]4-yl)phenyl phosphate, AMPPD) are appropriate substrates for ALP. The chemiluminescent substrate undergoes hydrolysis in the presence of ALP to yield an unstable intermediate with light emission at 470 nm.[28] In a comparison of both enzyme labels, ALP-labelling methods show two weak points when compared with HRP-labelling methods. On one hand, ALP-AMPPD system exhibits higher background in solutions containing proteins and small colloids. On the other hand, ALP has a higher molecular weight than HRP, meaning easier aggregation of ALP molecules in the case of complicated or long duration immunoreaction, which will yield a high background and unavoidable, disproportionate or false positive results in clinical usage.[29]

The signal could be increased by surface-enhanced CL. Metals like gold enhance the chemiluminescence signal, depending on distance. HRP labelled streptavidin was immobilized on a Au film and the distance was controlled by peptide chain to between 1 and 8 nm. With this experiment, CL signals were increased by more than one order of magnitude.[30] HRP-functionalized silica nanoparticles (NP) have shown a 61-fold increase in detection signals since the HRP–NP antibody conjugate binds to the immobilized analyte on a gold layered surface.[31] Previte *et al.* have developed a microwave-triggered metal-enhanced CL by means of a heating effect of the HRP reaction on a silver nanostructure.[32]

A novel and sensitive CLIA has been developed by employing a new CL enhancer, bromophenol blue (BPB), for the determination of α-fetoprotein (AFP) based on magnetic beads (MBs) and colloidal gold nanoparticles (AuNPs) modified with HRP-labelled anti-AFP antibodies.[33] BPB, as a chemical indicator, was found to act as a novel and effective signal enhancer of the peroxidase-catalysed CL reaction of luminol with hydrogen peroxide. The detection limits were one order of magnitude lower than that obtained without using AuNPs, and much lower than that typically achieved by ELISA.

The use of nanoparticle labels for catalytic reaction with luminol for CL generation has been examined as an alternative to enzyme labels.[34] Antibodies labelled with colloidal gold were detected with high sensitivity. Numerous Au^{3+} ions from each gold particle anchored on the surface of magnetic beads were released after oxidative gold metal dissolution (0.01 M HCl–0.1 M NaCl–0.25 Br_2 solution) and then quantitatively determined by a simple and sensitive Au^{3+}-catalysed luminol CL reaction. A concentration as low as 3.1×10^{-12} M human IgG (0.5 ng mL^{-1}) could be determined.[35] Gold nanoparticles dissolved to form $AuCl_4^-$ by employing a solution of 2% HNO_3 and 3.4% HCl had also a strong catalytic effect on the luminol–H_2O_2–Cl reaction. Antibody concentrations were detected down to a LOD of 1.5 ng mL^{-1}.[36] Hu *et al.* have dissolved gold nanoparticles at the solid support by using HCl–NaCl–Br_2. After one hour the CL could be started.[37] Antibodies in sera could be detected at $100\times$ lower dilution than with chromogenic ELISA. Wang *et al.* have found that the mixing of CdTe semiconductor nanocrystals (NCs) with luminol in the presence of $KMnO_4$ could induce a greatly sensitized effect on CL emission.[38] Duan *et al.* have developed an easily automated CL immunoassay based on gold nanoparticle labelled antibodies. In this assay, the CL reaction between luminol and $AgNO_3$ was triggered by gold nanoparticles.[39]

A few bioluminescence immunoassays have been published in the last ten years; HRP-based CLIA is the preferred CL label.[40] Lewis *et al.* have demonstrated that the bioluminescent label aequorin can be genetically modified to site-specifically attached nonpeptide molecules for the production of hapten–aequorin conjugates and could, therefore, enhance research in bioluminescence immunoassays.[41]

8.2.3   CLIA Procedures

Immunoassays can be classified into three different test formats: direct, indirect and sandwich, and additionally into competitive and non-competitive formats (Figure 8.1). Non-competitive immunoassays directly detect the immune complex at the surface, as opposed to competitive immunoassays where an analyte derivative is added that competes with the analyte for the binding site of the antibody. In direct formats the antibody is immobilized to the surface. Analytes and labelled analyte derivatives (tracer) are added. In contrast, an indirect format uses immobilized analyte derivatives. The latter format is a sandwich immunoassay where immobilized capture and detection antibodies are used. Competitive immunoassays are applied for the detection of small analytes that consist of one binding site for antibody recognition. Sandwich immunoassays are applied to high molecular weight molecules that consist of two or more binding sites.[42]

Instead of labelled detection antibodies, often labelled secondary antibodies (anti-human, anti-rabbit, anti-mouse, *etc.*) are used that react with unlabelled primary antibodies (human, rabbit, mouse, *etc.*) at the solid support. CLIAs, which contain secondary antibodies, need an additional reaction step and have therefore an extended assay time. However, as the labelling process reduces the activity of antibodies, high amounts of antibodies are needed, which is expensive, especially for sensitive detection antibodies. Other reasons for applying secondary antibodies are the detection of specific antibodies in samples such as human sera, which are identified with a secondary antibody (*e.g.* anti-human IgG). For multiplexed immunoassays, the advantage lies in the simple setting of the amount of secondary antibodies.

The principle of all immunoassays is the formation of highly selective antibody–analyte complexes and the differentiation of complexed and free antibodies. CL immunoassays are generally heterogeneous assays that use the solid support for the antibody–antigen reaction. Conversely, in homogeneous

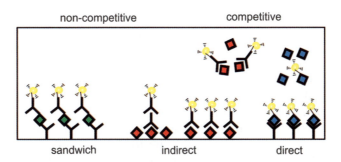

Figure 8.1   Test formats for chemiluminescence immunoassays (solid phase).

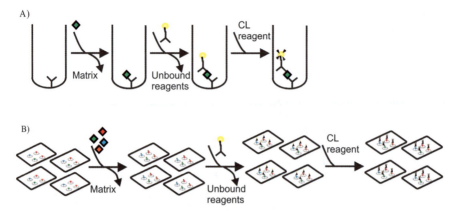

Figure 8.2   Single analyte (A) and multi-analyte (B) detection using static non-fluidic CL immunoassays.

immunoassays the whole reaction takes place in the liquid phase and the walls of the immunoanalytical system are inert. The CLIA procedure contains sequential reaction steps: immunoreaction by adding sample and CL-labelled reagents, removal of matrix and unbound reagent, addition of CL reagent, and finally CL detection. Matrix and unbound reagents are removed by aspirating (static immunoreaction) or by pumping through a flow channel (flow-through immunoreaction) as illustrated in Figures 8.2 and 8.3, respectively.

Multiplexed CL immunoassays can be used to quantify a set of analytes in a single immunoreactor by localized antibody immobilization in a microarray format (Figure 8.2b and Figure 8.3c.). The CL reaction occurs at the solid support and the generated light is detected in the liquid phase (Figure 8.3a) or directly at the surface (Figure 8.3b).

The removal of matrices and unbound reagents by washing steps as a process strategy increases the sensitivity in complex matrices and makes such a test system more robust. As consequence, up to 90% of these tests make use of this format in clinical diagnostics and environmental and food analysis.

Static immunoreactions are time consuming due to the needed equilibrium reaction under mass transport limitation. The heterogeneous immunoreaction in a flow cell is faster than in a static system. The diffusion layer is thinner and kept constant. Automated flow-through immunoreactions have the advantage that the immunoassay does not have to be in equilibrium if the conditions (flow rate and temperature) are constant[7] and are the fastest type of immunoassays. In flow-through immunoassays, the homogeneous equilibrium reaction with antibodies and analytes is carried out separately from the heterogeneous reaction. Some authors, therefore, define these immunoassays as non-competitive test formats, others as binding inhibition assays. In this chapter the phrase competitive immunoassay is used for test formats containing an analyte and an analyte derivative (immobilized on surfaces or conjugated to a label) that reacts with the binding places of their antibody.

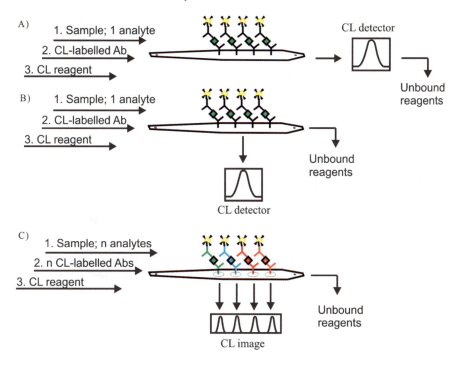

Figure 8.3 Flow-through chemiluminescence immunoreaction concepts: indirect detection (A), direct (B) and multiplexed analysis (C).

CL immunoassay can also be classified by the solid support. Microparticles as a dot have the smallest dimension, followed by lines in a flow-channel, surface areas of wells (microwell plates) or flat substrates (flow cells with glass, golden or plastic chips), and 3D immunoreaction in columns or membranes. In the last decade, the density of analytical information has been increased by developing new CL immunoassay processes. Parallel detection in multiwell plates allows the calibration and quantification of single analytes in one experiment. The formation of microarrays on solid supports increases the quantity of analytes that can be detected in a sample. The highest complexity is reached by parallelization of multiplexed microarray analysis. Figure 8.4 illustrates the choice of solid supports and the density of analytical information reached in a matrix.

The enforcement power of an analytical method often depends on practicability, simplicity of the assay procedure, low time consumption and low costs per analysis. Therefore, only these immunoassay procedures will become accepted for routine analysis that can be established on CL immunoassay platforms. Such an automated platform is a self-contained system, consisting of reagents for CL analysis, rinsing and cleaning, computer-controlled automated immunoassay procedure by addressing pumps and valves, CL signal detection, and data handling (Figure 8.5).

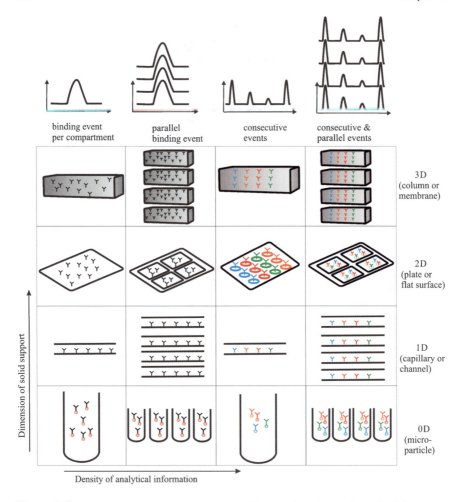

Figure 8.4   Overview of CL immunoassays with regard to dimension of solid support and analytical information.

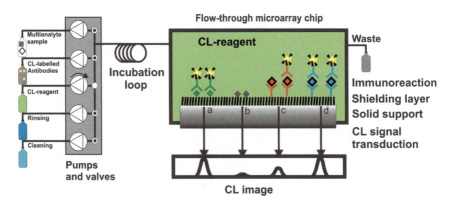

Figure 8.5   Example of CL immunoassay platforms (working principle).

8.3 Non-automated CLIA

8.3.1 CL-ELISA

CL-ELISA is performed in microwell plates (MWP). First, the MWP is prepared by coating each well with antibodies (direct format) or hapten-conjugates (indirect format) and by blocking to prevent unspecific binding. Sample and labelled-analyte derivative (direct competitive) or labelled antibodies (indirect competitive) are incubated in each well. After reaching equilibrium conditions, the CL reaction is initialized and detected with a luminometer. Washing steps are needed between each step to remove unreacted compounds. CL-ELISAs have been developed for many analytes in food, environmental or clinical samples. Table 8.1 lists CL-ELISAs published in 2000–2009.

8.3.2 Multiplexed CLIA

Microarray enzyme-linked immunoassays based on chemiluminescence were introduced by Joos *et al.* for multi-analyte detection in clinical diagnostics.[57] Other applications in diagnostics have been published for infectious diseases,[58] rheumatoid diseases,[59] and cytokines. Chemiluminescence antibody microarrays for simultaneous detection of multiple cytokines were introduced by Huang.[60] He had developed a sandwich multiplex CL immunoassay for 24 cytokines by adsorbing microspotted capture antibodies on membranes.[61] The membranes were incubated with biological samples and the bound proteins were recognized with biotin-conjugated antibodies and detected by HRP–streptavidin. On hybond membranes he could detect cytokines down to $5\,\text{pg}\,\text{mL}^{-1}$. The density of spotted anti-cytokine antibodies could be increased to 35 antibodies per chip by using poly(vinylidene difluoride) (PVDF) membranes.[62]

Moody *et al.* have used the bottom of 96-well polystyrene plates to form a 3×3 pattern of capture antibodies to cytokines in the bottom of the wells.[63] Urbanowska *et al.* have shown that cytokine biomarkers could analysed by such a multiplex sandwich CL immunoassay with an assay accuracy of between 70% and 130%, and an assay precision of less than 30%, which is comparable to classical single-analyte ELISAs.[64]

Marquette *et al.* have developed disposable screen printed chemiluminescence biochips for point of care diagnosis.[65] Four protein markers could be simultaneously determined in $25\,\text{min}$. The multiplex sandwich immunoassay was characterized by dynamic ranges of 0.5–50, 0.1–120, 0.2–20 and 0.67–$67\,\mu\text{g}\,\text{L}^{-1}$ for C-reactive protein, myoglobin, cardiac troponin I and brain natriuretic peptide, respectively. The capture antibodies were directly electrochemically grafted on the screen printed microarray.[66] The immobilization strategy was based on diazoated aniline derivatives, which could be electro-addressed, thus creating a covalent linkage with a conducting material. Nanostructured gold was used as conducting material, which enhanced the CL signal.[67]

Yang *et al.* have reported a one-step homogeneous bioluminescence immunoassay on a miniaturized multiwell plate.[68] The applied cloned enzyme donor

Table 8.1 Overview of CL-ELISA performed in microwell plates in the past ten years.

Matrix	Test format	Label	Analyte	LOD	IC_{50}; WR	Ref.
Milk	Indirect competitive	HRP	Gatifloxacin	$1\,pg\,mL^{-1}$	$0.4\,ng\,mL^{-1}$; –	43
Milk	Indirect competitive	HRP	Aflatoxine M_1	$0.25\,ng\,L^{-1}$	–; –	44
Skim milk	Direct competitive	HRP	Chloramphenicol	$0.05\,ng\,mL^{-1}$	–; 0.1–$10\,ng\,mL^{-1}$	45
Water, fruit juice, honeybee	Indirect competitive	HRP	Chlorpyrifos	1–$1.75\,ng\,mL^{-1}$	$3.5\,ng\,mL^{-1}$; –	46
Food samples	Direct competitive	HRP	Fumonisin B	$0.09\,\mu g\,L^{-1}$	$0.32\,\mu g\,L^{-1}$; 0.14–$0.9\,\mu g\,L^{-1}$	47
Water, soil, food	Indirect competitive	HRP	DDT	$0.2\,\mu g\,L^{-1}$	$0.6\,\mu g\,L^{-1}$; 0.1–$2\,\mu g\,L^{-1}$	48
Wastewater	Direct competitive	ALP	17β-Estradiol	$1.5\,pg\,mL^{-1}$	–; 2.5–$1600\,pg\,mL^{-1}$	49
Surface and tap water	Indirect competitive	HRP	Microcystin-LR	$0.032\,\mu g\,L^{-1}$	$0.2\,\mu g\,L^{-1}$; 0.062–$0.65\,\mu g\,L^{-1}$	50
Human serum	Direct competitive	ALP	Progesterone	$0.06\,ng\,L^{-1}$	–; 0.2–$125\,ng\,mL^{-1}$	51
Human serum	Sandwich	HRP/liposomes	PSA	$0.7\,pg\,mL^{-1}$	–; $0.74\,pg\,mL^{-1}$ – $0.74\,\mu g\,mL^{-1}$	52
Human serum	Sandwich	HRP	Human TSH	$0.01\,mU\,L^{-1}$	–; 0.1–$40\,mU\,L^{-1}$	53
Chicken and pig muscle	Indirect competitive	HRP	Sulfonamides	0.1–$0.43\,\mu g\,L^{-1}$	–; –	54
Chicken muscle	Indirect competitive	HRP	Chloramphenicol	$6\,ng\,L^{-1}$	–; 0.05–$5\,\mu g\,kg^{-1}$	55
Bovine urine	Direct competitive	HRP	Clenbuterol	$0.08\,\mu g\,L^{-1}$	–; –	56

immunoassay (CEDIA) is based on the bacterial enzyme β-galactosidase, which was genetically engineered into two inactive fragments: an enzyme donor and enzyme acceptor. When mixed together, the two fragments combined to form an active enzyme.[69] A conjugated analyte-derivate to ED competed for the binding site of an antibody. Only free analyte-ED could form the active enzyme, and a high analyte concentration correlated with a high bioluminescence signal. As bioluminescence reagent the Beta-Glo assay system was used: active β-galactosidase hydrolysis of 6-*O*-β-galactopyranosyl-luciferin to D-luciferin, and the enzyme firefly luciferase converted the produced D-luciferin into oxyluciferin and luminescent light.[70] The multiplexed assay for three different antiepileptic drugs (AED) has been performed on a miniaturized multiwell plate, the so-called ImmunoChip, which consisted of 1-μL wells in a 3×3 pattern. A serum sample (25 μL) containing all three AEDs was dispensed on the centre of nitrocellulose filter paper, which was placed above the wells of the array. The serum sample, applied to the middle, wicked along and through the filter paper and into each well, dissolving the lyophilized reagents and initiating the CEDIA and bioluminescent reactions. The signals from the bottom of the chip were imaged with a macro camera lens and a CCD camera. The assay performed on the ImmunoChip was fast (5 min), required small volumes of reagents and serum sample and had inter- and intraassay coefficients of variation of less than 10%.

Further application of non-fluidic CL microarray immunoassays systems have been published for food, environmental analysis and biosecurity by quantification of biotoxins[71–73] and microorganisms.[74] Magliulo *et al.* have presented a rapid multiplexed chemiluminescence immunoassay for the detection of *Escherichia coli* O157:H7, *Yersinia enterocolitica*, *Salmonella typhimurium* and *Listeria monocytogenes* in food samples.[75] The multiplexed CL-ELISA is performed on a 96×4 multiwell plate with four subwells in each well (Figure 8.6). A sandwich immunoassay was used as test format. Four different bacteria strains could be detected in parallel at each well. The standards for calibration curve and samples were simultaneously measured on the same platform. The assay was simple and fast, and the limit of quantification was of the order of 10^4–10^5 cfu mL^{-1}.

A multiplexed sandwich CL-ELISA for the detection of bacteria and viruses has been based on the ArrayTube System (ATS) from Clondiag Chip Technologies (Jena, Germany).[76] A single reagent tube was used for the incubation of sample and reagents. The microarray was located on a glass chip forming the bottom of a 1.5 mL plastic tube. Antibodies were deposited on the glass surface. A sandwich format was applied to simultaneously detect bacteria, viruses and toxins. The chemiluminescence signal was generated after an incubation step with streptavidin–poly-HRP. Washing steps and pipetting steps were manually operated. The incubation steps were carried out with a horizontal tube shaker. The detection limit for viruses (TCID$_{50}$ = 50% tissue culture infection dose) was 6×10^2–5×10^6 TCID$_{50}$ mL^{-1}, for toxins between 0.1 and 0.2 μg L^{-1} and for bacteria between 5×10^3 and 2×10^6 cfu mL^{-1}. The analytes were detected in 1 h 23 min.

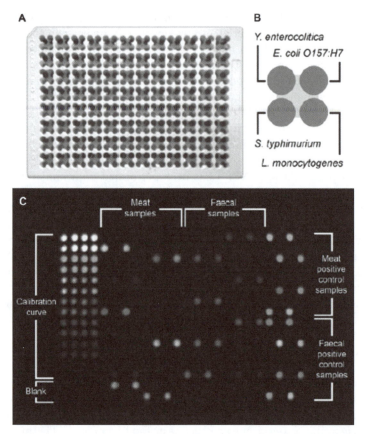

Figure 8.6 (A) Image of a microarray platform on a 96×4 well microtiter plate; (B) position of the immobilized antibodies within each subwell; and (C) CL image of the microtiter plate.[75] (Reproduced with permission from American Chemical Society, Copyright 2007.)

The nanodot array luminometric immunoassay (NALIA) system consists of conventional 96-well membrane-bottomed plates in which antigens or antibodies are adsorbed onto the underside of a nitrocellulose/cellulose acetate membrane.[77] A 5×5 format allowed the measurement of ten analytes in duplicates and was used for screening autoantibodies in systemic rheumatic disease. Sample, test sera and subsequent reagents were sucked through the membrane. The captured analytes were quantified by imaging CL with a CCD camera after an assay time of 95 min. Detection limits of $<20 \times 10^3 \,\text{IU}\,\text{L}^{-1}$ of anti-dsDNA were achieved (IU = international units of a WHO standard).

8.3.3 Microparticle and Magnetic Microparticle-based CLIA

Microparticle (MP) and magnetic microparticle (MMP)-based CLIA use microwell plates or tubes as immunoreaction compartment. MP and MMP serve as solid support for the immunoreaction and at the same time for separation of bound analytes from matrix and unbound reagents. MP-based CLIAs use centrifugal forces or sedimentation for separation. MMP-based CLIAs use magnetic forces by attaching a magnet under the microwell plate or at the wall of the tube. The immune reaction is carried out in solution, which reduces the incubation times. Sandwich CLIA can easily be performed in a single-step immunometric assay by incubating with a mixture of capture antibodies coated to MP, sample and labelled detection antibody. For C-reactive proteins a MP-based sandwich CL-ELISA has been developed.[78] MPs with a mean diameter of 8 mm were applied and acridinium ester was used as CL label. The assay covered a linear range of 0.01–50.00 mg L^{-1}. The calculated detection limit was 4 µg L^{-1}.

The tumour marker carbohydrate antigen 50 has been evaluated in human serum using a MMP-based CL-EIA.[79] A sandwich CLIA was performed by using FITC-labelled anti-CA50 antibodies as capture antibody and ALP-labelled anti-CA50 antibodies as detection antibody. After magnetic separation the CL substrate AMPPD was added and CL signals were measured during 30 min incubation at room temperature. The proposed method exhibited advantages of a lower minimum detectable concentration of 1.0 U mL^{-1}, in comparison to the commercially available immunoradiometric assay, and showed a larger linear range of 0–300 U mL^{-1}, as well as a shorter total assay time of only 50 min. With the same reagent setup α-fetoprotein was detectable down to 3 ng mL^{-1}.[80] The linear range observed was 3–1200 ng mL^{-1}. Through the use of MMPs, extremely high concentrations of α-fetoprotein could be measured in the samples without dilution. The application of MMPs has avoided the hook effect.

An ultrasensitive enhanced CL-EIA has been developed for determination of α-fetoprotein based on the combination of 4-(4-iodophenyl)phenol as enhancer and double-codified gold nanoparticles (DC-Au-NP) for further signal amplification. DC-Au-NPs labels were modified with HRP-conjugated anti-AFP; 100-times lower LODs were achieved (5 pg mL^{-1}) than that obtained using HRP-labelled anti-human AFP.[81]

Zhao *et al.* have developed a MMP-based CLIA in microwell plates to detect 17β-estradiol in seawater.[82] For their direct competitive assay format, they used MMPs coated with an antibody anti-fluorescein. These MMPs captured fluorescein-labelled antibodies in solution, which are then incubated with sample and ALP-conjugated estradiol-derivatives. The analytical performance was characterized with a working range of between 10 and 3000 ng L^{-1}, a LOD of 5.4 ng L^{-1} and an assay time of more than 60 min. Xin *et al.* have shown an improvement in LOD and detection time by using peroxidase-labelled estradiol derivatives as tracer and H$_2$O$_2$, luminol and *p*-iodophenol for enhanced CL reaction.[83] A detection limit of 2.5 ng L^{-1}, a working range of 15–1000 ng L^{-1} and a CL detection time of 10 min (30 min for ALP) were achieved.

In the literature a strategy for the separation of two different targets has been described that uses a combination of magnetic bead separation and precipitation. Antibodies to poly(*n*-isopropylacrylamide) (PNIP) and MMP were used to separate different targets by taking advantage of thermal response.[84] PNIP is known to aggregate and precipitate out of water when the temperature is raised above the lower critical solution temperature of 31 °C. Thus, it could be separated from supernatant by centrifugation. In addition, magnetic beads could be separated from PNIP by magnetic forces if the temperature was lower than the lower critical solution temperature. A homogeneous non-competitive ELISA was employed, formed by primary antibodies immobilized onto the surface of magnetic beads and PNIP, antigen as IgG and IgA in the sample and HRP-labelled second antibodies. Moreover, highly sensitive CL detection of HRP was applied, and the LOD of IgG and IgA was as low as 2.0 and $1.5 \, \text{ng} \, \text{L}^{-1}$.

The use of magnetic nanoparticles (MNPs) could improve the sensitivity and the reproducibility of fully automated CLIA because MNPs are completely dispersed in aqueous solution and allow accurate automatic handling.[85] However, the magnetophoresis is limited. Superparamagnetic poly(methacrylate divinylbenzene) microbeads with nanosized magnetic particles ($<10 \, \text{nm}$ in diameter) within their polymeric framework have been prepared to obtain higher magnetite contents and higher magnetophoresis ability.[86] The super-paramagnetic polymer microparticles had a diameter of $3 \, \mu\text{m}$ with many 8-nm magnetite nanoparticles distributed and immobilized in the inner polymeric frameworks. Free tumour marker hCGβ in serum was detected with a sandwich CLIA in 1 h by using ALP as CL label and AMPPD as CL-substrate. The detection limit was $0.22 \, \text{mIU} \, \text{mL}^{-1}$, which is one order lower than a chromogenic ELISA. A high-gradient magnetic field on steel ball columns achieved higher magnetophoresis by dealing with MNPs. Pappert *et al.* have developed a MNP-based sandwich CLIA for the combined enrichment and detection of *E. coli*,[87] which is especially interesting for quantification of bacteria in complex matrices like food or environmental samples. The immunomagnetic separation in combination with a sandwich CL-ELISA resulted in a reduction of the LOD by a factor of 20 compared to conventional sandwich CL-ELISA.

8.4 Platforms for CLIAs

8.4.1 Automated Single Analyte Detection

8.4.1.1 *Microparticle-based CLIA Platforms*

Microparticle-based CLIAs have been applied especially in clinical diagnostics. The first automated random-access platforms were introduced in the 1980s. A summary of existing platforms is shown in Table 8.2. Microparticles as solid phase material have advantages in the mass transfer of proteins and the rapid separation of bound and free immune complexes for fully automated applications in routine laboratories. The antibody–antigen binding equilibrium can be achieved more rapidly than when antibodies are immobilized on the planar surface, such as microwell plates. Zhang *et al.* have shown this by comparing

Table 8.2   Random access automated analysers.

Platform	Particles	CL-label	Throughput (tests h^{-1})
ACS:180 Chiron Diagnostics	MMP	Acridinium ester	180
ADVIA Centaur	MMP	Acridinium ester	200
Architect	MMP	Acridinium sulfona-mide ester	200
Immulite 2000	Polystyrene MP	Alkaline phosphatase	120
LIAISON	MMP	Aminobutyl-ethylisoluminol	140
LumiQuick	Polystyrene MP	Acridinium ester	150
Magic Lite	MMP	Acridinium ester	60
Nichols Advantage	MMP	Acridinium ester	80
SphereLight 180	Polystyrene MP	HRP	180

magnetic microparticles and coated tubes performing a sandwich CLIA for the quantification of α-fetoprotein.[29]

Two different principles are possible for separation of bound and free immune complexes by using microparticles (MP). MP-based CLIA uses the centrifugal forces to separate the particle complexes from the unbound labels. The magnetic attraction of magnetic microparticle (MMP) simplifies CL immunoassays platforms and increases the throughput.[88]

Microparticle-based CLIA.   Immulite is a random access automated analyser designed for MP-based CLIA to quantify various proteins by means of a sandwich immunoassay test format.[89] The automated separation of free and bound labels to polystyrene microparticles coated with antibodies is performed by spinning an assay tube about its longitudinal axis. Sample, excess reagent and wash solution are captured in a coaxial waste tank integral with the tube. The incubation carousel is heated to 37 °C and indexed as 30 s for a throughput of 120 tests h^{-1}. For the CL reaction the label alkaline phosphatase (ALP) is used. The CL substrate is a derivative of adamantyl 1,2-dioxetane phosphate, AMPPD. Counts are converted into analyte concentration by use of stored calibration curves. The curves are adjusted by lot-specific constants that are bar-coded on the reagents, and stored standard curves are periodically recalibrated with two calibrators that are loaded just like a sample.[89]

An automated immunoassay for cytokine measurement (IL-1, IL-6, IL-8, TNF-a, IL-2R) has been performed on the Immulite system. Polystyrene beads were either coated directly with monoclonal specific antibodies (IL-1β, TNF-α) or a biotin/streptavidin system with biotinylated antibodies (IL-6, IL-8, and IL2R) was used. The polyclonal detection antibodies were labelled with ALP. Patient serum and alkaline phosphatase-conjugated antibodies or, depending on the technique, a ligand-labelled antibody were incubated for 30 to 60 min at 37 °C. The increased incubation time gave a higher sensitivity, which was important for IL-1, IL-6 and TNF-α. Unbound conjugates were then removed by a centrifugal wash (3×), after which the chemiluminescent substrate was added, and the test unit was incubated for further 10 min. The bound complex, and thus also the generated photons, was measured by a luminometer.[90] Many other parameters in clinical diagnostics have been analysed with the Immulite instrument some of which have been published, such as: C-reactive protein (CRP),[91] folic acid,[92] pregnancy-associated plasma protein A,[93] troponin, CK-MP and myoglobin,[94] total testosterone,[95] anti-tissue-transglutaminase IgA, and antigliadin IgG and IgA,[96] vancomycin,[97] hepatitis B serology marker[88] and allergens.[98,99] For the measurement of circulating allergen-specific IgE on the Immulite system allergens were linked to biotin.[100] The ligand biotin was covalently linked to a soluble polymer matrix, which in turn attached to amino-, hydroxy-, carboxy or thiol-groups on the allergen. Biotinylated allergens and sIgE antibodies bound in the liquid phase. The formed allergen–antibody complex was subsequently captured by streptavidin-coated beads. A secondary AP-linked antibody against human IgE recognized bound IgE in sera.

Beside the Immulite platform only a few others exist that have used MP-based CLIA. Details have been published of the LumiQuick platform, which uses polystyrene MP with a mean diameter of 120 μm. A fully automated sandwich CLIA was developed to detect Anti-HTLV-I antibodies. Antigens were coated on the MP and, after reacting with serum antibodies, a mouse-anti human IgG conjugated with acridinium ester was added. The chemiluminescence generated was measured immediately for a period of 4 s by photon counting. Only 13 min per sample was needed to detect anti-HTLV-I antibodies, and 150 samples per hour could be tested.[101] The SphereLight 180 is also a fully automated assay system designed for point-of-care diagnostics.[102] PCT concentration could be measured with a linear response of up to $200 \, ng \, mL^{-1}$ and a LOD of $0.06 \, ng \, mL^{-1}$.

Magnetic Microparticle-based CLIA. Magnetic microparticle-based CLIAs have been used in many fully automated analyzing platforms. Chen *et al.* have compared the Immulite analyser with the ADVIA Centaur for determination hepatitis B serology markers. In both systems a qualitative CLIA assay was performed. The ADVIA Centaur employs one-step incubation with a total reaction time of 20 min, while the Immulite 2000 assay requires two incubation steps and a total assay time of 65 min.[88] Therefore, a higher throughput, of over 200 samples per hour, could be achieved. The fully automated random access ADVIA Centaur immunoassay system has offered testing for fertility, therapeutic drug monitoring, infectious disease, allergy, cardiovascular, anaemia, oncology, TDMs and thyroid, and has been specifically designed for use in large-volume laboratories.[103] The ADVIA Centaur uses magnetic particles for magnetic separation of bound and free labels. Acridinium ester is used as CL label. Published applications of the ADVIA Centaur are total homocysteine in plasma,[104] hepatitis B infection[105] and allergen-specific IgE in sera.[106] Figure 8.7 shows a typical assay for allergen diagnosis on the Advia Centaur.

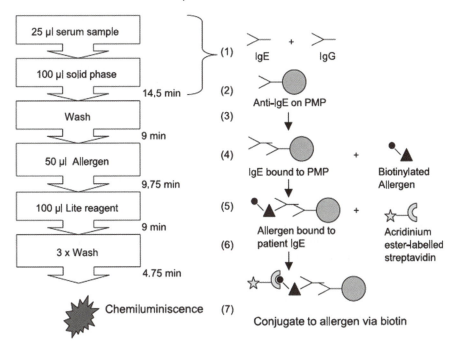

Figure 8.7 Specific IgE assay process and architecture of the ADVIA Centaur system.[106] (Reproduced with permission from Elsevier, Copyright 2004.)

The Magic Lite system was one of the first fully automated analysers to use magnetic particles for washing and acridinium ester as CL label.[107] The CLIA for thyrotropin included two anti-TSH monoclonal antibodies, one was immobilized on paramagnetic beads and the other was directly labelled with acridinium ester. Each antibody was specific to different epitopes on the intact TSH molecule. Serum was incubated with the detection antibody for 2.5 h, then the capture antibody was incubated for 30 min. Bound and free components were magnetically separated. Oxidation of acridinium ester by alkaline hydrogen peroxide initiated the CL emission. The light emission was expressed as photon counts accumulated during 2 s.[108] Applications for allergenes[109] and thyroxine[110] were also evaluated.

The ACS:180 from Chiron Diagnostics is also an automated analyser that uses paramagnetic beads and acridinium ester as CL label, with a throughput up to 180 tests h^{-1}.[111] On this system a competitive indirect immunoassay for deoxypyridinoline has been developed. The analyte derivative pyridinoline is covalently coupled to paramagnetic beads. The mAb anti-DPD antibodies are bound to pAb anti-mouse labelled with acridinium ester.[112] A competitive direct CL immunoassay on the ACS:180 has been developed for the detection of the pyrethroid insecticide 3-phenoxybenzoic acid in human urine. As tracer the 3-PBA-BSA-acridinium ester conjugate was prepared. A LOD of 0.01 µg L^{-1} and a working range 0.03–0.52 µg L^{-1} were achieved. The ACS:180

provided the first results within 15 min and produced up to 130 results in 1 h for this application.[113]

The LIAISON system is a fully automated system based on chemiluminescence and antigen bound to magnetic microparticles for the determination of infectious diseases. An anti-human IgG monoclonal antibody coupled to aminobutylethylisoluminol is added for chemiluminescence measurements. The system has allowed fast and precise measurement of *Toxoplasma*-specific immunoglobulin G (IgG) and IgM antibody levels and measurement of the IgG avidity index even at low levels of *Toxoplasma*-specific IgG antibodies in a single step without manual interference.[114] A LIAISON human cytomegalovirus (CMV) assay that determines the IgG avidity in serum has been evaluated. The system did not require sample dilution and was completed in 50 min. In the first incubation step, test sera were reacted with HCMV antigen coated onto magnetic particles in two separate wells. After washing, beads reacted with chaotropic solution or with buffer, respectively. After a further washing step, CL reaction was triggered and light was quantified.[115] For laboratory analysis of syphilis a recombinant antigen-based CLIA was developed.[116] The LIASON *Treponema pallidum* specific assay was an one-step sandwich qualitative CLIA performed on the LIAISON automated random access analyser that used the label isoluminol. Paramagnetic microparticles were coated with a recombinant treponemal antigen (TpN17). A patient's serum, the coated microparticles and an isoluminol-antigen conjugate were mixed and incubated. The isoluminol-antigen was the same antigen presented on the solid phase. It was a recombinant antigen of *T. pallidum*, and it was used to bind IgG and IgM antibodies in a sandwich manner. Specific antitreponemal antibodies presented in the specimens bound to the paramagnetic bead and the antigen conjugated during the incubation.[117]

The Nichols Advantage is a fully automated chemiluminescence immunoassay analyser that used streptavidin-coated MMP and acridinium ester as CL label. For the analysis of serum insulin-like growth factor I, the antibody to the C-terminal 62–70 amino acid sequences was biotinylated for capture and the antibody to the amino acid sequences of 1–23 and 42–61 was labelled with acridinium ester for the detection. Streptavidin magnetic particles were added to the reaction mixture. Free labelled antibody was then separated from the labelled antibody bound to the magnetic beads by aspiration and subsequent washing, while a strong magnetic force kept the magnetic particles in the well. An acidic hydrogen peroxide solution and a sodium hydroxide solution were added to the well to initiate the chemiluminescence reaction.[118] The thyroglobulin assay was a two-step chemiluminescence sandwich immunoassay that applied three monoclonal antibodies: two biotinylated and used for capture, and the third antibody labelled with acridinium ester for emitted light quantification. Throughput was up to 80 samples per hour with a time to first results of 51 min.[119] The aldesterone assay was a competitive one-site CLIA that used a biotinylated monoclonal antibody bound to streptavidin coated MMP. The acridinium ester labelled aldosterone competed with sample aldosterone for the limited amount of antibodies on the MPP.[120,121]

The Architect-i2000 is a random access analyser based on magnetic particle-based chemiluminescence immunoassay separation that uses acridinium-*N*-sulfonyl carboxamides as CL label. This label was expected to enhance the overall assay performance through higher quantum yield, stability and solubility. Published clinical applications are thyrotropin (TSH), free thyroxine (fT4) and testosterone,[122] salivary cortisol[123] and HTLV-I&II.[124] For estradiol a competitive one-step analysis was established that was completed in 30 min.[125]

MMP-based CLIAs performed in microwell plates were easily automated (Figure 8.8).[126] The fully automated sandwich CLIA for quantification of insulin used complexes of antibody, Protein-A and bacterial magnetic particle (BMP) to capture an alkaline phosphatase-conjugated detection antibody. Magnetic bacteria synthesized intracellular magnetite particles. Each particle was small (50–100 nm) and was covered with a stable lipid bilayer. Antibody–protein-A–BMP complexes had a mean diameter of 120 nm. A detection limit of $2 \mu U \, mL^{-1}$ of human insulin was observed with a linear range of 2–$254 \mu U \, mL^{-1}$. It was also observed that the antigen–antibody binding of recombinant antibody–protein-A–BMB complexes was two-times higher than that for chemically conjugated BMP-antibodies. With the same system 17β-estradiol (E2) analysis has been performed using anti-E2 monoclonal antibody immobilized on BMP and ALP-ES2 as tracer for a competitive direct immunoassay. A linear correlation between the luminescence intensity and E2 concentration was obtained between 0.5 and $5 \mu g \, L^{-1}$. The minimum detectable concentration of E2 was $20 \, ng \, L^{-1}$. All measurement steps were completed within 0.5 h.[127]

8.4.1.2 Flow-injection CLIA Platforms

Flow-through CLIA with Indirect CL Detection. Flow injection immunoassays were introduced to overcome the restrictions of ELISA techniques, which require trained persons and long assay times. They combine the advantages of, on the one hand, rapidity, precision and automation and, on the other hand, selectivity.[128] This method is based on automated injection of sample aliquots into a reagent-containing carrier stream. The residence time of the sample and reagents in the system are constant, which allows short assay times.[129]

Flow injection systems have been combined with CL immunoassays by using an immunoreactor that separated bound immune complexes from free components by a fluidic system as shown in Figure 8.3(A). In such a flow-through configuration enzyme labels reacted with the CL substrate in the reaction chamber and a flow-through CL detector measured CL signals behind the reaction chamber. Another point was that these immunoassays were regenerable, which allows the use of reusable columns.

A flow-injection CLIA has been developed for the detection of α-fetoprotein using Sepharose immunoaffinity columns.[130] AFP was detected by using a competitive indirect immunoassay as test format. AFP was immobilized on CNBr-activated Sepharose 4B. The analyte AFP and HRP-anti-AFP were

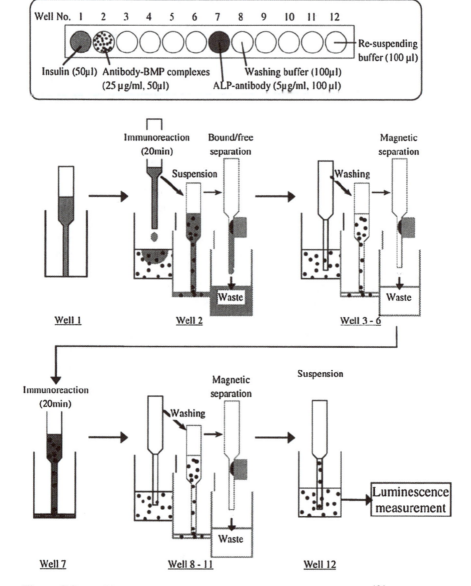

Figure 8.8 Fully automated CLIA for the quantification of insulin.[126] (Reproduced with permission from American Chemical Society, Copyright 2000.)

pre-incubated for 30 min and afterwards were flowed through the column. Bound immune complexes were detected behind the column with a PMT by reacting with a mixture of luminol, PIP and H_2O_2. The CL intensity decreased with increasing AFP concentration. A detection limit of $0.5 \, \mu g \, L^{-1}$ and a working range of 2–$75 \, \mu g \, L^{-1}$ was achieved. The immunoaffinity column could be repeatedly used 100 times without signal decrease. Carcinoembryonic antigen could be detected by similar instrumentation and assay procedure with a LOD of $0.5 \mu g L^{-1}$ and a working range of $1-25 \mu g L^{-1}$.[131] The hormone 17β-estradiol could be detected with a detection limit of $3 \mu g L^{-1}$ and a working range of $10-1000 \mu g L^{-1}$ (ref. 132) and estriol with a detection limit of $5 \mu g L^{-1}$ and working range $10-400 \mu g L^{-1}$.[133] A new strategy for immobilization of glycoproteins has been developed by using a boronate immunoaffinity column.[134] The interaction between a boronic acid Sepharose gel and the glycoprotein AFP was used for immobilization in the immunoaffinity column. Two linear ranges were observed: $5-120 \mu g L^{-1}$ and $300-1000 \mu g L^{-1}$, respectively. Glass beads have also been employed as material for immunoaffinity columns. An immunoreactor was fabricated by immobilization 3-aminophenylboronic acid on glass microbeads with γ-glycidylpropyltrimethoxy-silane as linkage.[135] AFP was directly immobilized on the immunoaffinity column, performing a competitive indirect immunoassay with pre-incubated analyte and antibody-HRP mixtures. Linear working ranges between 10 and $100 \mu g L^{-1}$ were achieved.

Magnetic microparticles (MMP) have been used for multiplexing a one-way flow-injection CLIA.[136] A mixture of sample, antibody-tagged MMPs and ALP-labelled detection antibodies were separately incubated in tubes to detect α-fetoprotein, carcinoma antigen 125, carcinoma antigen 199, and carcinoembryonic antigen. The immune complex was collected in a multichannel system by magnetic forces and unbound components were separated. With the help of two valves, the ALP substrate solution ABI was then sequentially mixed with the immune complexes in different channels for sequential triggering of the CL reaction in a time interval of 15 s. After triggering for 5 min, the mixtures were sequentially injected into a one-way detection channel in the same interval to form the analyte zones. With increasing concentrations of AFP, CA125, CA199 and CEA, the CL intensities increased linearly over the concentration ranges $1.0-80 \mu g L^{-1}$, $1.0-60 kU L$, $1.0-120 mU L^{-1}$ and $1.0-100 \mu g L^{-1}$, respectively. In a total assay time of 12 min the four tumour marker were detected in acceptable agreement with those from single-analyte tests of clinical sera.

A MMP-based immunosupported liquid membrane assay (m-ISLMA) based on CL detection of a HRP-labelled tracer that allowed sample clean up, analyte enrichment and detection has been developed.[137] A microporous polypropylene membrane was used for analyte extraction and separation of the sample channel and the detection channel in a m-ISLM unit. The detection channel contains antibodies immobilized on MMP, which were kept in motion during the extraction process by switching the magnets of both sides of the chamber. After extraction, antigen-tracer and HRP substrate were consecutively pumped through the detection channel and the CL product was detected with a CL flow-through detector. This method has improved the detection limit from $130 ng L^{-1}$ (ELISA) to $0.13 ng L^{-1}$ in a total assay time of 25 min. Progesterone has been detected down to $8.5 fg L^{-1}$.[138] The extraction strategy was also applied to simazine detection in mineral water, orange juice and milk.[139] A chemiluminescent based micro-immuno supported liquid membrane assay (μ-ISLMA) has been developed that enables clean up, enrichment and detection of simazine in a single miniaturized cartridge system. The μ-ISLM cartridge contains a supported liquid membrane (SLM) sandwiched between a donor and an acceptor plate, the latter being covered by a thin layer of gold onto which anti-simazine antibodies were covalently immobilized. Therefore, a direct competitive immunoassay was used by immobilizing the antibodies on a thin layer of gold. The μ-ISLMA was characterized by both a high apparent extraction efficiency and high apparent enrichment factor, which resulted in a very high sensitivity for simazine (LOD = 0.1 ng L^{-1}).

Flow-through CLIA with Direct CL Detection (Immunosensor). Flow-through CLIA platforms with direct CL detection consist of a flow injection system that delivers the analyte and reagents. In contrast to indirect detection methods, the detector is placed in front of the immunoreactor (Figure 8.3b).[140] Most authors define the direct detection principle as a CL immunosensor that is adapted to other optical biosensors consisting of a biorecognition element, a signal transducer and an electronic read-out. Jain *et al.* have compared both CL detection configurations by using a protein G column and a protein G monolithic disc, respectively.[141] Atrazine was quantified by applying a direct competitive immunoassay. The FI-CLIA detected the unbound atrazine tracers by reaction with CL substrate after the column, which caused an increase in CL signal with increasing the atrazine concentration. The CL immunosensor detected the bound atrazine tracer directly on the disc and the CL signal decreases at higher analyte concentrations. The systems have also shown further differences. The CL immunosensor was approximately 10-times more sensitive and matrix interferences were insignificant. The main drawback was that multiple assay steps were necessary, resulting in a lower throughput. Shellum *et al.* described the first flow-injection CL immunoassay with on-column CL detection using acridinium ester.[142] A competitive direct immunoassay was developed using a transparent PTFE tubing packed with immobilized antibodies on an Affi-prep gel.[143] Human IgG down to 7 fmol/20 μL was detected in 7 min. Triiodothyronine (T$_3$) had a detection limit of 0.4 ng mL^{-1} on this system.[144] An indirect competitive immunoassay has been used for quantification of T$_3$ and T$_4$,[145] as well as α-amino acids[146] by applying enantioselective antibodies.

A CL immunosensor with a transparent immunoaffinity reactor has been used for the quantification of AFP.[147] CL signals were detected on glass microbeads through the chamber by positioning a PMT in front of it. Anti-AFP antibodies labelled with HRP and AFP were incubated before flowing through the glass tube which contains AFP immobilized on epoxy-silanized glass microbeads. With increasing analyte concentrations the CL signal decreased. A detection limit of 2.7 μg L^{-1} and a working range of 5.0 and 100 μg L^{-1} for AFP was achieved in a total assay time of 36 min. Assay time and detection limit were improved to 16 min and 0.1 μg L^{-1} by applying a micro-bubble accelerated preincubation process to improve the mass transport of immunoreagents.[148] A three-minutes long CL IA has been achieved by dual acceleration of the immunoreaction by means of infrared heating and passive mixing.[149] The fluidic system combined a 3D helical glass tube for rapid mixing of immunoreagents with two spiral glass tubes for magnetic separation and CL detection, respectively. AFP could be detected, applying a sandwich test format, in a linear range of 0.2–$90\,\mu g\,L^{-1}$.

Other approaches of these types of immunosensors were the multiplexed analysis. By using two different CL labels (ALP and HRP) and a substrate zone-resolved technique, in 35 min CA 125 and CEA could be assayed in ranges 5.0–$100\,units\,mL^{-1}$ and 1–$120\,\mu g\,L^{-1}$.[150] Four tumour markers (CA153, CA155, CA 199 and CEA) were detected by using to fluidic channels and two CL labels. This assay was completed in 37 min and similar working ranges were achieved.[151] One other strategy for multiplexing has been the use of a moveable tubular optical shutter to resolve the CL signals produced in two channels.[152] In one channel, the capture antibodies are immobilized on a capillary and in the other channel the capture antibody are conjugated to paramagnetic beads. With this method CEA and AFP were assayed in ranges of 1–$60\,\mu g\,L^{-1}$ and 1.0–$80\,\mu g\,L^{-1}$, respectively, within 27 min.

Developments in the direction of miniaturization of CL immunosensors have been investigated to speed up the assay time, reduce reagent costs and perform multiplexed immunoassay in future.[153] A silicon microchip has been fabricated for the detection of atrazine in surface water and fruit juice samples.[154] The silicon microchip contains 42 porous flow channels where protein A and protein G were immobilized to perform a direct competitive test format with immobilized pAb and HRP conjugated atrazine-tracer. Detection limits down to $6\,ng\,L^{-1}$ were achieved on this regenerable CL immunosensor.

Table 8.3 presents further flow-through CLIA platforms. These platforms generally use HRP as CL label and luminol/H_2O_2 as CL reagent.

8.4.2 Automated Multi-analyte Detection

8.4.2.1 Non-fluidic CL Microarray Platforms

The first multi-analyte detection system based on CLIA was applied for the determination of total IgE and allergen-specific IgE in serum by the MAST chemiluminescence assay system (Figure 8.9)[163] and automated on a AP720S analyser.[164] Allergens were coated on cellulose threads and patient serum and anti-IgE labelled with HRP were incubated consecutively.

The first commercially available automated CL microarray platform was the Evidence biochip array analyser from Randox Laboratories (Figure 8.10). It was developed for the clinical diagnostic market. The platform consists of an automated dispensing station for reagent supply, incubation and washing steps, and sample introduction.[165] The biochip is secured in the base of a plastic well, which is then placed in a carrier holding nine biochips in a 3×3 format. Each biochip contains a 5×5 array. The capture antibodies are immobilized on an alumina substrate activated with silane. Light emission from the CL reaction on the surface of the biochip using HRP-labelled detection antibodies is imaged

Table 8.3 Overview of flow-through CLIA with direct CL detection.

Solid support	Test format	Analyte	LOD	WR	Assay time (min)	Ref.
MMP in a flow cell	Direct competitive	LAS	25 ppb	25–200 ppb	15	155
MMP in a flow cell	Direct competitive	APnEOs	10 ppb	10–1000 ppb	15	156
MMP in a flow cell	Sandwich	Vitellogenin	2 ng mL^{-1}	2–100 ng mL^{-1}	60	157
MMP in a flow cell	Indirect competitive	Clenbuterol		0.01–0.1 ng mL^{-1}	15	158
Glas chip	Indirect competitive	CA19-9	1 U mL^{-1}	2–25 U mL^{-1}	35	159
Gold microchip	Direct competitive	TNT	0.1 µg L^{-1}	0.1–10 µg L^{-1}	35	160
		Diuron	0.2 µg L^{-1}	0.1–10 µg L^{-1}		
		Atrazine	0.2 µg L^{-1}	0.1–10 µg L^{-1}		
Optical fibre; ultrabind membrane	Indirect competitive	2,4-D	4 µg L^{-1}	4–160 000 µg L^{-1}	20	161
Optical fibre; ultrabind membrane	Indirect competitive	Okadaic acid	0.1 µg L^{-1}	0.1–100 µg L^{-1}	20	162

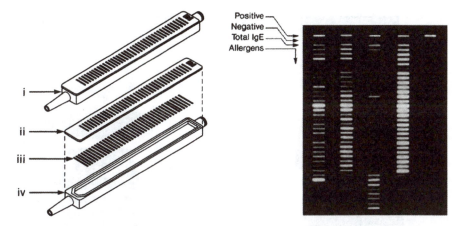

Figure 8.9 MAST chemiluminescence assay system for specific IgE detection in serum.[163] (Reproduced with permission from American Association Clinical Chemistry, Copyright 1985.)

by a CCD camera. With this platform an antibody microarray has been designed for the detection of 12 cytokines in parallel with a sensitivity of the assay in control samples spanning from $0.12\,pg\,mL^{-1}$ for IL-6 to $2.12\,pg\,mL^{-1}$ for IL-4 and an intra- and interassay precision expressed as %CV of typically < 12 for three multi-analyte control levels.[166] The Evidence analyser has enabled complete automation of assays for a test frequency of up to 2000 results per hour. This platform was the first fully automated CL microarray based on segmented microarrays. The assay process was automated with a spacious robotic station for dispensing, washing, incubation and readout. This system was well designed for multianalyte applications in the field of clinical diagnostics in the laboratory.

8.4.2.2 Fluidic CL Microarray Platforms

A CL flow-through protein microarray has been developed that uses PDMS as solid support for immobilized proteins and an integrated microfluidic system based on a SU-8/glass microfluidic reaction chamber (Figure 8.11).[167] The authors had shown that in a microfluidic system the injection volume and the flow rate had to be optimized. Diffusion was the mechanism for transporting the target toward the solid support, where the immunoreaction happened at distinct spots. Flowing samples at sufficiently high flow rates over the microarray surface could minimize the diffusion limitation and then increased the analytical signal. However, at too high flow rates a decrease of the CL signal occurred, related to a reduced number of interactions. For rapid analyzing of allergen-specific antibodies in serum samples incubation times of 6 min (flow rate $50\,\mu L\,min^{-1}$; $300\,\mu L$ sample volume) were found to be acceptable, with improvements to the detection limit of only one magnitude when compared to

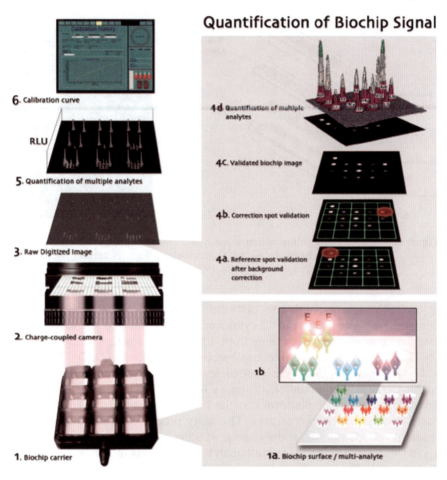

Figure 8.10 The Evidence biochip array analyser from Randox Laboratories.[165] (Reproduced with permission from American Association Clinical Chemistry, Copyright 2005.)

standard ELISAs on microtiter plates which needed a 60 min incubation step. For the direct modification of PDMS surfaces with spots of macromolecules the so-called "macromolecules to PDMS transfer" method has been applied. Beads or proteins were spotted and dried on a 3D master covered with Sylgard 184, cured and recovered, after peeling off, as spots of beads or proteins entrapped at the surface of the bar PDMS.[168,169] With this technique protein microarrays for allergen-specific antibody detection were integrated in microfluidic chips performing a multiplexed CLIA. Three different proteins, β-lactoglobulin, peanut lectin and human IgG, were immobilized and used as capturing agent for the detection of specific antibodies. Three specific antibodies were detected at the pM level in a 300-μL sample volume by using a 6-min sample incubation time. Sera from allergic patients were assayed using

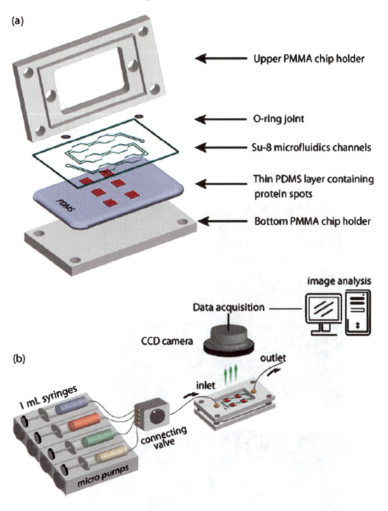

Figure 8.11 CL flow-through protein microarray on PDMS substrates.[167] (Reproduced with permission from Elsevier, Copyright 2008.)

the microfluidic device modified with apple hazelnut and pollen allergen. The results obtained compared favourably with those obtained with the classical Pharmacia CAP system. C-reactive protein has been detected on this microarray platform by performing a sandwich CLIA. Detection limits of 200 ng L^{-1} were achieved.[170]

Matsudaira *et al.* have developed an automated microfluidic assay system for the detection of autoantibodies related to autoimmune diseases like rheumatoid arthritis, multiple sclerosis and autoimmune diabetes.[171] Autoantigens were photoimmobilized on a polystyrene chip by microspotting a mixture of autoantigens, polymer of poly(ethylene glycol) methacrylate and photoreactive crosslinker. A transparent PDMS microfluidic chip was attached to the

(a)

(b)

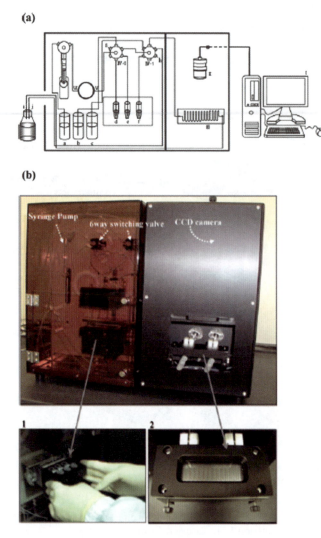

Figure 8.12 Schematic (a) and photograph (b) of an automated microfluidic assay system containing a polystyrene chip.[171] (Reproduced with permission from American Institute of Chemical Engineers, Copyright 2008.)

microspotted polystyrene plate. The setup of the automated assay system (Figure 8.12) consists of a CCD camera, syringe pumps and valves for processing the three flow-through steps of the automated CL immunoassay: sample, HRP-labelled secondary antibody and CL reagent. Multiple autoantigens could be detected in 30 min. Strong correlations between conventional ELISA and microarray assays were obtained.

Tai *et al.* have developed an automated microfluidic-based immunoassay cartridge for allergen screening (Figure 8.13).[172] Twenty Allergen extract

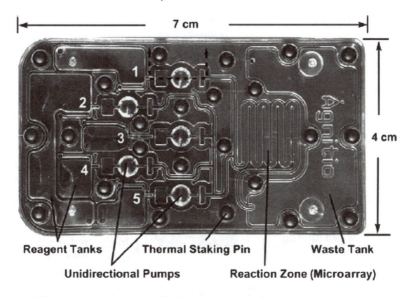

Figure 8.13 Automated microfluidic-based immunoassay cartridge for allergen screening.[172] (Reproduced with permission from Elsevier, Copyright 2009.)

targets, positive and negative controls, and IgE calibration standards were immobilized within the cartridge as a microarray. A computer-controlled array of solenoid valves provided the necessary actuation force for pumping by air. An automated CL indirect immunoassay was performed in a total time of 27 min. Detection limits of $2.4\,\mu g\,L^{-1}$ were achieved.

An automated ten-channel capillary immunodetector has been developed for the detection of *E. coli* O157:H7, SEB and bacteriophage M13 (Figure 8.14).[173] The multiplexed CL immunoassay was performed in ten parallel arranged glass capillaries in 29 min. The analytes were detected by a sandwich immunoassay using polyclonal capture antibodies immobilized on silanized glass capillaries and HRP labelled detection antibodies. CL signals of each capillary were detected by using a multianode-photomultiplier array. Limits of detection were $0.1\,\mu g\,L^{-1}$ for SEB, $10^4\,cfu\,mL^{-1}$ for *E. coli* O157:H7 and $5\times10^5\,pfu\,mL^{-1}$ for bacteriophage M13.

Karsunke *et al.* have performed a multichannel sandwich CL immunoassay on a ABS plastic chip for the rapid detection of pathogenic bacteria.[174] The simultaneous calibration and measurement were possible in one experiment by using six independently actuated flow-through microchannels. Five standard concentrations and one analyte sample were introduced in parallel. The capture antibodies were adsorptively immobilized in the microchannels of ABS by contact printing. Analytes, biotin-labelled detection antibodies, SA-HRP and CL reagent were pumped consecutively through the microchannel by using a six-channel peristaltic pump. The overall assay time for measurement and calibration was 18 min and was a solution for internal calibration of single-use

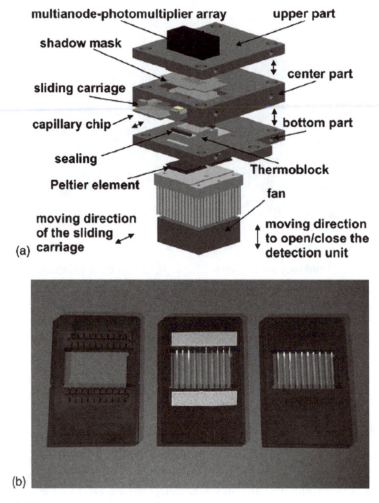

Figure 8.14 Automated immunodetector (a) with the corresponding ten-channel capillary chip (b).[173] (Reproduced with permission from Elsevier, Copyright 2007.)

microarrays. The detection limits were 1.8×10^4 cells mL^{-1} for *E. coli* O157:H7, 2×10^7 cells mL^{-1} for *S. typhimurium*, and 7×10^4 cells mL^{-1} for *L. pneumophila*.

The parallel affinity sensor array (PASA) has a CL flow-through microarray platform with an integrated fluidic system.[175] The platform consists of a CCD camera for readout, a fluidic system with pumps for automated reagent supply, and a microarray with spotted hapten-protein conjugates. The first application was a microarray chip for environmental contaminants in water. Triazines, 2,4-D and TNT could be detected in parallel. The PASA system was also convenient for automatically performed allergy diagnosis. Allergens and recombinant/purified allergens (24 preparations) have been used on the same epoxylated glass surface for the screening of allergen-specific IgE. A direct immunoassay format has applied been to detect 24 allergenic proteins in parallel.[176,177]

A second-generation CL microarray platform (Immunomat) has been used for the detection of multiple antibiotics in milk. The antibiotics chip included ten substances relevant in veterinary medicine. An indirect competitive test format was used. At first, the milk was incubated with a mixture of ten different antibodies. After a short incubation time, the immune reactants were flowed through the microarray. Free antibodies could bind to the distinct hapten spots and they were detected with a HRP-labelled secondary antibody. All liquid handling and sampling processing steps were fully automated, and one analysis was carried out in milk in less then 5 min. The detection limits ranged from 0.12 (cephairin) to $32 \, \mu g \, L^{-1}$ (neomycin).[178] An effective step-by-step chemistry has been developed generating a highly uniform PEG layer on silanized glass slides. High signal-to-noise ratios of more than 600 : 1 and low standard deviation of $<3\%$ were obtained detecting HRP on a microarray that contained immobilized anti-HRP antibodies.[179] Such an optimized protein microarray was introduced to detect pathogenic bacteria under flow through conditions. In 13 min *E. coli* O157:H7, *S. typhimurium* and *L. pneumophila* were detected on applying a sandwich immunoassay format. Detection limits down to 3×10^3 cells mL^{-1} were achieved.[180] It was also demonstrated that on this platform a capturing antibody based microarray could be used for hybridoma screening.[181] The third generation is the fully automated CL microarray chip reader, MCR 3 (Figure 8.15).

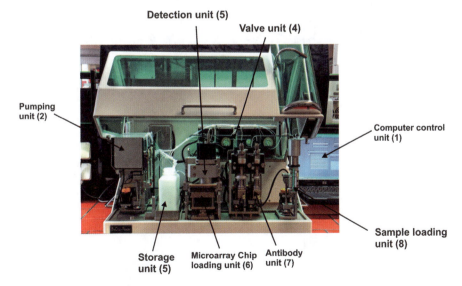

Figure 8.15   Fully automated microarray chip reader (MCR 3) for fast and regenerable CLIAs.[182] (Reproduced with permission from Elsevier, Copyright 2009.)

Table 8.4 Overview of existing flow-through CL microarray platforms.

Platform	Solid support	Immobilized molecules	Matrix	Analyte	LOD	Assay time (min)	Ref.
Microfluidic cartridge	Nitrocellulose coated silicon rubber	20 Allergenic proteins (extracts)	Serum	Allergen specific IgE	2.4 ng mL^{-1}	26.7	172
PASA	Silanized glass slides	BSA-conjugated analyte derivatives	Water	Atrazine Terbutylazine TNT	0.04 µg L^{-1} 0.02 µg L^{-1} 0.13 µg L^{-1}	29	175
PASA	Silanized glass slides	24 Allergenic proteins (extracts and recombinant)	Serum	Allergen specific IgE	0.16–1.9 µg L^{-1}	<60	176
Immunomat	Silanized glass slides	3 BSA-conjugated antibiotic derivatives	Milk	Ten antibiotics	0.12–32 µg L^{-1}	4 40 s	178
Immunomat	PEGylated glass slides	373 Hybridoma supernatants on antibody coated glass slides	Buffer	Aflatoxin B$_2$-HRP	Nn	7	181

Immunomat	PEGylated glass slides	Three polyclonal antibodies	Buffer	E. coli O157:H7 S. typhimurium L. pneumophila	3×10^3 cells mL^{-1} 1×10^5 cells mL^{-1} 3×10^6 cells mL^{-1}	15	180
MCR 3	PEGylated glass slides	13 Antibiotics	Milk	13 Antibiotics	0.05–135 µg L^{-1}	6	183
Multichannel microarray chip	ABS	Three polyclonal antibodies	Buffer	E. coli O157:H7 S. typhimurium L. pneumophila	2×10^4 cells mL^{-1} 2×10^7 cells mL^{-1} 8×10^4 cells mL^{-1}	18[a]	174
Automated micro-fluidic assay system	PEGylated poly-styrene plates	Photo-immobilized autoantigens	Serum	Autoantibodies of autoimmune diseases	Nn	30	171
10K-IDWG	Silanized glass capillaries	Three polyclonal antibodies	Buffer	SEB M13 E. coli O157:H7	0.1 µg L^{-1} 5×10^5 pfu mL^{-1} 1×10^4 cfu mL^{-1}	29	173
Microfluidic biochip	PDMS layer	Three allergenic proteins	Serum	Allergen-specific IgG antibodies	6.6–66 pM	51	167

[a]Total assay time of calibration and detection.

The MCR 3 was designed as a stand-alone platform, with the goal of quantifying multiple analytes in complex matrices of food and liquid samples, for field analysis or for routine analytical laboratories.[182] The CL microarray platform is a self-contained system for the fully automated multiplexed immunoanalysis: the microarray chip, the fluidic system and software module that enables automated process control, calibration and determination of analyte concentrations during operation.

Therefore, a regenerable microarray was required to avoid the replacement of a microarray after each measurement. The MCR 3 uses a flow-through microarray chip, which consists of two channels for parallel measurement and regeneration. To allow multiple analysis, a regenerable microarray chip was developed based on epoxy-activated PEG chip surfaces, onto which micro-spotted antibiotic derivatives like sulfonamides, β-lactams, aminoglycosides, fluoroquinolones and polyketides were coupled directly without further use of linking agents.[183] Using the chip reader platform MCR 3, this antigen solid phase was stable for at least 50 consecutive analyses. Some 13 antibiotics were quantified simultaneously in 6 min. Detection limits were mainly dependent on the antibody–analyte affinity and were determined as between 0.05 (tetracyclin) and $135\,\mu g\,L^{-1}$ (neomycin).

Table 8.4 depicts the performance of automated flow-through chemiluminescence microarrays.

8.5 Integrated Platform Concepts

An immunostrip with a chemiluminescence reader has been developed for point-of-care testing using the cross-flow chromatography principle to perform a sandwich immunoassay.[184] The strip consists of five different types of functional membrane pads consecutively connected by partial superimposition. They are, from the bottom, a glass fiber membrane for sample application, two glass membranes for the release of the detection antibody labelled with HRP and the biotinylated capture antibody, respectively, a NC membrane for signal generation, and a cellulose membrane for adsorption. The analytical protocol consists of a two-step procedure: initiation of sample flow to induce antigen/antibody binding coupled with biotin–streptavidin capture of the analyte and initiation of the flow of enzyme substrate for chemiluminometric signal generation (Figure 8.16). First, the sample containing analyte is absorbed from the bottom of an immuno-strip (vertical flow), inducing immune complex formation at each of two predetermined sites, indicated as analyte and control lines, on the signal generation pad. Second, two horizontally arranged pads are placed on each lateral side of the signal generation pad, and the substrate is then added onto the supply pad to initiate enzymatic signal generation (horizontal flow). Samples containing cardiac troponin I have been analysed in 15 min. A chemiluminescence signal proportional to the analyte concentration was produced by adding a luminogenic substrate to the tracer enzyme complexed with the analyte on the chip. The luminescence signal was detected for

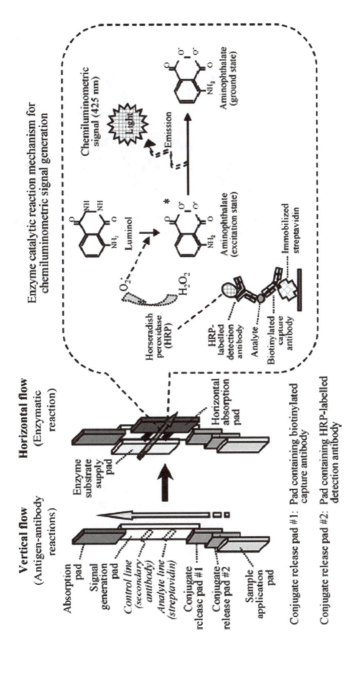

Figure 8.16 Immunostrip using the cross-flow chromatography principle to perform a sandwich immunoassay.[184] (Reproduced with permission from Elsevier, Copyright 2009.)

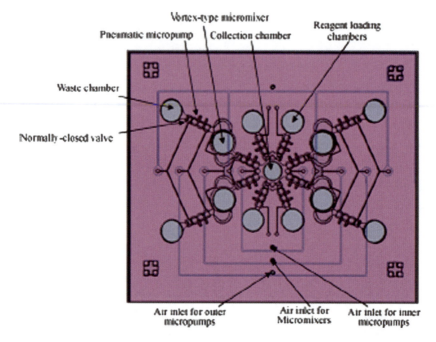

Figure 8.17 Principle of an integrated microfluidic chip performing CLIA.[185] (Reproduced with permission from Elsevier, Copyright 2009.)

30 s in a dark chamber mounted with a cooled charge-coupled device. This system can detect cardiac troponin I presented in serum at concentrations as low as 27 μg L^{-1}, which is 30-times lower than those measured using the conventional rapid test kit with colloidal gold as the tracer.

Yang *et al.* have developed a new microfluidic chip integrated with pneumatic micropumps, normally closed microvalves and vortex-type micromixers for C-reactive protein (CRP) measurement (Figure 8.17).[185] Measurements are performed by using a MMP-based immunoassay on the integrated microfluidic chip platform. The magnetic beads are coated with CRP-specific DNA aptamers which recognize, purify and enrich the target CRP. The entire process including sample pre-treatment and the interaction between the target CRP and anti-CRP antibody is performed automatically on a single chip. The chemiluminescence signal is measured using a luminometer to detect the concentration of CRP afterwards. The entire reaction time is less then 25 min and a detection limit of 12.5 μg L^{-1} has been achieved.

8.6 Conclusions

Chemiluminescence immunoassays are now established techniques in analytical chemistry for rapid analysis in clinical diagnostics, environmental investigations, food and feed control, and biological warfare agent detection.

Chemiluminescence immunoassays can flexibly be adapted to each application. Sandwich, indirect competitive and direct assay formats have been established for the detection of nearly all possible analytes. As CL detection is extremely sensitive, chemiluminescence immunoassays often achieve very low detection limits. CL signals are detected in solution or at solid supports like immunoreactors, microparticles or magnetic microparticles, glass slides, gold surfaces, plastic carriers, *etc*. The combination of solid support detection and separation of bound and unbound reagent is easy to adapt in automated immunoanalytical platforms by using microparticles, magnetic microparticles or flow-through systems. As the technique is cheap and easy to perform, fully automated platforms have been established in clinical chemistry for high-throughput analysis. Especially, multiplexed microarray analysis on automated chemiluminescence readout systems is challenging and will enlarge the analysis spectrum to food and environmental analysis to quantify multiple contaminants in one sample. The increase in density of analytical information reduces the costs of controlling regularly food, feed, water and other samples for safety reasons. The new platforms have now to be validated by all kinds of samples and the corresponding analysis. Antibiotics in milk, pathogenic bacteria in water and food, viruses in air, water or other samples, allergens in food, drug control in urine and antibody indication from autoimmune diseases are some of the cited application that should be highlighted for multiplexed analysis. A current drawback in CL immunoassay developments is the expensive development and production of antibodies – a spur to research and development of alternative recognition elements. Recombinant antibodies from large phage display libraries or aptamers are possible candidates for integration in new CL immunoassay analysers.

References

1. D. Knopp, *Anal. Chim. Acta*, 1995, **311**, 383.
2. D. S. Hage, *Anal. Chem.*, 1999, **71**, 294.
3. D. Knopp, *Anal. Bioanal. Chem.*, 2006, **385**, 425.
4. L. Gamiz-Gracia, A. M. Garcia-Campana, J. J. Soto-Chinchilla, J. F. Huertas-Perez and A. Gonzalez-Casado, *Trends Anal. Chem.*, 2005, **24**, 927.
5. R. L. Divi, F. A. Beland, P. P. Fu, L. S. Von Tungeln, B. Schoket, J. E. Camara, M. Ghei, N. Rothman, R. Sinha and M. C. Poirier, *Carcinogenesis*, 2002, **23**, 2043.
6. A. Roda, M. Guardigli, E. Michelini, M. Mirasoli and P. Pasini, *Anal. Chem.*, 2003, **75**, 462A.
7. H. A. H. Rongen, R. M. W. Hoetelmans, A. Bult and W. P. Vanbennekom, *J. Pharm. Biomed. Anal.*, 1994, **12**, 433.
8. A. Bange, H. B. Halsall and W. R. Heineman, *Biosens. Bioelectron.*, 2005, **20**, 2488.

9. W. R. G. Baeyens, S. G. Schulman, A. C. Calokerinos, Y. Zhao, A. M. G. Campana, K. Nakashima and D. De Keukeleire, *J. Pharm. Biomed. Anal.*, 1998, **17**, 941.

10. M. Seidel and R. Niessner, *Anal. Bioanal. Chem.*, 2008, **391**, 1521.

11. A. Roda, M. Guardigli, P. Pasini, M. Mirasoli, E. Michelini and M. Musiani, *Anal. Chim. Acta*, 2005, **541**, 25.

12. L. Zhao, L. Sun and X. Chu, *Trends Anal. Chem.*, 2009, **28**, 404.

13. C. M. Maragos, *Anal. Bioanal. Chem.*, 2009, **395**, 1205.

14. C. A. Marquette and L. J. Blum, *Anal. Bioanal. Chem.*, 2006, **385**, 546.

15. J. S. Woodhead and I. Weeks, *J. Biolumin. Chemilumin.*, 1989, **4**, 611.

16. E. P. Diamandis, *Clin. Chim. Acta*, 1990, **194**, 19.

17. L. J. Kricka, *Anal. Chim. Acta*, 2003, **500**, 279.

18. M. Adamczyk, P. G. Mattingly, J. A. Moore, Y. Pan, K. Shreder and Z. H. Yu, *Bioconjugate Chem.*, 2001, **12**, 329.

19. M. Adamczyk, Y. Y. Chen, J. R. Fishpaugh, P. G. Mattingly, Y. Pan, K. Shreder and Z. G. Yu, *Bioconjugate Chem.*, 2000, **11**, 714.

20. Y. Q. Lai, Y. Y Qi, J. Wang and G. N. Chen, *Analyst*, 2009, **134**, 131.

21. A. Roda, P. Pasini, M. Mirasoli, E. Michelini and M. Guardigli, *Trends Biotechnol.*, 2004, **22**, 295.

22. A. Roda, P. Pasini, M. Guardigli, M. Baraldini, M. Musiani and M. Mirasoli, *Fresenius' J. Anal. Chem.*, 2000, **366**, 752.

23. R. Creton and L. F. Jaffe, *BioTechniques*, 2001, **31**, 1098.

24. Y. Dotsikas and Y. L. Loukas, *Anal. Chim. Acta*, 2004, **509**, 103.

25. Y. Dotsikas and Y. L. Loukas, *Talanta*, 2007, **71**, 906.

26. J. X. Luo and X. C. Yang, *Anal. Chim. Acta*, 2003, **485**, 57.

27. R. S. Chouhan, K. V. Babu, M. A. Kumar, N. S. Neeta, M. S. Thakur, B. E. A. Rani, A. Pasha, N. G. K. Karanth and N. G. Karanth, *Biosen. Bioelectron.*, 2006, **21**, 1264.

28. I. Bronstein, J. C. Voyta, G. H. G. Thorpe, L. J. Kricka and G. Armstrong, *Clin. Chem.*, 1989, **35**, 1441.

29. Q. Y. Zhang, X. Wang, Z. J. Li and J. M. Lin, *Anal. Chim. Acta*, 2009, **631**, 212.

30. G. W. Lu, H. Shen, B. L. Cheng, Z. H. Chen, C. A. Marquette, L. J. Blum, O. Tillement, S. Roux, G. Ledoux, M. H. Ou and P. Perriat, *Appl. Phys. Lett.*, 2006, **89**, 223128.

31. Y. Wu, C. Chen and S. Liu, *Anal. Chem.*, 2009, **81**, 1600.

32. M. J. R. Previte, K. Aslan, S. N. Malyn and C. D. Geddes, *Anal. Chem.*, 2006, **78**, 8020.

33. S. Bi, Y. M. Yan, X. Y. Yang and S. S. Zhang, *Chem.–Eur. J.*, 2009, **15**, 4704.

34. A. P. Fan, Z. J. Cao, H. A. Li, M. Kai and J. Z. Lu, *Anal. Sci.*, 2009, **25**, 587.

35. A. P. Fan, C. W. Lau and J. Z. Lu, *Anal. Chem.*, 2005, **77**, 3238.

36. Z. P. Li, Y. C. Wang, C. H. Liu and Y. K. Li, *Anal. Chim. Acta*, 2005, **551**, 85.

37. D. H. Hu, H. Y. Han, R. Zhou, F. Dong, W. C. Bei, F. Jia and H. C. Chen, *Analyst*, 2008, **133**, 768.

38. Z. P. Wang, J. Li, B. Liu and J. H. Li, *Talanta*, 2009, **77**, 1050.
39. C. F. Duan, Y. Q. Yu and H. Cui, *Analyst*, 2008, **133**, 1250.
40. J. P. Gosling, *Clin. Chem.*, 1990, **36**, 1408.
41. J. C. Lewis and S. Daunert, *Anal. Chem.*, 2001, **73**, 3227.
42. C. A. Marquette and L. J. Blum, *Biosen. Bioelectron.*, 2006, **21**, 1424.
43. K. Ding, C. H. Zhao, Z. Z. Cao, Z. Q. Liu, J. T. Liu, J. H. Zhan, C. Ma and R. M. Xi, *Anal. Lett.*, 2009, **42**, 505.
44. M. Magliulo, M. Mirasoli, P. Simoni, R. Lelli, O. Portanti and A. Roda, *J. Agric. Food Chem.*, 2005, **53**, 3300.
45. S. Lin, S. Q. Han, Y. B. Liu, W. B. Xu and G. Y. Guan, *Anal. Bioanal. Chem.*, 2005, **382**, 1250.
46. C. Soler, S. Girotti, S. Ghini, F. Fini, A. Montoya, J. Manclus and J. Manes, *Anal. Lett.*, 2008, **41**, 2539.
47. Y. Quan, Y. Zhang, S. Wang, N. Lee and I. R. Kennedy, *Anal. Chim. Acta*, 2006, **580**, 1.
48. A. E. Botchkareva, S. A. Eremin, A. Montoya, J. J. Manclus, B. Mickova, P. Rauch, F. Fini and S. Girotti, *J. Immunol. Methods*, 2003, **283**, 45.
49. L. X. Zhao, J. M. Lin, Z. J. Li and X. T. Ying, *Anal. Chim. Acta*, 2006, **558**, 290.
50. F. Long, H. C. Shi, M. He, J. W. Sheng and J. F. Wang, *Anal. Chim. Acta*, 2009, **649**, 123.
51. S. Q. Ren, X. Wang, Z. Lin, Z. J. Li, X. T. Ying, G. N. Chen and J. M. Lin, *Luminescence*, 2008, **23**, 175.
52. Y. Zheng, H. Chen, X. P. Liu, J. H. Jiang, Y. Luo, G. L. Shen and R. Q. Yu, *Talanta*, 2008, **77**, 809.
53. Z. Lin, X. Wang, Z. J. Li, S. Q. Ren, G. N. Chen, X. T. Ying and J. M. Lin, *Talanta*, 2008, **75**, 965.
54. J. W. Liu, G. Z. Fang, Y. Zhang, W. J. Zheng and S. O. Wang, *J. Sci. Food Agric.*, 2009, **89**, 80.
55. S. X. Zhang, Z. Zhang, W. M. Shi, S. A. Eremin and J. Z. Shen, *J. Agric. Food Chem.*, 2006, **54**, 5718.
56. A. Roda, A. C. Manetta, F. Piazza, P. Simoni and R. Lelli, *Talanta*, 2000, **52**, 311.
57. T. O. Joos, M. Schrenk, P. Hopfl, K. Kroger, U. Chowdhury, D. Stoll, D. Schorner, M. Durr, K. Herick, S. Rupp, K. Sohn and H. Hammerle, *Electrophoresis*, 2000, **21**, 2641.
58. C. Whelan, E. Shuralev, G. O'Keeffe, P. Hyland, H. F. Kwok, P. Snoddy, A. O'Brien, M. Connolly, P. Quinn, M. Groll, T. Watterson, S. Call, K. Kenny, A. Duignan, M. J. Hamilton, B. M. Buddle, J. A. Johnston, W. C. Davis, S. A. Olwill and J. Clarke, *Clin. Vaccine Immunol.*, 2008, **15**, 1834.
59. Y. F. Feng, X. Ke, R. S. Ma, P. Chen, G. G. Hu and F. Z. Liu, *Clin. Chem.*, 2004, **50**, 416.
60. R. P. Huang, *J. Immunol. Methods*, 2001, **255**, 1.
61. R. P. Huang, R. C. Huang, Y. Fan and Y. Lin, *Anal. Biochem.*, 2001, **294**, 55.

62. Y. Lin, R. C. Huang, X. Cao, S. M. Wang, Q. Shi and R. P. Huang, *Clin. Chem. Lab. Med.*, 2003, **41**, 139.

63. M. D. Moody, S. W. Van Arsdell, K. P. Murphy, S. F. Orencole and C. Burns, *BioTechniques*, 2001, **31**, 186.

64. T. Urbanowska, S. Mangialaio, C. Zickler, S. Cheevapruk, P. Hasler, S. Regenass and F. Legay, *J. Immunol. Methods*, 2006, **316**, 1.

65. C. A. Marquette, F. Bouteille, B. P. Corgier, A. Degiuli and L. J. Blum, *Anal. Bioanal. Chem.*, 2009, **393**, 1191.

66. B. P. Corgier, C. A. Marquette and L. J. Blum, *Biosens. Bioelectron.*, 2007, **22**, 1522.

67. B. P. Corgier, F. Li, L. J. Blum and C. A. Marquette, *Langmuir*, 2007, **23**, 8619.

68. X. Yang, J. Janatova, J. M. Juenke, G. A. McMillin and J. D. Andrade, *Anal. Biochem.*, 2007, **365**, 222.

69. D. R. Henderson, S. B. Friedman, J. D. Harris, W. B. Manning and M. A. Zoccoli, *Clin. Chem.*, 1986, **32**, 1637.

70. X. Y. Yang, J. Janatova and J. D. Andrade, *Anal. Biochem.*, 2005, **336**, 102.

71. V. C. Rucker, K. L. Havenstrite and A. E. Herr, *Anal. Biochem.*, 2005, **339**, 262.

72. A. Y. Rubina, V. I. Dyukova, E. I. Dementieva, A. A. Stomakhin, V. A. Nesmeyanov, E. V. Grishin and A. S. Zasedatelev, *Anal. Biochem.*, 2005, **340**, 317.

73. M. H. Yang, Y. Kostov, H. A. Bruck and A. Rasooly, *Anal. Chem.*, 2008, **80**, 8532.

74. N. Karoonuthaisiri, R. Charlermroj, U. Uawisetwathana, P. Luxananil, K. Kirtikara and O. Gajanandana, *Biosens. Bioelectron.*, 2009, **24**, 1641.

75. M. Magliulo, P. Simoni, M. Guardigli, E. Michelini, M. Luciani, R. Lelli and A. Roda, *J. Agric. Food Chem.*, 2007, **55**, 4933.

76. B. Huelseweh, R. Ehricht and H. J. Marschall, *Proteomics*, 2006, **6**, 2972.

77. J. D. McBride, F. G. Gabriel, J. Fordham, T. Kolind, G. Barcenas-Morales, D. A. Isenberg, M. Swana, P. J. Delves, T. Lund, I. A. Cree and I. M. Roitt, *Clin. Chem.*, 2008, **54**, 883.

78. S. C. Shiesh, T. C. Chou, X. Z. Lin and P. C. Kao, *J. Immunol. Methods*, 2006, **311**, 87.

79. X. Wang, J. M. Lin and X. T. Ying, *Anal. Chim. Acta*, 2007, **598**, 261.

80. X. Wang, Q. Y. Zhang, Z. J. Li, X. T. Ying and J. M. Lin, *Clin. Chim. Acta*, 2008, **393**, 90.

81. X. Y. Yang, Y. S. Guo, S. Bi and S. S. Zhang, *Biosens. Bioelectron.*, 2009, **24**, 2707.

82. L. X. Zhao and J. M. Lin, *J. Biotechnol.*, 2005, **118**, 177.

83. T. B. Xin, S. X. Liang, X. Wang, H. F. Li and J. M. Lin, *Anal. Chim. Acta*, 2008, **627**, 277.

84. Y. Zhou, Y. H. Zhang, C. W. Lau and J. Z. Lu, *Anal. Chem.*, 2006, **78**, 5920.

85. X. M. Xie, N. Ohnishi, Y. Takahashi and A. Kondo, *J. Magn. Magn. Mater.*, 2009, **321**, 1686.

86. X. Y. Guo, Y. P. Guan, B. Yang, Y. N. Wang, H. L. Lan, W. T. Shi, Z. H. Yang and Z. H. Lu, *Int. J. Mol. Sci.*, 2006, **7**, 274.

87. G. Pappert, M. Rieger, R. Niessner and M. Seidel, *Microchim. Acta*, 2010, **168**, 1.

88. D. Chen, L. Kaplan and Q. Liu, *Clin. Chim. Acta*, 2005, **355**, 41.

89. A. L. Babson, D. R. Olson, T. Palmieri, A. F. Ross, D. M. Becker and P. J. Mulqueen, *Clin. Chem.*, 1991, **37**, 1521.

90. F. Berthier, C. Lambert, C. Genin and J. Bienvenu, *Clin. Chem. Lab. Med.*, 1999, **37**, 593.

91. W. L. Roberts, R. Sedrick, L. Moulton, A. Spencer and N. Rifai, *Clin. Chem.*, 2000, **46**, 461.

92. W. L. Roberts, E. L. Schwarz and L. Moulton, *Clin. Chem.*, 2000, **46**, 420.

93. Z. Bostanian, G. Hall, E. Whitters, J. D. Lei, K. Pregger, D. Sustarsic, E. Unver and A. S. El Shami, *Clin. Chem.*, 2003, **49**, A29.

94. J. Wen-Quinto, E. Whitters, J. D. Lei, N. Panosian-Sahakian, K. Jaggi, K. Pregger and A. S. El Shami, *Clin. Chem.*, 2004, **50**, A17.

95. G. Andree, B. Bachani, K. Pregger, A. P. Durham and A. S. El Shami, *Clin. Chem.*, 2004, **50**, A100.

96. A. Wang, E. Whitters, J. D. Lei, E. Unver, N. Panosian-Sahakian, D. Sustaric, J. Iagnemma, C. Chang and A. S. El Shami, *Clin. Chem.*, 2004, **50**, A53.

97. Z. Bostanian, N. Panosian-Sahakian, D. Chaturvedi, M. Chaturvedi, E. Whitters, J. Lei, D. Sustarsic and A. S. El Shami, *Clin. Chem.*, 2004, **50**, A131.

98. T. M. Li, P. Fu and V. Zic, *Clin. Chim. Acta*, 2005, **361**, 199.

99. Y. W. Lee, J. H. Sohn, J. H. Lee, C. S. Hong and J. W. Park, *Clin. Chim. Acta*, 2009, **401**, 25.

100. M. Ollert, S. Weissenbacher, J. Rakoski and J. Ring, *Clin. Chem.*, 2005, **51**, 1241.

101. S. Kamihira, S. Nakashima, S. Saitoh, M. Kawamoto, Y. Kawashima and M. Shimamoto, *Jpn. J. Cancer Res.*, 1993, **84**, 834.

102. H. Yamada, S. Matsuda, Y. Ushio, K. Nakamura, S. Kobatake, S. Satomura and S. Matsuura, *Clin. Chim. Acta*, 2008, **388**, 38.

103. F. Dati, *Clin Lab*, 2004, **50**, 53.

104. K. Demuth, W. Ducros, S. Michelsohn and J. L. Paul, *Clin. Chim. Acta*, 2004, **349**, 113.

105. F. Dati, G. Denoyel and J. van Helden, *J. Clin. Virol.*, 2004, **30**, S6.

106. A. B. Petersen, P. Gudmann, P. Milvang-Gronager, R. Morkeberg, S. Bogestrand, A. Linneberg and N. Johansen, *Clin. Biochem.*, 2004, **37**, 882.

107. M. P. Bounaud, J. Y. Bounaud, M. H. Bouinpineau, L. Orget and F. Begon, *Clin. Chem.*, 1987, **33**, 2096.
108. G. C. Zucchelli, A. Pilo, S. Masini, M. R. Chiesa and A. Masi, *J. Biolumin. Chemilumin.*, 1989, **4**, 620.
109. J. Kleinetebbe, M. Eickholt, M. Gatjen, T. Brunnee, A. Oconnor and G. Kunkel, *Clin. Exp. Allergy*, 1992, **22**, 475.
110. J. Beaman, J. S. Woodhead, K. Liewendahl and H. Mahonen, *Clin. Chim. Acta*, 1989, **186**, 83.
111. R. F. Dudley, *J. Clin. Immunoassay*, 1991, **14**, 77.
112. T. G. Rosano, R. T. Peaston, H. G. Bone, H. W. Woitge, R. M. Francis and M. J. Seibel, *Clin. Chem.*, 1998, **44**, 2126.
113. K. C. Ahn, P. Lohstroh, S. J. Gee, N. A. Gee, B. Lasley and B. D. Hammock, *Anal. Chem.*, 2007, **79**, 8883.
114. E. Petersen, M. V. Borobio, E. Guy, O. Liesenfeld, V. Meroni, A. Naessens, E. Spranzi and P. Thulliez, *J. Clin. Microbiol.*, 2005, **43**, 1570.
115. M. G. Revello, G. Gorini and G. Gerna, *Clin. Diagn. Lab. Immunol.*, 2004, **11**, 801.
116. A. Marangoni, V. Sambri, S. Accardo, F. Cavrini, A. D'Antuono, A. Moroni, E. Storni and R. Evenini, *Clin. Diagn. Lab. Immunol.*, 2005, **12**, 1231.
117. C. S. Knight, M. A. Crum and R. W. Hardy, *Clin. Vaccine Immunol.*, 2007, **14**, 710.
118. G. Brabant, A. von zur Muhlen, C. Wuster, M. B. Ranke, J. Kratzsch, W. Kiess, J. M. Ketelslegers, L. Wilhelmsen, L. Hulthen, B. Saller, A. Mattsson, J. Wilde, R. Schemer, P. Kann and K. B. German, *Horm. Res.*, 2003, **60**, 53.
119. A. C. M. Persoon, J. M. W. Van den Ouweland, J. Wilde, I. P. Kema, B. H. R. Wolffenbuttel and T. P. Links, *Clin. Chem.*, 2006, **52**, 686.
120. F. H. Perschel, R. Schemer, L. Seiler, M. Reincke, J. Deinum, C. Maser-Gluth, D. Mechelhoff, R. Tauber and S. Diederich, *Clin. Chem.*, 2004, **50**, 1650.
121. C. Schirpenbach, L. Seiler, C. Maser-Gluth, F. Beuschlein, M. Reincke and M. Bidlingmaier, *Clin. Chem.*, 2006, **52**, 1749.
122. A. Ognibene, C. J. Drake, K. Y. S. Jeng, T. E. Pascucci, S. Hsu, F. Luceri and G. Messeri, *Clin. Chem. Lab. Med.*, 2000, **38**, 251.
123. A. C. Heijboer, F. Martens and M. A. Blankenstein, *Ann. Clin. Biochem.*, 2009, **46**, 261.
124. X. X. Qiu, S. Hodges, T. Lukaszewska, S. Hino, H. Arai, H. Yamaguchi, P. Swanson, G. Schochetman and S. G. Devare, *J. Med. Virol.*, 2008, **80**, 484.
125. J. Taieb, D. H. M. Lozano, C. Benattar, C. Messaoudi and C. Pous, *Clin. Biochem.*, 2007, **40**, 1423.
126. T. Tanaka and T. Matsunaga, *Anal. Chem.*, 2000, **72**, 3518.
127. T. Tanaka, H. Takeda, F. Ueki, K. Obata, H. Tajima, H. Takeyama, Y. Goda, S. Fujimoto and T. Matsunaga, *J. Biotechnol.*, 2004, **108**, 153.

128. A. A. Arefyev, S. B. Vlasenko, S. A. Eremin, A. P. Osipov and A. M. Egorov, *Anal. Chim. Acta*, 1990, **237**, 285.
129. A. P. Osipov, A. A. Arefyev, S. B. Vlasenko, E. M. Gavrilova and A. M. Yegorov, *Anal. Lett.*, 1989, **22**, 1841.
130. J. H. Lin, F. Yan and H. X. Ju, *Appl. Biochem. Biotechnol.*, 2004, **117**, 93.
131. J. H. Lin, F. Yan and H. X. Ju, *Clin. Chim. Acta*, 2004, **341**, 109.
132. S. H. Wang, S. L. Lin, L. Y. Du and H. S. Zhuang, *Anal. Bioanal. Chem.*, 2006, **384**, 1186.
133. S. H. Wang, L. Y. Du, S. L. Lin and H. S. Zhuang, *Microchim. Acta*, 2006, **155**, 421.
134. Y. F. Wu and S. Q. Liu, *Analyst*, 2009, **134**, 230.
135. Y. F. Wu, Y. F. Zhuang, S. Q. Liu and L. He, *Anal. Chim. Acta*, 2008, **630**, 186.
136. H. Liu, Z. F. Fu, Z. J. Yang, F. Yan and H. X. Ju, *Anal. Chem.*, 2008, **80**, 5654.
137. M. Tudorache, M. Co, H. Lifgren and J. Emneus, *Anal. Chem.*, 2005, **77**, 7156.
138. M. Tudorache, I. A. Zdrojewska and J. Emneus, *Biosens. Bioelectron.*, 2006, **22**, 241.
139. M. Tudorache and J. Emneus, *Biosens. Bioelectron.*, 2006, **21**, 1513.
140. J. H. Lin and H. X. Ju, *Biosens. Bioelectron.*, 2005, **20**, 1461.
141. S. R. Jain, E. Borowska, R. Davidsson, M. Tudorache, E. Ponten and J. Emneus, *Biosens. Bioelectron.*, 2004, **19**, 795.
142. C. Shellum and G. Gubitz, *Anal. Chim. Acta*, 1989, **227**, 97.
143. A. Hacker, M. Hinterleitner, C. Shellum and G. Gubitz, *Fresenius' J. Anal. Chem.*, 1995, **352**, 793.
144. D. Dreveny, C. Klammer, J. Michalowsky and G. Gubitz, *Anal. Chim. Acta*, 1999, **398**, 183.
145. H. Silvaieh, R. Wintersteiger, M. G. Schmid, O. Hofstetter, V. Schurig and G. Gubitz, *Anal. Chim. Acta*, 2002, **463**, 5.
146. H. Silvaieh, M. G. Schmid, O. Hofstetter, V. Schurig and G. Gubitz, *J. Biochem. Biophys. Methods*, 2002, **53**, 1.
147. Z. F. Fu, C. Hao, X. Q. Fei and H. X. Ju, *J. Immunol. Methods*, 2006, **312**, 61.
148. Z. J. Yang, Z. F. Fu, F. Yan, H. Liu and H. X. Ju, *Biosens. Bioelectron.*, 2008, **24**, 35.
149. H. Liu, Z. J. Yang, F. Yan, Y. M. Xu and H. X. Ju, *Anal. Chem.*, 2009, **81**, 4043.
150. Z. F. Fu, H. Liu and H. X. Ju, *Anal. Chem.*, 2006, **78**, 6999.
151. Z. Fu, Z. Yang, J. Tang, H. Liu, F. Yan and H. Ju, *Anal. Chem.*, 2007, **79**, 7376.
152. Z. J. Yang, H. Liu, C. Zong, F. Yan and H. X. Ju, *Anal. Chem.*, 2009, **81**, 5484.
153. J. Yakovleva, R. Davidsson, A. Lobanova, M. Bengtsson, S. Eremin, T. Laurell and J. Emneus, *Anal. Chem.*, 2002, **74**, 2994.
154. J. Yakovleva, R. Davidsson, M. Bengtsson, T. Laurell and J. Emneus, *Biosens. Bioelectron.*, 2003, **19**, 21.

155. R. Q. Zhang, K. Hirakawa, D. Seto, N. Soh, K. Nakano, T. Masadome, K. Nagata, K. Sakamoto and T. Imato, *Talanta*, 2005, **68**, 231.
156. R. Q. Zhang, H. Nakajima, N. Soh, K. Nakano, T. Masadome, K. Nagata, K. Sakamoto and T. Imato, *Anal. Chim. Acta*, 2007, **600**, 105.
157. N. Soh, H. Nishiyama, Y. Asano, T. Imato, T. Masadome and Y. Kurokawa, *Talanta*, 2004, **64**, 1160.
158. Z. F. Li, L. R. Jian, H. R. Wang and Y. L. Cui, *Food Addit. Contam.*, 2007, **24**, 21.
159. J. H. Lin, F. Yan, X. Y. Hu and H. X. Ju, *J. Immunol. Methods*, 2004, **291**, 165.
160. I. M. Ciumasu, P. M. Kramer, C. M. Weber, G. Kolb, D. Tiemann, S. Windisch, I. Frese and A. A. Kettrup, *Biosens. Bioelectron.*, 2005, **21**, 354.
161. C. A. Marquette and L. J. Blum, *Talanta*, 2000, **51**, 395.
162. C. A. Marquette, P. R. Coulet and L. J. Blum, *Anal. Chim. Acta*, 1999, **398**, 173.
163. C. R. Brown, K. W. Higgins, K. Frazer, L. K. Schoelz, J. W. Dyminski, V. A. Marinkovich, S. P. Miller and J. F. Burd, *Clin. Chem.*, 1985, **31**, 1500.
164. S. Lee, H. S. Lim, J. Park and H. S. Kim, *Clin. Chim. Acta*, 2009, **402**, 182.
165. S. P. FitzGerald, J. V. Lamont, R. I. McConnell and E. O. Benchikh, *Clin. Chem.*, 2005, **51**, 1165.
166. S. P. FitzGerald, R. I. McConnell and A. Huxley, *J. Proteome Res.*, 2008, **7**, 450.
167. K. A. Heyries, M. G. Loughran, D. Hoffmann, A. Homsy, L. J. Blum and C. A. Marquette, *Biosens. Bioelectron.*, 2008, **23**, 1812.
168. C. A. Marquette, A. Degiuli, E. Imbert-Laurenceau, F. Mallet, C. Chaix, B. Mandrand and L. J. Blum, *Anal. Bioanal. Chem.*, 2005, **381**, 1019.
169. C. A. Marquette, M. Cretich, L. J. Blum and M. Chiari, *Talanta*, 2007, **71**, 1312.
170. K. A. Heyries, L. J. Blum and C. A. Marquette, *Langmuir*, 2009, **25**, 661.
171. T. Matsudaira, S. Tsuzuki, A. Wada, A. Suwa, H. Kohsaka, M. Tomida and Y. Ito, *Biotechnol. Progr.*, 2008, **24**, 1384.
172. L. W. Tai, K. Y. Tseng, S. T. Wang, C. C. Chiu, C. H. Kow, P. Chang, C. Chen, J. Y. Wang and J. R. Webster, *Anal. Biochem.*, 2009, **391**, 98.
173. E. Yacoub-George, W. Hell, L. Meixner, F. Wenninger, K. Bock, P. Lindner, H. Wolf, T. Kloth and K. A. Feller, *Biosens. Bioelectron.*, 2007, **22**, 1368.
174. X. Y. Z. Karsunke, R. Niessner and M. Seidel, *Anal. Bioanal. Chem.*, 2009, **395**, 1623.
175. M. G. Weller, A. J. Schuetz, M. Winklmair and R. Niessner, *Anal. Chim. Acta*, 1999, **393**, 29.
176. B. I. Fall, B. Eberlein-Konig, H. Behrendt, R. Niessner, J. Ring and M. G. Weller, *Anal. Chem.*, 2003, **75**, 556.
177. B. I. Fall and R. Niessner, *Methods Mol. Biol.*, 2009, **509**, 107.

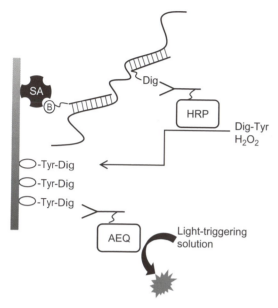

2 Enzyme-amplified aequorin-based bioluminometric hybridization assay. Denatured DNA is hybridized, in streptavidin (SA)-coated wells, with an immobilized biotinylated (B) probe and a digoxigenin (Dig)-labeled detection probe. Hybrids react with horseradish peroxidase (HRP) conjugated to anti-digoxigenin antibody. HRP catalyzes the oxidation of Dig-tyramine (Dig-Tyr) by H_2O_2, resulting in attachment of multiple Dig moieties to the solid phase. Aequorin (AEQ) conjugated to anti-digoxigenin antibody is then allowed to bind to the immobilized Dig. AEQ is determined by Ca^{2+} addition.

d to the expressed proteins encoded by the transgenic sequences, espe-
processed foods, and because DNA-based methods offer detectability
to that of protein methods.[8–10] Target DNA sequences frequently used
O screening are the 35S promoter of the cauliflower mosaic virus and the
synthase (NOS) terminator from *Agrobacterium tumefaciens* since these
most commonly used regulatory elements for the production of trans-
ants.[11] The challenge is to detect the low number of genetically modified
copies in a background of unaltered genome. Hence, DNA amplifica-
PCR constitutes an essential step of GMO detection methods. In one
target DNA was biotinylated through PCR and was added to strepta-
ated microtiter wells.[7] After removal of the non-biotinylated strand by
treatment, the target DNA was hybridized to the detection probe, an
cleotide containing a target-specific sequence and a poly(dT) tail.
ination of the hybrids was accomplished through the interaction of the
) tail with an aequorin–(dA)$_{30}$ conjugate (Figure 9.3).[11] Three targets
tected by this approach (each in a separate well): the 35S promoter, the
rminator and the endogenous, soybean-specific, lectin gene, which is a

178. B. G. Knecht, A. Strasser, R. Dietrich, E. Martlbauer, R. Niessner and M. G. Weller, *Anal. Chem.*, 2004, **76**, 646.
179. A. Wolter, R. Niessner and M. Seidel, *Anal. Chem.*, 2007, **79**, 4529.
180. A. Wolter, R. Niessner and M. Seidel, *Anal. Chem.*, 2008, **80**, 5854.
181. M. Rieger, C. Cervino, J. C. Sauceda, R. Niessner and D. Knopp, *Anal. Chem.*, 2009, **81**, 2373.
182. K. Kloth, R. Niessner and M. Seidel, *Biosens. Bioelectron.*, 2009, **24**, 2106.
183. K. Kloth, M. Rye-Johnsen, A. Didier, R. Dietrich, E. Martlbauer, R. Niessner and M. Seidel, *Analyst*, 2009, **134**, 1433.
184. I. H. Cho, E. H. Paek, Y. K. Kim, J. H. Kim and S. H. Paek, *Anal. Chim. Acta*, 2009, **632**, 247.
185. Y. N. Yang, H. I. Lin, J. H. Wang, S. C. Shiesh and G. B. Lee, *Biosens. Bioelectron*, 2009, **24**, 3091.

CHAPTER 9

Gene Assays Based on Bio(Chemi)luminescence

ELEFTHERIA LAIOS,[a] PENELOPE C. IOANNOU[b] AND
THEODORE K. CHRISTOPOULOS[c]

[a] General Hospital of Katerini, 60100, Katerini, Greece; [b] Department of
Chemistry, University of Athens, 15771, Athens, Greece; [c] Department of
Chemistry, University of Patras, 26500, Patras, Greece

9.1 DNA Hybridization Assays

One of the first reports describing the use of recombinant aequorin as a reporter
molecule in a bioluminometric nucleic acid hybridization assay was in 1996.[1]
Microtiter wells were coated with anti-digoxigenin antibody. The target DNA
was hybridized simultaneously with an immobilized digoxigenin-labeled capture probe and a biotinylated detection probe. The hybrids were determined
using an aequorin–streptavidin conjugate followed by the measurement of
luminescence in the presence of excess Ca^{2+} (Figure 9.1). The linearity of the
assay was in the range 0.1–200 pM (5 amol well^{-1} to 10 fmol well^{-1}) and the S/B
ratio at 0.1 pM (5 amol well^{-1}) was 5.3. A configuration in which the aequorin–
streptavidin conjugate was replaced by a preformed complex of biotinylated
aequorin to streptavidin resulted in equivalent signals and detectability. An
advantage of using the preformed complex is that it was conveniently prepared
by simply mixing the two components (streptavidin and biotinylated aequorin),
thus providing a practical alternative to the use of aequorin–streptavidin
conjugates prepared by covalent crosslinking techniques. The aequorin–biotin
and the aequorin–streptavidin conjugates in this study were obtained commercially (the same group later reported the construction of a plasmid suitable for
bacterial expression of *in vivo*-biotinylated aequorin, facilitating further the
development of highly sensitive hybridization assays,[2] as well as a method for

Chemiluminescence and Bioluminescence: Past, Present and Future
Edited by Aldo Roda

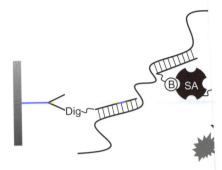

Figure 9.1 Bioluminometric hybridization assay bas[...]
is hybridized with an immobilized captu[...]
detection probe. Probe immobilization [...]
igenin (Dig)–anti-digoxigenin interactio[...]
(AEQ)–streptavidin (SA) conjugate. AEQ [...]

rapid conjugation of streptavidin to aequorin,[3] as [...]
proposed assay was applied to the detection and [...]
of the mRNA for prostate-specific antigen (PSA) [...]
staging of prostate cancer. PSA mRNA from a sin[...]
million non-PSA expressing cells was detected wit[...]

Through these studies it was shown that althoug[...]
not entail substrate turnover it does provide hig[...]
alkaline phosphatase (ALP) using chemilumino[g...]
tivity of aequorin-based hybridization assays wa[...]
mically introducing multiple aequorin labels pe[...]
DNA was hybridized simultaneously, in streptavid[...]
an immobilized biotinylated capture probe and a d[...]
probe. The hybrids reacted with horseradish pero[x...]
anti-digoxigenin antibody. A digoxigenin-tyrami[n...]
used as the hydrogen donor. In the presence of hy[...]
catalyzed the oxidation of Dig-Tyr, resulting in [...]
multiple Dig-Tyr molecules to the solid phase, th[...]
with the digoxigenin moiety remaining exposed.[6] Th[...]
were then reacted with aequorin conjugated to [...]
(Figure 9.2). As low as 20 fM (1 amol well^{-1}) target [...]
S/B ratio of 2.7 and the analytical range extende[d...]
assay was directly compared to an assay that used o[...]
anti-digoxigenin antibody, without the peroxidase [...]
ratio of 2 was obtained for 160 fM target DNA, i[...]
amplification resulted in an eight-fold increase in se[...]
assay that used only aequorin conjugated to anti-di[...]

Chemiluminometric hybridization assays for the [...]
are also applicable to testing for genetically-mo[d...]
DNA is the preferred analyte in GMO testing du[...]

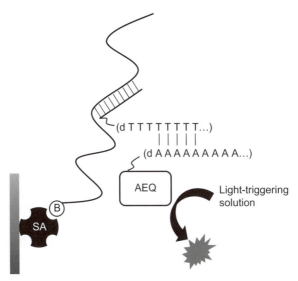

Figure 9.3 Bioluminometric hybridization assay based on a universal aequorin-labeled oligonucleotide probe. Biotinylated (B) target DNA is captured on streptavidin (SA)-coated wells and denatured by NaOH. Immobilized single-stranded DNA is then hybridized with a poly(dT) tailed-detection probe. Hybrids react with the universal aequorin (AEQ)–(dA)$_{30}$ conjugate. AEQ is determined by Ca^{2+} addition.

reference gene for confirmation of the integrity of extracted DNA. For all three targets, as low as 2 pM (100 amol well^{-1}) of amplified DNA was detected with an S/B ratio of about 2. The analytical range extended up to 2000 pM. As low as 0.05% GMO content in soybean was detectable with an S/B ratio of 8.2. This method offered 80-times higher detectability than agarose gel electrophoresis and ethidium bromide staining of PCR-amplified DNA, a commonly used GMO detection method. GMO analysis by real-time PCR has also been reported but the cost is much higher than the proposed assay.[12] Biosensors have also been proposed for GMO detection.[13,14] The simple bioluminometric assay is advantageous because it is based on a universal aequorin-labeled oligonucleotide probe, which can be used with other target sequences in combination with appropriate poly(dT)-tailed specific probes. The assay may be used for the semi-quantitative assessment of GMO content since luminescence increases with GMO content. But, to obtain accurate quantitative results, an internal standard should be used that contains the same primer binding sites and is distinguishable from the target by hybridization. This is addressed in Section 9.2 (quantitative PCR).[15]

Another application of hybridization assays as detection methods for PCR products is the area of molecular diagnostics. An assay using PSA mRNA as the model target has already been presented in this section.[1] The mRNAs for PSA and prostate-specific membrane antigen (PSMA) are the most commonly investigated markers for identification of prostate cancer cells in blood, lymph nodes and bone marrow by the use of reverse transcriptase PCR (RT-PCR). However, the routine clinical application of PSA mRNA or PSMA mRNA assays for the molecular diagnosis and monitoring of prostate cancer is prohibited by false-negatives and false-positives, *i.e.*, the marker may not be detected in patients with diagnosed metastatic cancer or may be detected in normal blood samples and in non-prostate cell lines. Combined screening of PSA and PSMA mRNAs has been proposed as a more useful marker than the separate PSA mRNA or PSMA mRNA markers. The method involves two separate nested RT-PCR steps employing primers specific for each target mRNA. The limitation of nested RT-PCR is its complexity and high risk of contamination. Detection of amplification products is usually performed by agarose gel electrophoresis and ethidium bromide staining. The result is confirmed by Southern transfer using target-specific probes carrying signal-generating labels. Enhanced sensitivity is achieved by either increasing the number of PCR cycles or using sensitive methods for the detection of the amplification products. The need for a rapid and simple assay for both PSA and PSMA mRNAs has been addressed with the development of a sensitive method for the simultaneous detection of both markers using duplex RT-PCR and a chemiluminometric microtiter well-based DNA hybridization assay. [16] Total RNA from peripheral blood was reverse-transcribed using oligo(dT)$_{20}$ primer followed by duplex PCR in the presence of two pairs of primers specific for PSA and PSMA. The biotinylated amplification products were determined through a hybridization assay. Specifically, after thermal denaturation, biotinylated PCR products were captured on microtiter wells coated with PSA- or PSMA-specific probes conjugated to bovine serum albumin (BSA). The captured single-stranded DNA was then detected by reacting with an ALP–streptavidin conjugate. Luminescence was measured with a chemiluminogenic substrate (Figure 9.4). Using the duplex PCR, 50 copies of PSA DNA and 5 copies of PSMA DNA were detected with S/B ratios of 9.7 and 22, respectively. The detectability of cancer cell equivalents was assessed by performing RT-PCR in samples containing total RNA corresponding to 0.04–400 LNCaP cells in the presence of 1 µg of total RNA from healthy subjects isolated from whole blood. The detectability was one cancer cell equivalent in 10 mL of blood and was comparable to that achieved by nested RT-PCR. The assay offers several advantages. Sensitivity is achieved without using nested PCR, therefore minimizing contamination problems, while specificity is accomplished through the use of gene-specific probes. The assay is suitable for high-throughput screening as opposed to the commonly used labor-intensive methods of electrophoresis and Southern blotting. Furthermore, the assay can be applied to the simultaneous mRNA detection of other marker combinations.

Another application of hybridization assays is the detection of parasites. A novel bioluminescence DNA hybridization assay has been developed for the detection of *Plasmodium falciparum*, the most prevalent and deadly species of malaria. [17] The gold standard for malaria detection is light microscopy but low-throughput, high cost and high skill limit its applicability, especially in

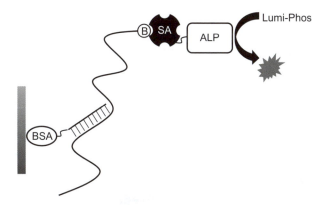

Figure 9.4 Chemiluminometric hybridization assay. Denatured biotinylated (B) DNA is hybridized with an immobilized probe [oligonucleotide conjugated to bovine serum albumin (BSA)]. Hybrids react with an alkaline phosphatase (ALP)–streptavidin (SA) conjugate. ALP is determined by adding a chemiluminogenic substrate.

developing regions where malaria detection is mostly needed. Promising alternatives to light microscopy, including hybridization assays involving PCR amplification, have emerged. However, there is still a need for assays suitable for the conditions typically encountered in developing areas with limited resources. An assay was developed that employed aequorin as a label and did not require PCR amplification of the sample.[17] Sensitivity was accomplished by using a genetically engineered mutant aequorin that contained a unique cysteine at position 5 and was characterized by greater bioluminescence activity and activity after conjugation to streptavidin.[18] The assay was based on the competition between the target DNA and the probe (B-probe) for hybridization with an immobilized probe. Specifically, the assay was performed by first immobilizing a probe on neutravidin-coated microtiter wells. Target DNA was then allowed to hybridize with the immobilized probe. Next, the B-probe (complementary to the immobilized probe) was added to the well to hybridize with any immobilized probe that was left unbound by the target DNA. Aequorin–streptavidin conjugate was then employed for interaction with the biotin of the B-probe and quantitation of the amount of B-probe, which in turn was related to the amount of the target *Plasmodium falciparum* DNA (Figure 9.5). The assay showed a detection limit of 3 pg μL^{-1} and was used for the detection of target DNA in spiked human serum samples.

A different direction in the detection of PCR products is the recent development of a flow-through chemiluminescence microarray read-out system not only for the detection but also for the quantification of *E. coli* DNA after PCR amplification.[19] The method has applications in the field of water monitoring and quality control since the most dangerous water contaminants are pathogenic microorganisms. Traditionally, these pathogens are detected by labor-intensive, time-consuming microbiological methods that require bacterial

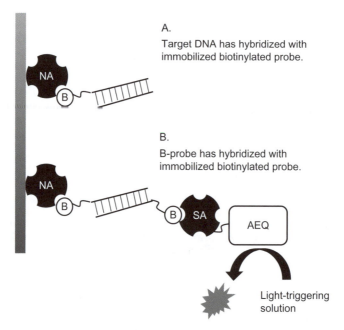

A.

Target DNA has hybridized with immobilized biotinylated probe.

B.

B-probe has hybridized with immobilized biotinylated probe.

NA
B

NA
B
B SA
AEQ

Light-triggering solution

Figure 9.5 Bioluminometric hybridization assay for the detection of *Plasmodium falciparum*. The assay is based on the competition between the target DNA and the probe (B-probe) for hybridization with an immobilized probe. Target DNA is hybridized in neutravidin-coated wells with an immobilized probe (schematic A). Next, the B-probe (complementary to the immobilized probe) is added to hybridize with any immobilized probe that was left unbound by the target DNA (schematic B). Aequorin–streptavidin (SA) conjugate is used for interaction with the biotin of the B-probe and quantitation. AEQ is determined by Ca^{2+} addition.

cultures. However, DNA microarrays can also be used for the quantification of microorganisms in water samples. In fact, fluorescence, electrochemical, chemiluminescence and label-free microarray read-out systems have been developed. Although sensitive, microarrays still require a preceding PCR amplification step. For this purpose, the authors have developed a stopped-PCR method that overcomes the drawback of the sigmoidal amplification curve of end-point PCR and allows the sensitive quantification of target DNA after PCR by means of a chemiluminometric DNA microarray.[19] The principle of stopped-PCR is to end amplification in the logarithmic phase of the reaction, at the point where the spread between different starting amounts of target DNA is at a maximum. As a consequence, the amplification product is strongly dependent on the initial concentration of the target DNA. The chemiluminometric hybridization assay involved microarrays constructed on a poly(ethylene glycol)-modified glass substrate. A NH_2-modified capture probe, which was covalently immobilized onto the microarray surface *via* the free amine, served for capturing the target and HRP–streptavidin conjugate was used as the reporter molecule. The target

DNA was biotinylated through PCR and was captured on the microarray surface by the immobilized probes. The hybrids were detected by HRP–streptavidin conjugate. HRP activity was determined by adding a chemiluminogenic substrate (luminol and H_2O_2). The chemiluminescence intensity was recorded by a sensitive charge-coupled device (CCD) camera. All assay steps, including the addition of denatured biotinylated target DNA, HRP–streptavidin conjugate and HRP substrates, were conducted with the flow-through chemiluminescence read-out system. The method was applied to the detection of the uidA gene (β-galactosidase) of *E. coli*. The detection limit for the uidA gene was 1.1×10^5 copies mL^{-1}. Detection and quantification of *E. coli* was feasible in the range from 10^6 to 10^9 copies mL^{-1}. The assay time was 7 h and the limiting factor was the double-stranded DNA analyte. The authors observed that the assay was very fast (15 min) and sensitive (detection limit 40 copies mL^{-1}) if the analyte was single-stranded DNA. This means that the two strands of the amplification product should be separated before the microarray hybridization assay, *i.e.*, not only denatured to form single strands but also physically separated such that only one strand is added to the microarray. One possible configuration would be biotinylation of the target DNA through PCR, removal of the biotinylated strand by using streptavidin-labeled microbeads, addition of the single-stranded target DNA to the microarray (this strand is now the analyte and it should be designed such that it carries a label incorporated through PCR) and detection of the hybrids through an HRP–antibody conjugate that recognizes the label on the single-stranded DNA analyte. This treatment might improve the sensitivity of the proposed assay by over three orders of magnitude and reduce assay time to 15 min, making the proposed flow-through chemiluminescence microarray read-out system extremely superior to traditional microbiological techniques, which take 18 h for *E. coli* detection and 10 days for *Legionella*.

A novel ultrasensitive flow injection chemiluminometric hybridization assay has been reported that is based on probes linked to copper sulfide (CuS) nanoparticles in combination with chemiluminescence of the luminol-H_2O_2–Cu^{2+} system.[20] The DNA probe was labeled with CuS nanoparticles and the target DNA was immobilized on a glass-carbon electrode. After hybrid formation, cupric ions (Cu^{2+}) were released from dissolution of the CuS nanoparticles on the probe, and the target was determined by the chemiluminescence intensity of luminol–H_2O_2–Cu^{2+}. The chemiluminescence was proportional to the concentration of the dissolved Cu^{2+}. To increase the sensitivity of the DNA biosensor, a Cu^{2+} preconcentration process was performed by using the anodic stripping voltammetry process, *i.e.*, the Cu^{2+} was retained temporarily by electrochemical preconcentration on a platinum electrode placed in an anodic stripping voltammetric cell. The assay gave a linear response curve in the range of 2–100 pM target DNA. The detection limit was estimated to be 0.55 pM. The selectivity of the DNA biosensor was investigated by allowing hybridization of the labeled DNA probe with the complementary target DNA, a two-base mismatched target DNA and a non-complementary target DNA. A well-defined chemiluminescence signal was obtained for the complementary sequence whereas the chemiluminescence intensity of the two-base mismatched sequence was significantly weaker, and the non-complementary sequence showed no response. The drawbacks of this method are the requirement of additional steps due to the dissolution of the nanoparticles and the Cu^{2+} preconcentration, as well as the relatively time-consuming labeling of the probe with the CuS nanoparticles (about 13 h). One way to simplify the assay and increase sensitivity is to investigate more suitable nanoparticle tags and chemiluminescent reactions.

In another publication the authors observed that the catalytic activity of gold nanoparticles (AuNPs) on luminol–H_2O_2 chemiluminescence is greatly enhanced after aggregation in the presence of 0.5 M NaCl.[21] This observation led them to the development of a homogeneous label-free chemiluminescence detection system for sequence-specific DNA hybridization. This is the first label-free chemiluminometric hybridization assay. The method is based on the fact that single- and double-stranded oligonucleotides have different tendencies for adsorption onto AuNPs. In the absence of target DNA (single-stranded DNA due to denaturation), the probe (single-stranded) is adsorbed on the surface of AuNPs and does not allow the aggregation of AuNP in 0.5 M NaCl. The dispersed AuNPs induce a weak chemiluminescence signal of the luminol–H_2O_2 system. In contrast, if hybridization occurs between the single-stranded target DNA and the single-stranded probe (resulting in a double-stranded conformation) then aggregation of AuNPs at 0.5 M NaCl occurs because the probe is no longer on the surface of the AuNPs. The aggregated AuNPs, in turn, induce a strong chemiluminescence signal of the luminol–H_2O_2 system. The detection limit of target DNA was estimated to be 1.1 fM. The sensitivity was more than six orders of magnitude higher than an AuNP-based colorimetric method. The method was also satisfactorily applied to human plasma samples. In this assay, both hybridization and detection occur in homogeneous solution, thereby eliminating the need for covalent attachment of the AuNP to the probe or the target DNA. The assay avoids the stripping procedure of metal nanoparticles, which would result in a high chemiluminescence background. However, it is not known whether longer target DNA molecules in the presence of excess unrelated DNA would interfere with the detection.

A different direction in the development of DNA hybridization assays is the use of magnetic beads. A magnetic bead-based DNA hybridization assay using either conventional chemiluminescence detection or chemiluminescence imaging has been developed.[22] In this method, a sandwich DNA hybridization assay was performed. The assay consisted of a NH_2-modified capture probe immobilized onto carboxylated magnetic beads, a biotinylated detection probe, and HRP–streptavidin conjugate as the reporter molecule. Hybridization occurred in one step by mixing the magnetic beads (which carry the capture probe) with the target DNA and the biotinylated detection probe. The hybrids were detected by adding HRP–streptavidin conjugate. HRP activity was determined by its chemiluminescent reaction (luminol/oxidant/enhancer system). The chemiluminescence intensity was measured both conventionally (with a luminometer), as well as with chemiluminescence imaging. The method has been applied to the detection of sequence-specific DNA related to the avian influenza A H1N1 virus; conventional chemiluminescence detection and chemiluminescence imaging had similar sensitivities (as low as 10 amol target DNA, *i.e.*, 0.1 pM). The assay with chemiluminescence detection was linear in the range 0.3–300 pM target DNA. The assay with chemiluminescence imaging had a wider linear range, from 0.1 to 1000 pM target DNA. The sensitivity of the magnetic bead-based assay is five-times better than that of the CuS nanoparticle assay,[20] but the AuNP assay remains the most sensitive (100-times better than the magnetic bead assay).[21] The magnetic bead-based assay with chemiluminescence detection was also used for genotyping of single nucleotide polymorphisms (SNP). The assay was able to distinguish between perfectly complementary sequences and single-base mismatched sequences by optimizing the stringency of the hybridization and washing steps.

Another recent application involving DNA hybridization has been the development of a multiplex chemiluminescence microscope imaging method for identification and classification of cervical intraepithelial neoplasia (CIN) lesions.[23] Classification of CIN lesions into low-grade (CIN1) or high-grade (CIN2-3) lesions is critical for optimal patient management. A shortcoming of current conventional histological diagnosis on biopsy samples is inter-observer variability. The authors developed a method for two complementary biomarkers that combined (a) immunohistochemical localization and quantitative detection of p16^{INK4A} (a protein marker of high-grade CIN lesions) and (b) *in situ* hybridization for the localization of human papillomavirus (HPV) DNA (in low-grade CIN lesions HPV DNA is an important indicator of the risk of progression to a higher grade lesion). Both determinations were performed in the same tissue biopsy section, sequentially. Different enzyme labels were employed to avoid any interference between the two assays. The enzyme label for detection of p16^{INK4A} was ALP and enzyme activity was measured by using a luminol/oxidant/enhancer system. The enzyme label for detection of HPV DNA was HRP and the enzyme activity was measured by using a dioxetane-based substrate. In fact, HPV DNA localization was performed by using a pool of digoxigenin-labeled DNA probes that recognize the 6, 11, 16, 18, 31, 33 and 35 HPV genotypes. The hybrids were detected by using HRP conjugated to anti-digoxigenin antibody. The light emission of each chemiluminescent reaction remained stable for 20–30 min, thus allowing optimal handling of the sample and the signal and enabling the acquisition of several images from different areas of the same section.

In the field of biosensors, a DNA detection approach that has received great attention is electrogenerated chemiluminescence; the electrogenerated chemiluminescence is obtained from the excited state of a luminophore generated at the electrode surface during an electrochemical reaction. CdS nanocrystals are an example of a semiconductor nanocrystal that provides electrogenerated chemiluminescence in the presence of co-reactant.

In a recent report, a simple electrogenerated chemiluminescence biosensor for the highly sensitive and specific detection of DNA was developed.[24] The biosensor featured CdS:Mn nanocrystals as the luminophore and gold nanoparticles (AuNPs) as both the electrogenerated chemiluminescence quencher and enhancer. AuNPs have made excellent building blocks for the construction of DNA biosensors because of the following features: (1) AuNP-enhanced Raman scattering and AuNP-enhanced surface plasmon resonance, (2) fluorescence quenching due to fluorescence resonance energy transfer (FRET) when the fluorophore and the AuNPs are in close proximity, and (3) AuNP-enhanced fluorescence due to interaction of the excited fluorophore with surface plasmons when the fluorophore and the AuNPs are largely separated.

The features of the biosensor and the assay steps are as follows: a CdS:Mn nanocrystal film (it functions as the luminophore) on a glassy carbon electrode – the nanocrystals have carboxylic acid groups due to modification with 3-mercaptopropionic acid; AuNPs (function as both electrogenerated chemiluminescence quencher and enhancer) covalently attached to one end of a DNA hairpin probe (specific for the target DNA) through 6-mercapto-1-hexanol; crosslinking of the hairpin probe–AuNP conjugates to the CdS:Mn film through N-(3-dimethylaminopropyl)-N'-ethylcarbodiimide hydrochloride (EDC) and N-hydroxysuccinimide (NHS) (through the second end of the DNA hairpin probe); addition of the target DNA.

There are three stages at which electrogenerated chemiluminescence was measured. In the first stage, the CdS:Mn nanocrystal film on the glassy carbon electrode exhibited electrogenerated chemiluminescence in the presence of the co-reactant $S_2O_8^{2-}$ ions. In the second stage, the hairpin probe–AuNP conjugates were brought close to the CdS:Mn nanocrystal film by EDC/NHS crosslinking, in which case the electrogenerated chemiluminescence peak height decreased by 25% in comparison with that of the CdS:Mn nanocrystal film before assembly; the decrease indicated the quenching of electrogenerated chemiluminescence due to FRET between the CdS:Mn nanocrystal film and the AuNPs (close proximity). In the third stage, the AuNPs were largely separated from the CdS:Mn nanocrystal film because the addition of target DNA (5 fM) caused unfolding of the hairpin probe due to hybridization between the target DNA and the probe. The peak height increased by 55% in comparison with that before assembly; the increase indicated the enhancement of electrogenerated chemiluminescence due to the interactions of the excited CdS:Mn nanocrystals with electrogenerated chemiluminescence-induced surface plasmon resonance in AuNPs (large separation). This combination of quenching and enhancement of electrochemiluminescence by AuNPs together in one assay provided high sensitivity for target DNA detection (50 aM). The assay was linear in the range of 50 aM to 5 fM target DNA. The specificity of the biosensor was studied by using one-base mismatched target DNA, three-base mismatched target DNA and a non-complementary target DNA.

DNA has been commonly used as a recognition molecule (probe) in hybridization assays but not as a signal-generating molecule (reporter). However, DNA is an excellent candidate for a reporter molecule since it is much more stable than enzyme reporters, which must maintain activity during (a) isolation from the corresponding cell/organism, (b) conjugation to molecules that are required for binding to the analyte and (c) storage. A novel direction in hybridization assays has been introduced by using a DNA fragment (DNA template) coding for an enzyme, as the reporter molecule. In these assays, after hybrid formation, the solid-phase-bound DNA template is expressed by *in vitro* transcription/translation and the activity of the synthesized enzyme is measured, resulting in a highly sensitive analytical system. Expressible DNA fragments encoding firefly luciferase as well as *Renilla* luciferase were used as labels. The application is not limited to enzymes, but has also been extended to photoproteins such as aequorin, in which case the expressible DNA label encodes apoaequorin.

A hybridization assay that utilizes as a label an expressible enzyme coding DNA fragment was designed for the first time in 1996.[25] The DNA label contained a firefly luciferase coding sequence downstream from a T7 RNA polymerase promoter. The DNA label was expressed by *in vitro* transcription/translation, leading to the production of luciferase. Because the T7 promoter sequence was present only in the DNA label, the transcription/translation process used the luciferase-coding DNA exclusively, and not the probe-target hybrids. In this assay configuration, the capture probe was tailed with digoxigenin-dUTP and was bound to microtiter wells coated with anti-digoxigenin antibody. Denatured target DNA was hybridized simultaneously with the immobilized capture probe and a detection probe tailed with biotin–dATP. The hybrids reacted with a preformed streptavidin–luciferase DNA complex. The complex was prepared by mixing biotinylated luciferase-coding DNA with a large excess of streptavidin. The assay was completed by expressing the solid-phase-bound DNA label by coupled (one-step) *in vitro* transcription/translation. The activity of the synthesized luciferase was then measured. Luciferase catalyzed the luminescent reaction of luciferin, O_2 and ATP to produce oxyluciferin, AMP, pyrophosphate and CO_2 (Figure 9.6).[26,27] As low as 5 amol target DNA was detected with an S/B ratio of 2. The luminescence was linearly related to the amount of target DNA up to about 5000 amol. The high sensitivity achieved was a result of the combined amplification due to transcription/translation and the substrate turnover. The proposed method was compared directly with a fluorometric and a chemiluminometric hybridization assay in which the streptavidin–DNA complex was replaced by an ALP–streptavidin conjugate for detection of the hybrids. The activity of solid-phase-bound ALP was then measured by using a fluorogenic substrate (4-methylumbelliferyl phosphate) or a chemiluminogenic substrate (dioxetane derivative, CSPD). The fluorometric assay detected 200 amol target DNA with an S/B ratio of 2.4. The chemiluminometric assay detected 100 amol target DNA with an S/B ratio of 3.2. Therefore, the assay utilizing the expressible streptavidin–DNA complex resulted in a 20- to 40-fold increase in sensitivity. Notably, luciferases have found only limited use as enzyme labels in DNA hybridization assays and immunoassays due to the significant loss of activity upon conjugation.[28]

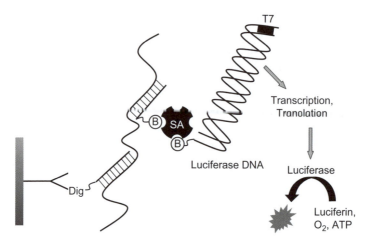

Figure 9.6 Bioluminometric hybridization assay based on an expressible DNA fragment encoding firefly luciferase as a label. Denatured DNA is hybridized with an immobilized capture probe and a biotinylated (B) detection probe. Probe immobilization is accomplished through digoxignenin (Dig)–anti-digoxigenin interaction. Hybrids react with a streptavidin (SA)–firefly luciferase DNA complex. Solid-phase-bound firefly luciferase DNA is measured by coupled *in vitro* transcription/translation. The activity of synthesized luciferase is measured.

   The luciferase-coding DNA was selected as the expressible enzyme coding DNA fragment because luciferase is a monomeric protein, it requires no post-translational modification and its activity can be readily measured in the transcription/translation mixture without prior purification. Because of these attractive properties, the luciferase cDNA has been used extensively as a reporter gene to monitor gene expression in various tissues. The proposed methodology was the first that used luciferase as a reporter molecule in hybridization assays and represents a novel approach for introducing multiple enzyme molecules in the system, because from a single expressible DNA fragment several enzyme molecules (12–14 luciferase molecules) can be synthesized by *in vitro* transcription/translation.[29]

   Subsequently, the investigation of novel hybridization assay configurations based on the *in vitro* expression of DNA reporter molecules was extended by exploiting the biotin–streptavidin interaction for the capture of hybrids to the solid phase.[30] This modification enhanced system versatility as it allowed two types of hybridization assay configurations to be performed. In configuration A (captured target hybridization assay) the target DNA was end-labeled with biotin (through PCR) and captured on streptavidin-coated wells (Figure 9.7). The one strand was removed by NaOH treatment and the remaining strand was hybridized with a poly(dA)-tailed oligonucleotide probe. Configuration B (sandwich-type hybridization assay) involved simultaneous hybridization of heat-denatured target DNA with a biotinylated capture probe (immobilized on

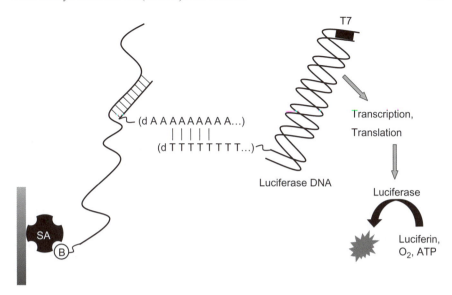

Figure 9.7   Bioluminometric hybridization assay based on an expressible DNA fragment encoding firefly luciferase as a label (captured target configuration). Biotinylated (B) target DNA is captured on streptavidin (SA)-coated wells and denatured by NaOH. Immobilized single-stranded DNA is then hybridized with a poly(dA) tailed-detection probe. Hybrids react with a poly(dT)-tailed DNA fragment encoding firefly luciferase. Solid-phase-bound luciferase DNA is measured by *in vitro* expression (either coupled or separate transcription/translation). The activity of synthesized luciferase is measured.

streptavidin-coated wells) and a poly(dA)-tailed detection probe (Figure 9.7). In both configurations the hybrids reacted with a poly(dT)-tailed luciferase-coding DNA fragment followed by *in vitro* expression of the bound DNA on the solid phase. Expression was accomplished either by using a commercially available coupled (one-step) transcription/translation or by performing sequential transcription and translation reactions that were optimized separately, leading to an increased expression yield (Figure 9.7). For configuration A, at the level of 0.93 fmol target DNA the S/B ratios were 2.6 and 16.7 with the coupled and the separate transcription/translation protocols, respectively. As low as 0.1 fmol target DNA was detected with the separate expression protocol, with an S/B ratio of 2.7, providing at least nine-times higher detectability than the coupled protocol. The reproducibility of configuration A was tested at the level of 3.1 fmol target DNA. For configuration B, at the level of 0.1 fmol target DNA the S/B ratios were 2.2 and 4.6 using the coupled and the separate transcription/translation protocols, respectively. In addition, the assays were directly compared to the sandwich hybridization assay in which the label was a streptavidin–luciferase DNA complex (Figure 9.6). The signals obtained with configuration B using the coupled and the separate expression protocols were

10- and 25-times, respectively, higher than those obtained with the configuration in Figure 9.6. An S/B ratio of 2.4 was obtained for 0.3 fmol target DNA using the streptavidin–luciferase DNA complex as the detection reagent. Configuration B offered higher sensitivity than configuration A but the latter was simpler to perform and potentially automatable because it avoided the heat denaturation step of the target DNA and had a shorter incubation time for hybridization.

A significant improvement of the assay involved the preparation of the firefly luciferase DNA label.[30] In the original use of an expressible enzyme coding DNA as a label,[25] the detection reagent was a complex of streptavidin with a biotinylated luciferase-coding DNA fragment. The preparation of the complex is tedious and time consuming, involving restriction enzyme digestion of the appropriate plasmid to produce three fragments with recessed 3' ends, a filling-in reaction with the Klenow fragment of DNA polymerase I in the presence of biotin–dATP to create DNA fragments that are biotinylated at both termini, purification of the DNA by ethanol precipitation and another digestion to remove the one biotinylated end, therefore, leaving a 2.1 kbp fragment labeled with biotin only at one terminus. After electrophoretic separation, the DNA is excised, purified and complexed with an excess of streptavidin. Finally, the complexes are purified by HPLC and concentrated. The yield of the entire procedure is 10–20%. In contrast, the subsequent procedure was simpler, involving a single digestion to linearize the plasmid (the entire 4.3 kbp plasmid is used as the label), purification by ethanol precipitation, enzymatic tailing of the DNA with dTTP and no additional purification.[30] The detection probe was a poly(dA)-tailed oligonucleotide. The luciferase-coding DNA fragment is attached through its 3' terminus to the detection probe to avoid steric hindrance during transcription.

In addition to firefly luciferase as a DNA label for hybridization assays,[25,30] an apoaequorin DNA label has also been reported.[31] The constructed label contained the T7 RNA polymerase promoter, the apoaequorin coding sequence and a downstream $(dA/dT)_{30}$. Two hybridization assay configurations were developed. In configuration A (captured target assay), biotinylated target DNA was captured on streptavidin-coated microtiter wells. The one strand was removed by NaOH treatment and the remaining strand was hybridized with a poly(dT)-tailed detection probe. In configuration B (sandwich-type assay), the target DNA was hybridized simultaneously with an immobilized capture probe (through biotin/streptavidin) and a poly(dT)-tailed detection probe. In both configurations, the hybrids reacted with poly(dA)-tailed apoaequorin DNA. The DNA label was subjected to coupled *in vitro* transcription/translation to produce multiple apoaequorin molecules in solution, which were converted into fully active aequorin in the transcription/translation reaction mixture. Generated aequorin was determined by its characteristic Ca^{2+}-triggered bioluminescence (Figure 9.8). Configuration A was linear in the range 0.5–7812 amol target DNA. The S/B ratio at the level of 0.5 amol was 1.9. In addition, the assay was directly compared to a hybridization assay in which the photoprotein aequorin was used as the label (Figure 9.10 top), as well as to a hybridization

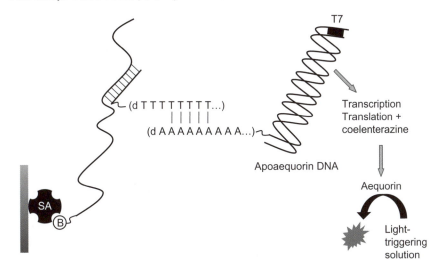

Figure 9.8 Bioluminometric hybridization assay based on an expressible DNA fragment encoding apoaequorin as a label (captured target configuration).
Biotinylated (B) target DNA is captured on streptavidin (SA)-coated wells
and denatured by NaOH. Immobilized single-stranded DNA is then
hybridized with a poly(dT) tailed-detection probe. Hybrids react with
poly(dA)-tailed apoaequorin DNA. Solid-phase-bound apoaequorin
DNA is measured by coupled *in vitro* transcription/translation to produce
apoaequorin molecules in solution, which are converted into fully active
aequorin in the reaction mixture. Generated AEQ is determined by Ca^{2+}
addition.

assay that uses poly(dT)-tailed luciferase DNA as the label (Figure 9.7 configuration A). The aforementioned assays were able to detect 25 amol (S/
B = 1.6) and 20.5 amol (S/B = 1.9) target DNA, respectively. Thus, there was
more than a 40-fold improvement in sensitivity when using the DNA encoding
apoaequorin as a label. The dramatic improvement in sensitivity observed in
configuration A, as compared to the assay in which aequorin was the label, was
due to the amplification introduced by the *in vitro* expression of apoaequorin
DNA into several active aequorin molecules. Each DNA label was estimated to
produce 156 aequorin molecules. Configuration B was linear in the range 0.25–
1562 amol target DNA. The S/B ratio at the level of 0.25 amol was 1.4.

The development of hybridization assays based on firefly luciferase-coding or
apoaequorin-coding DNA labels was followed by a novel dual-analyte expression hybridization assay. Two DNAs encoding firefly luciferase (FLuc) and
Renilla luciferase (RLuc) were used as labels for the development of a microtiter
well-based expression hybridization assay that allowed simultaneous determination of two target DNA sequences in the same well.[32] The RLuc DNA was
chosen because RLuc, like FLuc, is a monomeric protein, it requires no post-
translational modification and its activity can be readily measured in the tran-
scription/translation reaction without prior purification. The constructed FLuc

label contained the T7 RNA polymerase promoter, the firefly luciferase coding sequence and a single biotin at the 3′ terminus. Preparation of the streptavidin–FLuc complex involved a single digestion to linearize the FLuc plasmid (the entire 4.3 kbp plasmid was used as label), purification by ethanol precipitation, enzymic labeling with biotin-ddUTP, ethanol precipitation, complexing with streptavidin, and purification of the protein–DNA complex by electroelution. The yield was 40–60%, and the overall procedure was much simpler and faster than previously.[25] The constructed RLuc label contained the T7 RNA polymerase promoter, the *Renilla* luciferase coding sequence and a poly(dA) tail. Preparation of the poly(dA)-RLuc label involved a single digestion to linearize the RLuc plasmid (the entire 4.0 kbp plasmid was used as the label), purification by ethanol precipitation and enzymic tailing with dATP. In the model assay, the target DNAs (target DNA A and target DNA B) were heat-denatured and hybridized simultaneously with the specific immobilized capture probes and the detection probes. The capture probes were conjugated to BSA and used for coating the well. One detection probe was biotinylated (with a single biotin at the 3′ end) while the other detection probe was poly(dT)-tailed. The hybrids reacted with the streptavidin–FLuc DNA complex and the poly(dA)-tailed RLuc DNA. Subsequently, the DNA labels were expressed *in vitro* simultaneously and independently in the same transcription/translation reaction mixture. The activities of the generated firefly and *Renilla* luciferases were co-determined in the same sample based on the differential requirements of their characteristic bioluminescent reactions for Mg^{2+}. Firefly luciferase was measured first. The firefly luciferase reaction was then terminated by the addition of EDTA to measure *Renilla* luciferase with its coelenterazine substrate (Figure 9.9). The S/B ratio obtained for 66 amol of target A in the presence of 10 fmol of target B (a 150-fold excess) was 2.7. The S/B ratio obtained for 66 amol of target B in the presence of 10 fmol of target A was 6.

The following requirements must be fulfilled for the simultaneous determination of two target DNA sequences by an expression hybridization assay:

1 The reporter genes can be expressed independently in the same reaction mixture (the transcription and translation of one gene does not affect the expression of the other) and in a broad range of concentrations, despite the fact that they use the same promoter and share the same transcription and translation machinery.

2 The gene products can be co-determined in the same well, without splitting the sample (the presence of one enzyme and its substrate does not interfere with the assay of the other enzyme). In this work, enzyme determination is based on the fact that, contrary to FLuc, RLuc does not require the presence of Mg^{2+} for full activity.

3 There is no cross-reactivity between probes and target DNA sequences during hybridization.

The dual-analyte expression hybridization assay opens the way for the possible application of multiple enzyme-coding DNA labels for the simultaneous

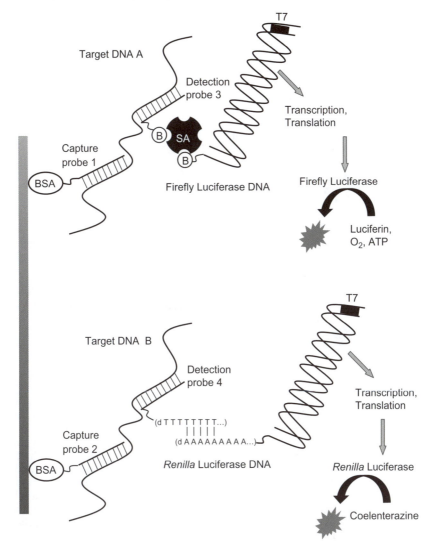

Figure 9.9 Dual-analyte bioluminometric hybridization assay based on expressible DNA fragments encoding firefly luciferase and *Renilla* luciferase as labels. Wells are coated with bovine serum albumin (BSA)-probe conjugates. Denatured DNAs (targets A and B) are hybridized with immobilized capture probes (probes 1 and 2) and detection probes [biotinylated (B) detection probe 3 and poly(dT)-tailed detection probe 4]. Hybrids react with a streptavidin (SA)–firefly luciferase DNA complex and a poly(dA)-tailed *Renilla* luciferase DNA. Solid-phase-bound luciferase DNAs are measured by coupled *in vitro* transcription/translation. The activity of synthesized luciferases is measured sequentially.

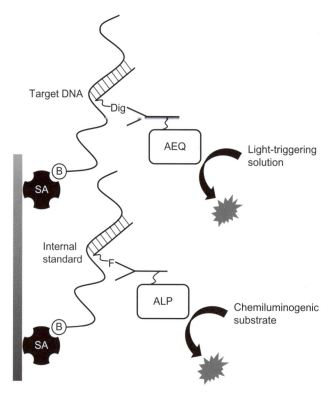

Figure 9.10   Dual-analyte bio(chemi)luminometric hybridization assay for quantitative competitive PCR. Biotinylated (B) target DNA and biotinylated internal standard (competitor) are captured on a single streptavidin (SA)-coated well and denatured by NaOH. Immobilized single-stranded target DNA and internal standard are simultaneously hybridized with their respective specific probes, a digoxigenin (Dig)-labeled probe and a fluorescein (F)-labeled probe. Hybrids react with a mixture of aequorin (AEQ) conjugated to anti-digoxigenin antibody and alkaline phosphatase (ALP) conjugated to anti-fluorescein antibody. AEQ is determined first by Ca^{2+} addition and ALP is determined by adding a chemiluminogenic substrate.

determination of many target DNAs in the same well. The color of emitted light can also serve as an extra parameter to distinguish between the synthesized luciferase reporters.

9.2   Quantitative PCR

In quantitative PCR the goal is to relate the signal obtained from the amplified DNA to the initial number of copies of the target sequence in the sample prior to amplification. The challenge is to compensate for sample-to-sample variation of the amplification efficiency due to the presence of inhibitors and the variability in reaction conditions. This is achieved by co-amplifying, in the same reaction tube, the target sequence with a synthetic DNA or RNA competitor (internal standard, IS) that has the same primer binding sites and similar size to the amplified target (competitive PCR). As a result, the ratio of the two amplification products (target/competitor) is linearly related to the initial amount of target in the sample.[33,34]

Competitive PCR requires highly sensitive assays for the amplification products. This is because the amplification of the target suppresses the amplification of the competitor to undetectable levels (if the target is much higher than the competitor) or *vice versa*, *i.e.*, the amplification of the competitor overpowers the target amplification (if the target is present in minute amounts with respect to the competitor). Thus, the use of an IS is only possible in assays that are highly sensitive in detecting the amplified products.[35,36]

Electrophoresis-based competitive PCR (slab gel or capillary electrophoresis) requires internal standards differing in size (insertion or deletion) to enable electrophoretic separation of the amplification products from target DNA and IS. However, sequence length is also a major determinant of the amplification efficiency. Differences in size cause differences in amplification efficiency.[37] Alternatively, the IS may have the same size as the target but contains a new restriction site, allowing product digestion prior to electrophoresis. The use of HPLC may facilitate the separation and quantification of the two products.[38] It has been observed, however, that co-amplification of substantially homologous DNA sequences, such as target and IS, yields heteroduplexes during PCR due to hybridization of target strands with IS strands (even if their sizes are different).[37] Heteroduplexes may cause errors in electrophoresis if they cannot be resolved from the homoduplexes. In the case of an IS having the same size as the target, but differing only in a restriction site, the heteroduplexes interfere because they are resistant to digestion.

Below, we discuss quantitative PCR methods based on the detection of amplification products by hybridization. Following PCR, the target DNA and the IS are quantified by hybridization to specific probes allowing confirmation of the amplified sequences and high sensitivity. The IS has the same size as the target DNA, thus circumventing the problem of variation of amplification efficiency. In addition, heteroduplex formation is not a concern because only one strand of DNA (target or IS) is captured on the microtiter well through hybridization.

In one report the authors developed a dual-analyte chemiluminescence hybridization assay for quantitative PCR, which allowed simultaneous determination of both amplified target DNA and IS in the same microtiter well.[39] The target DNA from the sample was co-amplified with a constant amount of a recombinant DNA IS that had the same size and primer binding regions as the target DNA, differing only by a 24-bp sequence, located between the primers. Biotinylated PCR products from target DNA and IS were captured on a single microtiter well coated with streptavidin. The non-biotinylated strands were removed by NaOH treatment. The immobilized single-stranded target DNA and IS were allowed to simultaneously hybridize with their respective specific probes, *i.e.* a digoxigenin-labeled probe and a fluorescein-labeled probe. The hybrids were determined by allowing aequorin conjugated to anti-digoxigenin antibody and ALP conjugated to anti-fluorescein antibody (in the same mixture) to bind to their corresponding haptens. Aequorin was determined by adding Ca^{2+}. ALP was measured by the subsequent addition of CSPD (without prior washing of the wells) (Figure 9.10). The ratio of the luminescence values obtained from the target DNA and IS amplification products was linearly related to the number of target DNA molecules present in the sample prior to amplification. The linear range of the assay extended from 430 to 315 000 target DNA molecules. The S/B ratios observed at 430 molecules of target DNA in the presence of 10 000, 20 000 and 40000 molecules of IS were 4.4, 2.4 and 2.7, respectively. Since only 5% of the initial PCR mixture was used in the assay, the luminescence signal was essentially obtained from amplification product corresponding to 22 target DNA molecules.

A bioluminescence hybridization assay using aequorin was applied to the detection of PSA mRNA as discussed in Section 9.1.[1] The detection of PSA mRNA was extended to the quantification of PSA mRNA in a subsequent paper. The authors reported a simple, rapid and sensitive assay protocol for the quantification of PSA mRNA in peripheral blood (a potential marker for molecular staging of prostate cancer) by using an IS, RT-PCR and a chemiluminometric hybridization assay.[40] The key to quantification was the recombinant RNA IS, which had the same primer binding sites and size as the amplified PSA mRNA but differed only in a 24-bp segment. Amplified sequences were labeled with biotin during PCR by using a 5' biotinylated upstream primer. The products were heat-denatured and hybridized with oligonucleotide-specific probes (for PSA and IS) that were immobilized in separate microtiter wells. The hybrids were measured using ALP–streptavidin conjugate and a chemiluminogenic substrate (Figure 9.4). The ratio of the luminescence values obtained for the PSA mRNA and the RNA IS was a linear function of the initial amount of PSA mRNA present in the sample prior to RT-PCR. As few as 50 copies of PSA mRNA were detected with an S/B ratio of 2. The linearity of the assay extended up to 500 000 copies of PSA mRNA. Samples containing total RNA from PSA-expressing LNCaP cells gave luminescence ratios that were linearly related to the number of cells in the range 0.04–400 cells.

Another marker for the molecular staging of prostate cancer is prostate-specific membrane antigen (PSMA) mRNA. PSMA mRNA expression is restricted to the prostate. This tissue-specificity renders PSMA mRNA an excellent candidate marker for molecular staging of prostate cancer. A quantitative competitive RT-PCR method has been reported, in which the biotinylated PCR products from amplified target and IS were captured on streptavidin-coated wells and detected by using conjugates of oligonucleotide probes with ALP.[41] The probes were specific for PSMA and IS. The use of probes that are directly labeled with the reporter molecule made the assay procedure shorter (streptavidin: biotinylated target: probe-reporter conjugate) as opposed to indirect labeling in which the reporter molecule is conjugated to a

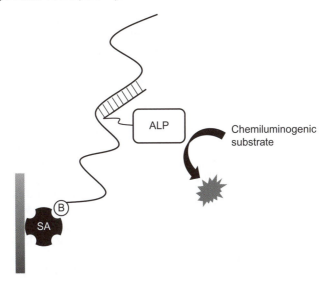

Figure 9.11 Chemiluminometric hybridization assay for quantitative PCR. Biotinylated (B) target DNA and biotinylated internal standard are captured in separate streptavidin (SA)-coated wells and denatured by NaOH. Immobilized single-stranded target DNA and internal standard are each hybridized with alkaline phosphatase (ALP) conjugated to target-specific probes.

specific antibody or a hapten (streptavidin: biotinylated target: probe: antibody-reporter conjugate) (Figure 9.11). The ratio of the luminescence values obtained for PSMA mRNA and the RNA IS was a linear function of the initial amount of PSMA mRNA copies initially present in the sample, before RT-PCR. As few as 500 copies of PSMA mRNA were detected with an S/B ratio of 5. The linear range extended from 500 to 5 000 000 PSMA mRNA copies. Samples containing total RNA from PSMA-expressing LNCaP cells gave luminescence ratios linearly related to the number of cells in the range 0.5–5000 cells.

An interesting application of the above principles of quantitative competitive PCR with chemiluminometric detection is in the field of quantification of genetically modified organisms (GMO) in food. A high-throughput double competitive quantitative PCR method has been developed for quantification of the transgene and a plant reference gene.[15] The latter compensates for differences in the total amount and integrity of DNA between samples. The advantage of double competitive PCR (compared to real-time PCR) is that any potential PCR inhibitors equally affect the amplification of target and IS, so that the ratio of their PCR products gives the ratio of their initial amounts in the sample, providing absolute quantification of the GMO-specific sequence and the reference gene. The conventional double competitive PCR method for GMO quantification entails the co-amplification of each target sequence with a competitive synthetic DNA IS that closely resembles the target DNA and shares the same primers. The IS also contains an insertion or deletion, large enough to allow separation from the target by slab gel electrophoresis. Each sample is titrated with the IS, *i.e.*, increasing, and known, amounts of IS are added to aliquots containing a constant amount of target, followed by PCR and electrophoresis. When the band densities are the same, the starting quantities of target and IS are equal. Internal standards are required for both transgene and reference gene. The conventional double competitive PCR is a low-throughput and labor intensive method due to multiple PCRs required for titration of each sample, electrophoresis and densitometry. A recent improvement in double competitive PCR is the introduction of capillary electrophoresis with laser-induced fluorescence detection for faster and automatable separation of the amplification products.

To address the drawbacks of conventional double competitive PCR, a high-throughput double quantitative competitive PCR (HT-DCPCR) method has been developed for GMO quantification.[42] In HT-DCPCR, electrophoresis and densitometry were replaced by a rapid, microtiter well-based bioluminometric hybridization assay and there was no need for titration of each sample. Instead, the target was co-amplified with a constant amount of IS. The determination of GM soya was chosen as a model. Internal standards (competitors) were constructed both for the transgene (35S promoter sequence) and the reference gene (lectin). Each IS had the same primer binding sites and size with the target sequences but differed in short internal segment. Each target sequence (35S and lectin) was co-amplified with a constant amount of the respective IS. For the hybridization assay, a universal solid phase coated with BSA–$(dT)_{30}$ conjugate was used. Each specific probe consisting of a 24-nt region complementary to its respective analyte and a poly(dA) tail was added to a separate well (four probes each in a separate well and each specific for one analyte). The four analytes (biotinylated amplified fragments of 35S target, 35S IS, lectin target and lectin IS) were denatured and hybridized with their respective specific probes. The hybrids were determined by using aequorin–streptavidin conjugate (Figure 9.12). The ratio of the luminescence values obtained for the target and the IS competitor was linearly related to the starting amount of target DNA. The detectability and linear range of each of the four hybridization assays were established. The limits of quantification were 4 pM amplified 35S target (200 amol well^{-1}), 6 pM amplified 35S IS (300 amol well^{-1}), 5 pM amplified lectin target (250 amol well^{-1}) and 13 pM amplified lectin IS (650 amol well^{-1}). The analytical range of the assays extended up to 1000 pM. For the quantitative competitive PCR of 35S promoter, the limit of detection and the limit of quantification were found to be 3 and 24 copies of 35S DNA, respectively. HT-DCPCR was evaluated by determining the GMO content of soybean powder certified reference materials. GMO contents determined by HT-DCPCR were in close agreement with the nominal values of the certified reference materials. In addition, HT-DCPCR was compared to real-time PCR (TaqMan assay) in various real samples. The results obtained by the two methods were in good agreement.

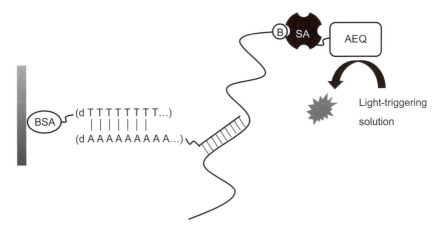

Figure 9.12 Bioluminometric hybridization assay for double quantitative competitive PCR. Wells are coated with bovine serum albumin (BSA)–(dT)$_{30}$ conjugate. For each analyte (35S target, 35S internal standard, lectin target or lectin internal standard), a specific probe consisting of a 24-nt region complementary to the analyte and a poly(dA) tail is added to a separate well. Each denatured biotinylated (B) target DNA is hybridized with its specific probe. Hybrids react with aequorin (AEQ)–streptavidin (SA) conjugate. AEQ is determined by Ca^{2+} addition.

The advantages of the HT-DCPCR assay are as follows. First, the high detectability that arises from the bioluminometric hybridization assay allows accurate and precise determination of the PCR products of each target and its respective IS despite the suppression of amplification due to their competition for the same primer set. This in turn enables quantification of the target sequence without the need for titration of each sample with various quantities of IS. Second, the hybridization assay is performed on a universal solid phase to ensure simplicity and high-throughput. The overall procedure (including PCR and hybridization assay) was completed in 2.5 h. Third, the amplified sequences for target and IS have identical sizes. This is an advantage over electrophoretic based double competitive PCR which requires internal standards differing in size, leading to differences in amplification efficiency. Fourth, heteroduplexes (as formed during co-amplification of substantially homologous DNA sequences) are not an issue because only one strand is captured on the microtiter well through hybridization.

The next challenge is the development of multi-analyte hybridization assays for simultaneous quantification of several target sequences in the same sample. Advantages include higher sample throughput, reduced consumption of reagents, smaller sample volume and lower cost of analysis compared to single-analyte assays. In principle, multi-analyte configurations require either the use of a single reporter along with spatial separation of the assays or the successful combination of several labels in a single assay. Spatial separation along with a single label has been employed in a multi-analyte hybridization assay for detection of various pathogens by constructing microtiter plates containing main wells with built-in sub-wells, with each sub-well corresponding to a single assay.[42] Few reports describe the combination of two labels in a single hybridization assay with applications to the quantification of PCR products.[4,32] These assays are based on the combination of two chemiluminescent labels (detection of two targets in the same sample and in one microtiter well).

A more recent advance is the development of a quadruple-analyte chemiluminometric hybridization assay for simultaneous quantification of four nucleic acid sequences.[43] As a model, the assay was applied to double quantitative competitive PCR for the determination of GMOs. The transgene is denoted NK, the reference gene is denoted IVR (invertase), and their respective internal standards (competitors) are NK-IS and IVR-IS. The four targets were amplified by a single PCR. The four biotinylated PCR products were then captured in the same microtiter well, which is coated with streptavidin. The non-biotinylated strands were removed by NaOH treatment, and the immobilized single-stranded DNA fragments were allowed to hybridize with a mixture of four specific probes. Each probe contained a sequence complementary to its respective target and a sequence or a hapten that allowed linkage with a unique chemiluminescent reporter. Specifically, NK and NK-IS probes were labeled at the 3' end with the haptens fluorescein and digoxigenin, respectively, to allow recognition by HRP-anti-fluorescein and ALP-anti-digoxigenin antibodies. The IVR-IS probe consisted of a segment at the 3' end complementary to IVR-IS and a sequence at the 5' end complementary to the oligonucleotide that was conjugated to β-galactosidase (GAL). In the middle of the IVR-IS probe, a 14-nt random spacer was introduced to avoid interference with hybridization. The IVR probe carried a poly(dA) tail at the 3' end, which was recognized by aequorin–(dT)$_{30}$ conjugate. Next, a mixture containing the four chemiluminescent reporters was added to the well (HRP conjugated to anti-fluorescein antibody, ALP conjugated to anti-digoxigenin antibody, GAL-oligo conjugate and aequorin–(dT)$_{30}$ conjugate). The four chemiluminescent reactions were triggered sequentially. The flash-type reaction of aequorin was triggered first by adding Ca^{2+} and decayed in a few seconds. The glow-type chemiluminescent reactions catalyzed by HRP, GAL and ALP were then triggered sequentially by adding appropriate substrates (Figure 9.13). Regarding the order in which HRP, GAL and ALP activities were measured, the authors observed that the GAL signal decreased considerably when measured after ALP (possible GAL inactivation due to exposure to the alkaline solution of the ALP substrate). The measurement order of HRP activity did not affect GAL. Based on these findings, the measurement order followed in the quadruple-hybridization assay was aequorin, HRP, GAL and ALP. A successful quadruple-analyte hybridization assay requires the absence of signal cross-talk between the reporters. A washing step was introduced after each measurement to remove one substrate before addition of the next. The signals were linearly related to the concentration of target sequences. The limits of quantification were 9 pM amplified IVR DNA (0.45 fmol well^{-1}), 3 pM of IVR-IS DNA (0.15 fmol well^{-1}), 6 pM of NK DNA (0.3 fmol well^{-1}) and 5 pM amplified NK-IS DNA (0.25 fmol well^{-1}). The analytical range extended up to 1000 pM DNA. For quantitative competitive

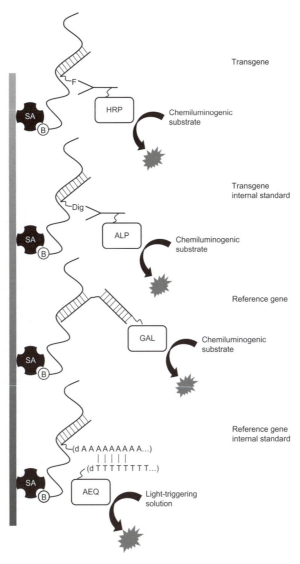

Figure 9.13 Quadruple-analyte bio(chemi)luminometric hybridization assay for double quantitative competitive PCR. The four biotinylated (B) target DNAs (transgene, transgene internal standard, reference gene and reference gene internal standard) are captured in a single streptavidin (SA)-coated well and denatured by NaOH. Immobilized single-stranded DNAs are hybridized with specific probes (a mixture of four probes is added to the well). Each probe contains a sequence complementary to its respective target and a sequence or a hapten [fluorescein (F) or digoxigenin (Dig)] that allows linkage with its unique reporter. Hybrids react with a mixture of horseradish peroxidase (HRP) conjugated to anti-fluorescein antibody, alkaline phosphatase (ALP) conjugated to anti-digoxigenin antibody, β-galactosidase (GAL)–oligonucleotide conjugate, and aequorin (AEQ)–$(dT)_{30}$ conjugate. The four reporters are determined sequentially in the same well.

PCR of NK or IVR, the S/B ratio at 50 copies of NK and IVR target DNA was 1.6 and 2.8, respectively. This assay combined the high sensitivity, wide dynamic range and simple instrumentation of chemiluminometric assays with the ability to simultaneously quantify four nucleic acid sequences in a single microtiter well. The assay is universal since the same reporter conjugates can be used for multi-analyte quantification of any sequences with properly designed sequence specific probes.

9.3   Genotyping of Single Nucleotide Polymorphisms (SNPs)

Single nucleotide polymorphisms (SNPs) constitute the most common form of human genetic variation. The method chosen for SNP detection depends on whether we are screening for unknown or known mutations. In this section, we focus on the detection of known mutations and particularly on methods that are suitable for the routine molecular diagnosis laboratory. DNA microarray technology plays an important role in genome-wide association studies of various SNPs with certain diseases because each chip enables parallel genotyping of thousands of SNPs in a single sample. However, it is expected that only a small number of SNPs will be routinely analyzed in the clinical setting. For instance, SNPs in disease-related genes will be tested for diagnosis and monitoring of disease; SNPs in genes encoding drug-metabolizing enzymes will be detected to design improved therapeutic strategies. Consequently, in the routine clinical laboratory high sample-throughput is much more useful than high SNP-throughput.

SNP genotyping methods, in general, include isolation of genomic DNA, exponential amplification (usually by PCR) of the region that spans the SNP, a genotyping reaction to distinguish the alleles and, finally, the detection of the genotyping reaction products. Sequencing of the amplified fragment is the reference method, but the relatively high cost of this technique prohibits its wide use in the clinical laboratory. Most genotyping techniques are based on one of the following principles. PCR combined with restriction fragment length polymorphism (RFLP) analysis is a commonly used method, but its low-throughput is a major drawback. Several genotyping methods for known mutations rely on the hybridization of allele-specific oligonucleotide probes (ASOs) under conditions that allow discrimination of a perfect match from a mismatch. ASOs may also be used as PCR primers because DNA polymerase amplifies only when the 3′ end of the primer perfectly complements the target. This approach, however, requires two PCRs per SNP. Enzyme-based genotyping assays, such as primer extension (PEXT), oligonucleotide ligation reaction (OLR) and invasive cleavage, have proven to be more robust and specific than ASO hybridization. The PEXT reaction is the most widely used because it is the simplest, it requires the least number of probes, and it can be easily optimized. Mini-sequencing, pyrosequencing, flow cytometry and mass spectrometry have been used for the detection of PEXT products. Here we discuss the development of cost-effective, rapid, automatable high-throughput bio(chemi)luminometric assays for SNP detection using PEXT or OLR.

One such method is based on the fact that the kinetics of light emission of chemiluminescent reactions varies considerably. The flash-type emission from an acridinium ester or the aequorin reaction has a decay half-life of about 1 s. The glow-type emission from enzyme-catalyzed chemiluminescent reactions may last several minutes or hours. The variation in emission kinetics allows the development of multi-analyte hybridization assays.

To this end, the difference in light-emission kinetics between the aequorin bioluminescent reaction and the ALP-catalyzed chemiluminescent reaction has been exploited for the analysis of bi-allelic polymorphisms in a single microtiter well.[44] The genotyping of the IVS-1-110 locus of the human β-globin gene was chosen as a model. Genomic DNA, isolated from whole blood, was first subjected to PCR using primers flanking the polymorphic site. A single OLR employing two allele-specific probes (normal probe, N; mutant probe, M) and a common probe (C) carrying a characteristic tail was then performed.[45] Probe N was labeled at the 5′ end with biotin whereas the 3′ end had a nucleotide specific for the normal allele. Probe M was labeled at the 5′ end with digoxigenin whereas the 3′ end had a nucleotide specific for the mutant allele. Probes N and M were designed to anneal to the target DNA at a position adjacent to probe C. Probes that are perfectly matched to the target sequence are covalently joined by the ligase. In contrast, a mismatch at the junction inhibits ligation. Thus, two different ligation products, depending on allele composition of the sample, may be formed. Biotin-labeled N-C was formed when the normal allele was present and digoxigenin-labeled M-C was formed when the mutant allele was present (Figure 9.14a). Next, the ligation products were captured in a single microtiter well through hybridization of the tail of the C probe with an immobilized complementary oligonucleotide (prior to capturing, the ligation products were heat-denatured to ensure separation of the non-ligated probes). The ligation products were detected by adding a mixture of biotinylated aequorin complexed to streptavidin and ALP conjugated to anti-digoxigenin antibody. The characteristic Ca^{2+}-triggered bioluminescence of aequorin (flash-type) was measured first, followed by the addition of the dioxetane aryl phosphate substrate for ALP (Figure 9.14b). The ratio of the luminescence signals obtained from ALP (signified mutant allele) and aequorin (signified normal allele) gave the genotype of each sample. A heterozygote for the mutation gave both signals. Free C probe (not ligated) was also captured to the well but not detected. The dual-analyte bio(chemi)luminometric genotyping assay provided clear distinction of the three genotypes with signal ratios differing by more than an order of magnitude. The assay is an excellent candidate for high-throughput genotyping of a large number of individuals for bi-allelic polymorphisms.

The previously mentioned assay used two different reporters for detection of the oligonucleotide ligation products in a single well.[44] An alternative genotyping assay was subsequently developed in which the high specificity of OLR was again combined with the simplicity and sensitivity of chemiluminometric

(a) Oligonucleotide ligation reaction

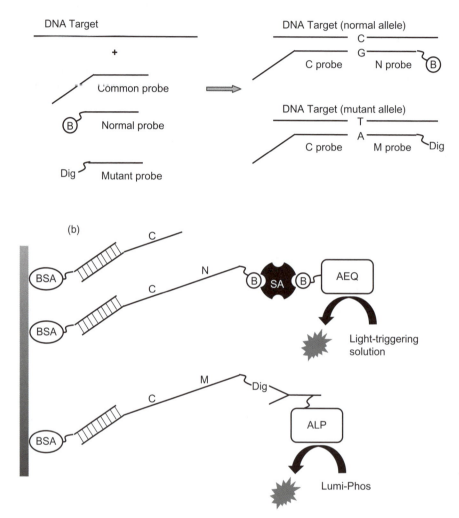

Figure 9.14   (a) Illustration of the principle of the oligonucleotide ligation reaction (OLR). The assay employs two allele-specific probes (normal probe, N; mutant probe, M) and a common probe (C) carrying a characteristic tail. N is 5′ labeled with biotin (B) and the 3′ end has a nucleotide specific for the normal allele. M is 5′ labeled with digoxigenin (Dig) and the 3′ end has a nucleotide specific for the mutant allele. N and M probes anneal to the target DNA at a position adjacent to C. Probes that are perfectly matched to the target sequence will be covalently joined by the ligase. (b) Dual-analyte bio(chemi)luminometric hybridization assay based on combined flash- and glow-type reactions for SNP genotyping by OLR. Denatured ligation products are hybridized in a single well with an immobilized probe [oligonucleotide conjugated to bovine serum albumin (BSA)]. Hybrids react with a mixture of biotinylated (B) aequorin (AEQ) complexed to streptavidin (SA) and alkaline phosphatase (ALP) conjugated to anti-digoxigenin antibody. AEQ is determined first by Ca^{2+} addition and ALP is determined by adding a chemiluminogenic substrate.

detection but this time the assay employed a universal detection approach that allowed simultaneous analysis of several samples for various SNPs in the same microtiter plate.[46] The method was applied to the genotyping of four SNPs within the genes of histamine H2 receptor (HRH2), serotonin receptor (HTR2A1 and HTR2A2) and b3 adrenergic receptor (ADRB3). SNPs in these neurotransmitter receptor genes form the basis of pharmacogenetic studies on the efficacy of anti-psychotic agents.

Genomic DNA, isolated from whole blood, was first subjected to PCR using primers flanking the polymorphic site. A single OLR employing two allele-specific probes (normal probe, N; mutant probe, M) and a biotinylated common probe (C) was then performed. Probe N was labeled at the 3′ end with digoxigenin whereas the 5′ end contained a nucleotide specific for the normal allele. Probe M was labeled at the 3′ end with fluorescein whereas the 5′ end contained a nucleotide specific for the mutant allele. Thus two different ligation products were formed. Digoxigenin-labeled N–C was formed when the normal allele was present and fluorescein-labeled M–C was formed when the mutant allele was present. Next, the ligation products were captured in streptavidin-coated microtiter wells through the biotin moiety of the C probe and denatured by NaOH treatment. The ligation products were detected by adding either ALP conjugated to anti-digoxigenin antibody or ALP conjugated to anti-fluorescein antibody (for each SNP to be analyzed there was one OLR whose products were then split into two wells for capturing) (Figure 9.15). The ratio of the luminescence signals obtained from ALP conjugated to anti-digoxigenin antibody (signified normal allele) and ALP conjugated to anti-fluorescein antibody (signified mutant allele) gave the genotype of each sample. A heterozygote for the mutation gave both signals. Free C probe (not ligated) was also captured in the well but not detected.

The use of aequorin as a reporter molecule in the detection of oligonucleotide ligation products as applied to SNP genotyping[44] was followed by the development of a hybridization assay for the detection of primer extension (PEXT) products by aequorin.[47] As discussed in the previous sections, conjugates of aequorin with antibodies or streptavidin had already been successfully used as reporter molecules for the detection of PCR products and for quantitative PCR.[4,42]

The developed assay was applied to SNP genotyping of the mannose-binding lectin 2 (MBL2) gene.[47] Since MBL2 SNPs have been associated with the functional deficiency of MBL (a key component of the innate immune system), there is a growing need to develop genotyping methods for screening MBL2 allelic variants. The method involves the following: (a) a single PCR to amplify the genomic region of interest encompassing all six variant nucleotide sites, (b) PEXT reactions (using unpurified PCR products) in the presence of biotin-dUTP and a DNA polymerase lacking 3′-5′ exonuclease activity and (c) a microtiter well-based assay of the extension products with an aequorin–streptavidin conjugate. Two PEXT reactions were performed for each site. For each variant, two primers were designed for PEXT (normal and mutant primer) and they were extended in two separate reactions. Each of these primers contained a poly(dA) segment at the 5′ end and an allele-specific nucleotide at the 3′ end.

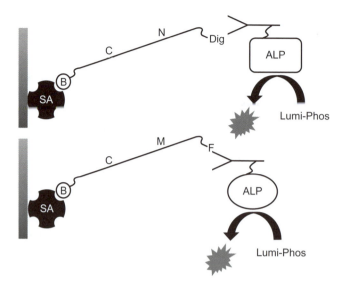

Figure 9.15 Chemiluminometric hybridization assay based on a universal detection approach for SNP genotyping by OLR (oligonucleotide ligation reaction). Following OLR, ligation products are captured on streptavidin (SA)-coated wells through the biotin moiety of the C probe and denatured by NaOH. Ligation products react with either alkaline phosphatase (ALP) conjugated to anti-digoxigenin antibody or ALP conjugated to anti-fluorescein antibody (for each SNP to be analyzed there is one ligation reaction whose products are then split into two wells for capturing).

The distinction between genotypes is based on the high accuracy of nucleotide incorporation by DNA polymerase. Biotin-dUTP (along with the other dNTPs) was incorporated in the extended primer only when the primer was a perfect match to an allele (Figure 9.16a). The products were captured by hybridization on the surface of microtiter wells that were coated with BSA–$(dT)_{30}$ conjugate. Only the extended primers (with the incorporated biotins) were detected by reaction with an aequorin–streptavidin conjugate (Figure 9.16b). Genotypes were assigned by the signal ratio of the normal-specific primer to the mutant-specific primer. The PEXT reaction was completed in 10 min and the detection of the products in less than 40 min. The assay is a cost-effective, rapid, robust and automatable method for detecting all known allelic variants of MBL2, introducing for the first time aequorin as a reporter in genotyping by PEXT reaction.

A method that combines the high detectability and dynamic range of chemiluminescence with the high allele-discrimination ability of PEXT has been developed for the simultaneous characterization of 15 SNPs in a high-throughput microtiter well-based assay in a dry-reagent format.[48] As a model for the development and validation of the method, the 15 most common β-hemoglobin (*HBB*) gene mutations found in the populations of the

(a) PEXT reaction with normal primer

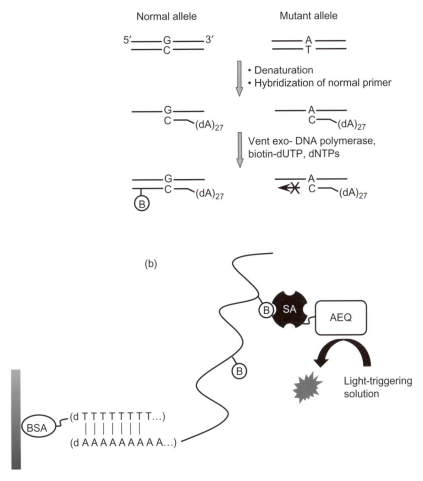

Figure 9.16 (a) Primer extension (PEXT) reaction with normal primer. Two primers (normal and mutant) are designed for PEXT and they are extended in two separate reactions. Each of these primers contains a poly(dA) segment at the 5' end and an allele-specific nucleotide at the 3' end. Biotin-dUTP (along with the other dNTPs) is incorporated by DNA polymerase in the extended primer only when the primer is a perfect match to an allele. (b) Bioluminometric hybridization assay based on aequorin for SNP genotyping by PEXT reaction. The products of each PEXT reaction are captured on wells coated with bovine serum albumin (BSA)–(dT)$_{30}$ conjugate. The extended primers [with the incorporated biotins (B)] are detected by reaction with an aequorin (AEQ)–streptavidin conjugate. AEQ is determined by Ca^{2+} addition.

Mediterranean basin were detected. Although various methods for *HBB* genotyping have been reported there is still a need for a cost-effective method (in terms of both equipment and reagents) with high sample-throughput, for simultaneous screening of several mutations. The method consist of: (a) duplex PCR to amplify the genomic region of interest producing two fragments encompassing all 15 mutations, (b) PEXT reactions (using unpurified PCR products) in the presence of fluorescein-dCTP and (c) a microtiter well based assay of extension products with HRP conjugated to anti-fluorescein antibody and a chemiluminogenic substrate. Two PEXT reactions were performed for each mutation. Two primers were designed for PEXT (normal and mutant primer) and they were extended in two separate reactions. Each of these primers contained a biotin at the 5′ end and an allele-specific nucleotide at the 3′ end. Fluorescein-dCTP was incorporated in the extended primer only when the primer was a perfect match to an allele. The products were captured by hybridization on the surface of microtiter wells coated with streptavidin. Only the extended primers (with the incorporated fluoresceins) were detected by reaction with HRP conjugated to anti-fluorescein antibody (Figure 9.17). Genotypes were assigned by the signal ratio of the normal-specific primer to the mutant-specific primer. A significant advantage is that the assay used lyophilized reagents for PCR and PEXT reactions, as well as dried streptavidin-coated wells. The lyophilized and dried reagents had a long shelf-life when

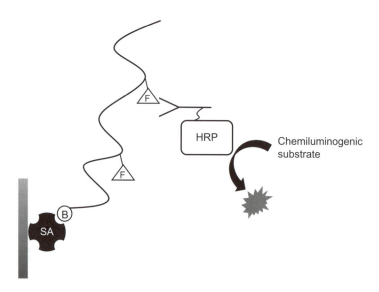

Figure 9.17 Chemiluminometric hybridization assay based on a dry-reagent format for genotyping of 15 SNPs by PEXT reaction. The products of each PEXT reaction are captured on streptavidin (SA)-coated wells. Only the extended primers [with the incorporated fluoresceins (F)] are detected by reacting with horseradish peroxidase (HRP) conjugated to anti-fluorescein antibody.

stored at 4 °C, and the use of pre-prepared lyophilized and dried reagents in a microtiter plate format made the entire assay much faster (about 3.5 h including PCR). This feature makes the method attractive for the routine molecular diagnostic laboratory.

The previously mentioned methods for SNP detection by PEXT require spatial separation during the capture of the PEXT products (two separate wells followed by two separate assays for detection).[47,48] A major step was the subsequent development of a dual-analyte bio(chemi)luminometric assay for SNP genotyping by PEXT.[49] The method consists of: (a) one PCR to amplify the genomic region of interest, (b) PEXT reactions in the presence of digoxigenin-dUTP (normal primer) and biotin-dUTP (mutant primer) and (c) a microtiter well-based assay of extension products with ALP conjugated to anti-digoxigenin antibody (for detection of the normal allele) and an aequorin–streptavidin conjugate (for detection of the mutant allele). For each SNP, two primers were designed for PEXT (normal and mutant primer) and they were extended in two separate reactions. Each of these primers contained a poly(dA) segment at the 5′ end and an allele-specific nucleotide at the 3′ end. The PEXT reaction of the normal primer was performed in the presence of digoxigenin-dUTP. The PEXT reaction of the mutant primer was performed in the presence of biotin-dUTP. The digoxigenin-labeled product (extension of normal primer) and the biotin-labeled product (extension of mutant primer) were captured by hybridization on the surface of a single microtiter well coated with BSA–(dT)$_{30}$ conjugate (for each SNP to be analyzed there were two separate PEXT reactions whose products were then captured on a single well). Detection was performed by adding a mixture of ALP conjugated to anti-digoxigenin antibody and an aequorin–streptavidin conjugate. First, aequorin was determined by its characteristic Ca^{2+}-triggered bioluminescence. ALP was then measured by adding a chemiluminogenic substrate (Figure 9.18). The ALP/aequorin signal ratio gave the genotype of the SNP.

In the above papers,[47–49] all assays used PEXT reactions for allele discrimination and, in all cases, both extended and non-extended primers were captured to the solid phase either through their poly(dA) segment[47,49] or their biotin moiety.[48] The difference in a subsequent paper is that only the extended primers were captured to the solid phase through the incorporated biotins.[50] The genotyping of two SNPs (A896G and C1196T) in the toll-like receptor 4 gene was chosen as a model. Toll-like receptors (TLRs) play a fundamental role in pathogen recognition and activation of innate immunity and SNPs in TLR have been associated with reduced host immune response to TLR ligands. The method consists of: (a) PCR to amplify the genomic region of interest, (b) PEXT reactions (using unpurified PCR products) in the presence of biotin-dUTP and (c) a microtiter well-based assay of extension products with an aequorin–(dT)$_{30}$ conjugate. Two PEXT reactions were performed for each SNP. Two primers were designed for PEXT (normal and mutant primer) and they were extended in two separate reactions. Each of these primers contained a poly(dA) segment at the 5′ end and an allele-specific nucleotide at the 3′ end.

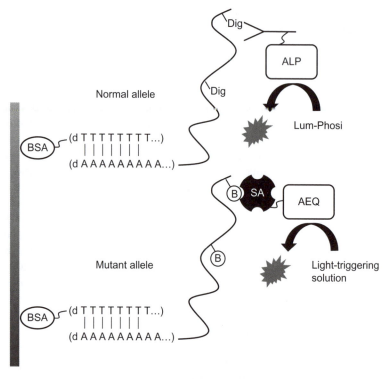

Figure 9.18 Dual-analyte bio(chemi)luminometric hybridization assay based on combined flash- and glow-type reactions for SNP genotyping by PEXT reaction. The products of both PEXT reactions [a digoxigenin (Dig)-labeled product from extension of normal primer and a biotin (B)-labeled product from extension of mutant primer] are captured in a single well coated with bovine serum albumin (BSA)–(dT)$_{30}$ conjugate. Hybrids react with a mixture of alkaline phosphatase (ALP) conjugated to anti-digoxigenin antibody and an aequorin (AEQ)–streptavidin conjugate. AEQ is determined first by Ca^{2+} addition and ALP is determined by adding a chemiluminogenic substrate.

Biotin-dUTP was incorporated in the extended primer only when the primer was a perfect match to an allele. The extended products were captured on streptavidin-coated wells and detected by using an aequorin–(dT)$_{30}$ conjugate (Figure 9.19).[13] Genotypes were assigned by the allelic ratio (AR), which was calculated from the equation $AR = L_N/ (L_N + L_M)$, where L_N and L_M are the luminescence signals obtained from PEXT reaction with the N and M primer, respectively. The theoretical values of AR for a normal sample (N/N), a heterozygote and a mutant homozygote are 1, 0.5 and 0, respectively. The authors also investigated and presented the possibility of screening pooled samples to reduce cost and increase throughput, a feature that makes this already cost-effective assay (compared to homogeneous fluorometric assays and mass spectrometry) even more attractive. The method is also rapid, as opposed to

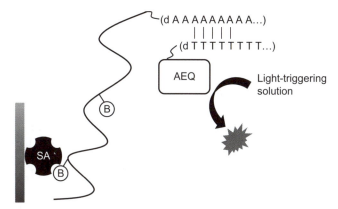

Figure 9.19 Bioluminometric hybridization assay based on aequorin for SNP genotyping by PEXT reaction. The products of each PEXT reaction are
captured on streptavidin (SA)-coated wells. Only the extended primers
[with the incorporated biotins (B)] are captured on the wells and detected
by adding an aequorin–(dT)$_{30}$ conjugate. AEQ is determined by Ca^{2+}
addition.

other time-consuming methods reported for TLR4 SNPs (RFLP analysis), and
amenable to automation.

The high-throughput PEXT-based bioluminometric assay for genotyping
A896G and C1196T SNPs in the TLR4 gene was followed by the development
of a quadruple-analyte PEXT-based bio(chemi)luminometric assay for the
simultaneous genotyping of the two TLR4 SNPs (simultaneous detection of
four alleles).[51] This became possible by successfully combining four reporters,
namely, aequorin, HRP, β-galactosidase (GAL) and ALP. A quadruple PEXT
reaction was first performed in the presence of two pairs of allele-specific
primers and biotin-dUTP (PCR products were used without purification). For
each SNP, two primers were designed for PEXT (normal and mutant primer,
N and M) with allele-specific nucleotides at the 3′ end. Each primer carried
a characteristic label (hapten or oligonucleotide) at the 5′ end to enable
detection through a specific reporter (enzyme or photoprotein). The N and M
primers for the 896 SNP were 5′-labeled with fluorescein and digoxigenin,
respectively. The N primer for the 1196 SNP contained a characteristic
oligonucleotide sequence at the 5′ end, which enabled detection through a
complementary oligonucleotide that is conjugated to GAL. The M primer for
the 1196 SNP contained a (dA)$_{21}$ sequence at the 5′ end that allowed detection through an aequorin–(dT)$_{30}$ conjugate. Biotin–dUTP was incorporated
in the extended primer. The biotin-labeled products were captured by hybridization on the surface of a single microtiter well coated with streptavidin
(for the two SNPs to be analyzed there was one PEXT reaction whose
products were then captured on a single well). Detection was performed by
adding a mixture of HRP conjugated to anti-fluorescein antibody (detection of
896 N allele), ALP conjugated to anti-digoxigenin antibody (detection of

896 M allele), GAL conjugated to a characteristic oligonucleotide (detection of 1196 N allele) and aequorin–$(dT)_{30}$ conjugate (detection of 1196 M allele). PEXT products were used without purification. The four reporters were determined in the same microtiter well (Figure 9.20). The HRP/ALP signal ratio gave the genotype of the 896 SNP whereas the GAL/aequorin signal ratio gave the genotype of the 1196 SNP. The PEXT reaction was completed in 15 min and the detection of the products in 75 min. The cost of the method was considerably reduced by performing four PEXT reactions in the same tube followed by detection of four PEXT products in the same well. The microtiter plate assay format facilitates automation of the genotyping method and provides high sample-throughput. PCR for the TLR4 gene was designed in such a way that both SNPs were present in the same amplified fragment. However, the loci of the two SNPs need not be contiguous, since a duplex PCR can be performed, providing two amplified fragments with the loci of interest. In this report, multi-analyte detection is accomplished through the use of four different reporters. A different chemiluminometric assay achieves multi-analyte detection through spatial separation.[42] The two concepts are not competitive but rather can be combined to further enhance the multiplicity of DNA targets.

9.4   Determination of Allele Burden

Thus far, we have discussed methods for detection of SNPs as applied to mutations that are inherited. Unlike inherited mutations, which are present in all cells, somatic (acquired) mutations occur only in certain cells of the body. Thus, somatic mutations are considered the primary cause of human cancer. An example is the somatic point mutation V617F of the JAK2 kinase, a recently discovered diagnostic marker for myeloproliferative neoplasms. The challenge with somatic point mutations is to develop sensitive, robust and practical methods to quantify the mutant allele while discriminating from a large excess of the normal allele that differs only in a single base-pair. Quantification of mutant allele burden (percentage of the mutant allele) is critical for diagnosis, therapeutic monitoring and detection of minimal residual disease. Sequencing has been used for allele burden studies but the detectability is low (15–20% of mutant allele). Current methods are based on real-time PCR. These homogeneous fluorometric assays are advantageous because amplification and detection are simultaneous. However, the assays require costly equipment and expensive reagents.

The bio(chemi)luminometric genotyping assays described in Section 9.3[44,46–51] have focused on the detection of alleles in the case of inherited mutations where the goal is the discrimination between the normal homozygote, mutant homozygote or heterozygote genotypes. The above methods provide no quantitative estimate of the percent of each allele (normal and mutant) in the DNA sample. However, in the case of somatic mutations, quantification of the allele burden is a crucial requirement. We next discuss a bioluminometric assay for

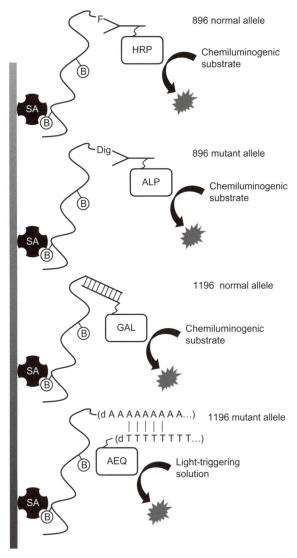

Figure 9.20 Quadruple bio(chemi)luminometric hybridization assay for SNP geno-typing by PEXT reaction. A quadruple PEXT reaction is performed. Normal and mutant primers for the 896 SNP are 5′-labeled with fluorescein (F) and digoxigenin (Dig), respectively. Normal primer for the 1196 SNP has a characteristic oligonucleotide sequence at the 5′ end. Mutant primer for the 1196 SNP has a $(dA)_{21}$ sequence at the 5′ end. Products of PEXT are captured on a single streptavidin (SA)-coated well. Only extended primers [with incorporated biotins (B)] are captured on the well and detected by reaction with a mixture of horseradish peroxidase (HRP) conjugated to anti-fluorescein antibody, alkaline phosphatase (ALP) conjugated to anti-digoxigenin antibody, β-galactosidase (GAL)–oligonucleotide conjugate and aequorin–$(dT)_{30}$ conjugate. The four reporters are determined sequentially in the same well.

quantification of allele burden as applied to JAK2 V617F mutation (model assay).[52]

The method involved a single PCR of the JAK2 genomic region of interest. PCR was followed by two PEXT reactions. The PEXT primers for the normal and mutant allele contained a poly(dA) segment at the 5' end and an allele-specific nucleotide at the 3' end (Figure 9.16a). Biotin–dUTP was incorporated in the extended primer only when the primer was a perfect match to an allele. The extended products were captured on streptavidin-coated wells and detected by using an aequorin–$(dT)_{30}$ conjugate (Figure 9.19). The assay was completed within 50 min after PCR. The allele burden was given by the ratio (expressed as percentage) of the signal due to the mutant allele over the total luminescence obtained from both alleles. The values of allele burden ranged from 0 (only the normal allele is present) to 100 (only the mutant allele is present). The authors demonstrated the linear relationship between the allele burden and the percent (%) luminescence signal due to the mutant allele. As value of as low as 0.85% of mutant allele was detected, and the linearity was extended to 100%. Therefore, the bioluminometric assay enabled the relative determination of the mutant allele burden.

9.5 Conjugation Strategies

In this section we discuss conjugation strategies aimed at reducing the cost of commercially available conjugates and/or improving the sensitivity, ease and practicality of assays.

In one report, the need for either costly commercial aequorin–biotin and aequorin–streptavidin conjugates or for chemical crosslinking of aequorin (to biotin or streptavidin) was replaced by the construction of a plasmid suitable for bacterial expression of *in vivo*-biotinylated aequorin.[53] The biotin tag facilitated both the isolation of aequorin from the crude cell extract as well as direct complexing of aequorin with streptavidin for utilization as a reporter molecule in the development of highly sensitive bioluminometric hybridization assays. The plasmid contained a biotin-acceptor coding sequence fused to the apoaequorin gene. The *birA* gene, encoding biotin protein ligase (BPL), was inserted downstream of the apoaequorin sequence. BPL post-translationally biotinylated the aequorin acceptor domain at a unique position. Functional aequorin was generated by incubating the lysate with coelenterazine followed by purification by affinity chromatography using immobilized monomeric avidin. After elution, the biotinylated aequorin was complexed with strepta-vidin. Purified aequorin was detected down to 1.6 amol with an S/B ratio of 4.4 and a linear range extending over three orders of magnitude. Moreover, the *in vivo* biotinylated aequorin was compared to a commercial aequorin and found to have identical performance. In addition, streptavidin-biotinylated aequorin complex was used as a reporter molecule in a microtiter well-based hybridization assay. Microtiter wells were coated with anti-digoxigenin antibody. A digoxigenin-labeled capture probe was then added to the well followed by hybridization of the heat-denatured biotinylated target DNA. The hybrids were determined using the streptavidin-biotinylated aequorin complex. The linearity of the assay ranged from 80 amol to 40 fmol target DNA. An S/B ratio of 2.1 was obtained at 80 amol. It was calculated that *in vivo*-biotinylated aequorin produced from 1 L of culture was sufficient for 300 000 hybridization assays. The entire process, including cell culturing, extraction of total soluble protein and generation and purification of fully active biotinylated aequorin, was completed in 2 days.

The availability of recombinant aequorin was further enhanced by developing a simple method for expression and purification of recombinant aequorin based on a commercially available starting vector and a common *E. coli* strain.[54] The paper reported a purification method for aequorin in one step based on immobilized metal-ion affinity chromatography, a method for purification of proteins with engineered poly-histidine tags. These tags form high affinity complexes with immobilized divalent metal ions, thereby allowing isolation of tagged proteins from a crude cellular extract. An appropriate plasmid was constructed in which a hexahistidine $(His)_6$-coding sequence was fused upstream of the apoaequorin cDNA. Overexpression of heterologous proteins in the *E. coli* cytoplasm is often accompanied by misfolding and segregation into inclusion bodies (insoluble and inactive aggregates). In the case of apoaequorin, inclusion bodies were solubilized by urea treatment and purification was accomplished in one step using a Ni^{2+}-nitrilotriacetic acid (Ni-NTA) agarose column in a way that proper protein refolding was ensured. Purified aequorin was detected down to 0.5 amol with an S/B ratio of 1.8 and a linear range extending over six orders of magnitude. The one-step purification procedure of $(His)_6$–aequorin lasted only a few hours.

The $(His)_6$- tag greatly facilitated the preparation of conjugates of aequorin to DNA probes as demonstrated in this next report.[55] Conjugates of aequorin to DNA probes represent the "direct labeling" approach in bioluminometric hybridization assays. This approach is advantageous over the "indirect labeling" approach because it eliminates an incubation step and a washing step, thus reducing the time required for assay completion. In the "indirect labeling" approach, a ligand is attached to the DNA probe, and the hybrids are detected by using a specific binding protein conjugated or complexed to aequorin.[1,4] However, even though direct labeling is preferred, the preparation of aequorin–DNA conjugates requires laborious chromatographic procedures followed by concentration steps, to remove the free (unreacted) DNA probe, which otherwise competes with the conjugate for hybridization to the target sequence. The authors reported a general procedure for the preparation of aequorin–DNA conjugates, and central to the conjugation protocols is the use of $(His)_6$–aequorin.[55] Conjugates were prepared by using either homobifunctional or heterobifunctional crosslinking reagents.

In one protocol, an amino-modified oligonucleotide was treated with a homobifunctional crosslinker carrying two N-hydroxysuccinimide ester groups, and the derivative was allowed to react with $(His)_6$–aequorin.

Following synthesis of the conjugate, the effective removal of the free oligo-nucleotide was crucial because it competes with the aequorin–oligo conjugate for hybridization to the complementary target DNA. The aequorin–oligo conjugates were purified cither by (a) both affinity capture on a Ni-NTA agarose column and anion-exchange HPLC or (b) only by anion-exchange HPLC. The performances of the purified aequorin–oligo conjugates were similar when tested in a hybridization assay, in which biotinylated target DNA was captured on streptavidin-coated wells. After removal of the non-biotiny-lated strand by NaOH treatment, the target DNA was hybridized to the aequorin–oligo conjugate. As low as 2 pM target DNA was detected using the conjugate purified both by affinity capture and HPLC (S/B ratio = 1.8). Moreover, it was found that conjugates purified by a single rapid affinity-capture step offered the same detectability and analytical range as those pur-ified both by affinity capture and HPLC. This simplified the preparation of conjugates by eliminating the need for HPLC purification.

In the second protocol, protected sulfhydryl groups were introduced into $(His)_6$–aequorin (to avoid aequorin inactivation due to derivatization of cysteine groups that play a role in the bioluminescent reaction) followed by reaction with a heterobifunctional crosslinker containing a N-hydroxysucci-nimide and a maleimide group. The aequorin–oligo conjugates were purified either by (a) both affinity capture on a Ni-NTA agarose column and anion-exchange HPLC or (b) only by affinity capture. The performances of the purified aequorin–oligo conjugates were similar when tested in hybridization assays. As low as 2 pM target DNA was detected using the conjugate purified both by affinity capture and HPLC (S/B ratio = 2.9). The linearity of the assay extended to 2000 pM target DNA for both conjugates. The conjugate obtained from a reaction of 10 nmol of $(His)_6$–aequorin was sufficient for about 5000 hybridization assays.

$(His)_6$–aequorin[54] has been exploited not only for the conjugation of oligo-nucleotides to aequorin[55] but also for the conjugation of streptavidin to aequorin.[56] The aequorin–streptavidin conjugate may be exploited as a uni-versal reporter molecule for bioluminometric DNA hybridization assays because streptavidin non-covalently binds to biotin with high affinity and biotin can be easily attached to practically any biomolecule. Protected sulfhydryl groups were introduced into $(His)_6$–aequorin whereas streptavidin was deri-vatized with maleimide groups. The conjugate was purified in a single step by immobilized metal-ion affinity chromatography. The performance of the aequorin–streptavidin conjugate was tested by using it as a reporter molecule in a microtiter well-based DNA hybridization assay. After thermal denaturation, biotinylated PCR products were captured on microtiter wells coated with target-specific probe conjugated to BSA. The captured single-stranded DNA was then detected by reacting with the aequorin–streptavidin conjugate. The limit of detection was found to be 0.29 pM (14.5 amol well^{-1}). The limit of quantitation was 0.50 pM (25 amol well^{-1}). The analytical range of the assay extended up to 500 pM. The conjugate obtained from 10 nmol of $(His)_6$–aequorin was sufficient for about 5000 hybridization assays.

Aside from aequorin, conjugation strategies have been employed for *Gaussia* luciferase. The cDNA for *Gaussia* luciferase (GLuc), the enzyme responsible for the bioluminescent reaction of the marine organism *Gaussia princeps*, was cloned in 2001.[57] The substrate for GLuc is coelenterazine. GLuc presented as an excellent candidate for a potential new bioluminescent reporter molecule for DNA hybridization assays since it is a monomeric protein and it gives high levels of light emission when transfected into mammalian cells. The first quantitative analytical study of GLuc was performed in 2002 and involved bacterial expression of *in vivo* biotinylated GLuc.[58] A plasmid encoding both a biotin acceptor peptide-GLuc fusion protein and the enzyme biotin protein ligase (BPL) was engineered. Purification of GLuc was then accomplished by affinity chromatography using immobilized monomeric avidin. Complex formation with streptavidin eliminated the need for chemical conjugation reactions, which are known to inactivate luciferases. Purified GLuc was detected down to 1 amol with an S/B ratio of 2 and a linear range extending over five orders of magnitude. Furthermore, the GLuc–streptavidin complex was used as a reporter molecule in a microtiter well-based DNA hybridization assay. Microtiter wells were coated with anti-digoxigenin antibody. A digoxigenin-labeled capture probe was then added to the well followed by hybridization of the heat-denatured biotinylated target DNA. The hybrids were determined using the GLuc conjugate (biotinylated GLuc complexed to streptavidin). Luminescence was measured in the presence of excess coelenterazine. The linearity of the assay ranged from 1.6 to 800 pM $(80\,\mathrm{amol\,well}^{-1}$ to $40\,\mathrm{fmol\,well}^{-1})$ target DNA. The S/B ratio at 80 amol was 1.4. This detectability was similar to the one exhibited in assays using *in vivo*-biotinylated aequorin.[2] It was calculated that *in vivo*-biotinylated GLuc produced from 1 L of culture was sufficient for 150 000 hybridization assays. The entire process, including cell culturing, extraction of total soluble protein and generation and purification of fully active biotinylated GLuc was completed in 2 days.

References

1. B. G. Galvan and T. K. Christopoulos, *Anal. Chem.*, 1996, **68**, 3545.
2. M. Verhaegen and T. K. Christopoulos, *Anal. Biochem.*, 2002, **306**, 314.
3. P. G. Zerefos, P. C. Ioannou and T. K. Christopoulos, *Anal. Chim. Acta*, 2006, **558**, 267.
4. M. Verhaegen and T. K. Christopoulos, *Anal. Chem.*, 1998, **70**, 4120.
5. E. Laios, P. C. Ioannou and T. K. Christopoulos, *Anal. Chem.*, 2001, **73**, 689.
6. M. N. Bobrow, T. D. Harris, K. J. Shaughnessy and G. J. Litt, *J. Immunol. Methods*, 1989, **125**, 279.
7. K. Glynou, P. C. Ioannou and T. K. Christopoulos, *Anal. Bioanal. Chem.*, 2004, **378**, 1748.
8. F. A. Ahmed, *Trends Biotechnol.*, 2002, **20**, 215.

9. E. Anklam, F. Gadani, P. Heinze, H. Pijnenburg and G. Van den Eede, *Eur. Food Res. Technol.*, 2002, **214**, 3.
10. J. W. Stave, *J. AOAC Int.*, 2002, **85**, 780.
11. K. Glynou, P. C. Ioannou and T. K. Christopoulos, *Bioconjugate Chem.*, 2003, **14**, 1024.
12. H. R. Permingeat, M. I. Reggiardo and R. H. Vallejos, *J. Agric. Food Chem.*, 2002, **50**, 4431.
13. G. Feriotto, M. Borgatti, C. Mischiati, N. Bianchi and R. Gambari, *J. Agric. Food Chem.*, 2002, **50**, 955.
14. E. Mariotti, M. Minunni and M. Mascini, *Anal. Chim. Acta*, 2002, **453**, 165.
15. A. K. Mavropoulou, T. Koraki, P. C. Ioannou and T. K. Christopoulos, *Anal. Chem.*, 2005, **77**, 4785.
16. E. Emmanouilidou, B. Tannous, P. C. Ioannou and T. K. Christopoulos, *Anal. Chim. Acta*, 2005, **531**, 193.
17. L. Doleman, L. Davies, L. Rowe, E. A. Moschou, S. Deo and S. Daunert, *Anal. Chem.*, 2007, **79**, 4149.
18. J. C. Lewis, J. J. Lopez-Moya and S. Daunert, *Bioconjugate Chem.*, 2000, **11**, 65.
19. S. C. Donhauser, R. Niessner and M. Seidel, *Anal. Sci.*, 2009, **25**, 669.
20. C. Ding, H. Zhong and S. Zhang, *Biosens. Bioelectron.*, 2008, **23**, 1314.
21. Y. Qi, B. Li and Z. Zhang, *Biosens. Bioelectron.*, 2009, **24**, 3581.
22. H. Li and Z. He, *Analyst*, 2009, **134**, 800.
23. M. Mirasoli, M. Guardigli, P. Simoni, S. Venturoli, S. Ambretti, M. Musiani and A. Roda, *Anal. Bioanal. Chem.*, 2009, **394**, 981.
24. Y. Shan, J. Xu and H. Chen, *Chem. Commun.*, 2009, **905**.
25. N. H. L. Chiu and T. K. Christopoulos, *Anal. Chem.*, 1996, **68**, 2304.
26. S. J. Gould and S. Subramani, *Anal. Biochem.*, 1988, **175**, 5.
27. D. W. Ow, J. R. De Wet, D. R. Helinski, S. H. Howell, K. V. Wood and M. Deluca, *Science*, 1986, **234**, 856.
28. L. J. Kricka, *Anal. Biochem.*, 1988, **175**, 14.
29. T. K. Christopoulos and N. H. L. Chiu, *Anal. Chem.*, 1995, **67**, 4290.
30. E. Laios, P. C. Ioannou and T. K. Christopoulos, *Clin. Biochem.*, 1998, **31**, 151.
31. S. R. White and T. K. Christopoulos, *Nucleic Acids Res.*, 1999, **27**, 19.
32. E. Laios, P. J. Obeid, P. C. Ioannou and T. K. Christopoulos, *Anal. Chem.*, 2000, **72**, 4022.
33. G. Gilliland, S. Perrin, K. Blanchard and H. F. Bunn, *Proc. Natl. Acad. Sci. USA*, 1990, **87**, 2725.
34. T. K. Christopoulos, in *Encyclopedia of Analytical Chemistry*, ed. R. A. Meyers, John Wiley & Sons, Ltd., Chichester, 2000, p. 5159.
35. J. Nurmi, H. Lilja and A. Ylikoski, *Luminescence*, 2000, **15**, 381.
36. J. Nurmi, T. Wikman, M. Karp and T. Lovgren, *Anal. Chem.*, 2002, **74**, 3525.
37. R. K. McCulloch, C. S. Choong and D. M. Hurley, *PCR Methods Appl.*, 1995, **4**, 219.

38. A. Hayward-Lester, P. J. Oefner and P. A. Doris, *BioTechniques*, 1996, **20**, 250.
39. M. Verhaegen and T. K. Christopoulos, *Anal. Chem.*, 1998, **70**, 4120.
40. E. Emmanouilidou, P. C. Ioannou, T. K. Christopoulos and K. Polizois, *Anal. Biochem.*, 2003, **313**, 97.
41. E. Emmanouilidou, P. C. Ioannou and T. K. Christopoulos, *Anal. Bioanal. Chem.*, 2004, **380**, 90.
42. A. Roda, M. Mirasoli, S. Venturoli, M. Cricca, F. Bonvicini, M. Baraldini, P. Pasini, M. Zerbini and M. Musiani, *Clin. Chem.*, 2002, **48**, 1654.
43. D. S. Elenis, P. C. Ioannou and T. K. Christopoulos, *Anal. Chem.*, 2007, **79**, 9433.
44. B. A. Tannous, M. Verhaegen, T. K. Christopoulos and A. Kourakli, *Anal. Biochem.*, 2003, **320**, 266.
45. U. Landegren, R. Kaiser, J. Sanders and L. Hood, *Science*, 1988, **241**, 1077.
46. D. K. Toubanaki, T. K. Christopoulos, P. C. Ioannou and C. S. Flordellis, *Anal. Biochem.*, 2009, **385**, 34.
47. P. G. Zerefos, P. C. Ioannou, J. Traeger-Synodinos, G. Dimissianos, E. Kanavakis and T. K. Christopoulos, *Hum. Mutat.*, 2006, **27**, 279.
48. K. Glynou, P. Kastanis, S. Boukouvala, V. Tsaoussis, P. C. Ioannou, T. K. Christopoulos, J. Traeger-Synodinos and E. Kanavakis, *Clin. Chem.*, 2007, **53**, 384.
49. J. Konstantou, P. C. Ioannou and T. K. Christopoulos, *Anal. Bioanal. Chem.*, 2007, **388**, 1747.
50. A. C. Iliadi, P. C. Ioannou, J. Traeger-Synodinos, E. Kanavakis and T. K. Christopoulos, *Anal. Biochem.*, 2008, **376**, 235.
51. D. S. Elenis, P. C. Ioannou and T. K. Christopoulos, *Analyst*, 2009, **134**, 725.
52. V. Tsiakalou, M. Petropoulou, P. C. Ioannou, T. K. Christopoulos, E. Kanavakis, N. I. Anagnostopoulos, I. Savvidou and J. Traeger-Synodinos, *Anal. Chem.*, 2009, **81**, 8596.
53. M. Verhaegen and T. K. Christopoulos, *Anal. Biochem.*, 2002, **306**, 314.
54. K. Glynou, P. C. Ioannou and T. K. Christopoulos, *Protein Expression Purif.*, 2003, **27**, 384.
55. K. Glynou, P. C. Ioannou and T. K. Christopoulos, *Bioconjugate Chem.*, 2003, **14**, 1024.
56. P. G. Zerefos, P. C. Ioannou and T. K. Christopoulos, *Anal. Chim. Acta*, 2006, **558**, 267.
57. B. J. Bryan and C. S. Szent-Gyorgyi, *U.S. Pat.* 6232107, May 2001.
58. M. Verhaegen and T. K. Christopoulos, *Anal. Chem.*, 2002, **74**, 4378.

CHAPTER 10
Biomolecular Interactions

ELISA MICHELINI, LUCA CEVENINI, LAURA
MEZZANOTTE, ANDREA COPPA AND ALDO RODA

Department of Pharmaceutical Sciences, University of Bologna, via
Belmeloro 6, 40126 Bologna, Italy

10.1 Introduction

The complexity of molecular interactions within living cells gives rise to the
need to measure different types of associations in a dynamic environment to
reveal the intricacies of highly orchestrated cellular events.

In the past few years, tools for the measurement of these interactions have
evolved greatly with the development of novel probes that facilitate in-depth
investigation of biomolecular interactions. Most of the strategies were first
developed *in vitro* and then adapted to imaging cellular events also in small
living animals, allowing the study of protein binding as well as their specific
inhibition *in vivo*. In particular, bioluminescent reporter genes have emerged as
powerful tools to develop such assays due to the minimal background signal of
bioluminescent proteins. These assays work not only as "on-off" sensors but
also as true quantitative bioanalytical tools. Moreover, the isolation of new BL
genes and their characterization has allowed us to exploit fully the potential of
such reporter for the development of innovative strategies to monitor protein
dynamics in living cells.

10.2 Strategies for Detecting Protein–Protein Interaction: an Overview

Several approaches have been developed to study protein–protein interactions
in living cells. Most strategies are based on fusion of the interacting molecules

Chemiluminescence and Bioluminescence: Past, Present and Future
Edited by Aldo Roda
© Royal Society of Chemistry 2011
Published by the Royal Society of Chemistry, www.rsc.org

to defined protein elements to reconstitute a biological or biochemical function. Besides the well-documented transcriptional strategies developed in recent years, several different techniques that use luminescent reporter proteins, such as fluorescence resonance energy transfer (FRET), bioluminescence resonance energy transfer (BRET), and protein fragment complementation, have been developed *in vitro* and *in vivo* in small animals. In particular, complementation strategies may provide a reversible and responsive system to detect transient events, which represent most protein–protein interactions.

While the detection of two protein interaction is now well documented, the demonstration of ternary complexes in cells is still a challenging task, and methods available for visualization and identification of ternary complexes in living cells are still limited and surely need further optimization.

More recently a non-transcriptional assay system based on molecular tension of a luciferase was described for the first time by Kim *et al.*[1] Based on the hypothesis that an artificially molecular tension to a full-length luciferase may affect the enzymatic activity through a modification of the active site, the authors developed a simple probe sensitive to estrogens in which a full-length *Renilla* luciferase 8 was sandwiched between the ligand binding domain of human estrogen receptor (ER-LBD) and the phosphorylation recognition domain of Src, a proto-oncogene tyrosine-protein kinase involved in the proteolysis pathway of estrogen receptors. This study provides new insights into the construction of a novel lineage of bioluminescent probes for estimating protein–protein interactions.

10.3   Transcriptional Strategies

Transcriptional strategies for imaging protein–protein interactions use reporter genes that encode for luminescent reporter proteins that can be detected after addition of specific substrates.

These approaches can detect transient or unstable interactions between proteins but the indirect readout of the reporter, due to the nuclear translocation of the released transcription factor and protein expression induction, results in a delay of signal visualization, thus limiting kinetic analysis.

10.3.1   Two-hybrid Systems

The yeast two-hybrid system has been widely used to detect protein interactions and most of the literature in this field refers to the use of this strategy. In a two-hybrid assay, two proteins of interest are expressed in yeast, one fused to the DNA-binding domain (BD) and the other fused to the activation domain (AD) of the transcription factor GAL4. When a protein interaction is induced, the fusion proteins bring together the two domains of the transcription factor; this event then induces reporter gene transcription.

Conventional two-hybrid systems are applicable to target protein interactions that are localized within the nucleus or that translocate into the nucleus because proximity to DNA and the transcriptional machinery is needed for reporter expression. This excludes membrane proteins and related partners. Furthermore, because the readout is based on transcriptional activation of a reporter gene, the measurement of these assays is inevitably delayed by several hours and do not allow us to evaluate protein dynamics in living cells.

Developed first in yeast, this classical two-hybrid system was later adapted for mammalian cells using different expression plasmids. Ray *et al.*[2] have developed an inducible mammalian two hybrid system by transfecting 293T cells in which the expression of both of the two fusion proteins (Gal4-ID and VP16-MyoD) is under the control of the promoter for NF-kB, a gene involved in apoptosis; by inducing NF-kB promoter with TNF-α, a pleiotropic cytokine secreted by macrophages, the system allows indirect monitoring of protein–protein interactions through VP16 mediated transactivation of a firefly luciferase reporter gene. The authors then performed imaging experiments in living mice implanted with transiently transfected cells to validate the ability to noninvasively and quantitatively image protein interactions *in vivo*.

The uses of the two-hybrid system, however, would not be to screen for potential protein interactions but, indeed, to take an existing protein–protein interaction and then study it in detail in the physiological cellular environment.

10.3.2   Split Ubiquitin

The split ubiquitin (Ub) system is another transcriptional strategy for imaging protein interaction that bypasses the nuclear localization or translocation requirement of protein complexes. In this strategy the ubiquitin, a small ubiquitous protein that binds target proteins and labels proteins for proteasomal degradation, is split into two fragments, one of which is fused to a transcription factor (TF) that remains tethered outside the nucleus under basal conditions; by taking advantage of ubiquitin-specific proteases (UBP) that recognize only intact Ub, upon protein interaction and cleavage by UBP, the TF transcolcates into the nucleus and induces reporter gene transcription. Nevertheless, split ubiquitin remains a transactivator reporting system and it does not allow real-time measurements of protein interactions.[3–6]

10.4   Resonance Energy Transfer (RET)

The Forster energy transfer is a non-radiative energy transfer that occurs between two luminescent molecules, a donor that transfers its energy to an acceptor, which then emits light. For an efficient energy transfer, donor and acceptor must be in close proximity (1–10 nm) and also be properly oriented. This strict dependence on the physical proximity of donor and acceptor is explained by Forster theory: "the RET process is due to the coupling between the dipoles of the excited donor and the acceptor." In fact the efficiency of RET

(E_{RET}) depends on the sixth power of the distance (r) between the two partners, Equation (10.1):

$$E = \frac{1}{\left[1 + (r/R_0)^6\right]} \tag{10.1}$$

where R_0 is the distance at which the RET efficiency is 50% and represents a characteristic parameter for a given donor–acceptor couple. In aqueous solution R_0 (in nm) is given by Equation (10.2):

$$R_0 = \left[2.8 \times 10^{17} \kappa^2 Q_D \varepsilon_A J(\lambda)\right]^{1/6} \tag{10.2}$$

where κ^2 is a parameter depending on the relative orientation of the donor and acceptor dipoles, Q_D is the donor emission quantum yield, ε_A is the maximal acceptor extinction coefficient (in $M^{-1} cm^{-1}$) and $J(\lambda)$ is the spectral overlap integral between the normalized donor fluorescence $F_D(\lambda)$ and the acceptor excitation spectra $E_A(\lambda)$.

The latter integral can be calculated as in Equation (10.3):

$$J = \int F_D(\lambda) E_A(\lambda) \lambda^4 d\lambda \tag{10.3}$$

All the parameters in Equations (10.2) and (10.3) depend on the photophysical properties of the donor and the acceptor, except the orientation factor (κ^2), whose value ranges from 4 (the donor and acceptor dipoles are aligned) to 0 (the dipoles are perpendicular to each other). In conclusion, for an efficient RET process the components of a RET couple must be matched to obtain a large R_0 value, which requires good spectral overlap and favorable dipole orientation, and the separation between the donor and the acceptor must be of the order of R_0 or less.[7]

Because most biological interactions such as receptor-ligand or protein–protein interactions occur in a distance of 1–10 nm, in theory the RET phenomenon can be used to monitor the activation state of any protein (*e.g.*, a receptor or a transcription factor) that undergoes association or conformational changes as a consequence of ligand binding.

10.4.1 FRET

The first application of RET process to the study of protein–protein interactions involved the use of fluorescence and has been named fluorescent resonance energy transfer (FRET), which occurs when both donor and acceptor are fluorescent molecules.[8–11]

The availability of several genes encoding for fluorescent (FL) proteins has allowed their use in living cells by fusing them with proteins of interest in different constructs for their expression. Fluorescent proteins, such as GFP and its variants, and fluorescent dyes have been widely investigated for their ability to act as donor or acceptor moieties. When they are fused or chemically bound to two protein of interest it is possible to evaluate the molecular association by measuring the fluorescent emission of acceptor. In fact, when the proteins interact, the donor and acceptor proteins are also in proximity and therefore the energy transfer can occur. This phenomenon can be observed *in vitro* using purified proteins or directly within living cells where the fusion proteins are expressed.[12] Although it has found wide application for the study of protein–protein interaction, the use of FRET presents serious limitations such as the requirement of an external light source, photobleaching of donor fluorophore, the simultaneous excitation of both the donor and the acceptor, and the autofluorescence that occurs within the cells or animal models. Most of these problems have been overcome by using another RET process, namely, bioluminescence resonance energy transfer (BRET).

10.4.2   BRET

Bioluminescence resonance energy transfer occurs when the donor molecule is a bioluminescent (BL) protein. This is a natural process found in some marine organisms such as the jellyfish *Aequorea victoria* and the sea pansy *Renilla reniformis*. These organisms are able to shift the color of their BL emissions to a longer wavelength; for example, in *R. reniformis* the donor luciferase emits blue light, and transfers energy to a GFP, which in turn emits green light.

The application of a BRET technique allows the monitoring of ligand-dependent interaction between two target proteins by genetically fusing them to a bioluminescent donor and a fluorescent acceptor; after addition of the luciferase substrate, two emissions are observed, one from the bioluminescent protein and another from the fluorescent acceptor. Since light emission from the donor takes place at a different wavelength to that of the acceptor, the energy transfer can be easily detected by measuring the ratio of the acceptor to the donor emission intensities. For example, if the maximum emission of the donor is at 490 nm and that of the acceptor is 530 nm, the BRET signal can be calculated by Equation (10.4):

$$\text{BRET signal} = [(\text{emission@530 nm}) - (\text{emission@490 nm}) \times C_f] / (\text{emission@490 nm})] \tag{10.4}$$

where C_f is a correction factor that is measured when only donor is present, under the same experimental conditions:

$$C_f = \text{emission@530 nm} / \text{emission@490 nm} \tag{10.5}$$

This ratiometric output allows us to compensate for well-to-well aspecific signal variations (*e.g.*, due to different cell numbers in each well or signal decay across the plate). Unlike FRET, BRET does not require a light source and,

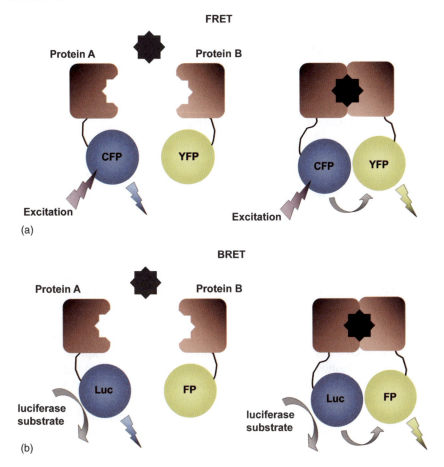

Figure 10.1 (a) Schematization of the FRET principle. One partner is fused to a fluorescent donor (*e.g.*, FP = CFP) and the other to a fluorescent acceptor (*e.g.*, YFP); after excitation of the donor, if interaction between the two partners occurs, light is emitted by the acceptor. (b) Schematization of the BRET principle. One partner is fused to a bioluminescent donor (*e.g.*, RLuc) and the other to a fluorescent acceptor (*e.g.*, EYFP); on addition of the substrate coelenterazine, BRET emission from EYFP is observed only when interaction between the two proteins occurs that brings Rluc and EYFP into close proximity.

therefore, there is an intrinsic low background in BRET assays, which should allow either detection of weak interactions or experiments to be performed with low concentrations of proteins (Figure 10.1).

The first application of BRET methodology goes back to 1999 and regarded the study of the dimerization of cyanobacteria circadian clock proteins.[13] BRET has been widely applied to the study of receptor dimerization, like the insulin receptor,[14] and to the evaluation of the homo- and hetero-dimerization of opioid receptors in live cells.[15] Estrogen receptor homodimerization has been studied with BRET and BRET[2], a BRET variant in which the energy transfer occurs between *Renilla* luciferase and a green fluorescent protein mutant (GFP[2]).[16,17] Extended BRET (eBRET) is also gaining popularity as a technique that allows the monitoring of protein–protein interactions in real time for many hours.[18]

10.4.3 Advances in BRET Technologies

Several BRET systems have also been used to create bioluminescent probes with emission at wavelengths that BL proteins alone cannot achieve. For example, semiconductor fluorescent nanocrystals or quantum dots (QDs) have been chemically bound to luciferase proteins to act as fluorescent acceptor in a BRET pair. The high separation of spectra between BL-donor and QD-acceptor makes these BRET pairs very amenable to study protein–protein interactions, especially *in vitro*. In particular, this technology has been widely exploited to monitor protease activity, to develop assays for metal ions detection, to label living cells and for live-imaging applications.[19] The main advantage of QDs-based BRET assays relies on the possibility of obtaining bioluminescence in the red region of the visible spectrum. This emission overcomes problems derived from strong attenuation of biological tissue to photons of wavelengths < 600 nm and can thus be applied for *in vivo* imaging. However, despite the high efficiency of emission of QDs and their tunable emission properties, their applications in *in vivo* imaging is limited by their large size, which can alter cell metabolism and influence QDs localization. This problem can be overcome by using organic dyes such as indocyanine and its derivatives as fluorophore acceptor for BRET systems; their reduced size allows their use as probe in small animals.

Recently, Wu *et al.* have conjugated a far-red fluorescent indocyanine derivative to *Cypridina* luciferase and linked this BRET system to an anti-human Dlk-1 monoclonal antibody to monitor localization of this tumor marker in animal models.[20]

The tetrameric red fluorescent protein (*DsRed*), cloned from *Discosoma* coral, has been mutated to create monomeric protein *mRFP1*, which can be fused to the proteins of interest to overcome the steric limitations of multimeric fluorescent proteins.[21] Moreover, *mRFP1* has been mutated to improve emission properties originating a class of fluorescent proteins assigned a wide range of fruit names, such as mPlum (649 nm), mRaspberry (625 nm), mCherry (610 nm), mStrawberry (596 nm), tdTomato (581 nm) and mOrange (565 nm), that have been successfully used as fluorescent acceptors.

Gammon *et al.* have used a fluorescent protein tdTomato as acceptor coupled with click beetle green luciferase (CBG) to obtain red bioluminescence. They developed a protease activity biosensor based on an intramolecular BRET system that leads to an emission peak at 580 nm. Careful characterization of this probe confirmed its potential application for high-throughput screening assays.[22]

Another intramolecular BRET assay with redshifted bioluminescence, defined BRET3, has been developed by De *et al.*, who used as BRET couple the red fluorescent protein mOrange and RLuc8. The fluorescent acceptor, coupled with donor RLuc8, and with the use of the substrate Ctz400, leads to an emission peak at 564 nm. This new BRET pair presents increased photon intensity and the best donor–acceptor spectral resolution (~ 85 nm) obtained to date. This optimal separation allows us to determine the luminescent signal also in a single cell, as demonstrated in HT1080 cells.[23]

Iglesias *et al.* have used a BRET-based biosensor for functional imaging of hypoxia in the cancer. By fusing fluorophore mCherry coding sequence to Rluc they obtained a new reporter protein that combines advantages of a luciferase system with a near-IR fluorescent emission.[24] They developed a hypoxia-sensing system based on a genetically encoded biosensor induced by the hypoxia transcription factor HIF-1α. In the presence of the genetically encoded biosensor, HIF-1α binds to the response element located in the regulatory module of the construction, inducing the transcription of the fusion protein firefly luciferase-NIR fluorophore. This dual tracer shows a proportional response to the quantity of HIF-1α.

To investigate the *in vivo* applicability of this system, two populations of cells, either with the activated system or 'basal state' cells (transfected only with the genetically encoded biosensor), were injected subcutaneously into SCID mice. The system remained active and its intensity was significantly higher in the cell population harboring the genetically encoded biosensor with the transcription factor HIF-1α than the cells displaying basal activation.

At present, several combinations of BRET formats and reagents are available for proteomics applications, including receptor research and mapping of signal transduction pathways.[25]

After recent improvements, the new focus in this field involves cell-based proteomics, which promise to give detailed information about the structure, organization and function of the complete set of human genes. BRET could be used in combination with high-throughput MALDI-MS and X-ray crystallography to study signaling events in living cells, without the need for expensive instrumentation.

Although BRET and FRET are particularly suitable for high-throughput screening and are therefore appealing to the pharmaceutical industry,[26] the techniques still show some drawbacks related to steric hindrance occurring between donor and acceptor and low efficiency of the energy transfer process. Steric hindrance of BRET proteins can hamper protein interaction. However, this problem may be overcome by using the recently isolated low-molecular weight (20 kDa) *Gaussia* luciferase.

10.4.4 SRET (Sequential BRET-FRET)

Recently, a new strategy that permits identification of heteromers formed by three different proteins, essential to decode the properties of molecular networks controlling intercellular communication, has been developed.[27]

This technique, called sequential BRET-FRET (SRET) is based on the combination BRET and FRET.

In SRET, the oxidation of a *Renilla* luciferase (Rluc) substrate by an Rluc fusion protein triggers acceptor excitation of a second fusion protein by BRET and subsequent FRET to a third fusion protein.

Moreover, by using different Rluc substrates it is possible to modulate the emission profile of the bioluminescent donor that could result in a shift of wavelength for SRET when paired to appropriate fluorescent acceptors. In particular, using DeepBlueC as a substrate, emission from Rluc (400 nm) allows energy transfer to a nearby GFP acceptor, which is detected (510 nm), and can result in a second energy transfer to YFP and concomitant emission of light at 530 nm (SRET^2). The use of coelenterazine h results in longer wavelength sequential BRET-FRET energy transfer (SRET^1), where the emission of Rluc (485 nm) allows energy transfer to YFP acceptor (530 nm) and a subsequent transfer to a DsRed with emission of light at 590 nm. SRET will only occur between these fusion proteins if the two partners, Rluc/GFP and GFP/YFP or Rluc/YFP and YFP/DsRed, are less than 10 nm apart.

The authors expressed a heterotrimeric G protein that consists of a G_α, G_β and G_γ subunit fused, respectively, to GFP, YFP and Rluc for SRET^2 or DsRed, YFP and Rluc for SRET^1. These results illustrate the usefulness of SRET to confirm that the stability of the $G_{\alpha\beta\gamma}$ complex is preserved for several heterotrimeric G proteins during the G-protein coupled receptor (GPCR) activation cycle. Both strategies were then applied to identify complexes of three different GPCRs (cannabinoid CB_1, dopamine D_2 and adenosine A_{2A} receptors), demonstrating for the first time that the $A_{2A}R$, D_2R and CB_1R can heterotrimerize in living cells by transfecting HEK-293T cells with plasmids encoding fusion proteins.

Indeed, the SRET technique, owing to its ability to identify direct interactions between three biomolecules in living cells, provides a valuable means of identifying the interaction of receptors within oligomeric complexes not only in horizontal molecular networks but also in any kind of protein–protein interaction.

10.5   Protein Fragment Complementation

The *in vivo* identification and characterization of protein–protein interactions (PPIs) are essential to understand cellular events in living organisms. The use of PCAs (protein complementation assays) are increasing, spanning different areas such as the study of biochemical networks, screening for protein inhibitors and determination of drug effects. The most exciting properties of PCAs are their ability to work *in vivo*, allowing the detection of even weak protein interactions in the endogenous environment and their applicability to high-throughput analysis of PPIs, thus becoming a very useful tool for proteomic and system biology studies.

The principle of the PCA strategy for detecting protein–protein interactions was first demonstrated by Pelletier *et al.* using the enzyme dihydrofolate reductase (DHFR), following the work of Johnsson and Varshavsky ("ubiquitin split protein sensor").[28]

Protein-fragment complementation assays (PCAs) are based in the fusion of the hypothetical binding partners to two rationally designed fragments of a reporter protein. The interaction between bait and prey proteins brings the split reporter fragments close enough to enable their non-covalent and specific reassembly followed by the recovery of the native structure and activity.

Different aspects have to be taken into account when designing a PCA experiment. A crucial requirement for the method is that the dissected fragments should not associate spontaneously in the absence of the binding proteins to avoid false positives. Thus, it is very important to perform appropriate controls in PCAs to ensure the specificity of the detected signal. For example, on studying the self-assembly of fluorescent proteins, it has been concluded that the fragments can self-associate with each other if they are expressed at high levels, regardless of the PPI, and so it is advisable to express the protein fusions at levels close to those of the endogenous counterparts.

Each protein requires specific breaking points that allow the non-covalent protein reconstitution while minimizing the spontaneous folding, and fusions constructs should be designed, sometimes with the introduction of a short amino-acidic linker, to minimize steric hindrance in the reconstitution of the active reporter protein. Moreover, the optimal dissection site may be different between inter- and intramolecular-format probes because the interactions between two separated proteins need a relatively strong affinity to recognize each other in the complex context of mammalian cells whereas two components inside a single fusion protein are adjacent.

Different types of proteins and enzymes have been used as reporter proteins, including DHFR, β-galactosidase, β-lactamase, fluorescent proteins and luciferases. Amongst the others, fluorescent or bioluminescent proteins are more suitable for studying specific protein interactions because they enable an easy imaging of binding *in vivo*.

10.5.1 Split Luciferase Complementation

One of the main advantages of the use of bioluminescent proteins for PCAs is their reversible reassembly due to the presence of structurally independent subdomains in the protein, allowing the study of protein interaction dynamics. The most commonly used Luciferases in optical imaging are *Renilla* luciferase, firefly luciferase and more recently *Gaussia* luciferase from a marine copepod *Gaussia princeps*.

A common feature of this type of enzymes is that they require a substrate to generate bioluminescence. This raises further considerations about the bioavailability of the BL substrate within the cells that may cause signal artifacts not related to the biological process. For instance, the substrate coelenterazine, used in *Renilla* and *Gaussia* luciferase, is unstable in air and is highly hydrophobic.

10.5.1.1   Renilla Luciferase Split Complementation

The split *Renilla* luciferase complementation method relies on the emission of luminescent signal upon protein–protein interaction induced complementation of the split *Renilla* luciferase. *Renilla* luciferase is a monomeric photoprotein with a molecular weight of 36 kDa and shares the conserved catalytic triad of residues employed by the dehalogenases, as confirmed by its crystallographic data.[29] The enzymatic reaction does not require ATP and its substrate coelenterazine is known to have a good penetration through cell membranes. The first investigation on *Renilla* luciferase split positions was carried out by Paulmurugan and Gambhir in 2003.[30] They validated complementation-based activation of split synthetic *Renilla* luciferase protein driven by the interaction of two strongly interacting proteins, MyoD and ID, in five different cell lines with transient transfection studies and identified the 687–688nt (amino acid 229) split site that produced relatively low background and high complementation-based signals. Next, Umezawa *et al.* used a split luciferase complementation method to analyze the protein interaction between the tyrosine-phosphorylated peptide (Y941) of IRS-1 (insulin receptor substrate 1) and the SH2 domain of PI3K (phosphoinositide 3-kinases) in living CHO cells overexpressing insulin receptor. When insulin receptor is activated, it phosphorylates IRS-1 that binds PI3K. They obtained good results by splitting *Renilla* luciferase between Ser91 and Tyr92.[31] In recent years split *Renilla* luciferase complementation methods have been applied for studying different protein–protein interaction. Of particular interest is the study of a ternary complex mediated by small molecules, like the complex composed of the human proteins FKBP12 (an immunosuppressant FK506 binding protein) and FRB (FKBP12 binding domain, the region of mammalian target of rapamycin mTor) and the immunosuppressant rapamycin.[32] With this approach it is also possible to follow cellular transport, like nuclear translocation, in cell cultures and *in vivo*. For example, the translocation of a particular protein (X) into the nucleus can be monitored by protein splicing of split-Rluc. RLuc-N (N-terminal portion of RLuc) connected with DnaE-N (N-terminal splicing domain of DnaE intein), which is predominantly localized in the nucleus, and DnaE-C (C-terminal splicing domain of DnaE) connected with RLuc-C (C-term of RLuc) and the protein X, which is localized in the cytosol. When the tandem fusion protein consisting of DnaE-C, Rluc-C and protein X translocates into the nucleus, the DnaE-C interacts with DnaE-N, reconstituting *Renilla* luciferase.[33] (Note: Inteins are protein segments implied in the protein splicing. They are able to excise themselves and rejoin the flanking regions, the exteins.)

In another work, a pair of genetically encoded indicators composed of cDNAs of glucocorticoid receptor (GR), split *Renilla* luciferase (RLuc) and a *Synechocystis* sp. DnaE intein has been developed. The GR fused with C-terminal halves of RLuc and DnaE is localized in the cytosol, whereas a fusion protein of N-terminal halves of RLuc and DnaE is localized in the nucleus. When GR translocation occur into the nucleus, the C-terminal RLuc meets the N-terminal one in the nucleus, and full-length RLuc is reconstituted by protein splicing with DnaE.[34]

These simple assays based on protein nuclear transports allow the selection of suitable drugs among candidates and has significant potential for risk assessments, such as carcinogenic chemical screening *in vitro*.

Another application is a new assay for monitoring protein–protein interaction in protoplasts. In this assay, the N- and C-terminal fragments of *Renilla reniformis* luciferase are translationally fused to bait and prey proteins, respectively. Split luciferase activity was measured by first transforming protoplasts with a DNA vector in a 96-well plate. DNA vector expressing both bait and prey genes was constructed. As proof of concept the authors detected the protein–protein interactions between the nuclear histones 2A and 2B, as well as between membrane proteins SYP (syntaxin of plant) 51 and SYP61, in *Arabidopsis* protoplasts.[35]

Regarding *in vivo* studies, *Renilla* luciferase is one of the most used photoproteins to date. In recent works *Renilla* luciferase split complementation methods based on the design of different strategies have been developed to study protein interaction in living subjects, too. A new intramolecular folding sensor encoding various hER–LBD fusion proteins that could lead to split *Renilla*-firefly luciferase reporter complementation in the presence of the appropriate ligands has been recently constructed and validated by Paulmurugan. In this system it is possible distinguish binding of agonists, SERMs (selective ER modulator), and pure antiestrogens. Dose–response curves for tamoxifen and 4-OHT antagonists were obtained with a dynamic range of 0.25–2 μM. Cells stably expressing the intramolecular folding sensors with wild-type and mutant hER–LBD were used for imaging ligand-induced intramolecular folding in living mice. This is the first hER–LBD intramolecular folding sensor suited for high-throughput quantitative analysis of interactions of hER with hormones and drugs using cell lysates, intact cells and molecular imaging of small living subjects.[36] Recently, interactions between the two most important components of the heat shock protein 90 (Hsp90) chaperone system, and the co-chaperone p23, using split RL reporters have been studied by bioluminescence imaging of intact cells in cell culture and living mice. Interaction with two different isoforms of Hsp90 chaperones, α and β, and the efficacy of different classes of Hsp90 inhibitors in living subjects have been studied using split *Renilla* complementation method, which should help in accelerating development of potent and isoforms of selective Hsp90 inhibitors.[37]

10.5.1.2 Firefly Luciferase Split Complementation

According to crystal structure data, firefly luciferase is composed of two distinct domains, a large N-terminal and small C-terminal domain, that encompass a large cleft within the active site for the binding of the substrate D-luciferin and cofactors, required for the bioluminescent reaction.

Paulmurugan *et al.*[38] have described the construction of a split firefly luciferase strategy by splitting Fluc between amino acid positions 437 and 438, obtaining N-FLuc(1–437) and C-FLuc (438–554) fragments, as previously demonstrated by Ozawa *et al.*[39] in a reconstitution strategy with inteins (DnaE).

This strategy was used to monitor the interaction of MyoD and ID proteins in mammalian cells and also in living mice implanted with transfected cells, confirming the possibility of using a bioluminescent reporter as a protein fragment complementation pair for the rapid imaging of protein interaction.

The authors studied other different sites by fusing N-Fluc and C-Fluc fragments with FRB- and FKBP12 interacting protein. Upon exposure of transfected cells to rapamycin only four fragments showed significant induction over untreated cells and the best non-overlapping complementation fragments were the previously described residues 1–437 and 438–550.[40]

In addition, Paulmurugan *et al.* have identified recently several combinations of firefly luciferase enzyme fragments that can self-complement without assistance of protein interactions. In particular, the fragment Nfluc (1–475)/Cfluc (265–550) can reconstitute 3–4% of the activity of intact luciferase. This self-associating split luciferase, however, has limited utility in the measurement of protein–protein interactions but, as proved by the authors with nuclear localization signals within fusion construct, could potentially be used to explore protein compartmentalization. Paulmurugan *et al.* further extended the research for optimal sites for fragment complementation by applying a combinatorial screening approach to a library of N- and C-terminal firefly luciferase fragments with the interacting proteins FRB/FKBP12.[41] The combinations with relatively high level of rapamycin mediated a protein–protein interaction associated bioluminescence signal with an extremely low background signal, and were further characterized by using different interacting protein partners and an intramolecular folding strategy to evaluate the optimal orientation of NFluc- and CFluc- reporter fragments and the interacting partners required for efficient protein–protein interaction mediated complementation. In particular, the combination of fragments Nfluc 398/Cfluc 394 (800-fold induction with an absolute signal of 1.7×10^8 RLU per µg of protein per min) was shown to be markedly superior, with a lower self-complementation signal and equal or higher post interaction absolute signal both in cells and in small living animals.

All published work about firefly luciferase complementation uses the wild-type enzyme characterized by the maximum emission spectra at 560 nm in the green-yellow region that is absorbed and scattered by tissue. Because the emission of this luciferase is modifiable from green to red ($\lambda = 615$ nm) by changing amino acid side chains within the active site and other areas of the protein, as described by Branchini *et al.*, further research on the use of these mutants for PCA imaging in small living animals will undoubtedly improve the signal output.[42,43]

10.5.1.3   Click Beetle Luciferase Split Complementation

Click beetle luciferase (CBluc) has different characteristics that make it suitable for multiple bioanalytical applications. It belongs to a class of pH insensitive luciferases, it is thermostable and produces stable light in different physiological conditions using the substrate D-luciferin. The luciferase from *Pyrophorus plagiophthalamus* has been engineered to produce mutants that are characterized by a maximum of emission at different wavelengths ranging from green to red (named CBred and CBgreen) and to have a sequence codon optimized for mammalian expression. These proteins can be split into two portions to obtain functionally inactive fragments. Although crystallographic data of CBluc are not available yet it can be assumed that its structure is similar to that of Fluc because of their sequence and functional homology. In fact they are both members of a superfamily of acyl-adenylate forming enzymes and there is a strong correspondence between the two hydrophobicity diagrams of the amino acids based on the scale of Kyte and Doolittle. In analogy with the split strategy applied for Fluc the dissection points were chosen inside or near the flexible loop region between 380 and 480 aa in recent work of Kim *et al.*[44] The authors used ten different dissection points and obtained good results in terms of signal-to-background ratio, splitting CBluc (the CBred mutant) into two parts, one containing the amino acids 1–439 and the other containing 440–542.They also tested two overlapping pairs of fragments in analogy with the studies performed for Fluc. They constructed a single-molecule format bioluminescence probe to evaluate the protein interaction between the ligand binding domain of androgen receptor (AR LBD) and a peptide sequence such as a conserved LXXLL motif of co-activators. The ARLBD is fused *via* a flexible linker to the conserved motif and sandwiched between the dissected fragments of CBluc (Figure 10.1). They tested the functionality of the probe in different cell lines and evaluated the dose–response curve of various ligands. The possibility of combining different fragments from the different mutants of CBluc has also been explored in subsequent work of Kim[45] and colleagues. In this case an integrated-molecule-format multicolor probe was created for simultaneous determination of multiple effects of a ligand. The probe contains the N-terminal fragment of CBred followed by the SH2 domain of SRc protein, the N-terminal fragment of CBgreen followed by a LXXLL motif and the ER LBD (estrogen receptor ligand binding domain); at the end of the probe the C-terminal fragment of CBred was inserted (Figure 10.2).

The addition of antagonists will cause the phosphorylation of ERLBD and the interaction with the SH2 domain of SRc, thus reconstituting the active CBred; meanwhile the addition of agonists will cause the interaction of ERLBD with the LXXLL motif and the reconstitution of an active CBgreen with consequent production of green light.

For optimal complementation of the split CB fragments the steric hindrance among the components during the intramolecular complementation and a special mismatch between the fragments has to be considered.

10.5.1.4 Gaussia Luciferase Split Complementation

The luciferase from the marine copepod *Gaussia princeps* is a monomeric protein of 185 amino acids (20 kDa) that does not require any cofactors to be active.[46] The humanized form of *G. princeps* luciferase (hGLuc) can generate

Protein fragment complementation

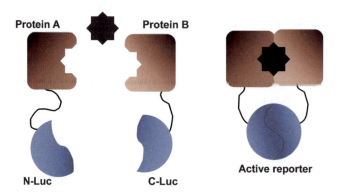

Figure 10.2   Split luciferase complementation principle. Proteins of interest are fused
to N-terminal and C-terminal domains of luciferase; after interaction of
the proteins, split luciferase becomes activated and emits luminescence.

over 100-fold higher bioluminescent signal than humanized forms of *Photinus
pyralis* (firefly; hFLuc) and *Renilla reniformis* (hRLuc) luciferases in cell lysates.
The small size of GLuc minimizes steric hindrance to the adjacent target proteins, making it an ideal candidate for the development of PCAs.

The molecular mechanisms of protein complementation between GLuc
fragments were demonstrated previously.[47] The authors generated a library of
fragments by PCR from the gene encoding hGLuc and they were fused to
GCN4 leucine zipper–coding sequences and co-expressed in HEK293 cells.
Fusions expressed alone did not result in detectable luminescence, indicating
that the PCA fragments alone have no activity while leucine zipper-induced
complementation of hGLuc fragments resulted in reconstitution of hGluc
activity, revealing that the constructs separated between Gly93 and Glu94, in
an unstructured region of the protein, gave the highest activity and were
selected for further hGLuc PCA. In particular, HEK293 cells expressing FKBP
and FRB fused to the hGLuc PCA fragments were treated with rapamycin
(inductor), FK506 (inhibitor) and cyclosporin A (negative control), with the
finding that the folding of hGLuc from fragments is completely reversible,
allowing drug induction and inhibition studies, and kinetic studies of protein
complex assembly and disassembly with the hGLuc PCA.

Kim *et al.* have used the complementation of split *Gaussia* luciferase to
develop a generally applicable bioluminescence template to visualize protein
dynamics related to cell signaling in living mammalian cells.[48] The authors
found an optimal dissection site for GLuc at Q105 within an hydrophilic region
(85–106 aa) of GLuc. The general applicability of this intramolecular luminescent template was proved by exploring ligand-activated dynamics of CaM,
ligand-binding domains (LBDs) of estrogen receptor (ER LBD; 305-550 AA),
androgen receptor (AR LBD; 672–910 AA) and glucocorticoid receptor (GR
LBD; 527–777 AA) in COS-7 cells. The use of *Gaussia* luciferase allows the

development of the smallest bioluminescent probe (total molecular weight is 41 kDa) for specific protein interaction monitoring in both cytosolic and nuclear compartments.

10.5.2 Split Luciferase Intein-mediated Protein Complementation

Intein-mediated split-luciferase assays were developed to overcome the limitations of two-hybrid and FRET systems to detect protein interactions occurring in the nucleus or the need for the partners to be in exacting close proximity.

Intein-mediated protein complementation relies on post-translational protein splicing reactions that lead to a precise excision of an intein (internal protein segment, DnaE) followed by ligation of flanking external proteins. The intein peptide is split into N- and C-terminal halves and fused in frame to each half of a reporter gene, which are in turn fused in frame to protein partners of interest; when the two interacting proteins come together, the intein is reconstituted and spliced out, leading to reconstruction of an active reporter protein (Figure 10.3).

Ozawa *et al.* 2000[39] have designed an intein-mediated split-firefly luciferase reporter system in which two interacting proteins were fused respectively to a fusion protein consisting of the N-terminal end fragment of DnaE and an N-terminal fragment of firefly luciferase (N-FLuc) and the other to a fusion

Split intein: protein reconstitution

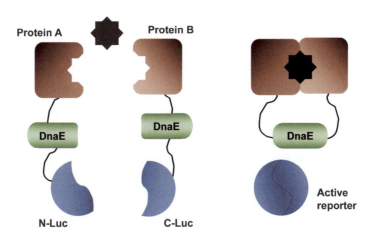

Figure 10.3 The intein peptide is split into N- and C-terminal halves and fused in frame to each half of a reporter gene; the two halves are in turn fused in frame to protein partners of interest. When the two interacting proteins come together, the intein is reconstituted and spliced out, leading to reconstruction of an active reporter protein.

protein consisting of the C-terminal fragment of DnaE and a C-terminal fragment of firefly luciferase (C-FLuc). Following interaction between the two proteins of interest, the formation of an intact DnaE leads to protein splicing and formation of a mature firefly luciferase that can be detected after addition of substrate D-luciferin. In particular, Fluc luciferase was split between amino acid positions 437 and 438 and used with inteins (DnaE) in a reconstitution strategy to detect the insulin-induced interaction of phosphorylated IRS 1 and SH2 domain of PI3K in a cell culture assay.

The potential of the reconstitution strategy to imaging cellular processes repetitively and non-invasively in living animals, by using a cooled charge-coupled device camera, was first explored by Gambhir *et al.*[38] The authors used the intein-mediated reconstitution of split firefly luciferase proteins (Fluc) driven by the interaction of two strongly interacting proteins, MyoD and Id, induced with TNF-α with cell implants (COS-1, 293T and N_{2A} cells) in living mice, demonstrating the ability to non-invasively image protein–protein interactions.

A limitation of methods based on split-reporter reconstitution is that the spontaneous interaction between DnaEs gives unwanted background signals and so appropriate control experiments, including measurements of the fluorescence or bioluminescence signals in the absence of a pair of interacting proteins and comparison of expression levels of the reporter proteins by Western blot analysis, should be performed. The major drawbacks of this technique are that the readout is an irreversible reaction and, due to the mechanism of splicing, there is a delay in the ability to reveal the interaction, making these strategies unsuitable for quantitative interrogation of reversible biochemical reactions as well as for the detection of protein–protein interactions in living cells or animals in real time.

10.6 Detection of Interactions involving more than Two Proteins: Combining SPLIT and RET Technologies

BRET, FRET and PCA, individually, are useful strategies for showing that two proteins are associated in living cells, but when these techniques are combined they can be used to demonstrate the interaction of three or more proteins in the same complex in living cells. In fact, as described before, either a bioluminescent or a fluorescent protein can be split into two fragments and fused to protein of interest that, if interacting, could reconstitute the functional donor or acceptor moiety for subsequent BRET or FRET.

As this strategy is relatively new, reports employing these techniques are few.

Most reports are based on bimolecular fluorescence complementation (BiFC) in which the reconstituted fluorescent proteins are the two moieties of a FRET couple.

Hu *et al.* have applied the BiFC-FRET to identify a ternary complex formed between Fos–Jun heterodimers and the NF-kB subunit, p65, and by quantifying the FRET signal it was also been possible to reveal the subnuclear localization of such ternary complexes. This finding reveals a cross-talk between AP-1 and NF-kB, which are two important families of transcription factors involved in cell proliferation, differentiation, immune and stress response.[49]

Using a combination of BiFC and BRET, Rebois *et al.* have demonstrated that the β_2AR, the G protein γ_2 subunit and adenylyl cyclize effector form a signaling complex, and combining GFP- and luciferase-based PCA and BRET they demonstrated the presence of four proteins in a single G-protein signaling complex.[50] This approach could be used to identify a large number of other proteins that are likely to be involved in the formation, trafficking, regulation and maintenance of G-protein mediated signal transduction pathways.

This strategy obviously suffers from the disadvantages of both PCA and RET techniques but, together with SRET, it allows the detection of more than two protein interactions.

10.7 Conclusion

The detection of protein interactions is still an ongoing task, and even more challenging is the real-time detection of protein interactions in living animals. The latter could provide useful information about the molecular basis of pathophysiological events, as well as the molecular response to therapeutic agents, thus representing a valuable tool also in the pharmaceutical field for developing new pre-clinical models.

There are few reports on BRET imaging in living mice, probably due to the intrinsic limitations caused by light scattering by tissues and absorption by hemoglobin, which can significantly affect short-wavelength (<600 nm) emission. In fact, typical BRET donor and acceptor peak emissions are less than 600 nm, thus complicating BRET imaging, especially in deep tissues.

The use of quantum dots, characterized by high quantum yields, large Stokes shifts and long wavelength emission, as BRET acceptors can partially overcome this problem. In addition, QDs have broad excitation spectra and size-tunable emission; these features can be exploited by pairing the same BL donor to different QDs for multiplexed BRET imaging. The only limitation is their incompatibility with genetic fusions, but several chemical methods are available for their conjugation to target proteins and for their delivery into cells.

Protein fragment complementation assays based on firefly or *Gaussia* luciferase, thanks to their great sensitivity due to minimal background association when co-expressed as individual proteins in the same cell, seem to be more suitable tools to move between cell-based and animal models.

Luciferase complementation imaging of protein interactions in cells and small animal models has been recently proposed to allow the rapid and repetitive measurement of target proteins interaction; nevertheless, many points remain to be clarified and new ways of understanding protein interactions will certainly emerge.

References

1. S. B. Kim, M. Sato and H. Tao, *Bioconjugate Chem.*, 2009, **20**(12), 2324.
2. P. Ray, H. Pimenta, R. Paulmurugan, F. Berger, M. E. Phelps, M. Iyer and S. S. Gambhir, *Proc. Natl. Acad. Sci.*, 2002, **99**(5), 3105.
3. I. Stagljar, C. Korostensky, N. Johnsson and S. te Heesen, *Proc. Natl. Acad. Sci. USA*, 1998, **95**, 5187.
4. N. Johnsson and A. Varshavsky, *Proc. Natl. Acad. Sci. USA*, 1994, **91**, 10340.
5. S. Thaminy, J. Miller and I. Stagljar, *Methods Mol. Biol.*, 2004, **261**, 297.
6. M. Fetchko and I. Stagljar, *Methods*, 2004, **32**, 349.
7. A. Roda, M. Guardigli, E. Michelini and M. Mirasoli, *Anal. Bioanal. Chem.*, 2009, **393**(1), 109.
8. P. Wu and L. Brand, *Anal. Biochem.*, 1994, **218**(1), 1.
9. A. B. Cubitt, R. Heim, S. R. Adams, A. E. Boyd, L. A. Gross and R. Y. Tsien, *Trends Biochem. Sci.*, 1995, **20**, 448.
10. R. Hovius, P. Vallotton, T. Wohland and H. Vogel, *Trends Pharmacol. Sci.*, 2000, **21**, 266.
11. D. A. Zacharias, G. S. Baird and R. Y. Tsien, *Curr Opin. Neurobiol.*, 2000, **10**, 416.
12. K. D. Pfleger and K. A. Eidne, *Nat. Methods*, 2006, **3**(3), 165.
13. Y. Xu, D. W. Piston and C. H. Johnson, *Proc. Natl. Acad. Sci. USA*, 1999, **96**, 151.
14. T. Issad, N. Boute and K. A. Pernet, *Biochem. Pharmacol.*, 2002, **64**, 813.
15. D. Wang, X. Sun, L. M. Bohn and W. Sadee, *Mol. Pharmacol.*, 2005, **67**, 2173.
16. E. Michelini, M. Mirasoli, M. Karp, M. Virta and A. Roda, *Anal. Chem.*, 2004, **76**, 7069.
17. K. L. Koterba and B. G. Rowan, *Nucl. Recept Signal*, 2006, **4**, 21.
18. K. D. Pfleger, J. R. Dromey, M. B. Dalrymple, E. M. Lim, W. G. Thomas and K. A. Eidne, *Cell Signal*, 2006, **18**, 1664.
19. U. Xia and J. Rao, *Curr. Opin. Biotechnol.*, 2009, **20**(1), 37.
20. H. Wu, K. Mino, H. Akimoto, M. Kawabata, K. Nakamura, M. Ozaki and Y. Ohmiya, *Proc. Natl. Acad. Sci. USA*, 2009, **106**(37), 15599.
21. R. E. Campbell, O. Tour, A. E. Palmer, P. A. Steinbach, G. S. Baird, D. A. Zacharias and R. Y. Tsien, *Proc. Natl. Acad. Sci. USA*, 2002, **99**(12), 7877.
22. S. T. Gammon, V. M. Villalobos, M. Roshal, M. Samrakandi and D. Piwnica-Worms, *Biotechnol. Prog.*, 2009, **25**(2), 559.
23. A. De, P. Ray, A. M. Loening and S. S. Gambhir, *FASEB J.*, 2009, **23**, 2702.
24. P. Iglesias and J. A. Costoya, *Biosens. Bioelectron.*, 2009, **24**, 3126.
25. L. I. Jiang, J. Collins, R. Davis, K. M. Lin, D. DeCamp, T. Roach, R. Hsueh, R. A. Rebres, E. M. Ross, R. Taussig, I. Fraser and P. C. Sternweis, *J. Biol. Chem.*, 2007, **282**, 10576.
26. K. J. Moore, S. Turconi, A. Miles-Williams, H. Djaballah, P. Hurskainen, J. Harrop, K. J. Murray and A. J. Pope, *J. Biomol. Screen*, 1999, **4**, 205.

27. P. Carriba, G. Navarro, F. Ciruela, S. Ferré, V. Casadó, L. Agnati, A. Cortés, J. Mallol, K. Fuxe, E. I. Canela, C. Lluís and R. Franco, *Nat. Methods*, 2008, **5**(8), 727.

28. B. Wang, J. Pelletier, M. J. Massaad, A. Herscovics and G. C. Shore, *Mol. Cell. Biol.*, 2004, **24**(7), 2767.

29. M. Loening, T. D. Fenn and S. S. Gambhir, *J. Mol. Biol.*, 2007, **374**(4), 1017.

30. R. Paulmurugan and S. S. Gambhir, *Anal. Chem.*, 2003, **75**(7), 1584.

31. A. Kaihara, Y. Kawai, M. Sato, T. Ozawa and Y. Umezawa, *Anal. Chem.*, 2003, **75**(16), 4176.

32. R. Paulmurugan and S. S. Gambhir, *Cancer Res*, 2005, **65**(16), 7413.

33. S. B. Kim, T. Ozawa, S. Watanabe and Y. Umezawa, *Proc. Natl. Acad. Sci. USA*, 2004, **101**(32), 11542.

34. S. B. Kim, T. Ozawa and Y. Umezawa, *Anal. Biochem.*, 2005, **347**(2), 213.

35. Y. Fujikawa and N. Kato, *Plant J.*, 2007, **52**(1), 185.

36. R. Paulmurugan and S. S. Gambhir, *Proc. Natl. Acad. Sci. USA*, 2006, **103**(43), 15883.

37. T. Chan, R. Paulmurugan, O. S. Gheysens, J. Kim, G. Chiosis and S. S. Gambhir, *Cancer Res.*, 2008, **68**(1), 216.

38. R. Paulmurugan, Y. Umezawa and S. S. Gambhir, *Proc. Natl. Acad. Sci. USA*, 2002, **99**, 15608.

39. T. Ozawa, A. Kaihara, M. Sato, K. Tachihara and Y. Umezawa, *Anal. Chem.*, 2001, **73**(11), 2516.

40. R. Paulmurugan and S. S. Gambhir, *Anal. Chem.*, 2005, **77**, 1295.

41. R. Paulmurugan and S. S. Gambhir, *Anal. Chem.*, 2007, **79**, 2346.

42. B. R. Branchini, D. M. Ablamsky, A. L. Davis, T. L. Southworth, B. Butler, F. Fan, A. P. Jathoul and M. A. Pule, *Anal. Biochem.*, 2010, **396**(2), 290.

43. H. Caysa, R. Jacob, N. Müther, B. Branchini, M. Messerle and A. Söling, *Photochem. Photobiol. Sci.*, 2009, **8**(1), 52.

44. S. B. Kim, Y. Otani, Y. Umezawa and H. Tao, *Anal. Chem.*, 2007, **79**(13), 4820.

45. S. B. Kim, Y. Umezawa, K. A. Kanno and H. Tao, *ACS Chem. Biol.*, 2008, **3**(6), 359.

46. B. A. Tannous, D. E. Kim, J. L. Fernandez, R. Weissleder and X. O. Breakefield, *Mol. Ther.*, 2005, **11**(3), 435.

47. I. Remy and S. W. Michnick, *Nat. Methods*, 2006, **3**(12), 977.

48. S. B. Kim, M. Sato and H. Tao, *Anal. Chem.*, 2009, **81**(1), 67.

49. Y. J. Shyu, C. D. Suarez and C. D. Hu, *Proc. Natl. Acad. Sci. USA*, 2008, **105**(1), 151.

50. T. E. Hébert, C. Galés and R. V. Rebois, *Cell. Biochem. Biophys.*, 2006, **45**(1), 85.

CHAPTER 11

Ultrasensitive Bioanalytical Imaging

MARA MIRASOLI,[a, b] SIMONA VENTUROLI,[c] MASSIMO GUARDIGLI,[a] LUISA STELLA DOLCI,[a] PATRIZIA SIMONI,[d] MONICA MUSIANI[c] AND ALDO RODA[a, b]

[a] Department of Pharmaceutical Sciences, University of Bologna, via Belmeloro 6, 40126 Bologna, Italy; [b] National Institute of Biostructure and Biosystems, N.I.B.B., Interuniversity Consortium, Rome, Italy; [c] Division of Microbiology, University of Bologna, via Massarenti 9, 40138 Bologna, Italy; [d] Department of Clinical Medicine, University of Bologna, via Massarenti 9, Bologna 40138, Italy

11.1 Introduction

The availability of low-light imaging devices based on high-sensitivity and high-resolution video cameras, such as intensified, or cooled, charge-coupled devices (CCD) or complementary metal oxide semiconductor (CMOS) image sensors, has allowed the development of bio- and chemiluminescence imaging methods, which rely not only on the detection of light emission down to the single-photon level, but also on the localization of the signal on the sample surface with excellent spatial resolution.

The possibility of performing signal localization has been exploited to assess the spatial distribution of a target analyte or biochemical process in macro- and microsamples, as well as to set up ultrasensitive bioanalytical methods that can simultaneously detect one analyte in several samples, and/or different analytes in each sample.

The self-illuminating feature of bio- and chemiluminescence reactions allows for analytically relevant emissions to be measured against a completely dark background. While absorption and fluorescence methods suffer from warm-up

Chemiluminescence and Bioluminescence: Past, Present and Future
Edited by Aldo Roda
© Royal Society of Chemistry 2011
Published by the Royal Society of Chemistry, www.rsc.org

and drift of light source, as well as interference from light scattering and from colored or fluorescent components present in the sample matrix, these effects are absent in bio- and chemiluminescence. Such techniques have already proved to be ultrasensitive, providing good spatial resolution, wide dynamic range and easy quantitative evaluation of the signal.[1]

Bio- and chemiluminescence measurements are characterized by high selectivity and, due to the absence of an excitation source, they require relatively simple and inexpensive instrumentation that is amenable to miniaturization. A wide number of applications have been described, for both macro- and microsamples in various fields, such as drug development, diagnostic applications, agrofood and environmental analysis and cultural heritage.[2–7]

Electrogenerated chemiluminescence (ECL) is a controllable form of chemiluminescence where light emission is initiated by an electron-transfer reaction occurring at an electrode surface. The most common system used for analytical purposes consists of the luminophore label $Ru(bpy)_3^{2+}$, or one of its derivatives, with tri-*n*-propylamine as a co-reactant. Electrogenerated chemiluminescence has been widely employed in clinical diagnostics applications, thanks to its high sensitivity and the possibility of turning the emission on and off by controlling the electrode potential. Recently, technological solutions to apply ECL in imaging applications are being envisaged.

Whole-body imaging in live animals represents a rapidly growing field of application of bio- and chemiluminescence imaging, which is making inroads into monitoring biological processes with clinical, diagnostic and drug-discovery applications. This subject is reported extensively in Chapter 12 and will not be treated here.

11.2 Instrumentation

Low-light imaging devices, such as high-sensitivity CCDs, are currently employed for bio- and chemiluminescence imaging in macro- and microsamples. Non-intensified cooled slow-scan CCDs, in which relatively long signal integration times (seconds to minutes) are employed to increase signal-to-noise ratio, are particularly suited for ultrasensitive and quantitative detection of steady-state weak bio- and chemiluminescence emissions. Intensified CCD detectors (either based on electron multiplying CCD detectors – EMCCD – or intensifier technology – ICCD) are, on the other hand, particularly suited for the real-time visualization of fast luminescent reactions and processes.

CMOS image sensors are expected to compete in the near future with CCDs, especially in the development of miniaturized integrated systems, thanks to their low power consumption, low voltage and possibility of integration on the same chip of all the functions required for image acquisition and processing.

Imaging devices and their technical characteristics and performance are described in Chapter 4.

Luminographs consist of an ultrasensitive video camera and an optical system enclosed in a light-tight box to prevent interference from ambient light.

The sample under investigation is placed in the luminograph and the pattern of light emission from its surface is recorded and converted into a digital image. The image is usually further processed employing suitable software, *e.g.*, by performing background subtraction and contrast enhancement procedures, to enhance the features of the bio-/chemiluminescence signal. The grayscale image can be converted into pseudo-colors to emphasize the differences in signal intensity, and then overlapped to the image acquired in transmitted light (live image) to obtain accurate analyte localization on the sample surface. To obtain quantitative information, the luminescence image can be analyzed by selecting the sample areas of interest and measuring the total number of photons emitted from within those areas as a function of time. Luminescence emission is usually expressed in photons per second and surface area or in relative light units (RLU; an arbitrary unit).

Resolution of luminescence imaging by employing standard or customs optics ranges from 100 to 200 µm for macrosamples analysis to a micrometer or sub-micrometer level when imaging measurements are performed by coupling the imaging detector to an optical microscope, thus enabling analysis at cellular and subcellular level.[2]

Recently, the contact imaging approach, characterized by the absence of optical elements (*e.g.*, lenses) between the sample and the imaging sensor, has been proposed for the development of low-cost imaging devices, suitable for implementation in miniaturized systems.[8] Because of the very short distance between sample and sensor surface, the light collection efficiency (η), which is calculated as the ratio between the light collection solid angle and the full solid angle, can be as high as 50%. In addition, loss of light that usually occurs in the optical system is avoided. On the other hand, the absence of optics leads to higher cross-talk and lower resolution, which can be improved for microscopic samples by reducing pixels size. In some cases, to gain accessibility to the surface of the sensor, a fiber optic face plate can be positioned between sample and sensor surfaces without loss of resolution, since the ordered arrangement of the fibers allows coherent transmission of the image from one face to the other.[9] In addition, the use of tapered face plates allows us to enlarge the useful analytical area of the sample up to the few cm^2 of the sensor surface.

11.3   Bioluminescent and Chemiluminescent Reagents for "*In Vitro*" Imaging Applications

In vitro chemiluminescence imaging often involves the use of enzyme labels, such as alkaline phosphatase (AP) or horseradish peroxidase (HRP), and suitable chemiluminescent substrates that allow for their detection with very high efficiency. The use of enzymes is generally preferred to that of organic chemiluminescence labels, due to the gain in sensitivity obtained thanks to enzyme amplification. Several substrates that produce steady-state light emission are commercially available and currently employed. In the presence of an excess substrate, the light emission is proportional to the enzyme amount, thus

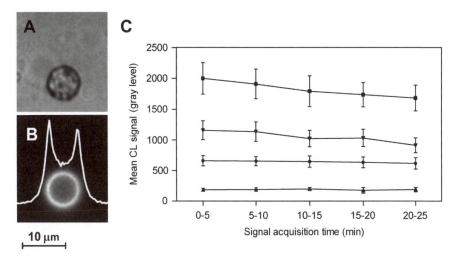

Figure 11.1 Microscope imaging (60×objective magnification) of HRP-coated microspheres (diameter 8 μm) using the Westar Supernova HRP CL substrate. (a, b) Live and chemiluminescence images of a microsphere, respectively. Panel (b) also shows the chemiluminescent signal profile obtained from the chemiluminescence image of the microsphere. (c) Chemiluminescent signal kinetic profiles measured for populations of microspheres coated by employing different amounts of HRP (data represent the mean chemiluminescent signal of the population of microspheres ± standard deviation). A glow-type kinetics is obtained for all HRP amounts. (Reproduced with permission from ref. 10.)

allowing for the quantification of the analyte, provided that a suitable calibration curve is produced.

Chemiluminescence systems suitable for imaging applications must be characterized by the production of short-lived reactions intermediates, which ensures accurate analyte localization, especially in microscope imaging applications (Figure 11.1).

Glow-type kinetics are preferred over flash-type kinetics, since they allow easy standardization of signal detection and integration of signal emission for a longer time to increase the signal/noise ratio. Furthermore, chemiluminescence substrates characterized by low background emission are required to achieve a high signal-to-noise ratio and thus improve the image quality and allow sensitive and specific quantification of the analyte.[11]

Bioluminescence imaging takes advantage of the high quantum yield of bioluminescence reactions and of the possibility of exploiting genetic engineering techniques to obtain functional labels that can be expressed in living cells and targeted to specific subcellular compartments. Bioluminescent proteins are most often employed as reporters in living cells, rather than labels in *"in vitro"* assays, mainly because of the difficulties of obtaining conjugates that retain bioluminescence activity upon chemical conjugation with the analyte or detection probe (*e.g.*, antibody or DNA fragment). Advances in molecular biology techniques and the cloning of new luciferases have expanded the range of application of bioluminescence imaging.[5]

Bio- and chemiluminescence analytical systems based on coupled enzyme reactions have also been developed, *e.g.*, by coupling ATP-involving reactions (kinases) with firefly luciferase, or NAD(P)H-involving reactions (dehydrogenases) with bacterial luciferase, or oxidase enzymes with the luminol–H_2O_2–HRP system.

11.4   Applications on Macrosamples

11.4.1   Imaging of Membranes and Gels

Because of its superior analytical performance and the possibility to perform direct and rapid quantitative evaluation of the signal over a wide dynamic range, chemiluminescence imaging in membranes and gels has found several applications.[6]

Southern, Northern or Western blot tests, and dot blot hybridization reactions with chemiluminescence detection rely on the detection of the target analyte (nucleic acids or proteins that are either blotted on filter membrane after separation by gel electrophoresis, or directly dotted on the membrane) by means of a labeled complementary gene probe or specific antibody, which can then be detected by enzyme-amplified chemiluminescence. Usually, the gene probe or antibody is indirectly labeled with the enzyme, *e.g.*, by employing the biotin/streptavidin or hapten/anti-hapten antibody systems. The use of chemiluminescence imaging detection offers the advantage of digital documentation and relatively large dynamic range of measured signals. Digital images can be easily analyzed to obtain quantitative information and stored in a computer to create archives and for exchange with other laboratories or with physicians for evaluation.

Chemiluminescence detection has been exploited to improve the sensitivity of Western blot (WB) for the serological diagnosis of infectious diseases. The immune response against bacterial or viral infections is conventionally assessed by ELISA, which detects antibodies directed against conformational bacterial or viral epitopes. On the other hand, the immune response against linear epitopes, which can be useful in determining the different phases of infection, can be detected by means of WB. Since the sensitivity of WB is not always as high as ELISA and the interpretation of results is subjective and may depend on the technical expertise of the operator, a WB assay with chemiluminescence detection was developed for the study of the IgG immune response against VP1 and VP2 linear epitopes of Parvovirus B19. Briefly, recombinant VP1 and VP2 proteins were separated under denaturing conditions by SDS-PAGE, and then transferred to a nitrocellulose membrane. After incubation with human serum, the membrane was incubated with (HRP)-conjugated anti human IgG, the enzyme activity of which was then detected by imaging upon addition of a HRP chemiluminescence substrate. The method, which was characterized by high reproducibility, provided an objective evaluation of the results and a semi-quantitative analysis of the presence of antibodies against VP1 and VP2 in human sera.[12]

To avoid the use of expensive specific antibodies, protein labeling with non-enzymatic tracers suitable for chemiluminescence detection, such as $Ru(bpy)_3^{3+}$,[13] metalloporphyrines[14] or Au(III), can be employed.[15] In this case, non-specific labeling of all the proteins in the gel is performed, rather than antibody-mediated specific target recognition.

11.4.2 Imaging of Microtiter Plates

Bio- and chemiluminescence reactions are particularly suited for the development of high-throughput screening (HTS) methods in microtiter plates or miniaturized formats, due to their high sensitivity in low volumes.[16] When imaging detection is performed, the signal is simultaneously measured from all the wells of the plate, thus increasing sample throughput and improving accuracy when analytical methods relying on the kinetic behavior of the chemiluminescence emission are employed. However, the lower sensitivity and narrower dynamic range of luminographs with respect to PMT-based luminometers has to be taken into account, as well as the absence of automated reagent dispensers suitable for the accurate measurement of rapid flash-type reactions. In addition, suitable correction optics (flat-field correction lenses) or algorithms must be used to avoid errors in the measurement of the signal intensity in the peripheral wells of the plate, due to the shadowing effects of the well walls.

High-density microtiter plates and bio- and chemiluminescence imaging detection have been used to develop various high-throughput enzyme assays, immunoassays, gene hybridization assays and whole-cell biosensor assays.[17–20]

A polymerase chain reaction chemiluminescence enzyme immunoassay based on a 384-well microtiter plate (384 PCR-CLEIA) has been developed for the semi-quantitative detection of the 15 high and low risk human *Papillomavirus* (HPV) genotypes more frequently associated with preneoplastic and neoplastic genital lesions.[21] Genotypization and quantification of HPV-DNA, which is useful in cervical screening programs, in the follow up post-surgical treatments and in monitoring the effectiveness of HPV vaccination programs, require high-throughput, flexible and semi-quantitative/quantitative molecular technologies. The use of such assays to better define the HPV genotype-specific prevalence and the association of viral load with the evolution of infection and with lesion progression is of great interest. The developed assay relied on PCR consensus amplification of HPV DNA and digoxigenin-labeling, product hybridization by means of type-specific biotin-labeled oligoprobes immobilized on the streptavidin-coated wells of a 384-well microtiter plate, and its quantification employing a HRP-labeled anti-digoxigenin antibody and chemiluminescence detection. The method, which allowed the detection, typing and quantification of 15 HPV genotypes in as many as 20 samples per assay, with a limit of detection of 10–50 DNA copies and high reproducibility, represents a rapid, convenient, high-throughput and low-cost diagnostic tool suitable for screening programs.

Chemiluminescence imaging of microtiter plates has also been exploited for the development of multiplexed quantitative assays, suitable for the simultaneous determination of several analytes in one sample. One possible approach is represented by the creation of capture probe (antibody or gene probe) mini-arrays at the bottom of a conventional 96-well microtiter plate to perform multiplexed detection of a small panel of analytes in a sample. The method, which was first reported for the development of a sandwich-type ELISA for measuring the concentration of seven different human cytokines in each sample,[22] is based on the capture of analytes by specific capture probes arrayed in the bottom of each well and their detection by means of a suitable enzyme-labeled detection reagent and a chemiluminescent substrate. The light emission was measured by imaging the entire plate with a CCD camera, and then quantified at each spot in the array. This technology combined high productivity and multiplexing ability of microarrays with the possibility to employ instrumentation required for standard 96-well microtiter plates.

Following a similar approach, multiplexed bioanalytical methods employing ECL imaging detection can be developed employing commercially available 24-, 96- and 384-well microtiter plates with up to 100 integrated electrodes arrayed at the bottom of each well.[23]

To avoid the need of expensive dedicated instrumentation to create mini-arrays of spots at the bottom of microtiter plate wells, custom-designed microtiter plates have been developed, in which each main well is internally divided into several sub-wells, thus allowing easy immobilization of several capture probes in different positions within the same main well. A multi-analyte binding assay for typing HPV by means of PCR-CLEIA was developed employing one such plate, in which each of the 24 main wells is divided into seven sub-wells.[24] Oligonucleotide probes specific for seven high-risk HPV genotypes were separately immobilized in each sub-well, then samples, previously digoxigenin-labeled during consensus PCR amplification, were added to the main wells. Each PCR product was able to diffuse within the main well and to hybridize to the immobilized probe corresponding to its genotype. Hybrids were subsequently detected and quantified by use of an HRP-labeled anti-digoxigenin antibody and chemiluminescence imaging (Figure 11.2). Assay of clinical samples gave results comparable to those obtained by conventional PCR-ELISA, with the advantage of the simultaneous determination of up to seven HPV DNAs in each sample.

Following a similar approach, a multiplexed immunoassay, suitable for the detection of *Escherichia coli* O157:H7, *Yersinia enterocolitica*, *Salmonella typhimurium* and *Listeria monocytogenes* pathogen bacteria in food samples has been developed employing a 96-well microtiter plate in which each main well was divided into four sub-wells.[25] In this case, four monoclonal antibodies,

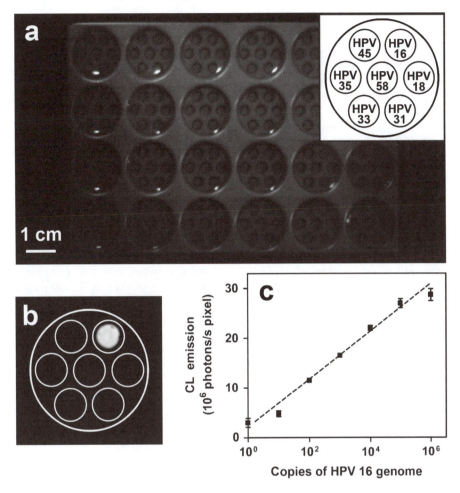

Figure 11.2 PCR-ELISA hybridization assay for HPV typing in a custom designed microtiter plate. (a) Image of the microtiter plate containing 24 main wells, each of them divided in seven sub-wells (inset: localization of the gene probes for the different HPV genotypes). (b) Chemiluminescence image obtained for a HPV 16-positive sample. (c) Calibration curve obtained by analyzing samples containing a known number of copies of the HPV 16 genome. (Reproduced with permission from ref. 6.)

each specific for one pathogen bacteria, were immobilized in the different sub-wells within the same main well. Upon addition of the sample to the main well, the bacteria were captured in the sub-well containing their specific antibody and then detected by adding a mixture of HRP-labeled specific polyclonal antibodies and by performing chemiluminescence imaging detection. The assay, which was characterized by high accuracy (recovery values ranging from 90 to 120%) and low limit of quantification (of the order of 10^4–10^5 CFU mL^{-1}) for all bacterial species, was suitable for screening procedures.

11.4.3 Imaging of Microarrays and Miniaturized Devices

Microarrays, which represent a powerful analytical tool for the rapid and simultaneous detection of multiple analytes on a single device, are essentially highly parallel, miniaturized surface-based assays in which numerous probes are immobilized in a spatially addressable manner. The main feature of microarrays is the ability to provide a huge amount of information with reduced time and samples/reagents consumption. Chemiluminescence is particularly suited as detection system in microarrays and miniaturized analytical formats, thanks to its high detectability in low volumes and simplicity of required instrumentation.

High-density microarrays allow the simultaneous detection of hundreds of analytes (nucleic acids or proteins) in each assay. Lower density arrays, possibly equipped with suitable microfluidic system and integrated detection apparatus, have found several applications in bioanalytical chemistry.

A multiplex chemiluminescence sandwich ELISA has been developed for the detection of *Escherichia coli* O157:H7, *Salmonella typhimurium* and *Legionella pneumophila* pathogen bacteria in water samples employing a flow-through antibody microarray platform.[26] A sandwich-type immunoassay was developed by capturing bacteria by means of a specific polyclonal antibodies microarray produced on poly(ethylene glycol)-modified glass substrates. The sandwich was completed by means of specific biotinylated antibodies that were detected employing streptavidin-HRP catalyzed chemiluminescent reaction and imaging with a CCD camera. The assay, which was completed in 13 min, displayed limits of detection in the range 10^3–10^6 cells mL^{-1}, which implies the necessity, in the future, of integrating the flow-through microarray with an online pre-enrichment module (*e.g.*, based on microfiltration or immunomagnetic capture).

Electrogenerated chemiluminescence is widely employed in clinical diagnostics as an ultrasensitive detection technique. A multiplex ECL bead-based platform has been developed to simultaneously quantify three antigens, VEGF, IL-8 and TIMP-1, in a sample.[27] In particular, three populations of polystyrene microspheres, internally encoded with fluorescent dyes to allow them to be distinguished from one another in the array, were coated with three different capture antibodies, each specific for one analyte. Pooled microspheres were allowed to self-assemble in the wells of a gold-coated etched fiber-optic bundle that acted as the working electrode for ECL measurements. The beads array was incubated with the sample containing the analytes, then with a pool of three biotinylated detection antibodies and finally with Ru(bpy)$_3^{2+}$. Upon performing a cyclic voltammogram while imaging ECL with an EMCCD coupled with a microscope, each microbead was associated to one analyte based on its fluorescent code and each analyte was quantified through the intensity of ECL emission. By increasing the number of fluorescent encoded beads, high multiplexing capacity can be obtained.

Multiplexed quantitative bioanalytical techniques relying on chemiluminescence imaging are an important diagnostic tool for the rapid and early diagnosis of infectious diseases and pathological conditions in general and for the investigation of the pathological pathway. The optimization of contact imaging techniques should facilitate the future development of portable, low-cost miniaturized analytical systems with integrated imaging devices.

11.4.4 Imaging of Whole Organs

Reactive oxygen species (ROS) include various extremely reactive compounds that play an important role in the defense mechanism against infection and in the pathogenesis of various diseases. Excessive production of ROS can compromise normal cellular functions, by damaging lipids, proteins and DNA. The investigation of ROS production in target organs and tissues as markers of functional significance for immune and inflammatory processes in health and disease is a challenging task, due to the high reactivity and instability of ROS. Chemiluminescence represents a very convenient and sensitive non-invasive technique that allows the direct monitoring of ROS production in biological systems based on the fact that every free radical reaction results in weak chemiluminescence emission, which can be detected by highly sensitive photodetectors. The use of light-amplifying substances, such as lucigenin, luminol or the luciferin analogue MCLA (2-methyl-6-[*p*-methoxyphenyl]-3,7-dihydroimidazo[1,2-*a*]pyrazin-3-one), increases the sensitivity of the detection process and, to a lesser extent, the specificity for some reactive species.

The possibility of performing chemiluminescence imaging, rather than performing batch measurements of chemiluminescence emission in organs perfusates or homogenates, allows us not only to assess the global luminescence emitted by the biological tissue but also to localize the emitted photons on the organ surface and thus to evaluate in real time temporal and spatial distribution of ROS generation.

A whole-organ chemiluminescence imaging method has been developed to perform real-time, quantitative localization of ROS in isolated and perfused rat livers exposed to oxidative damage due to ischemia-reperfusion.[28] Lucigenin was utilized as a chemiluminogenic probe to assess superoxide anion generation and a Saticon ultrasensitive video camera with image intensifier was employed to measure the tissue photons emission at a single photon level. For the first time, this method allowed evaluation in real time of the temporal and spatial distribution of ROS formation on the surface of an intact organ and investigation of the effect of specific scavengers. Employing this method, the effect of age, ethanol consumption[29] and steatosis[30] has also been investigated.

In recent years the availability of chemiluminescence and bioluminescence *in vivo* imaging applications has opened up a tremendous range of applications, providing the possibility of non-invasively investigating physiological and pathological processes in living animals. Luminol has been employed for non-invasive imaging of ROS/RNS production in a model of acute arthritis.[31] More recently, the compound L-012 (8-amino-5-chloro-7-phenylpyrido[3,4-*d*]pyridazine-1,4(2*H*,3*H*)dione), a luminol analogue, was proposed as a significantly more sensitive probe to visualize ROS in living mice for the study of dynamic processes in inflammation.[32] Although non-invasive *in vivo* imaging is an extremely powerful technique by which to investigate pathophysiological processes in intact organisms, as illustrated in Chapter 12 ("*in vivo*" molecular imaging) of this book, the investigation on isolated organs maintains its importance due to the limited ability to precisely localize the signal in specific regions of a target organ by means of whole body imaging.

11.4.5 Works of Art

Bio- and chemiluminescence imaging techniques have been recently proposed for the characterization of works of art and for evaluating their state of conservation. As artwork samples are generally very small, and the compounds to be determined are usually in low concentration, sensitive and selective analytical techniques are required.

It is well known that biodeteriogen agents (bacteria, fungi, yeast, algae and lichens) can cause physical–chemical damage to cultural heritage.[33] Analytical methods are required both to monitor the efficacy of biocide treatments aimed at removing biodeteriogens from the surface of artworks during the cleaning procedure and to detect early any new growth after cleaning. A bioluminescence low-light imaging technique has been developed for the evaluation of biodeteriogens spatial distribution, exploiting the adenosine triphosphate (ATP)-depending firefly luciferin–luciferase reaction.[34] More details are reported in Chapter 17.

11.5 Applications on Microsamples

The ultrasensitive localization and quantification of analytes inside tissues and cells is fundamental both for understanding physiopathological events and for the early diagnosis of pathologies. Low-light imaging detection devices can be easily connected to conventional optical microscopes to perform bio- and chemiluminescence imaging at the cellular level, provided that the microscope, or at least the sample, is enclosed in a light-tight container to prevent contact with external light. Bio- and chemiluminescent microscope low-light imaging is a valuable tool for the ultrasensitive localization of inorganic or organic molecules, enzymes, antigens and nucleic acids in living cells, fixed cells, tissue cryosections and paraffin-embedded tissue sections.

Bio- and chemiluminescence detection offers high sensitivity and easy and reliable evaluation of the amount of the probe, thus enabling the quantitative analysis of the target molecules. In addition, multiplexed detection can be performed by combining different detection principles and/or labels.

One of the main drawbacks of enzyme-catalyzed chemiluminescence imaging at the cellular level is the usually lower resolution with respect to colorimetric or fluorescent detection, due to the partial diffusion of the reaction intermediates, which causes the excited species responsible for the chemiluminescence reaction to emit photons at a site different from that where they were generated. In addition, since coverslips are usually not employed in chemiluminescence imaging to avoid internal reflection phenomena, the use of high-magnification oil immersion objectives is not possible. Nevertheless, resolution down to micrometer level can be reached, which is adequate for the localization of the biomolecules at the cellular and at the subcellular level.

Another disadvantage with respect to fluorescence detection is the limited number of bio- and chemiluminescence labels, which narrows down the multiplexing possibilities. In this respect, the recently developed transparent electrochemical cell optimized for ECL microscope imaging offers, for the first time, the possibility to exploit the sensitivity and specificity of ECL detection in single cell analysis.[35] Electrochemiluminescence detection combines the particular analytical performances in terms of high detectability of chemiluminescence with the possibility to control the time and position of the light-emitting reaction. In addition, ECL is characterized by improved resolution with respect to enzyme-catalyzed chemiluminescence since, in this case, the label itself [*e.g.*, $Ru(bpy)_3^{2+}$ complex] emits light and no diffusion phenomena may affect the imaging resolution. The developed electrochemical cell, which is based on a classic three-electrode configuration and is controlled though a conventional potentiostat, presents overall dimensions ($7.5 \times 2.5 \times 0.1$ cm) and a transparent field of view ($16\,mm^2$) that are compatible with conventional optical microscopy devices (Figure 11.3). Transparency, which was obtained employing fluorine-doped tin oxide (FTO)-coated glass as a support, allows acquisition of the transmitted light image, necessary for accurate localization of the ECL signal in the cell or tissue section, through the overlay of the ECL image to the transmitted light image. Optimization of the system employing $8\,\mu m$ diameter polystyrene beads coated with a $Ru(bpy)_3^{2+}$ complex demonstrated detectability of the $Ru(bpy)_3^{2+}$ complex down to $1 \times 10^{-19}\,mol\,\mu m^{-2}$ and spatial resolution of $0.4\,\mu m$ and also showed the suitability of the device for microscope ECL imaging of cells and tissue sections.

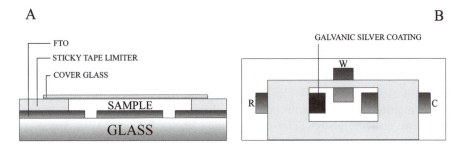

Figure 11.3   ECL optical imaging cell ($75 \times 25 \times 1$ mm). (A) side view; (B) top view (Ag quasi-reference electrode, $8\,mm^2$; FTO working electrode, $16\,mm^2$; FTO counter electrode, $20\,mm^2$). (Reproduced with permission from ref. 35.)

11.5.1 Metabolic Mapping and Enzyme Localization

Inorganic and organic molecules and enzymes can be detected with a spatial resolution at the cellular or subcellular level by bio- and chemiluminescence imaging by employing single or coupled enzyme reactions leading to photons emission.

The local concentration of metabolites, such as ATP, lactate, glucose, glucose-6-phosphate or the NAD redox state, have been detected down to femtomole levels by bioluminescence imaging using quantitative bioluminescence and single-photon imaging.[36] Enzymatic activity on a tissue section surface can be localized and quantified by adding a suitable enzyme cocktail able to produce coupled enzyme reactions, with the last being a bioluminescent one.[37]

To obtain a reliable snapshot of the actual analyte distribution in the biological tissue at the time of collection, the sample must be rapidly frozen and kept as such during the entire subsequent sample preparation. To perform metabolic mapping measurements, endogenous enzymes must be rapidly inactivated (*e.g.*, with a heat shock) before exposing the tissue section to the suitable enzyme cocktail; to perform measurements of enzyme activities, cryostat sections of the frozen tissue are exposed to a suitable enzyme solution, then brought to working temperature by means of a thermostated microscope stage. The resulting photon emission is then imaged and can be analyzed to obtain local metabolite or enzyme content upon proper method standardization and calibration.

A bioluminescence assay, based on coupled enzyme reactions terminating with $FMNH_2$-dependent bacterial luciferase reaction, has been developed recently for quantitative pyruvate imaging within sections of snap-frozen tissue, displaying a spatial resolution at the microscopic level.[38] The combination of this method with bioluminescence lactate mapping appears to be very promising for predicting the radio- and chemotherapy sensitivity of individual malignancies, which is known to be dependent on the lactate-to-pyruvate ratio of the tissue.

Imaging of Ca^{2+} genetically encoded indicators is widely used to measure transient Ca^{2+} increases in cells and subcellular compartments.[39] Several fluorescent indicators are employed, which offer high spatial resolution but suffer from photobleaching and sample autofluorescence. The photoprotein aequorin has been widely used as a reporter of Ca^{2+} physiology in various cell types following intracellular injection or gene expression; genetically encoded aequorin can be targeted to specific subcellular location.[40] Aequorin is nearly insensitive to changes in Mg^{2+} or pH and it covers a large dynamic range of measurable Ca^{2+} (from 10^{-7} to 10^{-3} M), which can be further extended by exploiting genetic modifications of the protein and/or use of different synthetic coelenterazines. The photoprotein obelin has also been proposed, which displays similar sensitivity to Ca^{2+} concentrations, with faster response kinetics.

The main limitation of bioluminescence imaging with photoproteins is the low photon yield, causing poor spatiotemporal resolution, which has made it difficult to perform studies at the single-cell level. To overcome this problem, in recent years fusion proteins of aequorin (or of the photoprotein obelin) and the green fluorescent protein (GFP) have been proposed that allow fluorescence labeling of expressing cells and bioluminescence Ca^{2+} imaging of single cultured cells, tissue slices and whole animals, with high signal-to-background ratio.[41,42] Recently, simultaneous and independent monitoring of Ca^{2+} fluxes in different subcellular compartments has been demonstrated by employing two spectrally distinct (green- or red-emitting) aequorins in different subcellular locations within the same cells.[43]

11.5.2 Immunohistochemistry and *In Situ* Hybridization

Localization of target antigens or nucleic acids in tissue sections and cells is routinely performed by immunohistochemistry (IHC) and *in situ* hybridization (ISH) techniques, usually relying on the colorimetric or fluorescent detection of an antibody or a DNA probe directly or indirectly labeled with an enzyme or a fluorescent molecule. Conventional prompt fluorescence is the most widely used detection principle, offering the possibility to perform multiplexed localization by using narrow band photoluminescent probes and spectral resolved analysis of the light emission. However, despite the use of combined techniques such as confocal or pseudo-confocal microscopy that improves the signal-to-noise ratio and the imaging resolution, photoluminescence is still affected by the autofluorescence of the sample matrix, which reduces detectability of the analyte specific signal. The high background signal can be efficiently minimized by the time-resolved fluorescence (TRF) analysis of long-lived fluorescence labels such as lanthanide chelates,[44] which, however, is not suited to obtain accurate quantitative information.

The need for high sensitivity and for objective quantitative information has prompted the development of IHC and ISH methods with chemiluminescence detection allowing spatial localization and quantitative evaluation of the labeled probe in tissue sections or single cells. Since in the presence of excess substrate the chemiluminescence signal intensity is proportional to the immobilized enzyme amount, quantitative ISH and IHC methods can be developed, once the system is optimized taking into account the non-specific chemical and instrumental background signal.

Various chemiluminescence IHC and ISH techniques have been developed for the localization of antigens or nucleic acids in single cells and tissue sections, by employing an antibody or a DNA probe labeled with HRP or AP and a suitable chemiluminescent substrate. Indirect labeling is usually performed by employing digoxigenin, biotin, or fluorescein non-isotopic labels or a two-step immunochemical reaction, with the secondary antibody being enzyme-labeled.

Chemiluminescent methods for the immunolocalization of tumoral markers, inflammatory mediators and viral antigens in infected cells have been developed. Viral nucleic acids were localized in infected cells and tissue sections by employing chemiluminescence ISH.[6]

Recently, the possibility of obtaining quantitative information from the photons emission was exploited to increase the diagnostic and prognostic potential of IHC and ISH techniques.

A direct, simple and rapid chemiluminescent IHC method has been developed for quantitative evaluation of the level of expression of the MRP2 transport protein (a member of the human multidrug resistance-associated protein family) in liver biopsy sections.[45] Alteration of MRP2 expression and/ or function is thought to play an important role in the pathogenesis of cholestasis and MRP2 levels could represent an indicator of primary biliary cirrhosis (PBC) progression. In the developed method, the MRP2 protein was localized by means of an anti-MRP2 monoclonal antibody, followed by a biotin-labeled anti-mouse antibody and a streptavidin–HRP conjugate, which was then detected by means of chemiluminescent imaging detection. Quantitative analysis of digital images allowed reliable and reproducible evaluation of the protein content of the tissue (Figure 11.4) and, when applied to the analysis of clinical samples from PBC patients under therapy with ursodeoxycholic acid, it provided results that were well correlated with the histological data and in line with those previously reported in the literature and obtained with conventional protein expression analysis techniques.

A quantitative chemiluminescence IHC method has been developed for evaluation of the level of expression of the p16^{INK4a} protein in cervical biopsy sections.[46] The p16^{INK4a} neoplastic marker protein is overexpressed in cervical intraepithelial neoplasias (CINs) caused by HPV. Since the positivity of p16^{INK4a} in the epithelium varies from low-grade CIN1 lesions (only the lower third of epithelium is positive) to high-grade CIN3 lesions (the whole epithelium is stained), the oncogenic risk of CINs could be evaluated by measuring the protein expression profile. In the developed method, the protein was localized by means of a monoclonal anti-p16^{INK4a} antibody, followed by a biotinylated anti-mouse antibody, streptavidin and a biotin–AP conjugate, with enzymatic activity measured by means of a chemiluminescence enzyme substrate and imaging detection. Quantitative analysis of the chemiluminescence signal showed a p16^{INK4a} concentration profile across the epithelium that changed depending on the degree of CIN lesion, according to data obtained by IHC with colorimetric detection. A signal elaboration procedure, based on the evaluation of both the chemiluminescence signal intensity and the fraction of epithelium area displaying a positive signal, allowed discrimination between low-risk (CIN1) and high-risk (CIN2 and CIN3) lesions with statistical significance. The newly developed method allowed objective evaluation of the risk of progression of CIN lesions, thus representing an advance with respect to the subjective evaluation obtained by means of histological analysis or p16^{INK4a} detection performed by IHC with colorimetric detection.

A CL ISH method based on peptide nucleic acid (PNA) probes has been developed for the ultrasensitive localization and the quantitative detection of parvovirus B19 nucleic acids in single infected cells.[47] The assay was based on the use of a biotin-labeled PNA probe detected by a streptavidin-linked AP and a chemiluminescent substrate. The developed ISH assay fulfilled all of the standard analytical requirements in terms of precision and accuracy with a

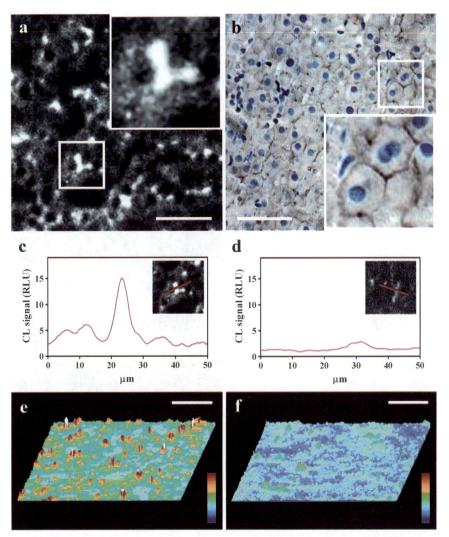

Figure 11.4 Immunohistochemical localization of MRP2 in liver tissue sections. (a,b) Comparison between images obtained using chemiluminescence (a) and colorimetric (b) detection (40× objective magnification); the enlarged view in the insets shows details of the signal distribution. (c,d) Profiles of the chemiluminescence emission across a canalicular membrane measured in tissue samples with high (c) and low (d) chemiluminescence signal, evaluated along the line shown in the insets. (e,f) Pseudo-color three-dimensional plots showing the spatial distribution of the chemiluminescence signal in samples with high (e) and low (f) signal intensity; the ruler shows the correspondence between colors and chemiluminescence signal. Bar = 50 μm. (Reproduced with permission from ref. 45.)

detectability superior to previously used ISH systems. In addition, thanks to the combination of sensitivity and quantitative approach, it was possible to study the kinetics of a virus cellular infection. In particular, the myeloblastoid cell line UT-7/EpoS1 was infected with Parvovirus B19, harvested at various times post-infection (2, 6, 12, 18, 24, 36 hpi) and analyzed by using the developed PNA-based chemiluminescence ISH assay. Under these experimental conditions, the earliest time allowing for the detection of B19 virus nucleic acids in infected cells was 12 h, while the fraction of infected cells reached its maximum at 24 h (Figure 11.5), and then decreased at longer times, due to the lysis

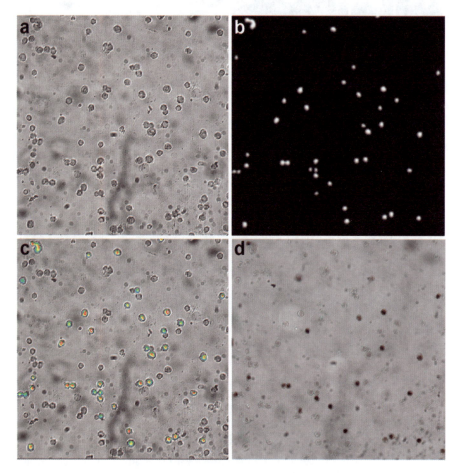

Figure 11.5   Images of the same field of a 24-hour-post-infection cell smear after ISH-PNA assays: (a) live image acquired in transmitted light; (b) chemiluminescence detection: signals higher than the threshold value, which are considered indicative of the presence of B19 virus nucleic acids, are shown; (c) overlay of the live image and the chemiluminescence signal elaborated in pseudo-colors; (d) results obtained with colorimetric detection. (Reproduced from ref. 47.)

of productively infected cells. The chemiluminescent ISH-PNA assay could thus represent a potent tool for the assessment of viral infections and for the quantitative evaluation of the virus nucleic acid load of infected cells in virus studies and diagnostics.

11.5.3 Combined *In Situ* Hybridization and/or Immunohistochemistry Techniques

The accurate diagnosis of a given pathology requires the simultaneous detection of multiple biomarkers that can complement each other, thus prompting the development of multiplexed bioanalytical methods. Multiplexed bio- and chemiluminescence imaging assays can be developed by combining different luminescence techniques and/or different probes within the same technique.

A multiplexed chemiluminescence ISH technique has been developed for the simultaneous detection of two viral DNAs, *i.e.*, Herpes simplex virus (HSV) and cytomegalovirus (CMV) DNAs in infected cells in the same specimen.[48] To obtain multiplex analysis, the two viral DNAs were detected by means of specific oligonucleotide probes indirectly labeled with either HRP (through the biotin–streptavidin system) or AP (though the digoxigenin–anti-digoxigenin antibody system) for HSV and CMV, respectively. By sequentially adding the chemiluminescence substrates specific for the two enzyme labels, it was possible to independently localize the two viral DNAs in the same specimen.

This double ISH assay, which combined the sensitivity of chemiluminescence detection with the possibility of performing multiple analyte localization in cells, opened the way to the development of assays combining IHC and ISH with chemiluminescence detection for the measurement of various biomarkers (proteins and nucleic acids) of a given pathology in cells or tissue sections.

As a first approach, a sensitive method that combined an enzyme-amplified fluorescent ISH (FISH) for the localization of high-risk HPV DNA with a CL-IHC method for the localization of the tumoral melanocytic marker HMB-45 in primary melanoma biopsy sections was developed to assess the association of mucosal HPVs with melanoma lesions.[49] HPVs have been recognized as the causal agents in cervical cancer and have been postulated as carcinogens in a range of other epithelial malignancies. The correlation between high-risk mucosal HPV and basal cell carcinoma and squamous cell carcinoma has been frequently reported but not conclusively demonstrated. In the developed method, the ISH detection was performed by means of a digoxigenin-labeled HPV-specific oligonucleotide probe, revealed with an AP-labeled anti-digoxigenin antibody and the enzyme-label fluorescence signal amplification technology, while the IHC localization used an anti-HMB-45 primary antibody, revealed by means of a biotinylated secondary antibody, a streptavidin/biotin/HRP detection system and a chemiluminescence substrate for HRP. Fluorescent ISH and chemiluminescence IHC reactions were performed sequentially on the same tissue section, then digital images of the fluorescence and chemiluminescence signals were separately recorded and the co-localization of the two signals was assessed using specific software for image analysis (Figure 11.6). The method was able to show the co-localization of both markers in the vast majority of cells, thus demonstrating the presence of HPV DNA within melanoma cells and stressing the association between HPV and melanoma cells.

To fully exploit the peculiar characteristics of chemiluminescence detection, such as high sensitivity and easy quantification of the target analyte, the chemiluminescence IHC method for quantitative evaluation of p16^{INK4A} overexpression in the epithelium of cervical biopsies reported above was combined with a chemiluminescent ISH method for the localization of HPV (generally accepted as a necessary but insufficient cause of cervical carcinoma).[10] Different label enzymes (AP and HRP) were employed for the chemiluminescent detection of the protein p16^{INK4A} and HPV DNA, respectively (Figure 11.7). Quantitative chemiluminescence image analysis was used to obtain objective evaluation of sample positivity. Results obtained with the multiplexed CL-IHC/CL-ISH localization procedure showed that the combined assay provides a better sample classification in negative, low-grade and high-grade lesions, with respect to the determination of the single biomarkers p16^{INK4a} and HPV DNA, thus offering an accurate and objective diagnostic tool providing important information for counseling, selection of therapy and follow up after surgical treatment.

These studies have demonstrated the potentiality of multiplexed chemiluminescence detection of proteins and nucleic acids in cells and tissue sections for diagnostic purposes and for the investigation of physiopathological processes. The development of methods able to simultaneously detect the overexpression of several tumor marker genes is gaining increasing interest. The availability of objective and reliable methods suitable for monitoring the state of expression of appropriately selected tumor biomarkers should enable prediction of the invasive and metastatic potential of a cancer, its ability to evade immune surveillance and its potential response to treatment. This would allow selection of the most appropriate treatment strategy for each individual patient, leading to personalized and predictive medicine. Chemiluminescence imaging represents a powerful tool, its main shortcomings being represented by the necessity to perform sequential measurements of the signal emitted by different enzyme labels, which increases the assay complexity and length, and the reduced multiplexing possibilities due to the limited availability of suitable labels detectable by chemiluminescence. The development of new chemiluminescent or bioluminescent labels suitable for imaging detection combined with optical microscopy is required to increase multiplexing possibilities.

11.5.4   BRET-based Methods

Resonance energy transfer (RET) is a proximity-based technology based on non-radiative energy transfer between donor and acceptor molecules according to the Forster mechanism. When the two molecules are in close proximity (1–10 nm) and the correct orientation, energy transfer is possible. Once the

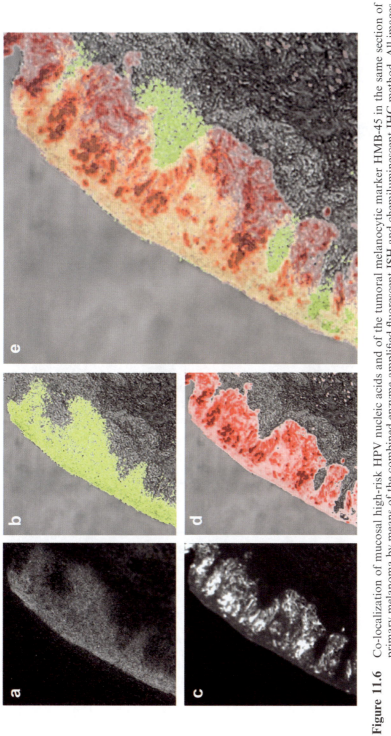

Figure 11.6 Co-localization of mucosal high-risk HPV nucleic acids and of the tumoral melanocytic marker HMB-45 in the same section of primary melanoma by means of the combined enzyme-amplified fluorescent ISH and chemiluminescent IHC method. All images were acquired at a 10× objective magnification. (a) Fluorescent signal obtained for the localization of HPV nucleic acids by means of the ISH procedure. (b) Overlay of the fluorescent signal, which was assigned the yellow color, and the bright-field transmitted-light image. (c) Chemiluminescent signal obtained for the localization of HMB-45 by means of the IHC procedure. (d) Overlay of the chemiluminescent signal, which was assigned the red color, and the bright-field transmitted-light image. (e) Final overlay image of the color processed fluorescent signal, the color processed chemiluminescent signal and the live image; co-localization of HPV DNA and HMB-45 marker is evidenced by the color combination, yielding an orange hue. (Reproduced with permission from ref. 49.)

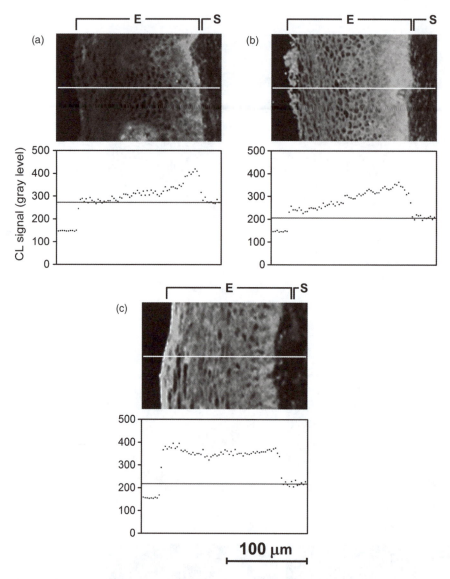

Figure 11.7 Localization of p16^{INK4A} by chemiluminescence-IHC in bioptic sections of representative (a) CIN1, (b) CIN2 and (c) CIN3 lesions. The upper panels show the chemiluminescence images, while the lower panels show the chemiluminescent signal profiles across the epithelium along the lines shown in the corresponding chemiluminescence image. The horizontal lines in the signal profiles indicate the mean background chemiluminescent signal measured in the stromal area. E = epithelium, S = stromal area. (Reproduced with permission from ref. 10.)

donor luminescence emission is elicited, the energy transfer results in a decrease of photons emitted by the donor and the appearance of photons emitted by the acceptor. The analytical signal is represented by the ratio between acceptor and donor emission intensities. The RET phenomenon has been widely exploited for monitoring bimolecular interactions, protein conformational changes or protease activities upon labeling the target proteins of interest (or different regions of the same protein) with the RET partners.[50,51]

While in fluorescence RET (FRET) donor and acceptor are both fluorescent molecules, bioluminescence RET (BRET) relies on a bioluminescent donor protein that emits energy upon addition of an organic substrate. Since BRET does not require photoexcitation of the donor, problems such as photodamage to cells, cell autofluorescence, direct excitation of the acceptor fluorophore and photobleaching are avoided.

Imaging BRET-based methods aimed at monitoring the dynamics of protein interactions within live mammalian cells are being increasingly proposed. The bioluminescent donor and fluorescent acceptor proteins can be genetically fused to proteins of interest and expressed in cells, with the possibility of targeting the fusion constructs to specific subcellular compartments. A cell-permeable substrate is subsequently added to initiate the bioluminescent reaction, which is detected by microscope low-light imaging.

The genetic constructs must be carefully designed in order both to ensure that the bioluminescent donor and the fluorescent acceptor are in the correct orientation for non-radiative energy transfer upon target proteins interaction and to avoid that the presence of the donor and acceptor proteins hinders or alters the specific protein–protein interaction under study.

As BRET is a ratiometric measurement, imaging instrumentation equipped with suitable filters for efficient and independent separate measurements of light output at the wavelengths corresponding to the emission maximum of the donor and acceptor must be employed.

Alternatively, the whole emission spectrum of the system under investigation can be recorded, from which the relative intensities donor and the acceptor emissions can be calculated. The required spectroscopic equipment has recently become available in FRET imaging microscopy. However, it must be taken into account that, since bioluminescence usually involves lower light intensities than fluorescence, filter-based approaches should be more efficient in BRET.

Despite the excellent signal/background ratio, the main limitation of BRET in microscopy imaging applications is the low intensity of light emission intrinsic to the bioluminescent luciferase reaction. However, in recent years the availability of enhanced sensitivity imaging detectors and improved bioluminescence probes and/or substrates has sustained the development of several BRET-based cellular imaging applications for the study of the spatiotemporal dynamics of protein interactions at the subcellular level.

Several BRET-based imaging methods have been developed to allow monitoring of protein–protein interactions in living cells and work is ongoing to obtain new BRET pairs with improved performance.[52–54] However, BRET is limited to the investigation of protein homo- or heterodimers and it does not allow demonstration of the existence of higher-order complexes involving more than two molecules. For this reason, a new approach, named sequential BRET-FRET technique (SRET), has been proposed to identify heteromers formed by the physical interaction of three different proteins in living cells.[55] In this system one protein was fused to Rluc (*Renilla* luciferase), the second protein was fused to fluorescent protein that works as a BRET acceptor (GFP2 or YFP) and the third protein with another fluorescent protein that works as a FRET acceptor (YFP or DsRed, respectively). With this approach, only when the three proteins form a complex does excitation of the BRET acceptor, and then of the FRET acceptor, occur by subsequent energy transfer upon addition of Rluc substrate. The method, which was employed to identify complexes of cannabinoid CB(1), dopamine D(2) and adenosine A(2A) receptors in living cells, will represent a fundamental tool for understanding the nature of protein assemblies and their interactions in cells.

Fluorescent proteins are common BRET acceptors, but organic molecules or fluorescent nanoparticles have also been employed. In particular, the use of semiconductor quantum dots (QDs) appears particularly promising, due to their characteristic large Stokes shift, which results in a much larger spectral separation of the acceptor emission from the donor emission, and thus higher sensitivity and accuracy. While producing bright and well-separated BRET emission, QDs present the disadvantage of their incompatibility with the gene fusion approach. Nevertheless, several chemical and biochemical strategies for binding QDs with the protein of interest and delivering the conjugate inside cells have been elaborated, as reviewed recently.[56]

BRET has been exploited for the development of various bioanalytical methods, such as immunoassays, and to investigate protein–protein interactions both *in vitro* and *in vivo*, as reported in other chapters of this book.

11.5.5 Works of Arts

The characterization and localization of inorganic and organic components of works of art is crucial for authentication studies, investigation of painting techniques and materials employed by the artist, and choice of the most suitable conservation practices. Various techniques have been developed to identify and localize organic materials (*e.g.*, pigment binders, adhesives and varnishes) in works of art, including immunological techniques to localize proteins. Recently, the use of immunochemistry techniques with chemiluminescence detection has been proposed for the localization of proteins in ancient painting layers.[57] Chapter 17 gives more details.

11.6   Conclusions

The possibility of localizing and quantifying the light emission on a target surface through bio- and chemiluminescence imaging represents a powerful tool for a wide range of applications. When macrosamples are analyzed, the distribution of the target analyte on the sample surface can be assessed, even on irregular surfaces. Employing high density analytical formats (*e.g.*, 384-well microtiter plates or microarrays), several samples and/or several analytes within each sample can be measured simultaneously. The high sensitivity of bio- and chemiluminescence measurements in small volumes and the relatively simple instrumentation are being increasingly exploited in the development of ultrasensitive assays in miniaturized formats, such as microarrays and lab-on-chip devices.

Coupling the imaging detector with a microscope also allows the localization and quantification of target biomolecules in tissue sections and single cells, taking the advantage of high detectability and possibility of quantification of the labeled probes, which is not possible with colorimetric or fluorescent detection. This has provided important advancements for the early and accurate diagnosis of various diseases.

The production of new improved bioluminescent and chemiluminescent probes, thanks to the powerful biotechnology techniques, and the continuous progress in light imaging instruments technology will allow further expansion of bio- and chemiluminescence imaging applications.

Chemiluminescence imaging provides an objective quantitative evaluation of the signal and, in addition, digital images can be stored in a computer for creating a real decision support system (DSS) for data exchange with other laboratories or with physicians. Decision support systems not only have a role to play in enhancing decision making but also in the study of diagnostic protocol, education, self-assessment and quality control. Finally, it is hoped that future quality control programs might benefit from the use of DSSs for assessing diagnostic performance in a more objective and reliable manner.

References

1. A. Roda, P. Pasini, M. Musiani, S. Girotti, M. Baraldini, G. Carrea and A. Suozzi, *Anal. Chem.*, 1996, **68**, 1073.
2. A. Roda, P. Pasini, M. Musiani, M. Baraldini, M. Mirasoli, M. Guardigli, M. Mirasoli and C. Russo, in *Chemiluminescence in Analytical Chemistry*, ed. A. M. García-Campaña and W. R. G. Baeyens, Marcel Dekker, New York, 2001, p. 473.
3. A. Roda, M. Guardigli, P. Pasini, M. Musiani and M. Baraldini, in *Luminescence Biotechnology: Instruments and Applications*, ed. K. Van Dyke, C. Van Dyke and K. Woodfork, CRC Press, FL, 2002, p. 481.
4. A. Roda, M. Guardigli, E. Michelini, P. Pasini and M. Mirasoli, *Anal. Chem.*, 2003, **75**, 462A.
5. A. Roda, P. Pasini, M. Mirasoli, E. Michelini and M. Guardigli, *Trends Biotechnol.*, 2004, **22**, 295.
6. A. Roda, M. Guardigli, P. Pasini, M. Mirasoli, E. Michelini and M. Musiani, *Anal. Chim. Acta*, 2005, **541**, 25.

7. A. Roda, M. Guardigli, E. Michelini and M. Mirasoli, *Trends Anal Chem.*, 2009, **28**, 307.
8. H. Ji, D. Sander, A. Haas and P. A. Abshire, *IEEE Trans. Circuits Syst. I, Reg. Papers*, 2007, **54**, 1698.
9. H. Eltoukhy, K. Salama and A. El Gamal, *IEEE J. Solid-St. Circuits*, 2006, **41**, 651.
10. M. Mirasoli, M. Guardigli, P. Simoni, S. Venturoli, S. Ambretti, M. Musiani and A. Roda, *Anal. Bioanal. Chem.*, 2009, **394**, 981.
11. A. Roda, P. Pasini, M. Baraldini, M. Musiani, G. Gentilomi and C. Robert, *Anal. Biochem.*, 1998, **257**, 53.
12. E. Manaresi, P. Pasini, G. Gallinella, G. Gentilomi, S. Venturoli, A. Roda, M. Zerbini and M. Musiani, *J. Virol. Methods*, 1999, **81**, 91.
13. L. Waguespack, A. Lillquist, J. C. Townley and D. R. Bobbit, *Anal. Chim. Acta*, 2001, **441**, 231.
14. X. Liu, L. Huang, W. R. G. Baeyens, J. Ouyang, D. He, G. Wan and L. Zhang, *Electrophoresis*, 2009, **30**, 3034.
15. J. Liu, X. Liu, W. R. G. Baeyens, J. R. Delanghe and J. Ouyang, *J. Proteome Res.*, 2008, **7**, 1884.
16. A. Roda, M. Guardigli, P. Pasini and M. Mirasoli, *Anal. Bioanal. Chem.*, 2003, **377**, 826.
17. M. Magliulo, M. Mirasoli, P. Simoni, R. Lelli, O. Portanti and A. Roda, *J. Agric. Food Chem.*, 2005, **53**, 3300.
18. A. Roda, M. Mirasoli, M. Guardigli, E. Michelini, P. Simoni and M. Magliulo, *Anal. Bioanal. Chem.*, 2006, **384**, 1269.
19. J. H. Lee, R. J. Mitchell, B. C. Kim, D. C. Cullen and M. B. Gu, *Biosens. Bioelectron.*, 2005, **21**, 500.
20. M. Guardigli, P. Pasini, M. Mirasoli, A. Leoni, A. Andreani and A. Roda, *Anal. Chim. Acta*, 2005, **535**, 139.
21. S. Ambretti, M. Mirasoli, S. Venturoli, M. Zerbini, M. Baraldini, M. Musiani and A. Roda, *Anal. Biochem.*, 2004, **332**, 349.
22. M. D. Moody, S. W. Van Arsdell, K. P. Murphy, S. F. Orencole and C. Burns, *BioTechniques*, 2001, **31**, 186.
23. www.mesoscale.com (accessed May 2010).
24. A. Roda, M. Mirasoli, S. Venturoli, M. Cricca, F. Bonvicini, M. Baraldini, P. Pasini, M. Zerbini and M. Musiani, *Clin. Chem.*, 2002, **48**, 1654.
25. M. Magliulo, P. Simoni, M. Guardigli, E. Michelini, M. Luciani, R. Lelli and A. Roda, *J. Agric. Food Chem.*, 2007, **55**, 4933.
26. A. Wolter, R. Niessner and M. Seidel, *Anal. Chem.*, 2008, **80**, 5854.
27. F. Deiss, C. N. LaFratta, M. Symer, T. M. Blicharz, N. Sojic and D. R. Walt, *J. Am. Chem. Soc.*, 2009, **131**, 6088.
28. A. Gasbarrini, P. Pasini, B. Nardo, S. De Notariis, M. Simoncini, A. Cavallari, E. Roda, M. Bernardi and A. Roda, *Free Radical Biol. Med.*, 1998, **24**, 211.
29. G. Addolorato, C. Di Campli, M. Simoncini, P. Pasini, B. Nardo, A. Cavallari, P. Pola, A. Roda, G. Gasbarrini and A. Gasbarrini, *Digest. Dis. Sci.*, 2001, **46**, 1057.

30. B. Nardo, P. Caraceni, P. Pasini, M. Domenicali, F. Catena, G. Cavallari, B. Santoni, E. Maiolini, I. Grattagliano, G. Vendemiale, F. Trevisani, A. Roda, M. Bernardi and A. Cavallari, *Transplantation*, 2001, **71**, 1816.
31. W. T. Chen, C. H. Tung and R. Weissleder, *Mol. Imag.*, 2004, **3**, 159.
32. A. Kielland, T. Blom, K. S. Nandakumar, R. Holmdahl, R. Blomhoff and H. Carlsen, *Free Radical Biol. Med.*, 2009, **47**, 760.
33. P. Fernandes, *Appl. Microbiol. Biotechnol.*, 2006, **73**, 291.
34. G. Ranalli, E. Zanardini, P. Pasini and A. Roda, *Ann. Microbiol.*, 2003, **53**, 1.
35. L. S. Dolci, S. Zanarini, L. Della Ciana, F. Paolucci and A. Roda, *Anal. Chem.*, 2009, **81**, 6234.
36. S. Walenta, T. Schroeder and W. Mueller-Klieser, *Biomol. Eng.*, 2002, **18**, 249.
37. P. Pasini, M. Musiani, C. Russo, P. Valenti, G. Aicardi, J. E. Crabtree, M. Baraldini and A. Roda, *J. Pharm. Biomed. Anal.*, 1998, **18**, 555.
38. U. G. A. Sattler, S. Walenta and W. Mueller-Klieser, *Lab. Invest.*, 2007, **87**, 84.
39. J. E. McCombs and A. E. Palmer, *Methods*, 2008, **46**, 152.
40. M. Brini, *Methods*, 2008, **46**, 160.
41. E. Drobac, L. Tricoire, A. F. Chaffotte, E. Guiot and B. Lambolez, *J. Neurosci. Res.*, 2010, **88**, 695.
42. V. Baubet, H. Le Mouellic, A. K. Campbell, E. Lucas-Meunier, P. Fossier and P. Brúlet, *Proc. Natl. Acad. Sci. USA*, 2000, **97**, 7260.
43. I. M. Manjarres, P. Chamero, B. Domingo, F. Molina, J. Llopis, M. T. Alonso and J. Garcia-Sancho, *Pflug. Arch. Eur. J. Phy.*, 2008, **455**, 961.
44. A. Roda, M. Guardigli, R. Ziessel, M. Mirasoli, E. Michelini and M. Musiani, *Microchem. J.*, 2007, **85**, 5.
45. M. Guardigli, M. Marangi, S. Casanova, W. F. Grigioni, E. Roda and A. Roda, *J. Histochem. Cytochem.*, 2005, **53**, 1451.
46. S. Venturoli, S. Ambretti, M. Mirasoli, D. Santini, M. Zerbini, A. Roda and M. Musiani, *Int. J. Gynecol. Pathol.*, 2008, **27**, 575.
47. F. Bonvicini, M. Mirasoli, G. Gallinella, M. Zerbini, M. Musiani and A. Roda, *Analyst*, 2007, **132**, 519.
48. G. Gentilomi, M. Musiani, A. Roda, P. Pasini, M. Zerbini, G. Gallinella, M. Baraldini, S. Venturoli and E. Manaresi, *BioTechniques*, 1997, **23**, 1076.
49. S. Ambretti, S. Venturoli, M. Mirasoli, M. La Placa, F. Bonvicini, M. Cricca, M. Zerbini, A. Roda and M. Musiani, *Br. J. Dermatol.*, 2007, **156**, 38.
50. A. Roda, M. Guardigli, E. Michelini and M. Mirasoli, *Anal. Bioanal. Chem.*, 2009, **393**, 109.
51. F. Ciruela, *Curr. Opin. Biotechnol.*, 2008, **19**, 338.
52. C. Wu, K. Mino, H. Akimoto, M. Kawabata, K. Nakamura, M. Ozaki and Y. Ohmiya, *Proc. Natl. Acad. Sci. USA*, 2009, **106**, 15599.
53. A. De, P. Ray, A. M. Loening and S. S. Gambhir, *FASEB J.*, 2009, **23**, 2702.

54. S. T. Gammon, V. M. Villalobos, M. Roshal, M. Samrakandi and D. Piwnica-Worms, *Biotechnol. Prog.*, 2009, **25**, 559.
55. P. Carriba, G. Navarro, F. Ciruela, S. Ferré, V. Casadó, L. Agnati, A. Cortés, J. Mallol, K. Fuxe, E. I Canela, C. Lluís and R. Franco, *Nat. Methods*, 2008, **5**, 7271.
56. Z. Xia and J. Rao, *Curr. Opin. Biotechnol.*, 2009, **20**, 37
57. L. S. Dolci, G. Sciutto, M. Guardigli, M. Rizzoli, S. Prati, R. Mazzeo and A. Roda, *Anal. Bioanal. Chem.*, 2008, **392**, 29.

CHAPTER 12

"In Vivo" Molecular Imaging

ERIC L. KAIJZEL,[a] THOMAS J. A. SNOEKS,[a] IVO QUE,[a]
MARTIN BAIKER,[b] PETER KOK,[b, c] BOUDEWIJN P.
LELIEVELDT[b, c] AND CLEMENS W. G. M. LÖWIK[a]

[a] Department of Endocrinology, Leiden University Medical Center,
Albinusdreef 2, 2300RC Leiden, The Netherlands; [b] Department of
Radiology, Division of Image Processing, Leiden University Medical Center,
Albinusdreef 2, 2300RC Leiden, The Netherlands; [c] Department of
Mediamatics, Delft University of Technology, Delft, The Netherlands

12.1 Whole Body Bioluminescent Imaging

Bioluminescence imaging (BLI) of luciferase reporters has been developed over
the last decade as a powerful tool for molecular imaging of small laboratory
animals. This technique provides a relatively simple, robust and extremely
sensitive means to study ongoing biological processes *in vivo* owing to exceptionally high signal-to-noise levels.[1] Various different luciferases have been
identified in nature with matching substrates available. The most commonly
used luciferase for molecular imaging purposes has been the one extracted from
the North American firefly (*Photinus pyralis*; FLuc) emitting light with a broad
emission spectrum and a peak around 560 nm, but other useful luciferases have
also been cloned from corals (*Tenilla*), jellyfish (*Aequorea*), several bacterial
species (*Vibrio fischeri*, *V. harveyi*) and red or green click beetle (*Pyrophorus
plagiophthalamus*), which have been optimized to produce green–orange
(544 nm) or red (611 nm) light after oxidizing luciferin.

Luciferases from the anthozoan sea pansy (*Renilla reniformis*) and the marine
copepod (*Gaussia princeps*) react with coelenterazine and are ATP-independent to
produce blue light with peak emission at approximately 480 nm. Despite the blue
emission wavelength of these enzymes, limited biodistribution and rapid kinetics
of coelenterazine in small animals, these luciferases have proven very useful for

Chemiluminescence and Bioluminescence: Past, Present and Future
Edited by Aldo Roda
© Royal Society of Chemistry 2011
Published by the Royal Society of Chemistry, www.rsc.org

in vivo applications for molecular imaging.[2–4] Because the substrates luciferin and coelentarazin for firefly luciferase (Fluc) and *Gaussia* luciferase (Gluc), respectively, show no cross reactivity, concomitant imaging of distinct cell populations that either express Fluc or Gluc can be performed within the same animal. This dual imaging technology has allowed monitoring of the *in vivo* trafficking of Gluc labeled T cells to an Fluc labeled tumor. Bioluminescent T cell imaging was achieved by modifying primary T cells with a membrane-anchored form of the *Gaussia* luciferase enzyme. In this way, concomitant imaging of T cells and tumor cells, which were modified to express firefly luciferase, was shown.[5] *Gaussia* luciferase has several advantages over other reporters commonly used for *in vivo* imaging in that it is 2000-fold more sensitive than firefly and *Renilla* luciferases and 20 000-fold more sensitive than the secreted alkaline phosphatase.[6,7] In addition, since *Gaussia* luciferase is secreted, its concentration in the blood correlates with expression level in biological processes in culture and *in vivo*[8] (Figure 12.1). Recently, the secreted GLuc was successfully evaluated as a biomarker for longitudinal monitoring of tumor burden and systemic metastasis in experimental metastases models of MDA-MB-231, a human breast adenocarcinoma cell line. In addition, secreted Gluc was measured to monitor treatment response to lapatinib, a tyrosine kinase inhibitor previously shown to reduce tumor outgrowth.[9] These studies showed that Gluc activity in the blood not only could track metastatic tumor progression by an accurate reflection of the amount of viable cancer cells in primary and metastatic tumors but also could serve as a longitudinal biomarker for tumor response to treatments.[10]

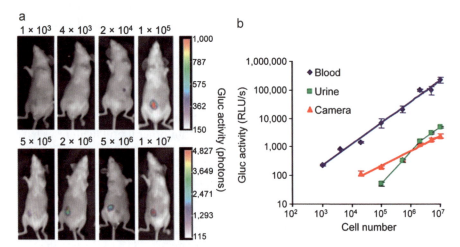

Figure 12.1 Monitoring tumor size with *Gaussia* luciferase. (a) Indicated numbers of Gli36-Gluc cells were implanted subcutaneously in mice ($n = 4$) and imaged with a CCD camera 3 days later. (b) Total relative light units (RLU) per second were calculated for tumors shown in (a). Gluc activity was measured in blood or urine using a luminometer. Results are presented as mean s.d. with $P < 0.001$ as calculated by Student's *t*-test ($n = 4$). (Adapted with permission from Macmillan Publishers Ltd: *Nat. Methods*, Wurdinger *et al.*, © 2008.)

Also, new redshifted variants of these marine luciferases, which have been accomplished recently for *Renilla* luciferase, will greatly improve their use in small-animal imaging.[11] These luciferases (RLuc and Gluc) also hold promise in *in vivo* imaging applications beyond reporter gene applications which require engineering of cells and animals to introduce the luciferases. By fusing a cancer-targeting engineered antibody to Rluc, the luciferase protein itself functions as an imaging probe, enabling the imaging of endogenous proteins and processes.[12] *Gaussia* and *Renilla* luciferases are explicitly suitable for the fusion to antibodies as their oxidation reactions are independent of ATP, the concentration of which is very low in the extracellular environment.

Despite the potential advantages of 611 nm emission for *in vivo* imaging, firefly luciferase remains the preferred enzyme and is the most frequently used for *in vitro* and *in vivo* bioluminescence molecular imaging. Its activity is ATP and O_2 dependent and, thus, only viable cells bioluminesce. Light from firefly luciferase peaks 10–12 min after injection of luciferin and decreases slowly over 60 min,[13] providing a broad time window for acquiring images. The combination of enzymatic amplification of signals from luciferase and the almost negligible background bioluminescence *in vivo* makes bioluminescence imaging with firefly luciferase a highly sensitive method for small-animal molecular imaging. It is also an excellent marker for kinetic and dynamic analyses of gene expression within short time frames because it lacks post-transcriptional modifications and has a relatively short half-life of approximately 3 h,[14,15] which makes it ideally suited for kinetic and dynamic analysis of gene expression within short time frames and, therefore, to identify circadian or even infradian rhythms of gene expression.

BLI is appealing for whole body imaging as mammalian tissues have low intrinsic bioluminescence and light is collected in the absence of external illumination sources; this results in almost no background activity, yielding in an exceptionally high signal-to-noise ratio (SNR), which makes it very sensitive and specific. Furthermore, the acquisition time of BLI measurements is short (seconds to a few minutes) compared to other imaging modalities and more animals can be analyzed at the same time.

12.2 BLI in Monitoring Cancer

BLI has become a routine modality for use in cancer biology; it is particularly suited for assessing tumor burden and metastatic spread. The most common use of BLI in cancer has been to assess mass and location of xenografted cells constitutively expressing luciferase, providing a robust strategy to monitor the effectiveness of anti-tumor drugs *in vivo*. Whole body BLI using firefly luciferase allows semi-quantitative measurements of tumor load and progression, metastasis and treatment response. Owing to the sensitivity of BLI luciferase-expression, tumor cells can be transplanted to any orthotopic site within a mouse or rat and subsequent tumor development, progression, and possible metastasis can be monitored in a rapid and time-sensitive manner. In addition,

BLI has proven very useful for the early detection of micro-metastases and minimal residual disease states in animal models.[1,16,17]

12.3   Monitoring Tumor Growth and Bone/Bone Marrow Metastases

A standard technique to induce bone metastasis is to inoculate tumor cells into the left heart ventricle to introduce tumor cells to the arterial circulation, leading to the colonization of cells to specific sites of the skeleton.[18] After this intracardiac injection of luciferase-expressing human MDA-231-B breast cancer cells (MDA-231-B/luc[+]), very small amounts of photon-emitting tumor cells can be detected in bone marrow/bone within a few days, mimicking micro-metastatic spread. A more straightforward method to induce local growth in bone marrow is the intra-osseous injection of tumor cells. Estimation of the lowest cell number detectable in bone after direct inoculation of these cells into the marrow cavity of the femur revealed that as low as 2×10^4 cells could be detected with a total volume of the estimated lesion of 0.5 mm.[3,17,19] Quantification at different time points of the bioluminescent signal localized over the site of implantation enables continuous monitoring *in vivo* of tumor growth and allows regular monitoring of the development and progression of experimental bone metastases in living animals with high sensitivity.[17,20] In animal models of xenotransplanted tumor growth and bone/bone marrow metastasis BLI not only allows the monitoring of very small metastatic deposits in bone marrow at a stage largely preceding tumor-induced osteolysis[17] but also to monitor the therapeutic efficacy of compound like bisphosphonates and BMP7 on tumor growth itself and bone metastasis.[21,22] This may help to better identify situations at risk for bone metastasis and develop novel therapeutic strategies that could be extended to the clinic.

Recent advances in understanding the molecular mechanisms of breast cancer metastasis to bone have provided initial insight into the role of TGFβ signaling in this process and revealed some of the molecular mechanisms involved in the transition of TGFβ from a tumor suppressor to a pro-oncogenic factor during tumorogenesis.[23,24] Using non-invasive bioluminescent imaging, the feasibility of monitoring the tumor-associated activity of the TGFβ/Smads proteins in bones has been shown using two related breast cancer cell lines that demonstrated different organ-specific metastatic potential.[25] In this study, the sites and expansion of metastases were visualized using a constitutive firefly luciferase reporter, while TGFβ signaling in metastases was monitored by microPET imaging and by non-secreted *Gaussia* luciferase that proved to be more sensitive and cost-effective than microPET. Concurrent and sequential bioluminescent imaging of metastases in the same animals has provided insight into the location and progression of metastases, and the timing and course of TGFβ signaling. This study nicely demonstrated the non-invasive imaging of TGFβ signal transduction pathway activity with high sensitivity and reproducibility, thereby providing the opportunity for an assessment of novel treatments that target TGFβ signaling.

Another advantage of BLI using firefly luciferase is that only metabolically active cancer cells contribute to bioluminescence production and, as such, the BLI signals are only derived from living cancer cells and neither from dead tumor cells in necrotic areas of the tumor nor from infiltrating host cells, tumor cell debris and peripheral tumor edema.[16,26] Therefore, BLI may serve as a surrogate quantitative measure of the number of metabolically active tumor cells.

12.4 BLI of Tumor-angiogenesis and Hypoxia

Furthermore, *in vivo* BLI has become indispensable for the imaging of processes like VEGF/VEGR-mediated (tumor) angiogenesis[27] or hypoxia that play important roles in cancer and cancer treatment. During tumor angiogenesis, key molecules like VEGF and VEGFR2 are locally upregulated[28] and VEGF-luc[29] and VEGFR2-luc[27] transgenic mouse models have been developed to image and quantify VEGF and VEGF receptor expression with BLI during (tumor) angiogenesis.

The VEGFR2-luc mouse has been validated using cutaneous wound healing models in which VEGFR2 expression was upregulated at the site of a punch wound marked by a clear localized increase of the luciferase signal. This increase in VEGFR2 gene activation reached a maximum after 7–10 days after the wound was inflicted.[27] Dexamethasone suppressed luciferase activity, concomitant with delayed healing and impaired angiogenesis. Using the VEGFR2-luc knock in mouse and the fluorescent SMF-mCherry breast cancer model, we were able to image and quantify angiogenesis and tumor growth real time *in vivo* (Figure 12.2).

The pVEGF-TSTA-fl mouse model also enabled the correlation between VEGF expression and BLI signal both *in vitro* and *in vivo*. This mouse model makes use of the GAL4-VP16 two-step transcriptional amplification (TSTA) system. In this system, the full length human VEGF promoter is placed upstream of the gene encoding the GAL4-VP16 fusion protein. Two GAL4 binding sites are placed upstream of the adenovirus E4 TATA minimal promoter driving the luciferase reporter gene, resulting in GAL4-induced luciferase expression. The TSTA system has been used to amplify prostate specific luciferase expression, leading to a 50-fold increase over the direct, one-step system.[30] This transgenic animal has also been used to demonstrate the correlation between VEGF expression and BLI signal both *in vitro* and *in vivo*. In addition, the authors show that this transgenic animal can be used to study the VEGF response in both wound healing assays as well as tumor growth.[29]

Key initiators of tumor angiogenesis like hypoxia and the subsequent stabilization of hypoxia inducible factor-1 (HIF-1) that binds hypoxia responsive elements (HREs) can also be analyzed with BLI-techniques. These HREs regulate the expression of pro-angiogenic signaling molecules like, among

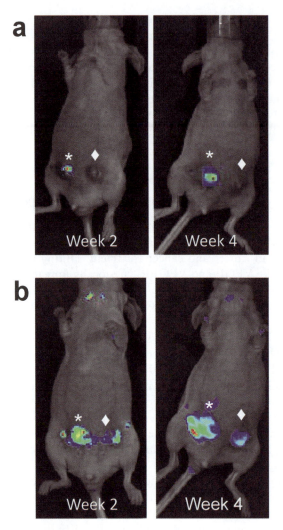

Figure 12.2   Multi-modality imaging of tumor growth and angiogenesis. VEGFR2-
luc knock-in mice were orthotopically, *i.e.* in the mammary fat pad,
inoculated with 1×10^6 SMF-mCherry cells (marked with *) and SMF
wild-type cells (marked with ◆), a murine breast cancer cell line. SMF-
mCherry cells constitutively express the far-red fluorescent protein
mCherry, which accumulates in the cytoplasm. Mice were imaged at
week 2 and week 4, both FLI (a) and BLI (b), using a IVIS Spectrum
(Caliper LifeSciences, USA) camera system. (a) Tumor growth of the
mCherry positive tumor could be followed over time using FLI. (b) The
BLI signal was present at the site of both the mCherry positive and
negative tumors, indicating a local upregulation of VEGFR2 expression
and tumor angiogenesis. (Kaijzel *et al.*, unpublished data.)

others, VEGF and its receptor. To be able to react to changes in oxygen tension, HIF-1 has an oxygen dependent degradation domain (ODD), which is hydroxylated in an oxygen dependent manner by prolyl hydroxylases. Under normoxic conditions, the ODD is hydroxylated and binds the Von Hippel–Lindau factor (VHL) targeting HIF-1 for proteasomal degradation. VHL is unable to bind to the ODD under hypoxic conditions, leading to stabilization of HIF-1 and the initiation of downstream signaling.[31]

To enable real time measurement of both *in vitro* as well as *in vivo* HIF-1 activity and stability, reporter constructs have been developed in which luciferase is driven by HREs.[32–34] The presence of HIF-1 leads to a significant upregulation of the reporter gene expression. HRE-Luciferase reporter constructs have been shown to be valuable tools to evaluate tumor hypoxia and the efficacy of hypoxia directed therapies *in vivo*.[33,34] In these studies, bioluminescence, a process requiring ATP and O_2, did not seem to be hampered by the low PO_2 in the tissue.

In another approach to visualise hypoxia and HIF-1 activity, a fusion protein consisting of the ODD of HIF-1 coupled to luciferase is expressed under a constitutively active promoter.[35] Similar to HIF-1, the ODD-Luc protein is targeted for proteasomal breakdown under normoxic conditions, but stabilizes under hypoxic conditions. These whole body BLI tools have proven very useful for studying the biology of hypoxia and mechanisms of response to experimental therapy.

12.5 BLI in other Disease Areas

Apart from cancer research, *in vivo* applications of BLI have been proven valuable in many other disease areas such as in central nervous system disorders,[36–38] cardiovascular disease,[39–41] infectious disease,[42–44] immunology and transplantation biology. BLI has been proven valuable in mouse models of graft *versus* host disease to non-invasively monitor the behavior, migration and fate of Fluc and labeled immune cells like T regulatory T-cells, cytotoxic T-cells or cytokine induced killer cells (CIKs).[45,46] By labeling T-cells genetically modified to express tumor–specific antigen receptors (CARs) with *Gaussia* luciferase, dual T cell Fluc + tumor cell imaging showed *in vivo* CAR-mediated T cells in tumors and concomitant imaging of T cells and tumor cells.[5]

Furthermore, the expression of luciferase can be controlled so that it is only expressed when a gene of interest is being transcribed. The use of optical reporter genes in either transplanted cells or in transgenic animals provides the opportunity to study gene-expression, -regulation and -function.

Another example of a disease area where BLI has enabled a whole new line of research is malaria. The life cycle of different species of malaria parasites, collectively called *Plasmodium*, has a blood stage and a liver stage. The liver is the main site of intracellular development of *Plasmodium* after infection. Furthermore, *Plasmodium* can stay clinically silent and dormant in the

liver over prolonged periods of time. Most drugs solely target the blood
stage of the parasite, leaving the parasites in the liver. Until recently it was
nearly impossible to specifically research treatment effects on liver stage
Plasmodium.

The luciferase expressing *P. berghei*[47–49] made it possible to image real time
blood distribution of the parasite and evaluate treatment efficacy *in vivo* over
time. The newly described *Pb*GFP-Luc_con (also a *P. berghei*) has been used to
study both blood and liver stage parasites *in vitro* and *in vivo*. Previously, it
would have been impossible to follow treatment efficacy on liver stage parasites
over time *in vivo*[50] (Figure 12.3).

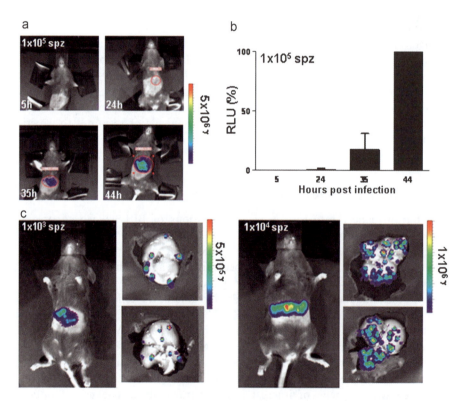

Figure 12.3   Bioluminescence imaging in livers of mice at different time points after
injection of sporozoites. (a) Non-invasive BLI in livers at different time
points. (b) Luminescence levels (photons s^{-1}) of livers in whole mice at
different time points after infection with 1×10^5 sporozoites ($n = 4$).
Photon counts from whole body imaging are expressed as the percentage
of the photon counts of mice at 44 h after infection ($= $RLU %). (c)
Distribution of luminescence signals in the livers of live mice and in
extracted livers of the same mice at 44 h after infection with 1×10^3 (left)
or 1×10^4 (right) of sporozoites. (Adapted from ref. 50.)

12.6 Transgenic Mouse Models Expressing Luciferase

Animal models modified to express luciferase are very useful to model disease pathways, providing information on where a compound or drug affects a particular pathway. In addition to utility in drug development and toxicology, induction of the luciferase reporter in a transgenic luciferase expressing animal model in response to a disease condition or a drug treatment may provide critical information about the role of that gene in a disease pathway. Nowadays, a growing number of luciferase expressing animal models are (commercially) available for drug metabolism and toxicology, disease areas like oncology/angiogenesis, metabolic and neurodegenerative diseases and inflammation. BLI can be used to monitor inflammation by driving luciferase with inflammation-specific regulatory sequences. For instance, Carlsen *et al.* have generated mice expressing luciferase under the control of the regulatory sequences of the NFκB gene, which in its turn is under the direct control of TNFα, a key cytokine produced during inflammation. Using NFκB-luciferase mice, Carlsen *et al.*[51] monitored osteoarthritic inflammation induced by injection of bacterial lipopolysaccharide, and quantified the therapeutic potential of dexamethasone treatment of the arthritic lesion. Using a similar approach, iNOS-luciferase mice were used to image zymosan-induced arthritis in the knee joint.[52] Most of these models consist of the mouse promoter driving the luciferase reporter and many of them are promoted by Xenogen (now Caliper LifeSciences, see www.caliperls.com).

Apart from these tissue-specific or inducible luciferase-based transgenics a growing number of transgenic animal models have been generated that represent spontaneous mouse models of cancer development that can be followed with whole body BLI. In these transgenics, a basal expression level of luciferase in the targeted tissues will strongly and steadily increase when tumor formation and progression starts. Similarly, the development of metastases in distant organs can also be detected by BLI. Examples of these models have been accomplished by crossing transgenics with tissue specific expression of luciferase, *i.e.*, in pituitary (POMC-luc)[53] or prostate (PSA-luc; luciferase expression targeted to prostate specific PSA expression),[54] with mice that spontaneously develop tissue specific cancers, like Cre-inducible Rb knock-out mice[53] that develop pituitary cancer and KIMAP mouse model of spontaneous prostate cancer development.[55,56] This knock-in mouse adenocarcinoma prostate model (KIMAP) was established by targeting the PSP94 gene with a tumor-inducer gene (SV40 Tag). The KIMAP model shows the applicability of the Gleason histological grading system, which is widely used in the clinical diagnosis and prognosis of human prostate cancer. First cross-breeds of PSA-luc and KIMAP have been generated recently and followed over time. Initial experiments are very promising as photon emission was found in the prostate region and increased consistently, starting at 20–24-weeks of age, which is coincident with the onset of spontaneous CaP tumors in mice. Dissection of the animals showed that luciferase expression was confined to the prostate; after removal of the prostate, local metastases could also be detected (Figure 12.4).

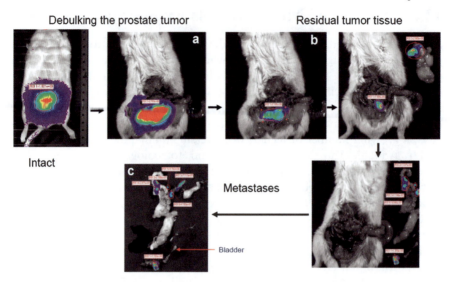

Figure 12.4 BLI of prostate tumor growth and metastasis in a spontaneous model of prostate cancer. PSA-luc transgenics were crossed with KIMAP transgenic mice and prostate tumor formation in male offspring was followed over time. After removal of the primary prostate tumor (a), residual tumor tissue (b) and metastatic deposits (c) could be detected with bioluminescence imaging. (Kaijzel *et al.*, unpublished data.)

12.7 Stem Cell Migration/Differentiation

Non-invasive BLI has also been widely utilized in stem cell research. Making use of different forms of luciferases, Wang *et al.*[57] have studied the engraftment and differentiation of mesenchymal stem cells (MSCs) in tumor-bearing mice. The fate and tumor tropism of injected Fluc-eGFP labeled MSCs were monitored by Fluc BLI while tumors were labeled with Rluc-mRFP to follow growth and metastasis and measured by Rluc BLI. This study elegantly showed that the engrafted MSCs selectively can localize, survive and proliferate in both subcutaneous tumor and lung metastasis by non-invasive bioluminescence imaging and *ex vivo* validation. Another study assessed the therapeutic efficacy and fate of engineered human mesenchymal stem cells for cancer therapy. When engineered to express secreted recombinant TRAIL, these Fluc-transduced cells could be tracked and were able to induce caspase-mediated apoptosis in gliomas, demonstrating the efficacy of diagnostic and therapeutic MSCs in preclinical glioma models. Such studies form a basis for developing stem cell-based therapies for different cancers.[58] In cardiovascular research, using a mouse model for myocardial infarction, the fate and survival of Fluc labeled embryonic stem cells were tracked with BLI for up to 8 weeks following injection into ischemic hearts, showing that the observed increased BLI signal correlated with improved neovascularization and cardiac function.[59] Further investigation into cardiovascular regenerative therapy was made by comparison of the *in vivo* fate and behavior of different cell types injected in the infarcted heart.[60,61] Using BLI, these studies showed that adipose stromal cells and bone marrow derived cells do not tolerate well the cardiac environment, resulting in acute donor cell death and a subsequent loss of cardiac function. BLI has also been employed in studies of the central nervous system *in vivo*. By engineering cells with different combinations of bioluminescent and fluorescent markers the fate of murine and human neural stem cells in glioma-bearing brains *in vivo* can be followed in real time with BLI.[62,63]

12.8 Going from 2D Qualitative to 3D Quantitative Whole Body Optical Imaging

In recent years, there has been a rapid growth of bioluminescent imaging applications in small animal models of disease driven by creative approaches to apply this technology in molecular imaging and the availability of BLI-instruments with advanced analysis software. Compared to other imaging modalities like MRI and PET/SPECT-imaging, advantages of bioluminescent imaging include the sensitivity and versatility of the technique and its low costs. Most current bioluminescent imaging techniques are performed in the 2D mode. However, this 2D planar imaging is primarily qualitative and hampered by relatively poor spatial resolution. Quantification of the photon emission is at best semi-quantitative as only relative changes in signal intensity can be measured over time. New developments in bioluminescence tomography (BLT) allow better quantification of photon emission[64,65] and localized analyses on a bioluminescent source distribution have become feasible in a mouse, which reveal molecular and cellular signatures critically important for numerous biomedical studies and applications. This quantitative 3D bioluminescent source information obtained by BLT can directly and more accurately reflect biological changes as opposed to 2D planar BLI. The localization and quantification accuracy of the reconstruction results achieved in optical tomographic methods depends on the depth and tissue dimensions and the optical properties of the tissues.

Combining bioluminescent imaging with other imaging modalities that are complementary can enhance the data sets by improving the optical image, and by correlating function and structure. Whole body optical imaging has been combined with X-ray[66] and magnetic resonance imaging (MRI) to yield informative images of mammalian biology.[67] In addition, fusing 3D optical images with 3D datasets obtained from the same animal using other MRI or CT will allow structural anatomic information to be obtained and will greatly enhance spatial resolution. Furthermore, this structural tissue information obtained by CT or MRI, in combination with a mouse tissue atlas, can be used in an attempt to correct for tissue-dependent photon scattering and absorption. In metastatic bone disease, there is a clear benefit in combining 3D BLI with CT. Co-registration of tomographic BLI data sets with CT images has provided information on tumor location from the optical imaging with high resolution

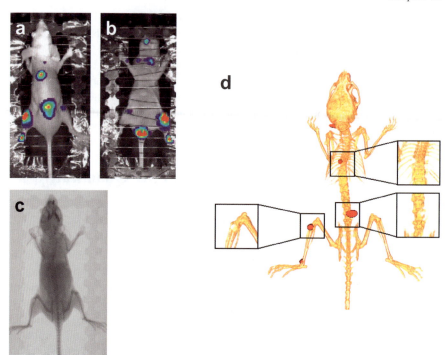

Figure 12.5 Co-registration of tomographic BLI data sets with CT. Fused bioluminescence and micro-computed tomography visualization of a mouse
intracardially injected with MDA-231-B/luc+ breast cancer cells that
developed extended bone metastases in spine, knee and ankle joints.
Three to four weeks after intracardially inoculation of the MDA-231-B/
luc+ cells in nude mice (*n* = 3), bone metastases where analyzed with an
IVIS 3D BLI Imaging system (Caliper LS, Hopkinton, MA, USA).
Depicted are the BLI images of a representative mouse in supine (a) and
prone position (b). The animal was subsequently scanned in a SkyScan
1178 µCT scanner (SkyScan, Kontich, Belgium) (c). The cut-outs of the
volume visualization of the fusion of the CT-images with the metastatic
tumors to bone (d) clearly show bone destruction at the location of the
metastatic breast cancer lesions in the computed tomography data,
whereas the bioluminescence sources (red) highlight the potential lesion
locations. Note that only the metastatic tumors to bone have been
depicted in this fusion. (Adapted from Kaijzel *et al.*, *Clin. Exp. Metastasis*, 2009, **26**, 371–379.)

structural details on the skeleton from CT imaging[68] (Figure 12.5). This enables
the direct study of the interaction between breast cancer metastasis and the
skeletal system from the combined imaging, which would not have been possible with any of the individual imaging modalities alone. Careful consideration
should be made when determining the number of CT-scans as these scanners
impose a relatively high ionizing radiation dose that may cause tissue damage
in longitudinal studies.[69] Fusing of tomographic optical images with CT or

MRI will provide structural anatomic information with enhanced spatial resolution. In addition, MR signals can localize sites of metastases prior to bone destruction or formation evident on μCT. Multimodality imaging is a promising way to register and relate different imaging data into a singular context.[70]

12.9 Drug Development and Therapeutic Intervention

Optical imaging, and bioluminescent imaging in particular, allows non-invasive, rapid and sensitive testing of (innovative) drugs and therapies for the treatment of cancer in animal models in relative high throughput compared to conventional drug testing.[71–73] The development of a fast growing number of BLI-based mouse models of disease, as discussed in an earlier section, enables the longitudinal monitoring of cytocidal effects of anti-neoplastic and antibiotic drugs in tumor burden,[74] metastatic dissemination,[22,75] as well as viral[76] and bacterial infections.[77] BLI allows spatiotemporal and quantitative analysis of tumor growth and, due to its sensitivity, is ideally suited to evaluate the effectiveness of therapeutic approaches that target both early stages of metastatic development and advanced metastatic disease. It gives detailed information on localization and growth of minimal metastatic deposits in the bone marrow of experimental animals at a stage that largely precedes tumor detection by other imaging modalities. Apart from accelerating drug development, the use of bioluminescent imaging will also lead to a faster optimization of new therapies. In bone metastatic studies, BLI can be used to assess the effect of drugs in the early phase of the disease, especially the localization and growth of tumor cells within the bone and even before osteolysis occurs, whereas radiography (X-ray and CT) will only monitor bone destruction by detecting osteolytic lesions.[21,22] Furthermore, the efficacy of drugs in bone metastatic studies can be assessed much faster by optical methods than by (conventional) X-ray staging, thus reducing the amount of animal suffering. Besides that, less laboratory animals are needed as, due to the non-invasive nature of the methods, repetitive measurements can be taken from the same animal, which also increases the reliability of observed effects. Using tomographic approaches it is now also possible to obtain better quantitative data. Owing to its sensitivity and simplicity it is now also widely used in drug development and drug screening.

12.10 Conclusion

BLI is a powerful and relatively simple non-invasive tool for longitudinal assessment of all kinds of biological processes in small laboratory animals that play a role in health and disease like cell migration, differentiation and fate as well as processes like hypoxia, angiogenesis, proteolysis, apoptosis, inflammation, infections, protein–protein interactions, gene-expression and signal transduction. BLI is easy to perform and multiple animals can be analyzed at the same time, yielding a very reasonable throughput. Furthermore, it also has an excellent sensitivity compared with other non-invasive imaging modalities. Therefore, BLI is perfectly suited to rapidly study the effects of new drugs and therapies. Two-dimensional planar BLI imaging does not provide information on anatomical location and has poor spatial resolution, but by using trans-illumination or BLI images from multiple angles 3D images can be reconstructed. When these 3D images are combined with information of a tissue atlas or combined with CT or MRI images the spatial resolution and anatomical localization strongly improves. We can conclude that bioluminescence imaging has become an important component of modern translational biomedical research and drug screening that will continue in the future.

References

1. C. H. Contag and M. H. Bachmann, *Annu. Rev. Biomed. Eng*, 2002, **4**, 235.
2. S. Bhaumik and S. S. Gambhir, *Proc. Natl. Acad. Sci. USA*, 2002, **99**, 377.
3. B. A. Tannous, D. E. Kim, J. L. Fernandez, R. Weissleder and X. O. Breakefield, *Mol. Ther.*, 2005, **11**, 435.
4. A. Pichler, J. L. Prior and D. Piwnica-Worms, *Proc. Natl. Acad. Sci. USA*, 2004, **101**, 1702.
5. E. B. Santos, R. Yeh, J. Lee, Y. Nikhamin, B. Punzalan, B. Punzalan, K. La Perle, S. M. Larson, M. Sadelain and R. J. Brentjens, *Nat. Med.*, 2009, **15**, 338.
6. C. E. Badr, J. W. Hewett, X. O. Breakefield and B. A. Tannous, *PLoS One*, 2007, **2**, e571.
7. B. A. Tannous, A. P. Christensen, L. Pike, T. Wurdinger, K. F. Perry, O. Saydam, A. H. Jacobs, J. Garcia-Anoveros, R. Weissleder, M. Sena-Esteves, D. P. Corey and X. O. Breakefield, *Mol. Ther.*, 2009, **17**, 810.
8. T. Wurdinger, C. Badr, L. Pike, R. de Kleine, R. Weissleder, X. O. Breakefield and B. A. Tannous, *Nat. Methods*, 2008, **5**, 171.
9. B. Gril, D. Palmieri, J. L. Bronder, J. M. Herring, E. Vega-Valle, L. Feigenbaum, D. J. Liewehr, S. M. Steinberg, M. J. Merino, S. D. Rubin and P. S. Steeg, *J. Natl. Cancer Inst.*, 2008, **100**, 1092.
10. E. Chung, H. Yamashita, P. Au, B. A. Tannous, D. Fukumura and R. K. Jain, *PLoS One*, 2009, **4**, e8316.
11. A. M. Loening, A. M. Wu and S. S. Gambhir, *Nat. Methods*, 2007, **4**, 641.
12. K. M. Venisnik, T. Olafsen, A. M. Loening, M. Iyer, S. S. Gambhir and A. M. Wu, *Protein Eng. Des. Sel.*, 2006, **19**, 453.
13. Z. Paroo, R. A. Bollinger, D. A. Braasch, E. Richer, D. R. Corey, P. P. Antich and R. P. Mason, *Mol. Imag.*, 2004, **3**, 117.
14. J. F. Thompson, L. S. Hayes and D. B. Lloyd, *Gene*, 1991, **103**, 171.
15. G. S. Lipshutz, C. A. Gruber, Y. Cao, J. Hardy, C. H. Contag and K. M. Gaensler, *Mol. Ther.*, 2001, **3**, 284.

16. M. Edinger, Y. A. Cao, Y. S. Hornig, D. E. Jenkins, M. R. Verneris, M. H. Bachmann, R. S. Negrin and C. H. Contag, *Eur. J. Cancer*, 2002, **38**, 2128.

17. A. Wetterwald, P. G. van der, I. Que, B. Sijmons, J. Buijs, M. Karperien, C. W. Lowik, E. Gautschi, G. N. Thalmann and M. G. Cecchini, *Am. J. Pathol.*, 2002, **160**, 1143.

18. F. Arguello, R. B. Baggs and C. N. Frantz, *Cancer Res.*, 1988, **48**, 6876.

19. G. van der Pluijm, B. Sijmons, H. Vloedgraven, M. Deckers, S. Papapoulos and C. Lowik, *J. Bone Miner. Res.*, 2001, **16**, 1077.

20. L. M. Kalikin, A. Schneider, M. A. Thakur, Y. Fridman, L. B. Griffin, R. L. Dunn, T. J. Rosol, R. B. Shah, A. Rehemtulla, L. K. McCauley and K. J. Pienta, *Cancer Biol. Ther.*, 2003, **2**, 656.

21. J. T. Buijs, N. V. Henriquez, P. G. van Overveld, H. G. van der, I. Que, R. Schwaninger, C. Rentsch, P. ten Dijke, A. M. Cleton-Jansen, K. Driouch, R. Lidereau, R. Bachelier, S. Vukicevic, P. Clezardin, S. E. Papapoulos, M. G. Cecchini, C. W. Lowik and P. G. van der, *Cancer Res.*, 2007, **67**, 8742.

22. G. van der Pluijm, I. Que, B. Sijmons, J. T. Buijs, C. W. Lowik, A. Wetterwald, G. N. Thalmann, S. E. Papapoulos and M. G. Cecchini, *Cancer Res.*, 2005, **65**, 7682.

23. R. Derynck, R. J. Akhurst and A. Balmain, *Nat. Genet.*, 2001, **29**, 117.

24. J. Massague, *Cell*, 2008, **134**, 215.

25. I. Serganova, E. Moroz, J. Vider, G. Gogiberidze, M. Moroz, N. Pillarsetty, M. Doubrovin, A. Minn, H. T. Thaler, J. Massague, J. Gelovani and R. Blasberg, *FASEB J.*, 2009, **23**, 2662.

26. S. K. Lyons, *J. Pathol.*, 2005, **205**, 194.

27. N. Zhang, Z. Fang, P. R. Contag, A. F. Purchio and D. B. West, *Blood*, 2004, **103**, 617.

28. T. T. Rissanen, I. Vajanto, M. O. Hiltunen, J. Rutanen, M. I. Kettunen, M. Niemi, P. Leppanen, M. P. Turunen, J. E. Markkanen, K. Arve, E. Alhava, R. A. Kauppinen and S. Yla-Herttuala, *Am. J. Pathol.*, 2002, **160**, 1393.

29. Y. Wang, M. Iyer, A. Annala, L. Wu, M. Carey and S. S. Gambhir, *Physiol Genomics*, 2006, **24**, 173.

30. M. Iyer, L. Wu, M. Carey, Y. Wang, A. Smallwood and S. S. Gambhir, *Proc. Natl. Acad. Sci. USA*, 2001, **98**, 14595.

31. J. M. Gleadle, *Nephrology (Carlton)*, 2009, **14**, 86.

32. T. Shibata, A. J. Giaccia and J. M. Brown, *Gene Ther.*, 2000, **7**, 493.

33. S. Lehmann, D. P. Stiehl, M. Honer, M. Dominietto, R. Keist, I. Kotevic, K. Wollenick, S. Ametamey, R. H. Wenger and M. Rudin, *Proc. Natl. Acad. Sci. USA*, 2009, **106**, 14004.

34. H. Harada, S. Kizaka-Kondoh and M. Hiraoka, *Mol. Imag.*, 2005, **4**, 182.

35. F. Li, P. Sonveaux, Z. N. Rabbani, S. Liu, B. Yan, Q. Huang, Z. Vujaskovic, M. W. Dewhirst and C. Y. Li, *Mol. Cell*, 2007, **26**, 63.

36. U. Abraham, J. L. Prior, D. Granados-Fuentes, D. R. Piwnica-Worms and E. D. Herzog, *J. Neurosci.*, 2005, **25**, 8620.

37. A. C. Liu, D. K. Welsh, C. H. Ko, H. G. Tran, E. E. Zhang, A. A. Priest, E. D. Buhr, O. Singer, K. Meeker, I. M. Verma, F. J. Doyle, III, J. S. Takahashi and S. A. Kay, *Cell*, 2007, **129**, 605.

38. J. Luo, A. H. Lin, E. Masliah and T. Wyss-Coray, *Proc. Natl. Acad. Sci. USA*, 2006, **103**, 18326.
39. Z. Li, J. C. Wu, A. Y. Sheikh, D. Kraft, F. Cao, X. Xie, M. Patel, S. S. Gambhir, R. C. Robbins, J. P. Cooke and J. C. Wu, *Circulation*, 2007, **116**, I46.
40. K. E. van der Bogt, A. Y. Sheikh, S. Schrepfer, G. Hoyt, F. Cao, K. J. Ransohoff, R. J. Swijnenburg, J. Pearl, A. Lee, M. Fischbein, C. H. Contag, R. C. Robbins and J. C. Wu, *Circulation*, 2008, **118**, S121.
41. M. Huang, D. A. Chan, F. Jia, X. Xie, Z. Li, G. Hoyt, R. C. Robbins, X. Chen, A. J. Giaccia and J. C. Wu, *Circulation*, 2008, **118**, S226.
42. M. A. Hutchens, K. E. Luker, J. Sonstein, G. Nunez, J. L. Curtis and G. D. Luker, *PLoS Pathog.*, 2008, **4**, e1000153.
43. I. J. Glomski, A. Piris-Gimenez, M. Huerre, M. Mock and P. L. Goossens, *PLoS Pathog.*, 2007, **3**, e76.
44. O. Disson, S. Grayo, E. Huillet, G. Nikitas, F. Langa-Vives, O. Dussurget, M. Ragon, A. Le Monnier, C. Babinet, P. Cossart and M. Lecuit, *Nature*, 2008, **455**, 1114.
45. V. H. Nguyen, S. Shashidhar, D. S. Chang, L. Ho, N. Kambham, M. Bachmann, J. M. Brown and R. S. Negrin, *Blood*, 2008, **111**, 945.
46. S. H. Thorne, R. S. Negrin and C. H. Contag, *Science*, 2006, **311**, 1780.
47. B. Franke-Fayard, D. Djokovic, M. W. Dooren, J. Ramesar, A. P. Waters, M. O. Falade, M. Kranendonk, A. Martinelli, P. Cravo and C. J. Janse, *Int. J. Parasitol.*, 2008, **38**, 1651.
48. B. Franke-Fayard, C. J. Janse, M. Cunha-Rodrigues, J. Ramesar, P. Buscher, I. Que, C. Lowik, P. J. Voshol, M. A. den Boer, S. G. van Duinen, M. Febbraio, M. M. Mota and A. P. Waters, *Proc. Natl. Acad. Sci. USA*, 2005, **102**, 11468.
49. B. Franke-Fayard, A. P. Waters and C. J. Janse, *Nat. Protoc.*, 2006, **1**, 476.
50. I. H. Ploemen, M. Prudencio, B. G. Douradinha, J. Ramesar, J. Fonager, G. J. van Gemert, A. J. Luty, C. C. Hermsen, R. W. Sauerwein, F. G. Baptista, M. M. Mota, A. P. Waters, I. Que, C. W. Lowik, S. M. Khan, C. J. Janse and B. M. Franke-Fayard, *PLoS One*, 2009, **4**, e7881.
51. H. Carlsen, J. O. Moskaug, S. H. Fromm and R. Blomhoff, *J. Immunol.*, 2002, **168**, 1441.
52. N. Zhang, A. Weber, B. Li, R. Lyons, P. R. Contag, A. F. Purchio and D. B. West, *J. Immunol.*, 2003, **170**, 6307.
53. M. Vooijs, J. Jonkers, S. Lyons and A. Berns, *Cancer Res.*, 2002, **62**, 1862.
54. S. K. Lyons, E. Lim, A. O. Clermont, J. Dusich, L. Zhu, K. D. Campbell, R. J. Coffee, D. S. Grass, J. Hunter, T. Purchio and D. Jenkins, *Cancer Res.*, 2006, **66**, 4701.
55. M. Y. Gabril, W. Duan, G. Wu, M. Moussa, J. I. Izawa, C. J. Panchal, H. Sakai and J. W. Xuan, *Mol. Ther.*, 2005, **11**, 348.
56. W. Duan, M. Y. Gabril, M. Moussa, F. L. Chan, H. Sakai, G. Fong and J. W. Xuan, *Oncogene*, 2005, **24**, 1510.
57. H. Wang, F. Cao, A. De, Y. Cao, C. Contag, S. S. Gambhir, J. C. Wu and X. Chen, *Stem Cells*, 2009, **27**, 1548.

58. L. S. Sasportas, R. Kasmieh, H. Wakimoto, S. Hingtgen, J. A. van de Water, G. Mohapatra, J. L. Figueiredo, R. L. Martuza, R. Weissleder and K. Shah, *Proc. Natl. Acad. Sci. USA*, 2009, **106**, 4822.

59. Z. Li, J. C. Wu, A. Y. Sheikh, D. Kraft, F. Cao, X. Xie, M. Patel, S. S. Gambhir, R. C. Robbins, J. P. Cooke and J. C. Wu, *Circulation*, 2007, **116**, I46.

60. K. E. van der Bogt, S. Schrepfer, J. Yu, A. Y. Sheikh, G. Hoyt, J. A. Govaert, J. B. Velotta, C. H. Contag, R. C. Robbins and J. C. Wu, *Transplantation*, 2009, **87**, 642.

61. K. E. van der Bogt, A. Y. Sheikh, S. Schrepfer, G. Hoyt, F. Cao, K. J. Ransohoff, R. J. Swijnenburg, J. Pearl, A. Lee, M. Fischbein, C. H. Contag, R. C. Robbins and J. C. Wu, *Circulation*, 2008, **118**, S121.

62. Y. Waerzeggers, M. Klein, H. Miletic, U. Himmelreich, H. Li, P. Monfared, U. Herrlinger, M. Hoehn, H. H. Coenen, M. Weller, A. Winkeler and A. H. Jacobs, *Mol. Imag.*, 2008, **7**, 77.

63. K. Shah, S. Hingtgen, R. Kasmieh, J. L. Figueiredo, E. Garcia-Garcia, A. Martinez-Serrano, X. Breakefield and R. Weissleder, *J. Neurosci.*, 2008, **28**, 4406.

64. A. J. Chaudhari, F. Darvas, J. R. Bading, R. A. Moats, P. S. Conti, D. J. Smith, S. R. Cherry and R. M. Leahy, *Phys. Med. Biol.*, 2005, **50**, 5421.

65. C. Kuo, O. Coquoz, T. L. Troy, H. Xu and B. W. Rice, *J. Biomed. Opt.*, 2007, **12**, 024007.

66. Q. Zhang, T. J. Brukilacchio, A. Li, J. J. Stott, T. Chaves, E. Hillman, T. Wu, M. Chorlton, E. Rafferty, R. H. Moore, D. B. Kopans and D. A. Boas, *J. Biomed. Opt.*, 2005, **10**, 024033.

67. L. S. Bouchard, M. S. Anwar, G. L. Liu, B. Hann, Z. H. Xie, J. W. Gray, X. Wang, A. Pines and F. F. Chen, *Proc. Natl. Acad. Sci. USA*, 2009, **106**, 4085.

68. P. Kok, J. Dijkstra, C. P. Botha, F. H. Post, E. L. Kaijzel, I. Que, C. W. G. M. Löwik, J. H. C. Reiber and B. P. F. Lelieveldt, *Integrated visualization of multi-angle bioluminescence imaging and micro CT*, in *Medical Imaging 2007: Visualization and Image-Guided Procedures*, ed. K. R. Cleary and M. I. Miga, SPIE, Bellingham, WA, 2007, Proceedings SPIE vol. 6509.

69. J. H. Waarsing, J. S. Day, J. C. van der Linden, A. G. Ederveen, C. Spanjers, N. De Clerck, A. Sasov, J. A. Verhaar and H. Weinans, *Bone*, 2004, **34**, 163.

70. M. Doubrovin, I. Serganova, P. Mayer-Kuckuk, V. Ponomarev and R. G. Blasberg, *Bioconjugate Chem.*, 2004, **15**, 1376.

71. M. Rudin, M. Rausch and M. Stoeckli, *Mol. Imag. Biol.*, 2005, **7**, 5.

72. W. S. El Deiry, C. C. Sigman and G. J. Kelloff, *J. Clin. Oncol.*, 2006, **24**, 3261.

73. S. Gross and D. Piwnica-Worms, *Curr. Opin. Chem. Biol.*, 2006, **10**, 334.

74. C. M. Shachaf, A. M. Kopelman, C. Arvanitis, A. Karlsson, S. Beer, S. Mandl, M. H. Bachmann, A. D. Borowsky, B. Ruebner, R. D. Cardiff, Q. Yang, J. M. Bishop, C. H. Contag and D. W. Felsher, *Nature*, 2004, **431**, 1112.

75. M. Deckers, M. van Dinther, J. Buijs, I. Que, C. Lowik, P. G. van der and P. ten Dijke, *Cancer Res.*, 2006, **66**, 2202.

76. G. D. Luker, J. P. Bardill, J. L. Prior, C. M. Pica, D. Piwnica-Worms and D. A. Leib, *J. Virol.*, 2002, **76**, 12149.

77. Y. A. Yu, S. Shabahang, T. M. Timiryasova, Q. Zhang, R. Beltz, I. Gentschev, W. Goebel and A. A. Szalay, *Nat. Biotechnol.*, 2004, **22**, 313.

CHAPTER 13

Biotechnological Improvements of Bioluminescent Systems

KRYSTAL TEASLEY HAMORSKY,[a] EMRE DIKICI,[a]
C. MARK ENSOR,[a] SYLVIA DAUNERT,[a] AUDREY L.
DAVIS[b] AND BRUCE R. BRANCHINI[b]

[a] Department of Chemistry, University of Kentucky, Lexington, Kentucky
KY, 40506-0055, USA; [b] Department of Chemistry, Connecticut College,
New London, Connecticut, CT 06320, USA

13.1 Designer Bioluminescent Systems in Biotechnology

13.1.1 Introduction

Biotechnology is a rapidly growing field with seemingly endless applications for environmental, biological and medical sciences. Genetic engineering, a focal point of modern biotechnology, is a sophisticated tool used for the creation of new biomolecules with unique and distinct properties. Genetic engineering has been instrumental in the creation of new molecular entities with unique characteristics, including genetically engineered medicines, knockout genes, analyte tracking and many more. Herein we focus on the use of genetic engineering for the development of modified bioluminescent proteins, specifically photoproteins, for detection and tracking of virtually any molecule of interest. The ability to improve the performance of such bioluminescent systems will be highlighted.

Chemiluminescence and Bioluminescence: Past, Present and Future
Edited by Aldo Roda
© Royal Society of Chemistry 2011
Published by the Royal Society of Chemistry, www.rsc.org

13.1.2 Bioluminescence

The term bioluminescence is a word that originates from the Greek word, *bios*, for "living" and the Latin word, *lumos*, for "light." It is a naturally occurring form of chemiluminescence where energy (40–60 kcal) is released from a chemical reaction in the form of visible light. There are many different types of organisms capable of emitting light ranging from bacteria and fungi, to mollusks, crustaceans, insects, fish and plants.[1,2] This curious phenomenon was first described by Aristotle (384–322 BC) as the light emitted from decaying wood.[1] Since then, scientists have studied bioluminescence in depth with the aim of understanding how it is produced.

There are two general types of bioluminescence. The first type involves a biochemical reaction where the total amount of the emitted light is directly proportional to the amount of an organic compound, known as luciferin, present in the organism. The light-emitting reaction is catalyzed by luciferase, an enzyme responsible for the oxidation of luciferin, resulting in the production of light. Scheme 13.1 depicts this biochemical reaction. Detailed information about the luciferin/luciferase type of bioluminescence can be found in Section 13.2. The second type of bioluminescence involves a different reaction mechanism and is used by the photoproteins. This type is discussed in Section 13.1.3.

13.1.3 Bioluminescent Photoproteins

In 1961, Osamu Shimomura and his colleagues isolated and identified a new type of protein that was luminescent. Until then, all known bioluminescence reactions involved some type of a luciferin–luciferase reaction. The protein Shimomura and colleagues discovered was capable of emitting light in aqueous

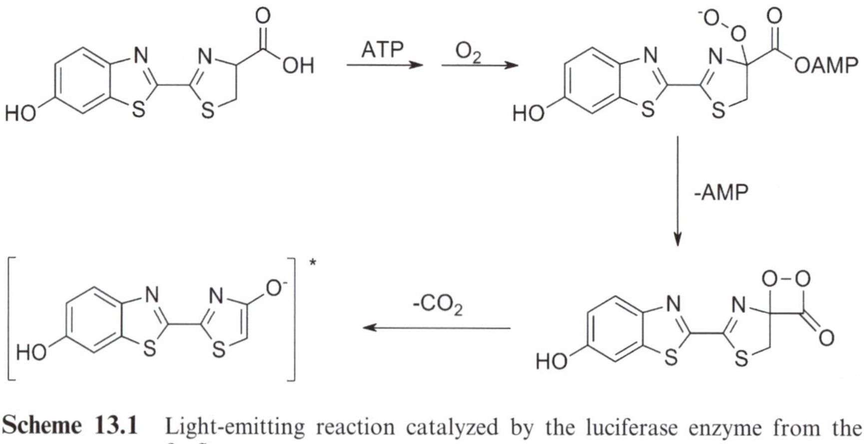

Scheme 13.1 Light-emitting reaction catalyzed by the luciferase enzyme from the firefly.

solutions when Ca^{2+} was added and, unlike the known luciferin–luciferase systems, the total amount of light emitted was proportional to the amount of the photoprotein present. The new protein, aequorin, was named after the genus of the jellyfish, *Aequorea*, from which it was isolated. Aequorin was initially thought to be an exceptional and perhaps unique protein but shortly after its discovery, Shimomura and Johnson unearthed another unusual bioluminescent protein in the parchment tubeworm *Chaetopterus*.[3] The total light emitted was again proportional to the amount of the protein present but, unlike aequorin, this newly found protein was activated by peroxide rather than calcium. Since these two proteins were not following any light-producing mechanism known at the time, a new term, "photoprotein," was introduced to denote the bioluminescent proteins capable of emitting light in proportion to the amount of the protein present.[4] The proportionality of the amount of photoprotein to total light emission makes a clear distinction between a photoprotein and a luciferin/luciferase system where the amount of light emitted is proportional to the amount of luciferin present and not to the amount of luciferase protein. There are at least three different types of photoproteins based on how they are induced to produce light. There are those such as aequorin and obelin that are calcium activated, photoproteins that are peroxide activated such as that from *Chaetopterus* and an ATP activated photoprotein from the *Luminodesmus* millipede.

Since the isolation of aequorin in 1961, several other photoproteins have been isolated, among these we can list obelin from *Obelia geniculata* [knotted thread hydroid (hydrozoa)],[5] clytin (a.k.a. phialidin) from *Phialidium gregarium* [hydrozoa],[6] thalassicolin from *Thalassicola* sp. [radiolarian],[7] pholasin from *Pholas dactylus* [common piddock, (mollusk)],[8] polynoidin from *Harmothoe lunulata* [polynoid worm],[9] mnemiopsin from *Mnemiopsis* sp. [cnidarian],[10] berovin from *Beroe ovata* [ctenephore],[10] symplectin from *Symplectoteuthis oualaniensis* [purpleback flying squid (Cephalopod)],[11] *etc.* There are also other unnamed photoproteins isolated from organisms such as *Chaetopterus variopedatus* [parchment worm],[3] *Ophiopsila californica* [Brittle star][12] and the only known photoprotein of terrestrial origin from the millipede *Luminodesmus sequoia*.[13]

13.1.4 Aequorin and Obelin

The calcium-regulated photoproteins aequorin and obelin have been studied more extensively than any of the other photoproteins identified so far. Both aequorin[14] and obelin[15] are composed of two distinct units, the apoprotein (apoaequorin or apoobelin) with an approximate molecular weight of 22 kDa, and a prosthetic group, coelenterazine (molecular weight of 472) (Figure 13.1). The apoproteins assemble spontaneously with coelenterazine in the presence of molecular oxygen to form the functional photoproteins. The photoproteins contain three conserved regions, called EF-hands, which serve as binding sites for Ca^{2+}. The Ca^{2+} binding to the EF-hands causes the photoprotein to

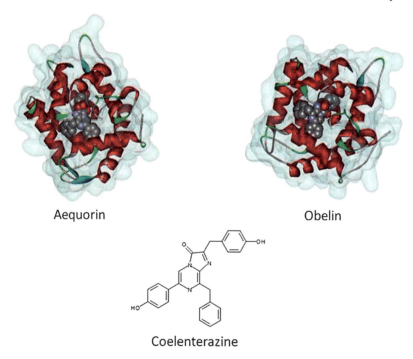

Aequorin   Obelin

Coelenterazine

Figure 13.1   X-Ray crystal structures of aequorin and obelin. The chemical structure
of coelenterazine is shown below. The protein structures are imaged
based on X-ray crystallography data (Aequorin: 1EJ3[14] Obelin: 1QV0[15])
using Accelrys DS Visualizer v2.0.1.7347.

undergo a conformational change resulting in the oxidation of coelenterazine
through an excited state to produce coelenteramide with a concomitant release
of CO_2 and emission of light. As the excited form of coelenterazine relaxes to
the ground state there is an emission of blue light at a λ_{max} of emission of
~470 nm for aequorin, and from 475 to 490 nm, depending on the source of the
protein, for obelin (Scheme 13.2).

13.1.5   Current Applications of Photoproteins

The bioluminescence characteristics of aequorin and obelin allow them to be
detected in the attomole range,[16] making them highly desirable reporters/labels
for many biological and analytical applications.[17] These proteins have found a
niche as reporters in certain applications where only radiolabels were previously used because they provide excellent sensitivity of detection while
avoiding the risks associated with exposure, handling and disposal of radiac-
tivity.[18] Both aequorin and obelin are quite stable, and under appropriate
storage conditions they can retain their activity for years.[19] Moreover, these
photoproteins offer significant advantages when employed as labels to detect

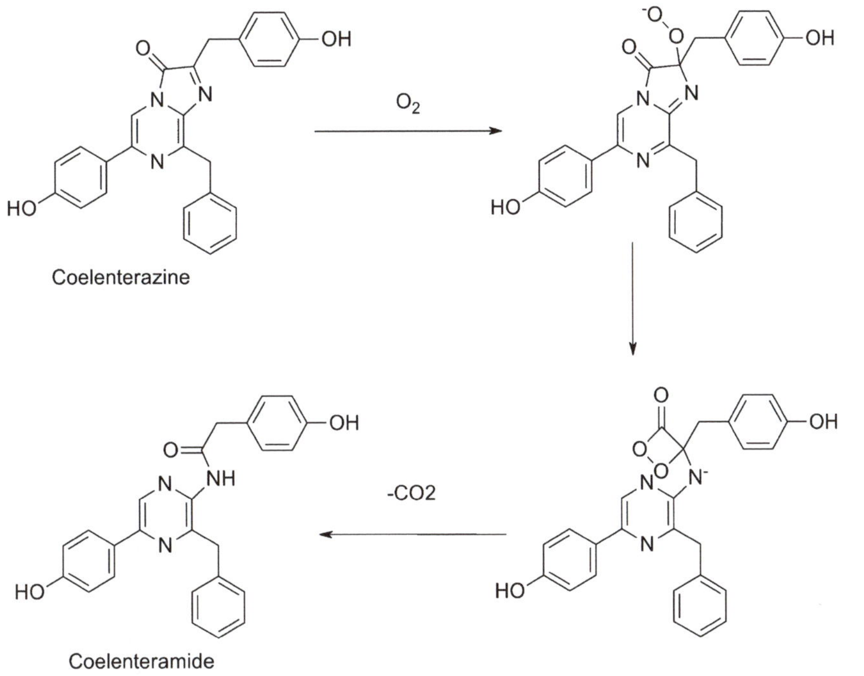

Scheme 13.2 Light-emitting reaction of aequorin and obelin.

biologically relevant molecules in biological fluids. Given that bioluminescence is quite rare in nature, and a light source for excitation is not required, any interference associated with the well-known background fluorescence of biological fluids is virtually eliminated.[20] In addition, unlike some chemiluminescent labels that require alkaline conditions of pH 9 or greater for maximum activity, the photoproteins are most active at physiological pH.[16]

These attractive properties have led to the use of the photoproteins in numerous applications. The cloning of aequorin in 1985 by Tsuji *et al.* and Prasher *et al.*[21,22] paved the way for its use as an intercellular Ca^{2+} indicator.[7,23,24] More recently aequorin and obelin have been employed as highly sensitive labels in competitive binding assays and in immunoassays for the detection of biologically important molecules[17,25] such as, thyroxine,[26,27] thyrotropin,[27] leu-enkephalin,[28] cortisol,[29,30] digoxin,[31] biotin,[16] angiotensin II,[32] methamphetamine,[33] prostacyclin[34] and serotonin.[35] The detection of pathogenic organisms such as *Salmonella*,[36] pseudorabies virus,[37] hepatitis B,[38] *Plasmodium falciparum*,[39] *Escherichia coli*[38] and *Shigella sonnei*[37] has also benefited from the use of aequorin as a reporter molecule as it yields more sensitive assays than those based on other types of luminescent labels.[40] Furthermore, the possibility of using aequorin in an homogeneous format makes miniaturization possible for use in field-studies and bedside-monitoring applications as well as for integration in microfluidics systems,[19] high-throughput screening systems[41–45] and for drug discovery.[42,46–48]

One factor that limits the more widespread use of photoproteins is the lack of range in the peaks of their emission wavelengths. Compared to the green fluorescent protein (GFP) and its variants, whose emission wavelengths range from blue all the way to red, the emission range of the photoproteins is much more limited. The bioluminescence emission maxima of the native photoproteins range between the 440 nm peak of thalassicolin to the 510 nm peak of polynoidin.[19] Even though a 70 nm difference in emission is enough to use in multi-analyte detection, such applications would greatly benefit if there was a wider range of emission to select from.

13.1.6 Designer Photoproteins

Green fluorescent protein (GFP)[49] from *Aequoria* and red fluorescent protein from *Discosoma* sp. (dsRed)[50] have been extensively studied, manipulated and mutated to create a multitude of variants capable of emitting fluorescent light in many different colors[51,52] and demonstrating varied stabilities[53–55] and maturation half-lives.[56,57] Unfortunately, the photoproteins have not gone through similar manipulation to the extent that the fluorescent proteins have, and, therefore, the varieties and their applications are more limited. To date, only a few reports have emerged that focus on designing photoproteins with altered bioluminescence emission. These include works that describe pairing different coelenterazine analogues with the photoproteins, ways of creating bioluminescence resonance energy transfer (BRET) reactions using photoproteins and employing random and rational mutagenesis. The following sections review recent advances in the design of photoproteins with fine tuned bioluminescence characteristics, such as emission maxima, stability, activity, reduced cross-reactivity and decay kinetics.

13.1.6.1   *Incorporation of Coelenterazine Analogues: Semi-Synthetic Aequorins*

Perhaps the easiest approach for tuning the emission characteristics of the bioluminescence reaction of aequorin is to substitute the native coelenterazine with synthetic coelenterazine analogues. Synthetic coelenterazines have been used to produce aequorins with different Ca^{2+} sensitivities,[58–60] varying regeneration times,[61] altered emission kinetics,[58–62] and different bioluminescence emission wavelengths.[58–62] Shimomura *et al.* tested 32 different coelenterazine analogues paired with recombinant aequorin and determined the time it takes to generate 50% of the maximum aequorin activity. The results showed that the regeneration time for these semi-synthetic aequorins varied from 8 min for coelenterazine-*e* to 300 min for coelenterazine-*n*.[61] In a similar work, Shimomura *et al.* tested the sensitivity of aequorin charged with each of the 32 synthetic coelenterazines towards Ca^{2+} ions and found that there was a

19 000-fold range in the Ca^{2+} sensitivities of the semi-synthetic aequorins. When luminescence was triggered with a low concentration of Ca^{2+} (0.1–1 μM Ca^{2+}) in a low ionic strength buffer, the intensity of light was proportional to the square of the Ca^{2+} concentration. The intensities of the tested coelenterazine analogues ranged from 0.01 relative light units (RLU) for coelenterazine-*n* to 190 RLU for coelenterazine-*hcp*.[60] The same semi-synthetic aequorins were also measured for their decay kinetics. Their half-lives were found to vary between 0.15 s for coelenterazine-*hcp* to 8 s for coelenterazine-*i*.[58–62] In addition, these studies showed that the bioluminescence emission maxima ranged from 438 nm for the coelenterazine analogue (**13.1**) to 476 for coelenterazine-*i*.[58–62]

13.1

13.1.6.2 *Bioluminescence Resonance Energy Transfer*

Scientists, inspired by nature, have tried to couple aequorin to natural or synthetic fluorophores in order to shift the emission wavelength towards the red end of the spectrum by employing bioluminescence resonance energy transfer (BRET). In jellyfish, the light emitted by the bioluminescence reaction of aequorin is transferred to green fluorescent protein (GFP) and the emission wavelength is shifted from blue (469 nm) to green (508 nm). By joining the genes for aequorin and GFP Gorokhovatsky *et al.* were able to create a fusion protein of aequorin and GFP joined by a cleavable 19 amino acid long peptide spacer.[63] Upon addition of Ca^{2+} to trigger the bioluminescence, they observed that the intensity of the bioluminescence emission decreased while a new peak corresponding to the fluorescence emission wavelength of GFP appeared. In addition, the reversal of this observation was reported when the spacer between the aequorin and GFP was cleaved. This was evidence that BRET was responsible for the emission shift in this system since the spatial proximity of the donor molecule (aequorin) to the acceptor molecule (GFP) was found to be essential for the energy transfer.[63]

Another example of the successful use of BRET was demonstrated by Shimomura *et al.* using a synthetic fluorophore, carboxyfluorescein.[62] Carboxyfluorescein succinimidyl ester (CFSE) was reacted with aequorin resulting in the conjugation of the carboxyfluorescein molecules to the aequorin through the protein's lysine residues. The resulting aequorin–CFSE conjugate had an emission maximum of 528 nm. The emission intensity was twice of that the aequorin bioluminescence and had a comparable decay half-life of between 0.6–1.2 s, demonstrating that much of the energy from the bioluminescence was being converted into fluorescence.

This conjugation of carboxyfluorescein to aequorin proved that through BRET the emission wavelength of aequorin bioluminescence could be shifted. However, the conjugation of the fluorophore to aequorin was not site-specific and since aequorin contains 14 solvent accessible lysine residues, this kind of approach results in more than one molecule of carboxyfluorescein conjugated to aequorin. This can produce batch-to-batch variation in the preparation of the conjugated aequorin, giving results that are variable or not reproducible. To overcome this drawback, Deo *et al.* have used a previously prepared mutant of aequorin that does not contain any cysteine residues and has higher bioluminescent activity as the template for site-directed mutagenesis. Starting with this cysteine-free mutant, Deo *et al.* prepared four different unique aequorin mutants containing single cysteine residues.[64] These mutants were A69C, G70C, G74C and G76C and the positions of these residues were selected due to their proximity to the location of the bound coelenterazine. These mutants were then individually conjugated to either *N*-[(2-(iodoacetoxy)ethyl]-*N*-methylamino-7-nitrobenz-2-oxa-1,3-diazole (IANBD ester) or an iodoacetamide derivative of Lucifer Yellow. The results showed that when the amino acid positions 74 and 76 were labeled with either fluorophore, no BRET was observed. The reason for this, as explained by the authors, is that even though the fluorophore was near enough to the chromophore for a successful energy transfer, the lack of fluorescence was due to the improper orientation of the fluorophore for an efficient dipole–dipole interaction. In contrast, when residues 69 and 70 were labeled, both conjugates showed an efficient BRET and an increase in the emission intensity corresponding to the emission maximum of the conjugated fluorophore ($\lambda_{max} = 536$ nm for IANDB and $\lambda_{max} = 531$ nm for Lucifer Yellow) was observed.[64]

13.1.6.3   *Random Mutagenesis*

Another logical approach for producing more desirable variants of aequorin or obelin is random mutagenesis.[65,66] Unfortunately, only aequorin has been subjected to these type of random mutagenesis studies. In one study, Tsuzuki *et al.* used random mutagenesis and *in vitro* evolution based on DNA shuffling to isolate aequorin mutants that showed a change in thermostability.[67]

Specifically, the cDNA of apoaequorin was digested with DNase I, and the DNA fragments were subjected to DNA shuffling. The bacterial colonies were screened in microtiter plates and colonies that showed luminescence activity were selected. The DNA of these colonies was amplified and the amplified DNA was used for a second round of DNA shuffling. After three rounds of DNA shuffling the brightest light emitting colonies were incubated at different temperatures and their activities were measured. The researchers found that out of 82 mutants that retained activity nine of them showed different thermostabilities. Two mutants, Q168R and L170I, exhibited an increase in protein lifetime at 37 °C. Further analysis showed that these mutants increased aequorin thermostability. Conversely, a mutant, F149S, was shown to decrease the thermostability of aequorin.

In another study, Tricoire *et al.* subjected aequorin to DNA shuffling and reported aequorin mutants with different decay kinetics and calcium sensitivities.[68] Following a procedure similar to the one outlined above, Tricoire isolated nine different aequorin mutants that have half-lives ranging from 0.7 s for a Q168R mutant to 39.9 s for E35G. They also demonstrated that each mutation that affected the decay-rate was located either within or in close proximity of one of the three calcium binding domains of aequorin. From these mutational studies, they found that even though each individual EF-hand is sufficient to trigger luminescence they exhibit different affinities towards Ca^{2+}. They also showed that the calcium EF1- and EF2-hands were more important for the kinetics of the bioluminescence decay than the EF3-hand and that the intensity of the luminescence was inversely proportional to the decay half-life of the same mutant.

13.1.6.4 Site-directed Mutagenesis

One of the most powerful tools in protein engineering is site-directed mutagenesis where one or more nucleotides in the protein's coding sequence is changed in order to alter the amino acid sequence in the protein. By using this powerful tool scientists can study the relationship between the structure and the function of a particular protein. This approach has been applied to both aequorin and obelin to introduce unique properties.

The early site-directed mutagenesis studies on aequorin were to gain information about its structure–function relationships. In 1986, Tsuji *et al.* performed site-directed mutagenesis and replaced the three cysteine residues, three glycine residues and one histidine residue.[69] They reported that the mutations of glycines within the EF-hand regions confirmed that a conformational change was required for bioluminescence emission. Furthermore, they reported that histidine 58 was involved within the active site. Other studies found that a C-terminal proline and the histidine residue at position 169[70] were both essential for its activity,[71] and the tryptophan residue at position 86 was involved in the bioluminescence emission.[72]

In 2000, when the crystal structure of aequorin was made available,[14] many researchers started performing site-directed mutagenesis to introduce desirable characteristics into aequorin. Armed with the knowledge of which particular amino acids are involved within the active site, scientists started engineering new aequorin and obelin mutants to suit their needs. In 2001, Deng *et al.* reported a shift in the bioluminescence emission of obelin from blue (485 nm) to violet (410 nm) when they changed W92 to phenylalanine.[73] They proposed that the bioluminescence spectral shift results from the removal of a hydrogen bond from the indole of W92 to a hydroxyl belonging to the 6-phenyl substituent of the coelenterazine. Then, in 2005, Stepanyuk *et al.* reported that after the mutation of the Y82 in aequorin to phenylalanine the H-bond that stabilizes the chromophore is removed. This resulted in a change of the spectral properties of aequorin to resemble that of obelin, in other words the emission peak was redshifted.[74] The reverse of this observation was also true for obelin. Obelin has a phenylalanine at the position that corresponds to the Y82 in aequorin. When this phenylalanine was changed to a tyrosine, the emission peak was blue-shifted. This was because the H-bonding to the oxygen atom of the 5-(*p*-hydroxy) phenyl group of coelenteramide is one of the important factors controlling the spectral properties of aequorin and obelin. In the Stepanyuk paper, a shift in the bioluminescence emission wavelength of aequorin from 469 to 501 nm was achieved, while the bioluminescence emission of obelin was shifted from 485 to 453 nm. This result was later confirmed by Dikici *et al.*[75] In this work, site-directed mutagenesis studies were performed on aequorin, generating several mutants with different emission wavelengths ranging from 466 to 494 nm.

13.1.7   Multi-analyte Detection

It is only logical to assume that these new mutants when coupled with different coelenterazine analogues should give even better separation in their bioluminescence emission spectra. This hypothesis was demonstrated by Dikici *et al.*[75] when mutants of aequorin paired with different coelenterazine analogues resulted in as much as 74 nm of separation between the emission maxima of two mutants, W86F mutant coupled with coelenterazine-*hcp* (445 nm) and Y82F mutant coupled with coelenterazine-*i* (519 nm) (Figure 13.2). Because of this separation these mutants could be employed in dual analyte detection based on wavelength resolution. In the same paper, the authors also described bioluminescence emission decay half-lives of the mutants ranging from 0.23 to 50.1 s. This combination of mutants with 0.23 s and 50.1 s half-lives represents an excellent opportunity for time-resolved dual analyte detection.

Today, scientists have developed bioluminescent proteins with emission maxima ranging from 390 nm (obelin W92F-H22E mutant coupled with native coelenterazine) to 519 nm (aequorin Y82F mutant coupled with coelenterazine-*i*). These new photoproteins have been used in dual analyte detections by two different groups. In 2008, Frank *et al.* reported a simultaneous detection of two different hormones.[76] They conjugated two different immunoglobulin-Gs to two different obelin mutants. Specifically, the Y138F mutant of obelin (λ_{max}

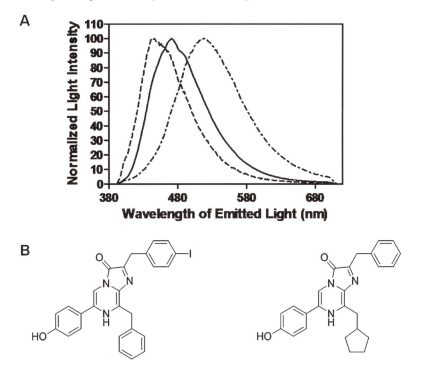

Figure 13.2 (A) Bioluminescence emission spectra of cysteine-free aequorin with native coelenterazine (—) and aequorin mutant W86F with coelenterazine-*hcp* (----), and aequorin mutant Y82F with coelenterazine-*i* (-·-·-·-). (B) Chemical structures of coelenterazine-*i* (left) and coelenterazine-*hcp* (right). (Reproduced with permission from ref. 75.)

498 nm) was conjugated to an anti-β-hFSH (human follicle stimulating hormone) and the W92F-H22E mutant of obelin (λ_{max} 390 nm) was conjugated to anti-β-hLH (human luteinizing hormone). These conjugates, having a 103 nm difference with little spectral overlap, were then used to detect the corresponding hormones by resolving both spectra with carefully selected transmission filters (Figure 13.3). This sandwich assay provided detection limits of 0.57 mIU mL^{-1} (milli International Units per milliliter) of hFSH and 1.1 mIU mL^{-1} of LH.

Rowe *et al.* have developed a dual analyte detection bioluminescent immunoassay employing genetically modified aequorins as labels. They demonstrated that this assay can simultaneously detect two cardiovascular markers, 6-keto-prostaglandin-F1-α and angiotensin-II, by employing luminescence time resolution.[77] In this assay, a 6-keto-prostaglandin-F1-α–aequorin conjugate and an angiotensin II–aequorin fusion protein were combined with two different coelenterazine analogues to form semi-synthetic aequorin variants. Time resolution was employed to resolve the signal of the two variants by the differences in their decay kinetics and half-lives. Figure 13.4 shows the time

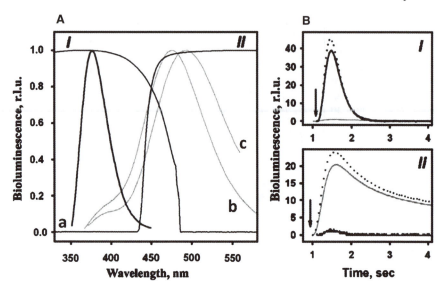

Figure 13.3 (A) Normalized bioluminescence spectra of wild-type (b), W92F-H22E (a) and Y138F (c) obelins; I, II indicate the transmission spectra of optical filters. (B) Transmission of bioluminescence signals through the optical filters built into channels I and II. Dotted lines show the signals without filters. The data were normalized to protein concentrations. Arrows show the moment of Ca^{2+} injection (r.l.u. = relative light units). (Reproduced with permission from ref. 76.)

resolution where the 6-keto-prostaglandin-F1-α signal was calculated from 0 to 6 s and the angiotensin II signal was calculated from 6.02 to 25 s. This method opens up new opportunities for multiplexing assays using the time resolution of aequorin signals to detect multiple analytes.

13.1.8 Protein Switches

Genetic engineering tools have been instrumental in the creation of new molecular entities with unique characteristics. The ability to rationally or non-rationally insert the gene of one protein into another has allowed for the creation of a new class of molecules called protein switches. Protein switches are defined as the fusion of two individual proteins creating a unique function that is regulated by an external signal. Protein switches contain both the receptor and transducer in a single polypeptide chain,[78] creating a much simpler system. The system is homogeneous in nature with no immobilization or washing steps necessary. Protein switches must contain a protein that will bind to an analyte causing a conformation change and a protein capable of emitting a detectable signal. Protein switches are being studied for purposes of understanding natural switches, to help explain protein form and function, to

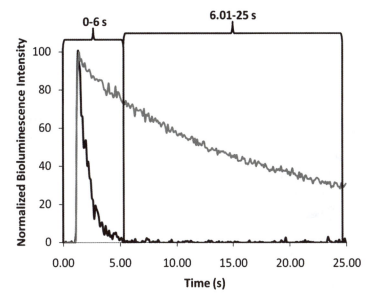

Figure 13.4 Bioluminescence decay kinetics profiles of 6-keto-PGF1R–aequorin–coelenterazine (native) (dark line) and angiotensin II–aequorin–coelenterazine-*i* (light line) conjugates, showing the flash *versus* glow-type decay kinetics of the two conjugates. The graph demonstrates how all the bioluminescent signal of the 6-keto-PGF1R–aequorin–coelenterazine (native) conjugate expires by the end of the 0–6 s time channel, leaving only the angiotensin II–aequorin–coelenterazine-*i* signal in the 6.01–25 s time channel. (Reproduced with permission from ref. 77.)

develop tools for elucidating cellular function and behavior and to create switches for sensing and biomedical applications.[79]

Of interest here are optical protein switches capable of detecting and monitoring specific molecules. Employing optical proteins as the signaling molecule is superior to using other labels, *i.e.*, radioactive labels, because optical proteins are derived from nature, non-hazardous, sensitive and biocompatible. Protein switches based on split fluorescent proteins have been studied for over ten years.[78,80–85] These fluorescent based systems have several advantages, including sensitivity and biocompatibility; however, they suffer from photobleaching and background scattering. To overcome these limitations bioluminescent proteins can be used to generate the optical output, yielding lower background, enhanced sensitivity and less expensive instrumentation. To that end, it was recently discovered that the photoprotein aequorin could be used to develop functional protein switches. Using molecular biology techniques, the gene for aequorin was rationally split and the glucose binding protein (GBP) gene was inserted, creating a gene that when expressed produces a protein switch capable of detecting glucose. The switch works in a way that simulates a light switch. In the presence of glucose, GBP binds the sugar and undergoes a conformational change bringing the two halves of aequorin in proximity,

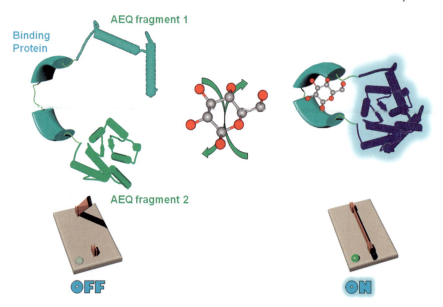

Binding Protein

AEQ fragment 1

AEQ fragment 2

OFF

ON

Figure 13.5 Depiction of the protein switch, showing the binding component and the optical component. In the absence of glucose the switch is "off;" however, in the presence of glucose the switch is "on."

creating an increase in bioluminescence, in essence turning the switch "on." In the absence of glucose the two halves of aequorin do not interact tightly for full bioluminescence and the switch is considered "off" (Figure 13.5). This aequorin–GBP protein switch is capable of selectively detecting glucose concentrations ranging from 10^{-2} to 10^{-7} mM (Figure 13.6), allowing for the determination of glucose at physiological levels in human plasma. The tertiary circular dichroism spectra of the aequorin–GBP protein switch demonstrated that the addition of glucose to the switch introduced a change in the tertiary structure, indicating protein structure rearrangement upon binding glucose. This structure change verifies that the change in bioluminescence activity is due to glucose binding to GBP causing a rearrangement of the aequorin–GBP protein switch. Moreover, this demonstrates that aequorin can be employed as the signaling mechanism in molecular switches.[86] Additionally, Scott *et al.* have developed a protein switch by inserting the cyclic AMP receptor protein into aequorin, creating a switch that responds selectively and linearly to cyclic AMP over several orders of magnitude as well as to detect changes in intracellular cAMP of intact cells in the presence of different external stimuli.[87]

Being able to split aequorin provides a gateway to expand the applications of the photoproteins. Discovering that larger proteins can be inserted into a photoprotein could allow for the development of molecular switches with enhanced characteristics compared to other types of sensing systems. Current use of photoproteins as labels requires chemical conjugation or end-to-end gene

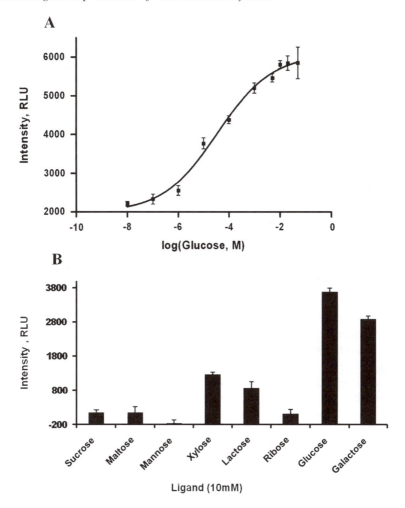

Figure 13.6 (A) Aequorin–GBP molecular switch dose-dependent response. Data points are an average of three measurements ±1 standard deviation. (B) Aequorin–GBP molecular switch selectivity study. Glucose and galactose were the only sugars that turned the switch "on." Data points are an average of three measurements ±1 standard deviation (RLU = relative light units).

fusion. Chemical conjugation techniques are difficult, can cause a loss in protein function and can result in variable performance. End to end gene fusion has the advantage of creating one to one conjugates with enhanced sensitivity and performance. However, end to end fusions may not be the optimal way of linking two individual proteins to maximize the coupling of their functions. Protein switches generate fusions that can have enhanced characteristics because one gene is inserted into the other. Kin *et al.* have provided evidence that protein switches can improve kinetic stability over end to end fusion.[88] In addition, protein switch design allows a more versatile combination of functional fusion proteins[78] and conformational changes in split fusion systems can show more of a dramatic change in structure and function. Rational design of protein switches allows for what researchers believe may be the most optimal switch; however, random fusion combinations may provide another avenue to optimal and enhanced systems. Random switch design can be accomplished *via* circular permutation (domain insertion) to create the best switch. Domain insertion has been studied in the creation of various molecular switches.[78–81,84,88,89] Researchers hypothesize that using circular permutation in the creation of random bioluminescent protein switches will expand their versatility and create superior systems.

13.2 Genetic and Chemical Engineering of Luciferases for Bioanalytical Applications

13.2.1 Introduction

For many light-based bioanalytical applications, including reporter gene assays, metabolite and toxin detection assays, and *in vivo* molecular imaging, wild-type bioluminescence systems have been used to achieve high-quality results. Nonetheless, researchers have taken advantage of chemical, genetic and protein engineering methods to make striking enhancements to natural luciferases. This section reviews the ways in which various luciferases – including enzymes from the sea pansy *Renilla reniformis*, the copepod *Gaussia princeps*, bacteria, and several families of fireflies, click beetles and railroad worms – have been altered genetically and chemically to better suit bioanalytical applications. In past decades, genetic enhancements, including codon-optimization,[90–93] addition of Kozak consensus sequences,[94,95] altered GC content[96] and elimination of peroxisomal targeting signals,[97,98] cryptic splice sites[93,99,100] and cryptic regulatory sequences[102–108] have been established as effective methods for improving luciferase expression in given cell types, and so these methods will not be described in detail. Instead, the principal focus will be protein engineering, followed by a discussion of recent approaches using chemical modification. Because this writing follows several excellent reviews already published[102–108] we will primarily cover work performed on this side of the century.

13.2.2 Firefly Luciferase Mutants

Firefly luciferase, particularly the enzyme from the North American firefly *Photinus pyralis*, has received the most attention of all the cloned luciferases, a trend that emerged partly because this enzyme was the first to be isolated and cloned,[109] and because it bioluminesces with an impressive quantum yield ($41.0 \pm 7.4\%$ compared to 15–30% for most luciferases).[103,110] The *Photinus pyralis* enzyme (Luc) became even more popular after its commercialization by the Promega Corporation, whose luciferase-encoding pGL3- and pGL4-basic vectors have been used for reporter assays,[111–113] bioluminescence imaging studies[114,115] and for constructing whole-cell biosensors.[116] While the luciferase genes in pGL3 and pGL4 have been extensively altered to improve expression in host organisms, they share an amino acid sequence that is essentially wild-type but for the removal of the Ser-Lys-Leu peroxisome targeting sequence.[117–119] Though native firefly luciferase has proven adequate for many bioanalytical applications, many useful changes to the protein have been made (Table 13.1).

13.2.2.1 Color Shifts

The bioluminescence color of firefly luciferase, which is yellow-green for Luc (560 nm at pH 7.8), has been shifted to shades of green, orange and red using several point mutations.[103,108] Still, despite the extraordinary efforts of research scientists worldwide, the structural determinants of beetle luciferase bioluminescence color have remained enigmatic. While it is not our intent to review this topic, we would like to note four particularly significant publications regarding the issue. Nakatsu and coworkers have reported a systematic crystallographic study of *Luciola cruciata* luciferase-inhibitor complexes that included a red-emitting point mutant (Ser286Asn) and concluded that the absence of a local conformational change observed only in the wild-type enzyme is the key determinant of green light because it controls the "molecular rigidity of the excited state of oxyluciferin."[120] Unfortunately, the crystallographic results with wild-type Luc and the red-emitting Ser284Thr mutant do not support the occurrence of this key conformational change.[121] Next, in a well-conceived experimental approach, Ando *et al.* remeasured the Luc quantum yield with respect to the consumption of substrate luciferin and reported its variation with pH.[110] According to their results, it appears that the quantum yield of "green" bioluminescence is approximately twice that of the "red" emission. Recently, Hirano and Naumov and coworkers in their respective laboratories have produced excellent summaries of the mechanisms that have been proposed to explain the basis of luciferase color determination. Based on the spectroscopic chemical model and theoretical calculation results using dimethyloxyluciferin, Hirano *et al.* proposed a "light-color modulation mechanism."[122] Naumov *et al.* produced an impressive crystallographic study of the elusive emitter oxyluciferin and, combining evidence from spectroscopic and theoretical studies, proposed that the phenolate ion of the enol form of oxyluciferin is the likely green emitter of firefly bioluminescence.[123] This conclusion is consistent with E. H. White's original proposition.[124] Certainly these interesting mechanisms will provide an important framework for future investigations.

Even in the absence of a full explanation of the bioluminescence color biochemistry, a host of color-shifted luciferases variants have been created (Table 13.1). Developed by random and rational mutagenesis techniques, these luciferase mutants can be used in multiplexed reporter systems for monitoring two

Table 13.1 Firefly luciferase variants engineered for bioanalytical applications.

Enzyme[a]	Mutations[b]	Properties[c]	Potential applications	Ref.
Luc	Q283R, S284G	λ_{max} at ~605 nm in *Escherichia coli*	*In vivo* imaging	132
	S293P	λ_{max} at ~600 nm (560 nm shoulder) in *E. coli*; blue-shift at pH 8.0 (~560 nm), redshift at pH 5.5 (~610 nm) in *E. coli*	Monitoring intracellular pH	132
	F14R/L35Q/V182K/I232K/F465R	6× more thermostable than Luc at 43°C	*In vivo* applications, pyrosequencing	143
	T214A/I232A/F295L/E354K	2× more thermostable than Luc at 35°C	*In vivo* applications	139
	T214A/A215L/I232A/V241I/G246A/F250S/F295L/E354K (Ppy GR-TS)	λ_{max} at 546 nm; 40× more thermostable than Luc at 37°C	Multi-color reporter assays	128
	T214A/A215L/I232A/S284T/F295L/E354K (Ppy RE-TS)	λ_{max} at 610 nm; 34× more thermostable than Luc at 37°C	Multi-color reporter assays and *in vivo* imaging	128
	I423L//D436G/L530R	10× greater luminescence output than Luc when assayed with low ATP levels	ATP detection assays	167
	R337Q, R337M	Enhanced thermostability and resistance to trypsin degradation	*In vivo* applications	149
	S239T/D357Y/A532T	Stability in chloroform, surfactants, ethanol	Cytotoxicity screenings	165

	Mutations[b]	Characterization[c]	Application	Reference
Lit	G216A/T217L/S234A/V243I/G248A/F252S/E356K/K547G/M548G (Lit GR-TS)	λ_{max} at 550 nm; expresses well at 37 °C	Multi-color reporter assays	138
	G216A/T217L/S234A/S286T/E356K/K547G/M548G (Lit RE-TS)	λ_{max} at 613 nm; expresses well at 37 °C	Multi-color reporter assays and *in vivo* imaging	138
Lat	A217L/E490K	Active in 10–25× more BAC than has been used in Luc assays	Intracellular ATP assays	163,164
Luc/Lcr	Residues 1–449 of Lcr containing T217I, T219I and V239I, residues 447–550 of Luc	>12× increased thermostability at 50 °C compared to Luc and Lcr, slightly greater catalytic efficiency for ATP, pH resistant bioluminescence spectrum	ATP detection assays Reporter assays	173
TLuc	H245N	λ_{max} at 617 nm, minimal activity loss from color shift	In vivo imaging	134
	Arg inserted at position 356	λ_{max} at 618 nm (shoulder 558 nm), minimal activity loss from shift	In vivo imaging	133,134

[a]Luciferases from the North American firefly *Photinus pyralis* (Luc), Italian firefly *Luciola italica* (Lit), Japanese fireflies *Luciola lateralis* (Lat) and *Luciola cruciata* (Lcr), and Iranian firefly *Lampyris turkestanis* (TLuc).

[b]Mutations are given with respect to the wild-type primary sequences.

[c]Characterization performed with pure proteins at pH 7.0–7.8 unless stated otherwise.

or more gene activities. Contrasting bioluminescence colors, most often red and green, can be detected simultaneously and separated using optical filters (Figure 13.7).[125–129] Red-emitting luciferases can be used alone for *in vivo* imaging studies, since long-wavelength red light transmits more easily than green light through live tissue.[93,130,131] An orange or red shift in firefly bioluminescence color can be achieved with few amino acid changes, as there are many point mutations that have been reported to shift the emission maximum to ~600–620 nm at pH 7. Unfortunately, bathochromic shifts in firefly luciferases are nearly always accompanied by activity loss,[93,128] and this compromise must be considered when selecting an enzyme for reporter or imaging applications.

Two red Luc variants (Gln283Arg and Ser284Gly both emit at ~605 nm) and two orange mutants (Ser293Pro and Leu287Ile both emit at ~600 nm with a shoulder at 560 nm) have been presented as options for biosensing and imaging (Table 13.1).[132] Though these proteins were observed only in *E. coli*, rather than purified and characterized, it was apparent that the variants lost some bioluminescence intensity relative to Luc. Recent mutants from the Iranian firefly *Lampyris turkestanis* were more thoroughly investigated; enzymes containing His245Asn or an Arg inserted at position 356 displayed emission maxima at 617 and 618 nm (shoulder at 558 nm) respectively (Table 13.1), and reportedly retained 77% and 81% specific activity compared to the native enzyme, although this result does not seem consistent with Ando's quantum yield findings for Luc.[110,133,134]

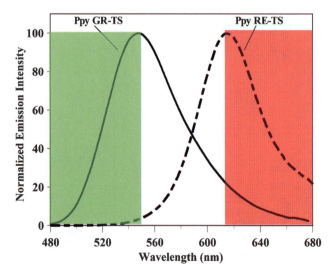

Figure 13.7   Thermostable Luc variants designed for dual-color assays. Normalized bioluminescence emission spectra produced by the Ppy GR-TS and Ppy RE-TS luciferases (Table 13.1) at pH 7.8 were obtained with purified proteins using a Veritas Microplate Luminometer, and the red and green shadings indicate the spectral regions transmitted through the green and red filters. (Originally published in ref. 128.)

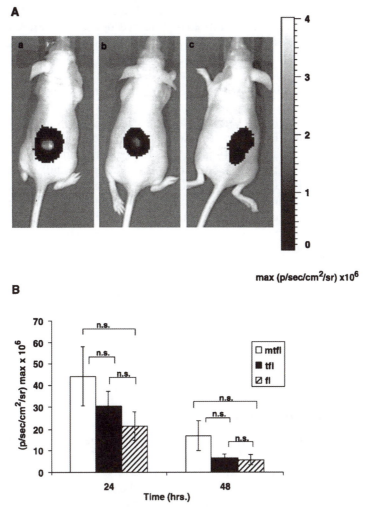

Figure 13.8 Improvement of *in vivo* bioluminescence imaging using a thermostable Luc mutant with the Ser-Lys-Leu peroxisome targeting sequence replaced by Ile-Ala-Val. Firefly luciferases were expressed under the control of a cytomegalovirus promoter in transiently transfected 293T cells, which were implanted in living mice. (A) Bioluminescence imaging of mice implanted with cells expressing a thermostable Luc mutant (T214A/I232A/E354K/F295L) with the Ser-Lys-Leu peroxisome targeting sequence replaced by Ile-Ala-Val (mtfl) (a), the same mutant with the peroxisome targeting sequence intact (tfl) (b) and unmodified Luc (fl) (c). (B) Comparison of mean light output from 293T cells expressing mtfl, tfl and fl as calculated from the ROIs drawn over the sites of cell implantation from three groups ($n = 3$) of living mice at 24 and 48 h. At both 24 and 48 h, mtfl showed a two-fold higher signal than fl and a 1.5-fold higher signal than tfl luciferase. (Adapted with permission from ref. 98; Figure 2.)

For dual-color assays and *in vivo* imaging, Branchini *et al.* developed a set of red- and green-emitting mutants that produced optically separable signals in bacterial cell lysates (Table 13.1).[127] The red-emitting mutant (Ser284Thr) was shown to be more effective than Luc for *in vivo* bioluminescence imaging.[135] These variants were later fortified with additional mutations for thermostability, at the expense of a shift in the red signal from 615 to 610 nm (Figure 13.7).[128] The resulting variants, named Ppy GR-TS and Ppy RE-TS, were recently used to visualize intestinal colonization of *E. coli* in mice,[136] and Ppy RE-TS was multiplexed with Luc and *Gaussia* luciferases to monitor transcriptional regulation of a cholesterol degradation enzyme in human cells.[137] When further engineering was performed to shift the emission spectrum of Ppy RE-TS from 610 to 617 nm, creating a mutant named Ppy RE9, a human codon-optimized version produced 100-fold greater luminescence in mammalian cells compared to Promega's red-emitting click beetle enzyme (CBR).[93,125]

Similar red- and green-emitting mutants were produced using luciferase from the Italian firefly *Luciola italica* (Lcr). These enzymes emitted 550 and 613 nm light when all homologous mutations from Ppy RE-TS and GR-TS were introduced, with the addition of Lys547Gly and Met548Gly (Table 13.1).[138] The variants, named Lit RE-TS and Lit GR-TS, were not used in dual-color assays together, but Lit RE-TS was multiplexed with Luc and *Gaussia* luciferases to monitor two pathways of bile acid biosynthesis in cell culture.[138]

13.2.2.2   Thermostability

The thermostability mutations used in these -TS enzymes were adapted from the work of Tisi, Baggett and Law *et al.*, who have performed extensive work toward improving the longevity of Luc at 37 °C.[139–143] Luc has a half-life of only 3–4 h in mammalian cells,[144] and although the cellular environment offers some protection against thermal denaturation,[142] the enzymes unfold quickly if they are exposed to heat even before cell lysis.[145] Thermostable luciferases may produce brighter signals, since they accumulate more quickly in cells,[142] and those with stability at temperatures even greater than 37 °C may be used in heat-inducible expression vectors or for pyrosequencing.[146]

One thermostable mutant created by Tisi *et al.* contained the mutations Thr214Ala, Ile232Ala, Phe295Leu and Glu354Lys and exhibited 2× greater stability than Luc when assayed at 35 °C (Table 13.1).[139] This variant was imaged *in vivo*, and showed modest improvement in signal compared to Luc, but performed significantly better when the Ser-Lys-Leu peroxisome targeting sequence was replaced by Ile-Ala-Val (Figure 13.8).[98] A more recent mutant by Law *et al.* containing Phe14Arg, Leu35Gln, Val182Lys, Ile232Lys and Phe465Arg exhibited a half-life 6× greater than that of Luc when assayed at 43 °C (Table 13.1).[143] Tisi and colleagues have applied this research to BART (bioluminescence assay in real time), a patented technology used for quantifying nucleic acids that requires thermostable enzymes because of the severe reaction conditions.[147,148]

A different research team, Riahi-Madvar and Hosseinkhani, found that Luc variants Arg337Gln and Arg337Met not only preferred slightly higher temperatures compared to Luc, but also demonstrated increased resistance to proteolysis by trypsin (Table 13.1).[149] Arg337Gln and Arg337Met remained 39% and 41% active after 3 h trypsin digestion at 23 °C, while native Luc was only 12% active. The same experiment at 37 °C demonstrated the thermostability of the mutants; Arg337Gln and Arg337Met retained 26% and 17% activity, while native Luc dropped to less than 1% activity.

Luciferases from other firefly species, including the Japanese firefly *Hotaria parvula*[150] and the Southern Russian firefly *Luciola mingrelica*,[151] have been altered for increased thermostability as well. One of the most striking examples is Promega's Ultra-Glo™ recombinant luciferase, which is derived from the luciferase of the North American firefly *Photuris pennsylvanica*. Available in the CellTiter-Glo® Luminescent Cell Viability Assay, the Ultra-Glo™ enzyme has a half-life of over 5 h in mammalian cells,[152] an improvement attributable to as many as 34 mutations that were discovered by directed evolution.[153,154] In addition, compared to Luc, Ultra-Glo™ demonstrates increased resistance to chemical inhibitors, a favorable quality for high-throughput screening applications.[155]

Notably, *in vitro* enzyme thermostability is highly dependent on buffer choice, ionic strength and osmolyte concentration.[156–158] These variables make it difficult to compare any of these reported thermostable luciferases, since the heat-inactivation assays were performed under different conditions. Furthermore, inactivation assays using pure proteins in arbitrary buffer solutions may not accurately predict enzyme performance in the intended application.

We must also point out that these thermostable mutants are ineffective for applications monitoring transient expression in real time, since enzymes that linger too long in cells usually mask changes in promoter activity. Even wild-type Luc is too stable for some reporter assays. Though destabilized Luc,[90,159] bacterial luciferase[160] and *Renilla* luciferase[90] genes have been developed, thermolabile proteins have not been mutagenically engineered.

13.2.2.3 Resistance to pH Color Shifts and Chemical Deactivation

Another potential limitation of *P. pyralis*, *H. parvula* and *L. turkestanicus* luciferases and other firefly luciferases is the shift in the bioluminescence spectra from yellow-green to red when the pH of the reaction is lowered.[135,150,161] This instability could render data unreliable when collected from a variable pH environment using a color-sensitive charge-coupled device or photomultiplier tube. Conveniently, some of the mutations reported to enhanced thermostability were also found to increase pH tolerance of the bioluminescence color (Figure 13.9).[143,148,150]

Stability against chemical deactivation is another favorable quality in luci-ferases. Benzalkonium chloride (BAC), an extractant used to remove intra-cellular ATP for quantification, inhibits luciferase activity.[162] Of several luciferases, that from *Luciola lateralis* was found to be the most tolerant of BAC, so it was used as a template for random mutagenesis.[163] Colony-level screenings yielded two beneficial mutations, Ala217Leu and Glu490Lys, which were combined to create a strongly BAC resistant mutant (Table 13.1).[163] When used in an ATP detection assay to estimate microbial biomass, the double mutant remained active in cell extracts containing 0.2% BAC, whereas previous experiments employing wild-type firefly luciferases had used 0.008–0.02% BAC.[164] The freedom to use more BAC allowed the experimenters to achieve outstanding detection sensitivity, such that picomolar levels of ATP could be quantified.

Intracellular ATP assays are also used for cytotoxicity testing, which requires luciferases to report cell viability without denaturing in the test che-micals. A Luc variant (Ser239Thr/Asp357Tyr/Ala532Thr) was engineered with increased stability in chloroform, surfactants, ethanol and other chemicals commonly used in cytotoxicity screenings, and displayed impressive activity and thermostability as well (Table 13.1).[165] However, when substituted for the thermostable luciferase variant included in the Vialight cell proliferation/cytotoxicity kit, the triple mutant only provided a slight improvement in the detection of chloroform and ethanol toxicity in normal human epidermal keratinocyte (NHEK) cells.[166] The authors propose that components in the Vialight reagent may have protected both proteins from chemical denaturation. Therefore, to assess the advantage of this triple mutant it must be evaluated in other cytotoxicity tests and should be compared with wild-type luciferase.

13.2.2.4   Altered Kinetics

To obtain a brighter bioluminescent signal, stabilizing firefly luciferase for experimental conditions is one approach, but an alternative route is to alter its signal kinetics. In one study, the combined mutations Ile423Leu, Asp436Gly and Leu530Arg were shown to increase the luminescence intensity of wild-type Luc by ten-fold when low levels of ATP were present (Table 13.1).[167] This mutant, capable of producing a measurable signal in the presence of just a single bacterium,[168] was used to detect *Salmonella* at a sensitivity 1000-times greater than a standard immunochromatographic lateral flow assay.[169] The improved luminescence of Ile423Leu/Asp436Gly/Leu530Arg was not attribu-table to increased specific activity, but rather to increased enzyme turnover. In the firefly luciferase reaction, enzyme turnover is quickly slowed by the gen-eration of the byproduct dehydroluciferyl-adenylate (L-AMP), which inhibits bioluminescence and thereby causes a flash-like signal with a rapid decay.[170] Enzyme turnover can be sustained by decreasing either the rate of L-AMP production or the enzyme's affinity for the inhibitor, resulting in glow kinetics rather than flash kinetics. It appears that the Ile423Leu/Asp436Gly/Leu530Arg

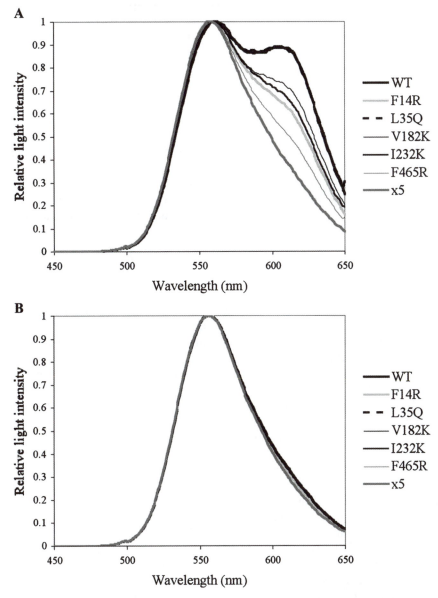

Figure 13.9 Mutations that enhance thermostability are combined to create a Luc variant (×5) that resists redshift at pH 6.5. Normalized bioluminescence spectra of luciferases at pH 6.5 (A) and 7.8 (B). Enzyme (0.31 nmol) was assayed with 1 mL of TEM containing 200 μM D-LH$_2$, 1 mM ATP, 270 μM CoA and 2 mM DTT, adjusted to pH 6.5 and 7.8. Spectra were recorded from 450 to 650 nm and scanned at 200 nm min^{-1} with a slit width of 10 nm. PMT voltage was 900 V. Spectra obtained were corrected for the baseline and sensitivity of the PMT. Experiments were subject to a pH accuracy of ± 0.05 unit and timing accuracy of ± 5 s. The spectrum for L35Q in (A) is identical with that of WT. [Reproduced with permission, from ref. 143 © The Biochemical Society (http://www.biochemj.org).]

variant possesses altered properties to decrease L-AMP interference, although the authors did not explicitly draw this conclusion.[167] In addition, the Ile423Leu, Asp436Gly and Leu530Arg single mutants each had a decreased ATP K_m, which may account for the triple mutant's ability to detect trace quantities of ATP.[167]

To detect larger amounts of ATP, the ATP K_m may need to be raised. Schneider and Gourse found that wild-type Luc could not report fluctuations of ATP in *E. Coli*, since even the lowest endogenous ATP concentrations saturated the enzyme.[171] The His245Phe mutant, which has a five-fold elevated ATP K_m,[172] was sensitive to ATP changes in the relevant physiological range and was successfully used to track *E. coli* growth rates.

While mutagenesis was an effective method in these two studies, it is also possible to alter the kinetics of firefly luciferases by using specialized reagents. Coenzyme A, for example, is known to sustain the bioluminescence output of luciferases over time.[170]

13.2.2.5   Chimeric and Fusion Proteins

The firefly luciferase variants discussed thus far have been altered with relatively few point mutations, but more intricate constructs have also been designed. Chimeric approaches allow researchers to obtain desired properties from a combination of luciferases. In one example, a luciferase variant (Thr217Ile) derived from Lcr was combined with Luc and was mutated at Thr219Ile and Val239Ile to create a chimeric protein variant with positive attributes from both parent enzymes (Table 13.1).[173] The impressive thermostability of LcrThr217Ile (over 12× more stable than both Luc and Lcr at 50 °C) was retained with minimal loss. The superior catalytic efficiency of Luc ($\sim 2.9 \times 10^{15}$ RLU mg^{-1} s^{-1} mM^{-1}) over LcrThr217Ile ($\sim 1.3 \times 10^{15}$ RLU mg^{-1} s^{-1} mM^{-1}) was not only retained but also exceeded ($\sim 3.0 \times 10^{15}$ RLU mg^{-1} s^{-1} mM^{-1}), an improvement due to the chimerization and/or the Thr219Ile/ Val239Ile mutations. These changes were also responsible for stabilizing the bioluminescence color; the chimeric protein possessed a pH stable bioluminescence spectrum whereas Luc and LcrThr217Ile did not.

Mutant and wild-type firefly luciferases have been further modified by fusion with biotin acceptor peptides (BAPs). Luc-BAP constructs lose little activity and thermostability compared to the wild-type enzyme, and can be localized using the biotin/avidin reaction.[174] For example, a biotinylated thermostable Luc variant (Glu354Lys) was immobilized on biotinylated cell membranes using streptavidin, and was used to detect ATP released upon hypotonic stress.[175] Zhang *et al.* modified this method by using streptavidin-coated beads, rather than free streptavidin, to increase the concentration of localized Luc and avoid interference with cell membrane proteins.[176] While Zhang *et al.* did not use an enhanced Luc, the method would likely be improved by employing a variant with improved thermostability and/or kinetic properties.

Beyond the luciferase constructs discussed here, there are great multitudes of Luc split-reporters, circular constructs and fusion proteins, but these exceed the scope of this chapter and are addressed in Chapter 10.

13.2.3 Click Beetle and Railroad Worm Luciferase Mutants

Certain members of the Elateridae family, more commonly known as click beetles, glow in a remarkable variety of bioluminescence colors, all of which appear to be pH independent over the 6–9 range.[177,178] In fact, the Jamaican click beetle *Pyrophorus plagiophthalamus* flaunts two sets of bioluminescent organs, the ventral set emitting in wavelengths from yellow-green to orange, depending on the individual, and the dorsal set emitting from green to yellow-green (Figure 13.10).[179–181] The availability of several bioluminescence colors from a single species lends itself to multicolor reporter systems, since few mutations are needed to produce enzymes with two distinct signals. Also convenient is that the *P. plagiophthalamus* luciferases share 95–99% amino acid sequence identity,[182] and are therefore likely to have similar expression levels and half-lives in cells.

Figure 13.10 Bioluminescence in several colors from the Jamaican click beetle *P. plagiophthalamus*. (A) Paired dorsal bioluminescent organs emitting green light. (B) Ventral bioluminescent organ emitting yellow light. The dorsal organ is also visible but does not bioluminesce during flight and is seen here due to the "irritated" state of the specimen. (Adapted from ref. 181 © 2009 National Academy of Sciences, U.S.A.)

Accordingly, Promega has developed a pair of vectors encoding red- and green-emitting *P. plagiophthalamus* luciferase variants, which are available in the Chroma-Glo™ luciferase assay system.[125] The enzymes, named click beetle red (CBR) and green (CBG) produce signals with maximum intensities at 613 and 537 nm.[183] They contain numerous primary sequence changes to enhance their stability at temperatures up to 40 °C, and to increase the redshift and luminescence brightness of CBR.[183] CBR and CBG have proven suitable for cell-sensor assays,[184] and for dual-reporter assays in mammalian cells[185] and plant cells.[186] Further manipulations of the enzymes include a CBG fusion with the fluorescent protein tdTomato for a BRET-based protease assay,[187] and a CBR fusion with Luc and other luciferases for a split-reporter assay.[188]

A Brazilian click beetle larva luciferase from *Pyrearinus termitilluminans* was found to have the greenest light emission of any native beetle luciferase (538 nm).[189] For dual-reporter and imaging applications, a multitude of red-emitting *P. termitilluminans* enzymes with emission maxima up to 619 nm have been developed and patented.[190] The ~80 nm separation between these red- and green-emitting enzymes is larger than those achieved using *P. pyralis* and *L. italica* mutants, and could make for a highly sensitive dual-color assay.[128,138]

Similar to click beetle luciferases, railroad worm luciferases are found in various natural colors and have the fascinating ability to bioluminescence in two distinct colors with two different sets of organs.[191] *Phrixothrix* railroad worm luciferases emit colors ranging from red to green and offer the same advantages as the click beetle enzymes – a set of multicolored reporters with high sequence homology and pH resistant spectra.[191] Optimized genes encoding red-emitting *Phrixothrix hirtus* luciferase (622 nm) and green-emitting *Phrixothrix vivianii* luciferase (549 nm) have been used in assays for studying circadian rhythms in whole cyanobacteria[192] and in mammalian cells.[193] *P. hirtus* luciferase was also paired with a green-emitting luciferase from *Rhagophthalmus ohbai* (550 nm) to monitor circadian rhythms in Rat-1 cells.[194]

To our knowledge, no protein engineering has been performed with railroad luciferases to tailor them for applications, but research has been performed to elucidate the bioluminescence color determinants.[195–197] There is evidence that red-emitting *P. hirtus* luciferase, despite its previous use in mammalian cells, is quite unstable at 37 °C as a pure protein.[93] Developing a thermostable version of this luciferase and a corresponding green-emitter could increase reporter assay sensitivity, though the outstanding bioluminescence color might be altered in the process.

13.2.4   Enhanced Coelenterazine/Luciferase Systems

Luciferases from the marine copepod *Gaussia princeps* (GLuc) and the sea pansy *Renilla reniformis* (RLuc) are unlike the beetle luciferases in that they react with coelenterazine, rather than beetle luciferin, to emit deep blue light

($\sim$480 nm) (Scheme 13.3).[198,199] Exploiting the two different substrates, Promega pioneered a dual-assay kit combining Luc and RLuc, which are expressed simultaneously but quantified using sequential reactions.[126] Also, while firefly luciferase requires the presence of ATP and magnesium, *Gaussia* and *Renilla* luciferases are useful in environments lacking said cofactors (Scheme 13.3). For example, RLuc bioluminescence has been measured in the murine bloodstream, where the ATP concentration is too low for a signal to be obtained from beetle luciferases.[200] GLuc ($\sim$20 kDa) and RLuc ($\sim$34 kDa) are also markedly smaller than firefly luciferases ($\sim$62 kDa), and are therefore more appropriate for applications requiring small vectors and/or small proteins, such as gene therapy studies and *in vivo* bioluminescent tags.

Gaussia, the brighter and smaller of the two marine luciferases, has required little alteration to provide excellent imaging and reporter data. A human

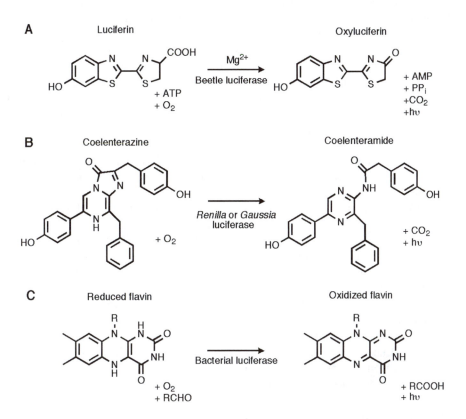

Scheme 13.3   Reactions catalyzed by luciferases from bioluminescent beetles, *Renilla*, *Gaussia* and bacteria. The structurally unrelated substrates shown produce light by diverse biochemical reactions with the shared requirement of molecular oxygen. The bacterial luciferase reaction (C) is coupled *in vivo* with a FMN:NADH oxidoreductase reaction to replenish the supply of reduced flavin.

codon-optimized GLuc produced a signal 200-times greater than that of humanized RLuc and similar to that of Luc under typical *in vivo* imaging conditions.[91] Furthermore, GLuc displayed impressive thermostability, providing a strong signal in a heat-inducible algal reporter system where RLuc was quickly denatured.[201] When expressed in mycobacteria, the native protein was affected neither by 48 °C heat shock nor by lowered pH.[202] Such innate thermostability appears to be shared by other recently cloned copepod luciferases.[203,204]

Considering only luminescence intensity, thermostability and size, GLuc seems to be a superior version of RLuc. But the distinguishing feature of *Gaussia* luciferase is that it is naturally secreted from eukaryotic cells,[205] a characteristic that can be advantageous or detrimental depending on the application. Luciferase secretion is useful for non-invasive assays in which the extracellular media is tested for bioluminescence; and, conveniently, the signal peptide on GLuc is surprisingly effective in mammalian cells, even better at inducing protein secretion than are signal peptides from human proteins.[206] On the downside, the secretion of GLuc can be a problem when the bioluminescent signal must be contained. An attempt to block secretion by removing the signal peptide caused GLuc expression in cells to drop, and adding an endoplasmic reticulum retention signal did not prevent luciferase from escaping.[91] But a membrane-anchored version of GLuc was successfully used to image T-cell accumulation in murine tumors, and performed much better than an anchored RLuc.[207]

Although GLuc appeared to perform adequately *in vivo*, one drawback both *Gaussia* and *Renilla* luciferases share is that they exhibit flash-like kinetics in cells.[91] The *in vivo* kinetic behavior of beetle luciferases vary, but sustained intensity has frequently been engineered into thermostable variants.[93] A Met43Ile variant of GLuc had a slowed signal decay, but only in the presence of the detergent Triton X-100, largely limiting the potential uses of this enzyme.[208] Greater improvements have been made to *Renilla* luciferase, which was changed at eight residues to produce an enzyme (RLuc8) with four-fold enhanced light-output, 200-fold increased stability in murine serum and a 5 nm redshift.[209] A triple-mutant RLuc (Met185Val/Lys189Val/Val267Ile) was discovered to have a prolonged luminescence signal, three-fold improved k_{cat} and greater intensity when expressed in *Arabidopsis*.[209]

To further advance their *in vivo* imaging applicability, it is desirable to redshift GLuc and RLuc, but few mutagenesis studies have been conducted to this end. In one report, six amino acid changes added to the variant RLuc8 increased the emission maximum from 480 to 535 nm, a shift that doubled the amount of detectable light from murine lungs.[210] But the vast majority of *Gaussia* and *Renilla* bioluminescence color alterations have been achieved using BRET-based approaches, which are discussed in Chapter 10.

An additional future direction for enhancing the GLuc and RLuc systems may be to alter their substrate specificity. Several problems associated with coelenterazine, such as its auto-chemiluminescence,[211] activation of MDR1 P-glycoprotein (Pgp),[212] binding to serum proteins and elimination from blood[2] have driven the development of coelenterazine analogs, of which there are over ten commercially available.[213] While some of these analogs improve RLuc performance under certain conditions, there are instances for which an appropriate analog may be difficult to find. For example, when assaying in blood, coelenterazine-*cp* is useful in that it is not a substrate for Pgp, but it exhibits increased background luminescence and decreases the signal from RLuc.[210,213] The Met185Val mutation was seen to quadruple the signal-to-noise ratio when coelenterazine-*cp* was used, evidencing an increased affinity for this analog.[200] This example suggests that further mutagenesis studies could be performed to enhance RLuc or GLuc affinity for particular coelenterazine analogs, tailoring the enzymes for specific applications. Finally, as is the case with firefly luciferase, RLuc and GLuc have been used in fusion-protein and split-reporter studies, which are discussed in detail in Chapter 10.

13.2.5 Bacterial Luciferases

Luciferases from *Vibrio harveyi*, *Vibrio fischeri*, *Xenorhabdus luminescens*, *Photorhabdus luminscens* and other bacteria perform a unique bioluminescence reaction, coupled with FMN:NADH oxidoreductase, to produce blue-green light ($\sim$490 nm) (Scheme 13.3).[214–216] The advantage of these bioluminescent systems is that the lux operons (luxCDABE) encode not only the luciferases (heterodimers of luxA and luxB) but also the enzymes for substrate synthesis (luxCDE).[217,218] Because substrate injection is not necessary, lux operons are useful in bacterial gene regulation studies[219] and whole cell biosensors,[220] and for use in small animals to monitor pathogenesis,[221,222] track tumor-targeting bacteria[223] and investigate bacteria-mediated gene transfer.[224] Furthermore, lux bioluminescence is quite bright; luciferase from *P. luminescens* produced a similar signal intensity to firefly luciferase when identical expression constructs were used in *E. coli*.[225]

Although a mammalian codon-optimized gene of *P. luminscens* has been synthesized,[226] lux operons are generally used in bacteria.[100] The endogenous reduced flavin mononucleotide in bacteria contributes a significant amount of substrate, an advantage that is not shared by eukaryotic cells.[106] In addition, the lux A and B subunits are expressed dicistronically, and unequal synthesis of the two peptides can render quantitative reporter assays inaccurate.[126] The cistrons for *V. harveyi* luciferase were combined into one coding region with success, but the resulting fusion protein was highly sensitive to 37 °C expression conditions, and would fare poorly in a mammalian system.[227,228] Several alterative enzymes – wild-type[229] and monocistronic *P. luminescens* luciferase[230] and wild-type *V. campbellii* luciferase[231] – are more thermostable at 37 °C than wild-type *V. harveyi* luciferase, but these enzymes have not been enhanced further for mammalian applications.

For applications in bacteria, several primary sequence alterations have been made to lux enzymes. A missense translation assay was designed by mutagenesis; the activity of *V. harveyi* luciferase was decreased by introducing a mutation at position 45 on the alpha subunit, such that translational errors might cause a more active form of the enzyme to be produced.[232] In addition, Hosseinkhani *et al.* showed that error-prone PCR of the alpha subunit can be used to alter the bioluminescence kinetics. They discovered that the Glu175Gly mutation speeds the signal decay of the *X. luminescens*, a property that might be useful for real-time monitoring of genetic regulation.[233] Color-shifted *V. harveyi* luciferase variants have been developed, such as the αAla75Gly/

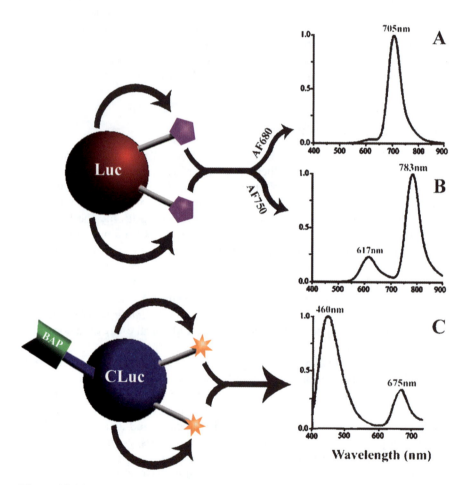

Figure 13.11   Bioluminescence resonance energy transfer (BRET) of luciferases tagged with fluorescent dyes. A red-emitting Luc mutant (617 nm) was covalently modified at two introduced non-native cysteine residues with the maleimide-based dye AF-680 (A) or AF-750 (B). (Unpublished results by B. R. Branchini, D. M. Ablamsky and J. C. Rosenberg.) Luciferase from *Cypridina noctiluca* (CLuc, 460 nm) was expressed as a fusion protein with a biotin–acceptor peptide (BAP) and was modified at two residues with an indocyanine-based organic dye (C). (Adapted from figures originally published in ref. 243.)

Cys106Val/Val173Cys and αAla75Gly/Cys106Val/Val173Ser mutants, which display emission maxima at 504 and 506 nm (from ~490 nm).[234] Unfortunately, color-shifting mutations in *V. harveyi* typically cause 80–99% activity loss,[234] and the shifts would need to be increased dramatically to confer any significant advantage for imaging applications. Aside from these examples, protein engineering of bacterial luciferases is an arena that remains largely unexplored. Currently, the primary focus has been the genetic tailoring of lux cassettes.[96,235,236]

13.2.6 Chemical Modifications to Luciferases

In addition to the myriad of mutagenesis studies, luciferases have also been modified by chemical methods using reactive cysteine and lysine residues. Several preliminary reports have suggested that cysteine modification disrupts bioluminescence activity. Luc bioluminescence was impaired by various cysteine-blocking agents,[237–239] and a cysteine-free variant lost 90% activity.[240]

Nevertheless, Branchini *et al.* were able to attach fluorescent maleimide-based dyes to a red-emitting Luc variant (Thr214Ala/Ala215Leu/Ile232Ala/Ser284Thr/Phe295Leu/Arg330Gly/Ile351Val/Phe465Arg) without disrupting enzyme activity.[241] Because the dyes did not access any of the four native cysteine residues, it was possible to react the dyes with two cysteines added at positions Thr169 and Ser399. The fluorescent dyes acted as BRET acceptors to the red bioluminescence signal of the luciferase (617 nm), producing near-infrared signals at either 705 nm (when Alexa Fluor® 680 dye was used) or 783 nm (when Alexa Fluor® 750 dye was used) that may be useful for imaging experiments (Figure 13.11A and B).

A more recently cloned luciferase from the marine ostracod *Cypridina noctiluca* (CLuc)[242] was likewise altered to create a BRET source of far-red light.[243] Glycol-chains were used to add the indocyanine-derived organic dye HiLyte Fluor™ 647 to a luciferase/biotin acceptor peptide fusion protein. The finished construct produced a BRET signal at 675 nm, though the bioluminescence signal at 460 nm was stronger (Figure 13.11C). Using the biotin/avidin reaction, this chemically-modified luciferase was conjugated to an antibody and used for cancer cell imaging.

13.3 Conclusions and Future Perspectives

Advancements in biotechnology have allowed for a rapid expansion of the applications of bioluminescent proteins and photoproteins in several fields, including clinical diagnostics and environmental monitoring. The greatest advantages of photoprotein labels are the sensitivity of detection afforded in physiological samples due to the lack of background bioluminescence, the simplicity and cost-effectiveness of the instrumentation employed to measure bioluminescence given that there is no need for an excitation source and the non-hazardous nature of the labels. The current limitation of photoproteins is the lack of a large enough array of photoprotein labels capable of emitting at different wavelengths; however, their diversity is growing. Throughout this chapter we have highlighted examples of photoprotein improvements that are capable of detecting one or more molecules as well as using aequorin as the signaling molecule in protein switch sensing systems. It is envisioned that research in this arena will provide investigators and clinicians with new diverse bioluminescence labels comparable to the existing fluorescence ones (*e.g.*, GFP, DSRed, *etc.*) that should enable a host of new bioanalytical and clinical methods. For example, the availability of a palette of bioluminescent labels should allow for the design and development of multiplex assays in clinical samples, the real-time imaging of more than one target compound in a specific location, as well as for the implementation of new gene probes and protein–protein interaction methods for use in genomics, proteomics and drug discovery.

While the field of analytical bioluminescence is currently dominated by wild-type *Photinus pyralis* luciferase, researchers should not overlook the many luciferase variants that have been engineered for specific applications. Laboratories that study luciferases are working hard to tailor enzymes for dual-color assays, *in vivo* imaging, nucleotide sequencing, and toxicology studies, and there is abundant evidence that many of these enzymes do improve experimental results. From variants containing simple but effective point mutations to complex fusions and chimeric proteins, the possibilities for altering luciferases are numerous and the outcomes are powerful. Enzymes with brightened signals and enhanced stability can illuminate phenomena that are too discreet for wild-type luciferases to expose. As new bioluminescent species and systems are discovered, such as the sponge *Suberities domuncula*[244] and the *Mycena* fungi,[245] we can expect still more luciferases to become available as templates for modification. So long as there is close collaboration between laboratories engineering the luciferases and those utilizing them, the reach of analytical bioluminescence applications will continue to extend.

Acknowledgements

S. D. would like to acknowledge support of a grant from the National Institutes of Health (GM 047915), and a Gill Professorship from the University of Kentucky. K. T. H. is grateful to the National Institutes of Environmental Health for a Pre-doctoral Fellowship.

The writing and the original, unpublished work presented by A. L. D. and B. R. B. were supported by the Air Force Office of Scientific Research under Grant FA9550-07-1-0043, the National Science Foundation under Grant MCB 0842831, and the Hans & Ella McCollum '21 Vahlteich Endowment.

References

1. E. N. Harvey, *Bioluminescence*, Academic Press, New York, 1952.
2. P. J. Herring, *Bioluminescence in Action*, Academic Press: London, New York, 1978.

3. O. Shimomura and F. H. Johnson, *Biolumin. Prog., Proc. Conf.*, 1966, **495**.
4. O. Shimomura, *Symp. Soc. Exp. Biol.*, 1985, **39**, 351.
5. A. K. Campbell, *Biochem. J.*, 1974, **143**, 411.
6. L. D. Levine and W. W. Ward, *Comp. Biochem. Physiol., B: Biochem. Mol. Biol.*, 1982, **72**, 77.
7. A. K. Campbell, M. B. Hallett, R. A. Daw, M. E. T. Ryall, R. C. Hart and P. J. Herring, *Biolumin. Chemilumin. [Int. Symp. Anal. Appl. Biolumin. Chemilumin.], 2nd*, 1981, **601**.
8. J. P. Henry, M. F. Isambert and A. M. Michelson, *Biochimie*, 1973, **55**, 83.
9. M. T. Nicolas, J. M. Bassot and O. Shimomura, *Photochem. Photobiol.*, 1982, **35**, 201.
10. W. W. Ward and H. H. Seliger, *Biochemistry*, 1974, **13**, 1500.
11. F. I. Tsuji and G. B. Leisman, *Proc. Natl. Acad. Sci. USA*, 1981, **78**, 6719.
12. O. Shimomura, *Photochem. Photobiol.*, 1986, **44**, 671.
13. J. W. Hastings and D. Davenport, *Biol. Bull. (Woods Hole, MA)*, 1957, **113**, 120.
14. J. F. Head, S. Inouye, K. Teranishi and O. Shimomura, *Nature*, 2000, **405**, 372.
15. Z. J. Liu, E. S. Vysotski, L. Deng, J. Lee, J. Rose and B. C. Wang, *Biochem. Biophys. Res. Commun.*, 2003, **311**, 433.
16. A. Witkowski, S. Ramanathan and S. Daunert, *Anal. Chem.*, 1994, **66**, 1837.
17. J. C. Lewis and S. Daunert, *Fresenius' J. Anal. Chem.*, 2000, **366**, 760.
18. A. K. Campbell, *Chemiluminescence Principles and Applications in Biology and Medicine*, Horwood Series in Biomedicine, VCH, New York, 1988.
19. S. Daunert and S. K. Deo, *Photoproteins in Bioanalysis*, Wiley-VCH Verlag GmbH, Weinheim, 2006.
20. R. J. Jackson, K. Fujihashi, H. Kiyono and J. R. McGhee, *J. Immunol. Methods*, 1996, **190**, 189.
21. S. Inouye, M. Noguchi, Y. Sakaki, Y. Takagi, T. Miyata, S. Iwanaga and F. I. Tsuji, *Proc. Natl. Acad. Sci. USA*, 1985, **82**, 3154.
22. D. Prasher, R. O. McCann and M. J. Cormier, *Biochem. Biophys. Res. Commun.*, 1985, **126**, 1259.
23. M. Brini, *Methods*, 2008, **46**, 160.
24. P. Pinton, A. Rimessi, A. Romagnoli, A. Prandini and R. Rizzuto, *Methods Cell Biol.*, 2007, **80**, 297.
25. S. K. Deo, J. C. Lewis and S. Daunert, *Bioconjugate Chem.*, 2001, **12**, 378.
26. J. C. Lewis and S. Daunert, *Anal. Chem.*, 2001, **73**, 3227.
27. L. A. Frank, A. I. Petunin and E. S. Vysotski, *Anal. Biochem.*, 2004, **325**, 240.
28. S. K. Deo and S. Daunert, *Anal. Chem.*, 2001, **73**, 1903.
29. M. Mirasoli, S. K. Deo, J. C. Lewis, A. Roda and S. Daunert, *Anal. Biochem.*, 2002, **306**, 204.
30. L. Rowe, S. Deo, J. Shofner, M. Ensor and S. Daunert, *Bioconjugate Chem.*, 2007, **18**, 1772.

31. S. Shrestha, I. R. Paeng, S. K. Deo and S. Daunert, *Bioconjugate Chem.*, 2002, **13**, 269.
32. X. Qu, S. K. Deo, E. Dikici, M. Ensor, M. Poon and S. Daunert, *Anal. Biochem.*, 2007, **371**, 154.
33. H. S. Kim and I. R. Paeng, *Bull. Korean Chem. Soc.*, 2006, **27**, 407.
34. U. A. Desai, S. K. Deo, K. V. Hyland, M. Poon and S. Daunert, *Anal. Chem.*, 2002, **74**, 3892.
35. J. Walstab, S. Combrink, M. Bruess, M. Goethert, B. Niesler and H. Boenisch, *Anal. Biochem.*, 2007, **368**, 185.
36. J. Wang, C. M. Ensor, G. J. Dubuc, S. A. Narang and S. Daunert, *Anal. Chim. Acta*, 2001, **435**, 255.
37. N. V. Rudenko, L. L. Sinegina, M. A. Arzhanov, V. N. Ksenzenko, T. V. Ivashina, O. S. Morenkov, L. A. Shaloiko and L. M. Vinokurov, *J. Biochem. Biophys. Methods*, 2007, **70**, 605.
38. L. Frank, S. Markova, N. Remmel, E. Vysotski and I. Gitelson, *Luminescence*, 2007, **22**, 215.
39. L. Doleman, L. Davies, L. Rowe, E. A. Moschou, S. Deo and S. Daunert, *Anal. Chem.*, 2007, **79**, 4149.
40. S. Inouye and J.-i. Sato, *Biosci., Biotechnol., Biochem.*, 2008, **72**, 3310.
41. S. Bovolenta, M. Foti, S. Lohmer and S. Corazza, *J. Biomol. Screen.*, 2007, **12**, 694.
42. S. K. Deo and S. Daunert, *Fresenius' J. Anal. Chem.*, 2001, **369**, 258.
43. E. Le Poul, S. Hisada, Y. Mizuguchi, V. J. Dupriez, E. Burgeon and M. Detheux, *J. Biomol. Screen.*, 2002, **7**, 57.
44. V. Menon, A. Ranganathn, V. H. Jorgensen, M. Sabio, C. T. Christoffersen, M. A. Uberti, K. A. Jones and P. S. Babu, *Assay Drug Dev. Technol.*, 2008, **6**, 787.
45. M. Suzuki, S. Furukawa, C. Kuramori, C. Sawa, Y. Kabe, M. Nakamura, J.-i. Sawada, Y. Yamaguchi, S. Sakamoto, S. Inouye and H. Handa, *Biochem. Biophys. Res. Commun.*, 2008, **368**, 600.
46. R. M. Eglen and T. Reisine, *Assay Drug Dev. Technol.*, 2008, **6**, 659.
47. H. Hoshino, *Expert Opin. Drug Discovery*, 2009, **4**, 373.
48. B. Schnurr, T. Ahrens and U. Regenass, *Comp. Med. Chem. II*, 2006, **3**, 577.
49. M. Ormo, A. B. Cubitt, K. Kallio, L. A. Gross, R. Y. Tsien and S. J. Remington, *Science*, 1996, **273**, 1392.
50. R. E. Campbell, O. Tour, A. E. Palmer, P. A. Steinbach, G. S. Baird, D. A. Zacharias and R. Y. Tsien, *Proc. Natl. Acad. Sci. USA*, 2002, **99**, 7877.
51. R. Heim and R. Y. Tsien, *Curr. Biol.*, 1996, **6**, 178.
52. R. Heim, D. C. Prasher and R. Y. Tsien, *Proc. Natl. Acad. Sci. USA*, 1994, **91**, 12501.
53. M. Ishii, J. S. Kunimura, H. T. Jeng, T. C. Penna and O. Cholewa, *Appl. Biochem. Biotechnol.*, 2007, **137–140**, 555.
54. M. Ishii, J. S. Kunimura, T. C. Penna and O. Cholewa, *Int. J. Pharm.*, 2007, **337**, 109.

55. C. B. Smith, J. E. Anderson, R. L. Fischer and S. R. Webb, *Environ. Pollut.*, 2002, **120**, 517.
56. A. Muller-Taubenberger and K. I. Anderson, *Appl. Microbiol. Biotechnol.*, 2007, **77**, 1.
57. H. Fukuda, M. Arai and K. Kuwajima, *Biochemistry*, 2000, **39**, 12025.
58. O. Shimomura, S. Inouye, B. Musicki and Y. Kishi, *Biochem. J.*, 1990, **270**, 309.
59. O. Shimomura, B. Musicki and Y. Kishi, *Biochem. J.*, 1988, **251**, 405.
60. O. Shimomura, B. Musicki and Y. Kishi, *Biochem. J.*, 1989, **261**, 913.
61. O. Shimomura, Y. Kishi and S. Inouye, *Biochem. J.*, 1993, **296**(Pt 3), 549.
62. O. Shimomura, B. Musicki, Y. Kishi and S. Inouye, *Cell Calcium*, 1993, **14**, 373.
63. A. Y. Gorokhovatsky, V. V. Marchenkov, N. V. Rudenko, T. V. Ivashina, V. N. Ksenzenko, N. Burkhardt, G. V. Semisotnov, L. M. Vinokurov and Y. B. Alakhov, *Biochem. Biophys. Res. Commun.*, 2004, **320**, 703.
64. S. K. Deo, M. Mirasoli and S. Daunert, *Anal. Bioanal. Chem.*, 2005, **381**, 1387.
65. A. V. Shivange, J. Marienhagen, H. Mundhada, A. Schenk and U. Schwaneberg, *Curr. Opin. Chem. Biol.*, 2009, **13**, 19.
66. W. P. Stemmer, *Proc. Natl. Acad. Sci. USA*, 1994, **91**, 10747.
67. K. Tsuzuki, L. Tricoire, O. Courjean, N. Gibelin, J. Rossier and B. Lambolez, *J. Biol. Chem.*, 2005, **280**, 34324.
68. L. Tricoire, K. Tsuzuki, O. Courjean, N. Gibelin, G. Bourout, J. Rossier and B. Lambolez, *Proc. Natl. Acad. Sci. USA*, 2006, **103**, 9500.
69. F. I. Tsuji, S. Inouye, T. Goto and Y. Sakaki, *Proc. Natl. Acad. Sci. USA*, 1986, **83**, 8107.
70. Y. Ohmiya and F. I. Tsuji, *FEBS Lett.*, 1993, **320**, 267.
71. M. Nomura, S. Inouye, Y. Ohmiya and F. I. Tsuji, *FEBS Lett.*, 1991, **295**, 63.
72. Y. Ohmiya, M. Ohashi and F. I. Tsuji, *FEBS Lett.*, 1992, **301**, 197.
73. L. Deng, E. S. Vysotski, Z. J. Liu, S. V. Markova, N. P. Malikova, J. Lee, J. Rose and B. C. Wang, *FEBS Lett.*, 2001, **506**, 281.
74. G. A. Stepanyuk, S. Golz, S. V. Markova, L. A. Frank, J. Lee and E. S. Vysotski, *FEBS Lett.*, 2005, **579**, 1008.
75. E. Dikici, X. Qu, L. Rowe, L. Millner, C. Logue, S. K. Deo, M. Ensor and S. Daunert, *Protein Eng. Des. Sel.*, 2009, **22**, 243.
76. L. A. Frank, V. V. Borisova, S. V. Markova, N. P. Malikova, G. A. Stepanyuk and E. S. Vysotski, *Anal. Bioanal. Chem.*, 2008, **391**, 2891.
77. L. Rowe, K. Combs, S. Deo, C. Ensor, S. Daunert and X. Qu, *Anal. Chem.*, 2008, **80**, 8470.
78. R. M. Ferraz, A. Vera, A. Aris and A. Villaverde, *Microb. Cell Fact.*, 2006, **5**, no pp given.
79. C. M. Wright, R. A. Heins and M. Ostermeier, *Curr. Opin. Chem. Biol.*, 2007, **11**, 342.

80. G. S. Baird, D. A. Zacharias and R. Y. Tsien, *Proc. Natl. Acad. Sci. USA*, 1999, **96**, 11241.
81. N. Doi and H. Yanagawa, *FEBS Lett.*, 1999, **453**, 305.
82. S. Lindman, I. Johansson, E. Thulin and S. Linse, *Protein Sci.*, 2009, **18**, 1221.
83. R. H. Newman and J. Zhang, *Mol. BioSyst.*, 2008, **4**, 496.
84. M. Ostermeier, *Protein Eng. Des. Sel.*, 2005, **18**, 359.
85. M. S. Siegel and E. Y. Isacoff, *Neuron*, 1997, **19**, 735.
86. K. Teasley Hamorsky, C. M. Ensor, Y. Wei and S. Daunert, *Angew. Chem., Int. Ed.*, 2008, **47**, 3718.
87. D. F. Scott, K. T. Hamorsky, C. M. Ensor, K. W. Anderson and S. Daunert, *Abstract of Papers, 238th ACS National Meeting*, Washington, DC, United States, August 16–20, 2009, 2009, ANYL.
88. C.-S. Kim, B. Pierre, M. Ostermeier, L. Looger and J. R. Kim, *Abstract of Papers, 238th ACS National Meeting*, Washington, DC, United States, August 16–20, 2009, 2009, BIOT.
89. G. Guntas, S. F. Mitchell and M. Ostermeier, *Chem. Biol.*, 2004, **11**, 1483.
90. K. V. Wood, *Cell Notes*, 2004, **8**, 2–6.
91. B. A. Tannous, D. E. Kim, J. L. Fernandez, R. Weissleder and X. O. Breakefield, *Mol. Ther.*, 2005, **11**, 435–443.
92. V. Gooch, A. Mehra, L. F. Mehra, L. Larrondo, J. Fox, M. Touroutoutoudis, J. Loros and J. Dunlap, *Eukaryotic Cell*, 2008, **7**, 28–37.
93. B. R. Branchini, D. M. Ablamsky, A. L. Davis, T. L. Southworth, B. Butler, F. Fan, A. P. Jathoul and M. A. Pule, *Anal. Biochem.*, 2010, **396**, 290–297.
94. A. L. Bonin, M. Gossen and H. Bujard, *Gene*, 1994, **141**, 75–77.
95. Y. Nakajima, T. Kimura, C. Suzuki and Y. Ohmiya, *Biosci. Biotechnol. Biochem.*, 2004, **68**, 948–951.
96. A. Craney, T. Hohenauer, Y. Xu, N. K. Navani, Y. Li and J. Nodwell, *Nucleic Acids Res.*, 2007, **35**, e46.
97. S. Gould, G. Keller and S. Subramani, *J. Cell Biol.*, 1988, **107**, 897–905.
98. P. Ray, R. Tsien and S. S. Gambhir, *Cancer Res.*, 2007, **67**, 3085–3093.
99. B. A. Rabinovich, Y. Ye, T. Etto, J. Q. Chen, H. I. Levitsky, W. W. Overwijk, L. J. N. Cooper, J. Gelovani and P. Hwu, *Proc. Natl. Acad. Sci. USA*, 2008, **105**, 14342–14346.
100. Promega, *Protocols & Applications Guide*, revised 2009, 8.0–8.15.
101. H. Mziaut, M. Trajkovski, A. Altkruger and M. Solimena, *Promega Notes*, 2007, **95**, 5–7.
102. K. E. Luker and G. D. Luker, *Antiviral Res.*, 2008, **78**, 179–187.
103. A. Roda, M. Guardigli, E. Michelini and M. Mirasoli, *Trends Anal. Chem.*, 2009, **28**, 307–322.
104. B. Binkowski, F. Fan and K. Wood, *Curr. Opin. Biotechnol.*, 2009, **20**, 14–18.
105. F. R. Leach, *J. Appl. Biochem.*, 1981, **3**, 473–517.
106. E. A. Meighen, *FASEB J.*, 1993, **7**, 1016–1022.
107. K. V. Wood, *Photochem. Photobiol.*, 1995, **62**, 662–673.

108. V. R. Viviani, *Cell. Mol. Life Sci.*, 2002, **59**, 1833–1850.
109. J. R. de Wet, K. V. Wood, D. R. Helinski and M. DeLuca, *Proc. Natl. Acad. Sci. USA*, 1985, **82**, 7870–7873.
110. Y. Ando, K. Niwa, N. Yamada, T. Enomot, T. Irie, H. Kubota, Y. Ohmiya and H. Akiyama, *Nat. Photon.*, 2008, **2**, 44–47.
111. L. Zhang, W. Li, X. Hong and H. Lin, *Mol. Cell. Endocrinol.*, 2009, **311**, 87–93.
112. S. V. Shenvi, E. J. Smith and T. M. Hagen, *Pharmacol. Res.*, 2009, **60**, 229–236.
113. F. Y. Zeng, J. Cui, L. Liu and T. Chen, *Cancer Lett.*, 2009, **284**, 157–164.
114. S. Matsumoto, F. Tanaka, K. Sato, S. Kimura, T. Maekawa, S. Hasegawa and H. Wada, *Lung Cancer*, 2009, **66**, 75–79.
115. H. Yao, S. S. Ng, W. O. Tucker, Y. K. T. Tsang, K. Man, X. M. Wang, B. K. C. Chow, H. F. Kung, G. P. Tang and M. C. Lin, *Biomaterials*, 2009, **30**, 5793–5803.
116. M. N. Kim, H. H. Park, W. K. Lim and H. J. Shin, *J. Microbiol. Methods*, 2005, **60**, 235–245.
117. B. A. Sherf and K. V. Wood, *Promega Notes*, 1994, **49**, 14.
118. A. Paguio, B. Almond, F. Fan, P. Stecha, D. Garvin, M. Wood and K. Wood, *Promega Notes*, 2005, **89**, 7–10.
119. Promega Corporation, *pGL3 Luciferase Reporter Vectors Technical Manual No. TM033*, revised 2007.
120. T. Nakatsu, S. Ichiyama, J. Hiratake, A. Saldanha, N. Kobashi, K. Sakata and H. Kato, *Nature*, 2006, **440**, 372–376.
121. A. M. Gulick and B. R. Branchini, unpublished results.
122. T. Hirano, Y. Hasumi, K. Ohtsuka, S. Maki, H. Niwa, M. Yamaji and D. Hashizume, *J. Am. Chem. Soc.*, 2009, **131**, 2385–2396.
123. P. Naumov, Y. Ozawa, K. Ohkubo and S. Fukuzumi, *J. Am. Chem. Soc.*, 2009, **131**, 11590–11605.
124. E. H. White, E. Rapaport, H. H. Seliger and T. A. Hopkins, *Bioorg. Chem.*, 1971, **1**, 92–122.
125. Promega Corporation, *Chroma-Glo™ Luciferase Assay System Technical Manual No. TM062*, revised 2009.
126. Promega Corporation, *Dual-Luciferase® Reporter Assay System Technical Manual No. TM040*, revised 2009.
127. B. R. Branchini, T. L. Southworth, N. F. Khattak, E. Michelini and A. Roda, *Anal. Biochem.*, 2005, **345**, 140–148.
128. B. R. Branchini, D. M. Ablamsky, M. H. Murtiashaw, L. Uzasci, H. Fraga and T. L. Southworth, *Anal. Biochem.*, 2007, **361**, 253–262.
129. Y. Nakajima, T. Kimura, K. Sugata, T. Enomoto, A. Asakawa, H. Kubota, M. Ikeda and Y. Ohmiya, *BioTechniques*, 2005, **38**, 891–894.
130. B. F. Eames, D. A. Benaron, D. K. Stevenson and C. H. Contag, *Proc. SPIE Int. Soc. Opt. Eng.*, 1999, **3600**, 36–39.
131. B. W. Rice, M. D. Cable and M. B. Nelson, *J. Biomed. Opt.*, 2001, **6**, 432–440.

132. E. Shapiro, C. Lu and F. Baneyx, *Protein Eng. Des. Sel.*, 2005, **18**, 581–587.
133. N. Tafreshi, S. Hosseinkhani, M. Sadeghizadeh, M. Sadeghi, B. Ranjbar and H. Naderi-Manesh, *J. Biol. Chem.*, 2007, **282**, 8641–8647.
134. N. K. Tafreshi, M. Sadeghizadehi, R. Emamzadeh, B. Rawbar, H. Naderi-Manesh and S. Hosseinkhani, *Biochem. J.*, 2008, **412**, 27–33.
135. H. Caysa, R. Jacob, N. Muther, B. Branchini, M. Messerle and A. Soling, *Photochem. Photobiol. Sci.*, 2009, **8**, 52–56.
136. M. L. Foucault, L. Thomas, S. Goussard, B. R. Branchini and C. Grillot-Courvalin, *Appl. Environ. Microbiol.*, 2010, **76**, 264–274.
137. E. Michelini, L. Cevenini, L. Mezzanotte, D. Ablamsky, T. Southworth, B. R. Branchini and A. Roda, *Photochem. Photobiol. Sci.*, 2008, **7**, 212–217.
138. E. Michelini, L. Cevenini, L. Mezzanotte, D. Ablamsky, T. Southworth, B. Branchini and A. Roda, *Anal. Chem.*, 2008, **80**, 260–267.
139. L. C. Tisi, P. J. White, D. J. Squirrell, M. J. Murphy, C. R. Lowe and J. A. H. Murray, *Anal. Chim. Acta*, 2002, **457**, 115–123.
140. L. C. Tisi, C. Lowe and J. A. H. Murray, in *Proceedings of the 11th International Symposium on Bioluminescence and Chemiluminescence*, ed. J. F. Case, P. J. Herring, B. H. Robison, S. H. D. Haddock, L. J. Kricka and P.E. Stanley, World Scientific, Singapore, 2001, pp. 189–192.
141. G. H. E. Law, O. A. Gandelman, L. C. Tisi, C. R. Lowe and J. A. H. Murray, in *Proceedings of the 12th International Symposium on Bioluminescence and Chemiluminescence*, ed. P. E. Stanley and L. J. Kricka, World Scientific, Singapore, 2002, p. 37–40.
142. B. Baggett, R. Roy, S. Momen, S. Morgan, L. Tisi, D. Morse and R. J. Gillies, *Mol. Imag.*, 2004, **3**, 324–332.
143. G. H. E. Law, O. A. Gandelman, L. C. Tisi, C. R. Lowe and J. A. H. Murray, *Biochem. J.*, 2006, **397**, 305–312.
144. J. F. Thompson, L. S. Hayes and D. B. Lloyd, *Gene*, 1991, **103**, 171–177.
145. E. M. Harrison, O. J. Garden, J. A. Ross and S. J. Wigmore, *J. Immunol. Methods*, 2006, **310**, 182–185.
146. J. Eriksson, B. Gharizadeh, T. Nordström and P. Nyrén, *Electrophoresis*, 2004, **25**, 20–27.
147. O. A. Gandelman, V. L. Church, C. A. Moore, C. Carne, H. Jalal, J. Murray and L. C. Tisi, in *Proceedings of the 14th International Symposium on Bioluminescence and Chemiluminescence*, ed. A. A. Szalay, P. J. Hill, L. J. Kricka and P. E. Stanley, World Scientific, Singapore, 2007, p. 95–98.
148. L. C. Tisi, G. H. E. Law, O. Gandelman, J. Murray and H. Augustus, pH Tolerant Luciferase, World Intellectual Property Organization, 2007, *WO/2007/017684*.
149. A. Riahi-Madvar and S. Hosseinkhani, *Protein Eng. Des. Sel.*, 2009, **22**, 655–663.
150. A. Kitayama, H. Yoshizaki, Y. Ohmiya, H. Ueda and T. Nagamune, *Photochem. Photobiol.*, 2003, **77**, 333–338.

151. G. Lomakina, Y. Modestova and N. Ugarova, *Moscow Univ. Chem. Bull.*, 2008, **63**, 63–66.

152. Promega, *CellTiter-Glo*® *Luminescent Cell Viability Assay Technical Manual*, revised 2009.

153. K. V. Wood, M. P. Hall and M. G. Gruber, Thermostable luciferases from Photuris pennsylvanica and pyrophorus plagiopthalamus and methods of production, World Intellectual Property Organization, 2001, *WO 01/20002 A1*.

154. M. P. Hall, M. G. Gruber, R. R. Hannah, M. L. Jennens-Clough and K. V. Wood, in *Bioluminescence and Chemiluminescence - Perspectives for the 21ˢᵗ Century*, ed. A. Roda, *et al.*, John Wiley & Sons, Ltd., Chichester, UK, 1998, pp. 392–395.

155. D. S. Auld, Y. Q. Zhang, N. T. Southall, G. Rai, M. Landsman, J. Maclure, D. Langevin, C. J. Thomas, C. P. Austin and J. Inglese, *J. Med. Chem.*, 2009, **52**, 1450–1458.

156. J. Eriksson, T. Nordström and P. Nyrén, *Anal. Biochem.*, 2003, **314**, 158–161.

157. B. N. Dominy, D. Perl, F. X. Schmid and C. L. Brooks, *J. Mol. Biol.*, 2002, **319**, 541–554.

158. K. Ikegaya, *J. Biochem.*, 2005, **137**, 349–354.

159. G. M. Leclerc, F. R. Boockfor, W. J. Faught and L. S. Frawley, *BioTechniques*, 2000, **29**, 590–591594–596, 598 *passim*.

160. M. S. Allen, J. R. Wilgus, C. S. Chewning, G. S. Sayler and M. L. Simpson, *Syst. Synth. Biol.*, 2007, **1**, 3–9.

161. H. H. Seliger and W. D. McElroy, *Proc. Natl. Acad. Sci. USA*, 1964, **52**, 75–81.

162. N. Hattori, K. Yajitate, M. Nakajima and S. Murakami, Method for analyzing intracellular components, Kikkoman Corporation, 2001, *US Pat.* 6238857.

163. N. Hattori, N. Kajiyama, M. Maeda and S. Murakami, *Biosci. Biotechnol. Biochem.*, 2002, **66**, 2587–2593.

164. N. Hattori, T. Sakakibara, N. Kajiyama, T. Igarashi, M. Maeda and S. Murakami, *Anal. Biochem.*, 2003, **319**, 287–295.

165. E. Kim-Choi, C. Danilo, J. Kelly, R. Carroll, D. Shonnard and I. Rybina, *Luminescence*, 2005, **21**, 135–142.

166. E. Kim-Choi, C. Danilo, J. Kelly, R. Carroll, D. Shonnard and I. Rybina, *Toxicol. In Vitro*, 2006, **20**, 1537–1547.

167. H. Fujii, K. Noda, Y. Asami, A. Kuroda, M. Sakata and A. Tokida, *Anal. Biochem.*, 2007, **366**, 131–136.

168. K. Noda, T. Matsuno, H. Fujii, T. Kogure, M. Urata, Y. Asami and A. Kuroda, *Biotechnol. Lett.*, 2008, **30**, 1051–1054.

169. M. Urata, R. Iwata, K. Noda, Y. Murakami and A. Kuroda, *Biotechnol. Lett.*, 2009, **31**, 737–741.

170. H. Fraga, D. Fernandes, R. Fontes and J. C. G.E. d. Silva, *FEBS J.*, 2005, **272**, 5206–5216.

171. D. A. Schneider and R. L. Gourse, *J. Biol. Chem.*, 2004, **279**, 8262–8268.

172. B. R. Branchini, R. A. Magyar, M. H. Murtiashaw, S. M. Anderson, L. C. Helgerson and M. Zimmer, *Biochemistry*, 1999, **38**, 13223–13230.
173. K. Hirokawa, N. Kajiyama and S. Murakami, *Biochim. Biophys. Acta-Protein Struct. Mol. Enzym.*, 2002, **1597**, 271–279.
174. J. Eu and J. Andrade, *Luminescence*, 2001, **16**, 57–63.
175. M. Nakamura, M. Mie, H. Funabashi, K. Yamamoto, J. Ando and E. Kobatake, *Anal. Biochem.*, 2006, **352**, 61–67.
176. Y. Zhang, G. J. Phillips, Q. Li and E. S. Yeung, *Anal. Chem.*, 2008, **80**, 9316–9325.
177. V. R. Viviani and E. J. H. Bechara, *Photochem. Photobiol.*, 1995, **62**, 490–495.
178. G. D. Kutuzova, R. R. Hannah and K. V. Wood, in *Proceedings of the 9th International Symposium of Bioluminescence and Chemiluminescence*, ed. J. W. Hastings, L. J. Kricka and P. E. Stanley, John Wiley & Sons, Ltd., Chichester, 1997, pp. 248–252.
179. H. H. Seliger, J. B. Buck, W. G. Fastie and W. D. McElroy, *J. Gen. Physiol.*, 1964, **48**, 95–104.
180. W. H. Biggley, J. E. Lloyd and H. H. Seliger, *J. Gen. Physiol.*, 1967, **50**, 1681–1692.
181. U. Stolz, S. Velez, K. V. Wood, M. Wood and J. L. Feder, *Proc. Natl. Acad. Sci. USA*, 2003, **100**, 14955–14959.
182. K. V. Wood, *Photochem. Photobiol.*, 1995, **62**, 662–673.
183. B. Almond, E. Hawkins, P. Stecha, D. Garvin, A. Paguio, B. Butler, M. Beck, M. Wood and K. Wood, *Promega Notes*, 2003, **85**, 11–14.
184. R. E. Davis, Y. Q. Zhang, N. Southall, L. M. Staudt, C. P. Austin, J. Inglese and D. S. Auld, *Assay Drug Dev. Technol.*, 2007, **5**, 85–103.
185. J. J. Hawes, J. D. Nerva and K. M. Reilly, *J. Biomol. Screen.*, 2008, **13**, 795–803.
186. R. Ogura, N. Matsuo, N. Wako, T. Tanaka, S. Ono and K. Hiratsuka, *Plant Biotechnol.*, 2005, **22**, 151–155.
187. S. T. Gammon, V. M. Villalobos, M. Roshal, M. Samrakandi and D. Piwnica-Worms, *Biotechnol. Prog.*, 2009, **25**, 559–569.
188. N. Hida, M. Awais, M. Takeuchi, N. Ueno, M. Tashiro, C. Takagi, T. Singh, M. Hayashi, Y. Ohmiya and T. Ozawa, *PLoS ONE*, 2009, **4**, e5868.
189. V. R. Viviani, A. C. R. Silva, G. L. O. Perez, R. V. Santelli, E. J. H. Bechara and F. C. Reinach, *Photochem. Photobiol.*, 1999, **70**, 254–260.
190. Y. Ohmiya, Y. Nakajima, V. Viviani, S. Nishii, T. Asai, A. Sugiyama and K. Masuda, Preparation of *Pyrearinus* luciferase variant with shifted maximum luminescence wavelength for altered color signal presentation, Jpn. Kokai Tokkyo Koho, 2008, *JP 2008289475*.
191. V. R. Viviani, E. J. H. Bechara and Y. Ohmiya, *Biochemistry*, 1999, **38**, 8271–8279.
192. Y. Kitayama, T. Kondo, Y. Nakahira, H. Nishimura, Y. Ohmiya and T. Oyama, *Plant Cell Physiol.*, 2004, **45**, 109–113.

193. Y. Nakajima, M. Ikeda, T. Kimura, S. Honma, Y. Ohmiya and K. Honma, *FEBS Lett.*, 2004, **565**, 122–126.

194. T. Noguchi, M. Ikeda, Y. Ohmiya and Y. Nakajima, *BMC Biotechnol.*, 2008, **8**, 40.

195. V. R. Viviani and Y. Ohmiya, *Photochem. Photobiol.*, 2000, **72**, 267–271.

196. V. R. Viviani, A. J. da Silva Neto and Y. Ohmiya, *Protein Eng. Des. Sel.*, 2004, **17**, 113–117.

197. V. R. Viviani, F. G. C. Arnoldi, F. T. Ogawa and M. Brochetto-Braga, *Luminescence*, 2007, **22**, 362–369.

198. W. W. Ward and M. J. Cormier, *J. Biol. Chem.*, 1979, **254**, 781–788.

199. B. T. Ballou, C. Szent-Gyorgyi and G. Finley, *11ᵗʰ Int. Symp. on Biolumin. & Chemilumin.*, 2000, Abstract p34. Asilomar, CA.

200. A. M. Loening, T. D. Fenn, A. M. Wu and S. S. Gambhir, *Protein Eng. Des. Sel.*, 2006, **19**, 391–400.

201. N. Shao and R. Bock, *Curr. Genet.*, 2008, **53**, 381–388.

202. S. Wiles, K. Ferguson, M. Stefanidou, D. B. Young and B. D. Robertson, *Appl. Environ. Microbiol.*, 2005, **71**, 3427–3432.

203. V. V. Borisova, L. A. Frank, S. V. Markova, L. P. Burakova and E. S. Vysotski, *Photochem. Photobiol. Sci.*, 2008, **7**, 1025–1031.

204. Y. Takenaka, H. Masuda, A. Yamaguchi, S. Nishikawa, Y. Shigeri, Y. Yoshida and H. Mizuno, *Gene*, 2008, **425**, 28–35.

205. C. Szent-Gyorgyi, B. T. Ballou, E. Dagnal and B. Bryan, *Proc. SPIE*, 1999, **3600**, 4–11.

206. S. Knappskog, H. Ravneberg, C. Gjerdrum, C. Trosse, B. Stern and I. F. Pryme, *J. Biotechnol.*, 2007, **128**, 705–715.

207. E. B. Santos, R. Yeh, J. Lee, Y. Nikhamin, B. Punzalan, B. Punzalan, K. L. Perle, S. M. Larson, M. Sadelain and R. J. Brentjens, *Nat. Med.*, 2009, **15**, 338–344.

208. C. A. Maguire, N. C. Deliolanis, L. Pike, J. M. Niers, L-A. Tjon-Kon-Fat, M. Sena-Esteves and B. A. Tannous, *Anal. Chem.*, 2009, **81**, 7102–7106.

209. J. Woo and A. G. von Arnim, *Plant Methods*, 2008, **4**, 23.

210. A. M. Loening, A. M. Wu and S. S. Gambhir, *Nat. Methods*, 2007, **4**, 641–643.

211. M. Verhaegen and T. K. Christopoulos, *Anal. Chem.*, 2002, **74**, 4378–4385.

212. A. Pichler, J. L. Prior and D. Piwnica-Worms, *Proc. Natl. Acad. Sci. USA*, 2004, **101**, 1702–1707.

213. H. Zhao, T. C. Doyle, R. J. Wong, Y. Cao, D. K. Stevenson, D. Piwnica-Worms and C. H. Contag, *Mol. Imag.*, 2004, **3**, 43–54.

214. J. W. Hastings, *CRC Crit. Rev. Biochem.*, 1978, **5**, 163–184.

215. J. W. Hastings, C. J. Potrikus, S. C. Gupta, M. Kurfurst and J. C. Makemson, *Adv. Microb. Physiol.*, 1985, **26**, 235–291.

216. J. Lee, I. B. C. Matheson, F. Muller, D. J. O'Kane, J. Vervoort and A. J. W. G. Visser, in *Chemistry and Biochemistry of Flavoenzymes*, ed. F. Muller, CRC Press, Boca Raton, 1990, pp. 109–151.

217. C. M. Miyamoto, M. Boylan, A. F. Graham and E. A. Meighen, *J. Biol. Chem.*, 1988, **263**, 13393–13399.
218. C. Y. Lee, R. B. Szittner and E. A. Meighen, *Eur. J. Biochem.*, 1991, **201**, 161–167.
219. B. K. Hammer and B. L. Bassler, *J. Bacteriol.*, 2009, **191**, 169–177.
220. B. C. Kim, C. H. Youn, J. M. Ahn and M. B. Gu, *Anal. Chem.*, 2005, **77**, 8020–8026.
221. J. L. McAuley, F. Hornung, K. L. Boyd, A. M. Smith, R. McKeon, J. Bennink, J. W. Yewdell and J. A. McCullers, *Cell Host & Microbe*, 2007, **2**, 240–249.
222. M. N. Seleem, M. Ali, S. M. Boyle and N. Sriranganathan, *FEMS Microbiol. Lett.*, 2008, **286**, 124–129.
223. J. J. Min, V. H. Nguyen, H. J. Kim, Y. Hong and H. E. Choy, *Nat. Protocols*, 2008, **3**, 629–636.
224. M. D. Larsen, U. Griesenbach, S. Goussard, D. C. Gruenert, D. M. Geddes, R. K. Scheule, S. H. Cheng, P. Courvalin, C. Grillot-Courvalin and E. W. Alton, *Gene Ther.*, 2008, **15**, 434–442.
225. K. Hakkila, M. Maksimow, M. Karp and M. Virta, *Anal. Biochem.*, 2002, **301**, 235–242.
226. S. S. Patterson, H. M. Dionisi, R. K. Gupta and G. S. Sayler, *J. Ind. Microbiol. Biotechnol.*, 2005, **32**, 115–123.
227. O. Olsson, A. Escher, G. Sandberg, J. Schell, C. Koncz and A. Szalay, *Gene*, 1989, **81**, 335–347.
228. A. Escher, D. J. O'Kane, J. Lee and A. A. Szalay, *Proc. Natl. Acad. Sci. USA*, 1989, **86**, 6528–6532.
229. P. J. Hill, C. E. Rees, M. K. Winson and G. S. Stewart, *Biotechnol. Appl. Biochem.*, 1993, **17**(1), 3–14.
230. A. Westerlund-Karlsson, P. Saviranta and M. Karp, *Biochem. Biophys. Res. Commun.*, 2002, **296**, 1072–1076.
231. C. Suadee, S. Nijvipakul, J. Svasti, B. Entsch, D. P. Ballou and P. Chaiyen, *J. Biochem.*, 2007, **142**, 539–552.
232. B. C. Ortego, J. J. Whittenton, H. Li, S. C. Tu and R. C. Willson, *Biochemistry*, 2007, **46**, 13864–13873.
233. S. Hosseinkhani, R. Szittner and E. A. Meighen, *Biochem. J.*, 2005, **385**, 575–580.
234. L. Y. Lin, R. Szittner, R. Friedman and E. A. Meighen, *Biochemistry*, 2004, **43**, 3183–3194.
235. E. J. Nelson, H. S. Tunsjo, P. M. Fidopiastis, H. Sorum and E. G. Ruby, *Appl. Environ. Microbiol.*, 2007, **73**, 1825–1833.
236. L. R. Mesak, G. Yim and J. Davies, *Plasmid*, 2009, **61**, 182–187.
237. R. Lee and W. D. McElroy, *Biochemistry*, 1969, **8**, 130–136.
238. S. C. Alter and M. DeLuca, *Biochemistry*, 1986, **25**, 1599–1605.
239. B. R. Branchini, R. A. Magyar, M. H. Murtiashaw, N. Magnasco, L. K. Hinz and J. G. Stroh, *Arch. Biochem. Biophys.*, 1997, **340**, 52–58.
240. J. R. Kumita, L. Jain, E. Safroneeva and G. A. Woolley, *Biochem. Biophys. Res. Commun.*, 2000, **267**, 394–397.

241. B. R. Branchini, D. A. Ablamsky and J. C. Rosenberg, *unpublished results*.
242. Y. Nakajima, K. Kobayashi, K. Yamagishi, T. Enomoto and Y. Ohmiya, *Biosci., Biotechnol., Biochem.*, 2004, **68**, 565–570.
243. C. Wu, K. Mino, H. Akimoto, M. Kawabata, K. Nakamura, M. Ozaki and Y. Ohmiya, *Proc. Natl. Acad. Sci. USA*, 2009, **106**, 15599–15603.
244. W. E. Muller, M. Kasueske, X. Wang, H. C. Schroder, Y. Wang, D. Pisignano and M. Wiens, *Cell. Mol. Life Sci.*, 2009, **66**, 537–552.
245. K. Mori, S. Jkojima, S. Maki, T. Hirano and H. Niwa, *J. Biol. Chem. Lumin.*, 2008, **23**, 86.

CHAPTER 14

Chemiluminescent and Bioluminescent Biosensors

CHRISTOPHE A. MARQUETTE AND LOÏC J. BLUM

Laboratoire de Génie Enzymatique et Biomoléculaire, Institut de Chimie et Biochimie Moléculaires et Supramoléculaires, Université Lyon 1-CNRS 5246 ICBMS, Bâtiment CPE 43, bd du 11 novembre 1918, 69622 Villeurbanne, Cedex, France

14.1 Introduction

Luminescence is the emission of light from an electronically excited compound returning to the ground state. The source of excitation energy serves as a basis for a classification of the various types of luminescence. Chemiluminescence occurs in the course of some chemical reactions when an electronically excited state is generated. Bioluminescence is a special case of chemiluminescence occurring in some living organisms and involves a protein, generally an enzyme.

Bio- or chemiluminescence measurements consist of monitoring the rate of production of photons and, thus, the light intensity depends on the rate of the luminescent reaction. Consequently, light intensity is directly proportional to the concentration of a limiting reactant involved in a luminescence reaction. With modern instrumentation, light can be measured at a very low level, and this allows the development of very sensitive analytical methods based on these light-emitting reactions. Bioluminescence-based and chemiluminescence-based sensors have been then developed with the aim of combining the sensitivity of light-emitting reactions with the convenience of sensors. Fibre-optics associated with a sensitive light detector appeared to be convenient transducers for designing biosensors involving these kinds of luminescent reactions. In

Chemiluminescence and Bioluminescence: Past, Present and Future
Edited by Aldo Roda
© Royal Society of Chemistry 2011
Published by the Royal Society of Chemistry, www.rsc.org

addition to these fibre-optic-based sensors several luminescence analytical systems including immobilized reagents but not fibre-optics have been described. More recently, chemi- and electrochemiluminescence detections have also been used instead of fluorescence for the development of biochips and microarrays.

14.2 Chemiluminescence

Chemiluminescence (CL) reactions are generally oxidoreduction processes and the excited compound, which is the reaction product, has a different chemical structure from the initial reactant. Several hundreds of organic and inorganic compounds are at the origin of chemiluminescence reactions, which can occur in liquid or solid phases, or at solid–liquid or solid–gas interfaces.[1-3] This chapter will mainly focus on liquid-phase light-emitting reactions based on the oxidation of 5-amino-2,3-dihydrophthalazine-1,4-dione (luminol) (Scheme 14.1) but other label such as alkaline phosphatase could also lead to chemiluminescent signal emission. Luminol oxidation leads to the formation of an aminophthalate ion in an excited state, which emits light when returning to the ground state. The quantum yield of the reaction is low (≈ 0.01) and the emission spectrum shows a maximum at 425 nm.[4] As suggested above, this reaction can be triggered through a wide range of catalysts, more or less specific for a particular oxidizing species and with varying efficiencies.

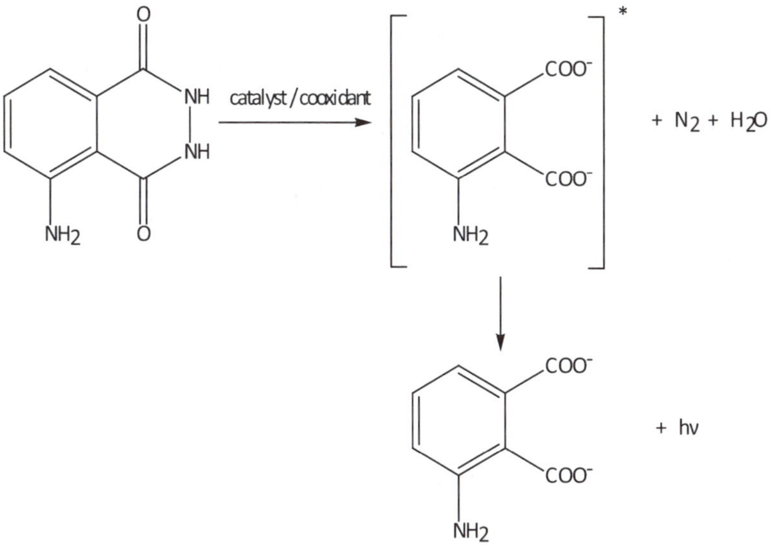

Scheme 14.1

14.2.1   Reaction Triggering

14.2.1.1   Bio-catalysis and Chemical Catalysis

The bio-catalysis of the luminol chemiluminescence is considered to be the most powerful way of triggering the light emission reaction. Heme-containing proteins, particularly horseradish peroxidases (EC 1.11.1.7), can catalyse the chemiluminescent reaction of luminol in the presence of hydrogen peroxide. The use of this enzyme has the advantage over other catalysts in that the chemiluminescent reaction can proceed at near-neutral pH (8–8.5). This chemiluminescence production system is then the basis of biochemistry-based analytical systems such as biosensors, immunoassays, immunosensors and microarrays.[5] Despite numerous studies,[6–11] the complex mechanism of the peroxidase-catalysed reaction and the stoichiometry remain hypothetical. Roughly speaking, the reaction sequence leading to light generation can be divided into two main processes (Figure 14.1a). First, during a series of enzymatic steps and in the presence of luminol (LH^-) and hydrogen peroxide, horseradish peroxidase (HRP) is successively converted into intermediary complexes (complex I and complex II) before being regenerated into free peroxidase. These enzymatic steps produce luminol radicals ($L^{-\bullet}$, $LH^\bullet$) that then enter a complex chemical pathway to finally generate luminol hydroperoxide (LO_2H^-), the precursor of the light emitter (excited 3-aminophthalate ion). This second part of the mechanism can be slightly modified, in terms of the species involved and their relative concentrations, when the enzyme to hydrogen peroxide molar ratio changes: when this ratio is high, a superoxide pathway is involved; for low ratios the predominant pathway is the diazaquinone route.

This reaction can be used for detecting peroxidase-labelled molecules, hydrogen peroxide generating labelled enzymes and hydrogen peroxide enzymatic precursors, *i.e.* substrates for oxidase enzymes generating hydrogen peroxide.

From a commercial point of view, the composition of the reagent solutions used for chemiluminescent immunoassays has been a field of great improvements and patenting. The main investigators here are Pierce (USA), Amersham (UK) and Covalab (France), who all commercialize measuring solutions composed of luminol in addition with an oxidation system and a specially designed chemiluminescent enhancer. This enhancer is the most important part of each kit since it enables the signal to be intensified while keeping the background luminescence low.

Finally, transition metal cations (Co^{2+}, Cu^{2+}, Cr^{2+}, Fe^{2+}, Fe^{3+}, Hg^{2+}, Mn^{4+}, Ni^{2+}) and their complexed forms (*e.g.* ferrocene, ferricyanide) can be used as catalyst with mitigated performances that are linked to the relatively low signal-to-noise ratio obtained and the required elevated pH of the reaction. When these transition metal cations are used to catalyse the reaction, they are involved in the production of luminol radicals, as the HRP is, but under harsher pH conditions. Analytical applications using such catalysts will

a) Bio-catalysed chemiluminescence

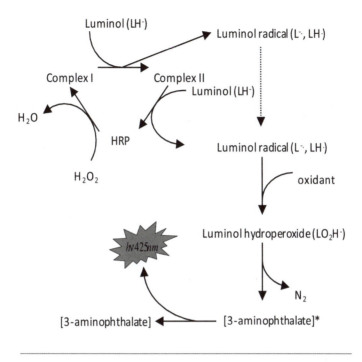

b) Electro-catalysed chemiluminescence

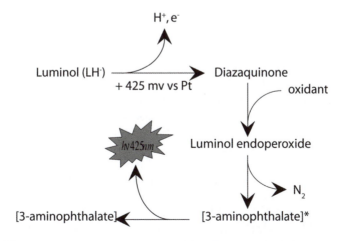

Figure 14.1 Schematic representation of (a) horseradish peroxidase catalysed reaction and (b) electro-catalysed chemiluminescent reaction.

therefore focus preferably on the detection of molecules affecting the rate or efficiency of the reaction and the detection of hydrogen peroxide enzymatic precursors.

14.2.1.2   Electro-catalysis

An original and unusual way to obtain highly sensitive hydrogen peroxide detection is the electrogenerated chemiluminescence or electro-chemiluminescence of luminol (ECL). The electrochemical oxidation of luminol is usually considered as the second most efficient way of triggering the reaction. The scheme is similar to the peroxidase-catalysed one with hydrogen peroxide as co-oxidant and a large range of working pH (Figure 14.1b).

In a mechanistic study of this ECL reaction, Sakura et al.[12] had proposed that luminol was first oxidized at the electrode surface and then reacted, mole to mole, with hydrogen peroxide (Figure 14.1b). The theoretical ratio (photon produced)/(H_2O_2 consumed) is then 1, whereas it is only 0.5 for the peroxidase-catalysed reaction. Moreover, avoiding the use of fragile enzymes for the catalysis of the chemiluminescent reaction could lead to more stable and reproducible sensors. Consequently, regarding the sensitivity of hydrogen peroxide detection, the electrogenerated chemiluminescence of luminol will be more efficient than the peroxidase-catalysed reaction.[13] Because of this high sensitivity for hydrogen peroxide most applications of ECL are dedicated to the detection of hydrogen peroxide, generating label enzymes and hydrogen peroxide enzymatic precursors. Less often, this electrogenerated reaction is used for the detection of luminol-labelled molecules.

The main competitors to luminol label in the field of analytical chemistry are ruthenium complexes.[14] Such molecules, which can be regenerated after having emitted their photon(s), appeared to be more appealing electroluminescent labels.

14.3   Bioluminescence

Although numerous luminous organisms are known, only a few of them has been studied and really exploited. Analytical applications of bioluminescence concern mainly the detection of ATP with the firefly luciferase and of NADH with some marine bacteria systems. Luciferase from the North American firefly, i.e. *Photinus pyralis*, has been studied extensively[15–17] and, subsequently, attention has been paid to the luciferase from *Luciola mingrelica*, i.e. the North Caucasus firefly.[18–20]

In some bioluminescent organisms, light is produced without the intervention of a luciferase, directly from a protein–luciferin complex, called a photoprotein, where the luciferin is tightly or covalently bound to the protein. These systems are able to release energy in the form of light emission, independently of a chemical or enzymatic reaction. This energy "discharge" occurs in the presence of a triggering compound, generally H^+ or Ca^{2+} ions depending on the bioluminescent systems.

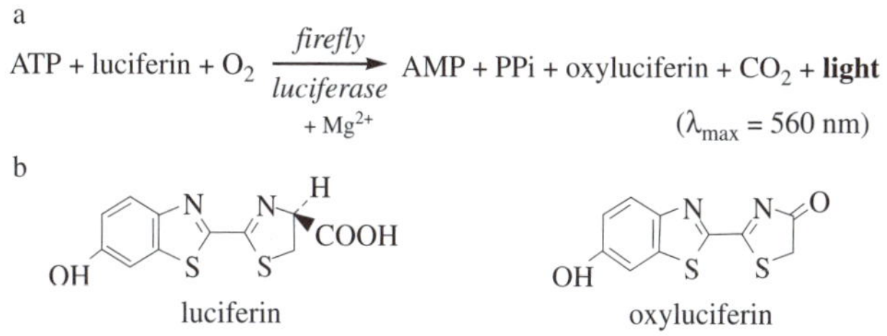

a

$$ATP + luciferin + O_2 \xrightarrow[\substack{\textit{luciferase} \\ + Mg^{2+}}]{\textit{firefly}} AMP + PPi + oxyluciferin + CO_2 + \textbf{light}$$

$$(\lambda_{max} = 560 \text{ nm})$$

b

luciferin   oxyluciferin

Figure 14.2   (a) Firefly luciferase bioluminescence reaction; (b) structure of the specific substrate luciferin and the corresponding reaction product oxyluciferin.

For example, the jellyfish *Aequorea* contains a photoprotein called aequorin of molecular weight about 20 000 and with a heterocyclic compound called coelenterazine covalently linked to it. The protein contains bound oxygen and three calcium binding sites and, upon addition of calcium ions, a blue light is produced.[21] This bioluminescence system can be used for imaging the Ca^{2+} content in living cells. However, it has not been exploited for sensor development since the protein does not turnover and consequently is efficient only once for the production of light.

14.3.1   Firefly Bioluminescence

The firefly luciferase (EC 1.13.12.7) catalyses the emission of light in the presence of ATP, Mg^{2+}, molecular oxygen and firefly luciferin, a specific natural substrate (Figure 14.2).

The light emission is yellow-green with a maximum at 560 nm. The quantum yield of the firefly luciferase bioluminescence reaction is close to 1 under optimum conditions of temperature and pH and in the presence of saturating luciferin concentration.[22]

Synthetic luciferin, as well as purified preparations of native and recombinant firefly luciferases, is now commercially available, allowing the bioluminescent determination of ATP to be used as a routine analysis technique in some laboratories.

14.3.2   Bacterial Bioluminescence

The bacterial bioluminescent reaction is also catalysed by a luciferase (EC 1.14.14.3) isolated from marine bacteria. The four most studied types are *Vibrio harveyi*, *V. fischeri*, *Photobacterium phosphoreum* and *P. leiognathi*.[23,24] In these different luminescent bacteria the same light emission mechanism is involved and the luciferases are similar.[25] The substrates of the bacterial luciferase reaction include reduced flavin mononucleotide ($FMNH_2$), molecular oxygen and a long-chain aldehyde (R-CHO). *In vitro*, decanal is generally used as the aliphatic aldehyde and light emission occurs with a peak at 490 nm. $FMNH_2$ is

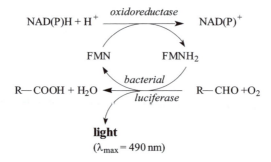

Figure 14.3   Coupled bacterial bioluminescent reaction allowing the detection of NADH or NADPH.

only a transient intermediate and is produced during an oxidoreduction reaction, catalysed by an oxidoreductase, which involves the oxidation of NADH or NADPH concomitantly with the reduction of FMN.

This reductase was isolated from various strains of bioluminescent bacteria as well as in several species of non-luminous aerobic and anaerobic bacteria. The two most useful light-emitting systems were isolated from *V. harveyi* and *V. fischeri*. Distinct oxidoreductases for NADH and NADPH have been identified in extracts of *V. harveyi*[26] whereas *V. fischeri* appears to have only one oxidoreductase acting on both NADH and NADPH. Thus, the bacterial bienzymatic system allows NAD(P)H to be assayed (Figure 14.3). In addition, the use of coupled reactions makes it easy to determine various substrates and enzymes involved in NAD(P)H producing or consuming reactions.

14.4   Biosensors Generalities

The first biosensor described, even if the term was not used at the time, was the combination of the Clark amperometric oxygen electrode serving as transducer and the enzyme glucose oxidase as sensing element for glucose monitoring. In 1962 Clark and Lyons[27] took advantage of the fact that an analyte like glucose could be enzymatically oxidized with, in parallel, consumption of the co-reagent O_2 or the appearance of a product, H_2O_2, which could be electrochemically monitored. The enzyme, retained by a permselective membrane, thus added to the amperometric detector a high selectivity that could not be obtained without the bio-recognition element.

In 1967 Updike and Hicks[28] gave the name "enzyme electrode" to a device consisting of a polyacrylamide gel with entrapped glucose oxidase coating an oxygen electrode for the determination of glucose. Besides amperometry, potentiometric electrodes were also proposed by Guilbault and Montalvo in 1969.[29]

Since the early 1970s various combinations of biological material associated with different types of transducers have given birth to the larger concept of "biosensor." A biosensor associates a bioactive sensing layer with any suitable transducer giving a usable output signal. Biomolecular sensing can be defined as the possibility of detecting analytes of biological interest, like metabolites, but also including drugs and toxins, using an affinity receptor (enzymes being the simplest and historically the first employed), which can be a natural system or an artificial one mimicking a natural one, able to recognize a target molecule in a complex medium among thousands of others.

To obtain a usable output signal that can be correlated with the amount or concentration of analyte present in the medium, multiple events must take place sequentially. Briefly, a first chemical or physical signal consecutive to molecular recognition by the bioactive layer is converted by the transducer into a second signal, generally electrical, with a transduction mode that can be electrochemical, thermal, optical, or based on mass variation. The selective molecular recognition of the target molecule can theoretically be achieved with various kinds of affinity systems, for example (but not exclusively), enzyme for substrate, antibody for antigen, lectin for sugar and nucleic acid for complementary sequence.

The first problem we must face is the degree of bioamplification obtained when molecular recognition occurs. If the bioactive molecule present in the sensing layer is a biocatalyst, a reaction takes place in the presence of the specific target analyte, and a variable amount of co-reagent or product may be either consumed or produced, respectively, in a short time depending on the turnover. Bio-catalysis thus corresponds to an amplification step, generating a chemical signal.

Another key point to which attention must be paid is the intrinsic specificity of the biological material involved in the recognition process. Some enzymes, for instance, may be strictly specific, like urease, or highly specific, like glucose oxidase. Others, like alcohol oxidase or amino acid oxidases, recognize a large spectrum of alcohols and amino acids, respectively. This intrinsic specificity is difficult to modify and problems of interferences may arise.

14.5 Immunosensor Generalities

Theoretically, all immunochemical techniques could be used to design immunosensors. Nevertheless, because the immunosensors are proposed as rapid testing systems, the reaction conditions, and particularly the reagent concentrations, have to be adapted. Indeed, the sensitivity of the detection system used enables the lowering of those concentrations, leading to high sensitivity in the measurement of the immune complex variations.

Most of the developed immunosensors are based either on competitive or sandwich assay, when applied to the detection of low (herbicides, toxins) and high (proteins, cells) molecular weight molecules, respectively (Figure 14.4). Two approaches could be considered when dealing with competitive immunosensors.

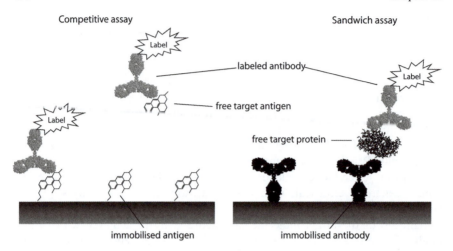

Figure 14.4 Schematic representations of the competitive and sandwich type immuno-tests.

In the first one, an immobilized antibody reacts with free antigens in competition with labelled antigens. In the second, the use of immobilized antigen and labelled antibody is generally preferred and avoids all problems related to antibody immobilization (lost of affinity, orientation).

A few theoretical aspects are presented here since, compared to immuno-tests on micro-plate, immunosensors rarely use incubation times long enough to reach the steady state. Under steady state conditions, the amount of labelled antibodies (Ab*) (or antigen) linked to the immobilized antigens (Ag) (or antibodies) depends only of the affinity constant (K) [Equation (14.1)]. Conversely, when steady state can not be reach because of time or diffusion limitations, the amount of immobilized antigen will have a great influence on the performance of the test. Competitive immunosensors could be considered as a system that measures the amount of "still-free" labelled antibodies after a definite incubation time with the antigen of the sample.[30] Under such conditions, the larger the amount of immobilized antigens, the faster the reaction with the antibodies to be measured and the better the performances of the immunosensor.

These considerations are useful in understanding why immobilization supports with highly specific surfaces, such the porous surfaces of macrometrics membranes and polymeric gels, are usually preferred:

$$Ab^* + Ag \underset{k_d}{\overset{k_a}{\rightleftarrows}} Ab^*:Ag$$

$$K = \frac{k_a}{k_d} = \frac{[Ab^*:Ag]}{[Ab^*][Ag]}$$

(14.1)

14.6 Analytical Applications

14.6.1 Chemiluminescent Biosensors

Interesting approaches to the specific detection of particular compounds in a complex mixture are based on the sequential use of highly specific enzymatic reactions and chemiluminescent detection. Chemiluminescent systems requiring hydrogen peroxide for the light emission are of particular interest since in addition to the determination of H_2O_2, several other compounds can be analysed through the coupling of the CL reaction with H_2O_2-generating enzymatic reactions.[31–37] This makes it possible to detect specific compounds, including, for example, glucose, lactate, urate, cholesterol, xanthine and choline, in complex media, taking advantage of the specificity of the oxidase enzyme used.

For this type of biosensor, FIA fibre optic sensors associated with membrane- or polymer-based immobilized enzymes (Figure 14.5a) have been used extensively; however, these appear to be declining in popularity (according to the number of publications) since the introduction of miniaturization. Nevertheless, the immobilization of both the oxidase enzyme and the peroxidase in millimetre-sized sensing layers, in contact with a fibre optic, has afforded sensitive, specific and stable analytical systems.[3,38,39] Interesting results were also obtained based on ferricyanide or electrochemically catalysed chemiluminescence (Figure 14.5b).

Table 14.1 presents typical detection performances of such macrosystems. They exhibit satisfactory detection limits and ranges and are in most cases

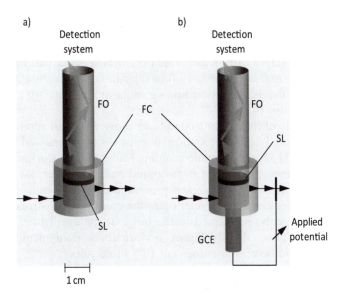

Figure 14.5 Flow injection analysis systems for enzyme-based chemiluminescent (a) and electrochemiluminescent (b) specific systems. FO = optical fibre, FC = flow cell, GCE = glassy carbon electrode and SL = sensing layer.

Table 14.1   Analytical characteristics of enzyme-based chemiluminescent and electrochemiluminescent sensors for bio-specific detection in FIA systems.

Compound	CL system	Detection limit	Range	Ref.
L-Glutamate	Luminol/peroxidase	40 nM	40–1 000 nM	37
	Luminol/peroxidase	10 µM	20 nM–5 µM	36
L-Lysine	Luminol/peroxidase	50 nM	50–1200 nM	68
Lactate	Luminol/ferricyanide	0.2 µM	1–1000 µM	31
	Luminol/electrochemical	1 µM	1 µM–3 mM	13
Uric acid	Luminol/ferricyanide	4 nM	10 nM–30 µM	32
Choline	Luminol/electrochemical	0.3 µM	0.3–1000 µM	43
	Luminol/peroxidase	15 nM	15 nM–300 µM	33
Glucose	Luminol/OH$^-$	4 µM	10 µM–1 mM	34
	Luminol/peroxidase	0.12 mM	0.2–2 mM	35
	Luminol/electrochemical	2 µM	2 µM–30 mM	13
Cholesterol	Luminol/electrochemical	20 µM	20 µM–2.5 mM	42, 44

validated in complex samples such as human sera. These systems benefit from the high sensitivity of light measurement systems such as photomultiplier tubes.

14.6.2   Electrochemiluminescent Biosensors

As mentioned above, an original and unusual way to obtain highly sensitive hydrogen peroxide detection is the electrogenerated chemiluminescence of luminol (ECL). Based on this electro-optical process, flow injection analysis H$_2$O$_2$ sensors have been developed.[13,40,41] The electrochemiluminescence was generated using a glassy carbon electrode polarized *versus* a platinum pseudo-reference electrode and integrated in a flow injection analysis system that could take advantage of the use of optical fibres to separate the detector and the flow system[13,42–44] (Figure 14.5b).

The optimization of the reaction conditions showed that an applied potential of + 425 mV *versus* a platinum pseudo-reference electrode enabled the realization of a sensitive H$_2$O$_2$ sensor while avoiding fouling of the working electrode. An optimum pH measurement of 9 was found and, moreover, the pH dependence of the ECL sensor appeared less pronounced than when using immobilized HRP as the sensing layer. Under optimum conditions, hydrogen peroxide measurements could be performed in the range 1.5 pmol–30 nmol. This ECL H$_2$O$_2$ sensor exhibited slightly higher performances than membrane-based horseradish peroxidase chemiluminescent FIA biosensors.[45]

For the development of glucose and lactate ECL FIA biosensors,[13] the hydrogen peroxide ECL sensor could be associated with the catalytic action of glucose oxidase and lactate oxidase. The oxidases were immobilized on synthetic preactivated membranes brought into contact with the glassy carbon electrode. The glucose or lactate electro-optical biosensor was then able to detect the target analyte with detection limits of 150 and 60 pmol, respectively. In each case, glucose and lactate measurements could be performed over four decades of concentration (Table 14.1).

These biosensors were tested for glucose and lactate measurements in sera, and for lactate measurements in whey solutions. Good agreements were obtained between the present method and reference methods. For glucose analysis in serum, the coefficient of variation for 53 repeated measurements performed over a 10 h period was 4.8% while for lactate analysis 80 assays performed over a 15 h period gave a coefficient of variation of 6.7%. Thus, the ECL-based biosensors gave the possibility of sensitively detecting glucose and lactate in complex matrices without pre-treatment of the samples.

A flow injection fibre optic electro-chemiluminescent biosensor for choline was also developed.[43,44] Choline oxidase was immobilized by physical entrapment in a photo-crosslinkable poly(vinyl alcohol) polymer (PVA-SbQ) after adsorption on weak anion-exchanger beads (DEAE-Sepharose). In this way, the sensing layer was directly created at the surface of the working glassy carbon electrode. Optimization of the reaction conditions and of the physicochemical parameters influencing the FIA biosensor response allows the measurement of choline concentration with a detection limit of 10 pmol. The DEAE-based system also exhibited a good operational stability since 160 repeated measurements of 3 nmol of choline could be performed with a variation coefficient of 4.5%.

A cholesterol flow injection analysis biosensor has also been described as an application of the H_2O_2 ECL sensor.[42] In that work, the luminol electrochemiluminescence, previously studied in aqueous media, was implemented in Veronal buffer with added 0.3% triton X-100 (v/v), 0.3% PEG and 0.4% cholate, to enable the solubilization of the cholesterol and then its efficient oxidation catalysed by the immobilized cholesterol oxidase. The ECL reaction thus occurred in a micellar medium and the performances of the H_2O_2 ECL sensor were investigated.

The calibration curve obtained for hydrogen peroxide exhibited a detection limit of 30 pmol and ranged over three decades at least. These performances compared well with those previously obtained in non-micellar media.[40] The presence of surfactant compounds in the ECL measurement buffer appeared thus to have little effect on the H_2O_2 ECL sensor performances. Under optimized conditions, the determination of free cholesterol could be performed with a detection limit of 0.6 nmol and a calibration curve ranging over at least two decades of concentration.

14.6.3   Chemiluminescent Immunosensors

When the compound of interest is not a substrate for oxidase enzymes or could not be converted into such a substrate, the chemiluminescence reaction can still be used in a biospecific system through the use of immunoreactions with

Table 14.2 Analytical characteristics of antibody-based chemiluminescent sensors for bio-specific detection.

Compound	CL system	Detection limit	Range	Ref.
Immunoglobu-lin G	Luminol/peroxidase	$2\,\mu g\,L^{-1}$	2–$60\,\mu g\,L^{-1}$	50
Salmonella typhimurium	Luminol/peroxidase	$2\times10^3\,CFU\,mL^{-1}$	2×10^3 to 2×10^6 $CFU\,mL^{-1}$	51
2,4-D	Luminol/peroxidase	$4\,\mu g\,L^{-1}$	$4\,\mu g\,L^{-1}$–$160\,mg\,L^{-1}$	46
	Luminol/electrochemical	$0.2\,\mu g\,L^{-1}$	$0.2\,\mu g\,L^{-1}$–$200\,mg\,L^{-1}$	69
Okadaic acid	Luminol/peroxidase	$0.1\,\mu g\,L^{-1}$	$0.1\,\mu g\,L^{-1}$–$100\,\mu g\,L^{-1}$	46
TNT	Luminol/peroxidase	$0.1\,\mu g\,L^{-1}$	0.1–$1000\,\mu g\,L^{-1}$	47
Serologic IgG	Luminol/peroxidase	1 : 800 000 titre	–	49
Alkylbenzene sulfonates	Luminol/peroxidase	50 ppb	–	48

peroxidase- or luminol-labelled antibodies. In such cases the specifically detected targets can be pesticides,[46] toxins,[46] explosives,[47] detergent,[48] proteins[49,50] or living cells[51] (Table 14.2).

Peroxidase-labelled antibodies are the most widely used chemiluminescent tools in the bio-analytical field. Indeed, these protein complexes are used for immunoassay on standard 96-well microplates, for immunosensors detection and for on-chip immunodetection (see below).

Numerous immunosensors based on chemiluminescent detection were described. Antibodies or antigens are then immobilized either at the surface of a macrometric membrane,[52] brought into contact with a fibre optic, or directly at the end of this fibre.[49] This latter enables setup the transfer of photons, produced during the catalysed reaction, to a light measurement system (usually a photomultiplier tube).

For particular application, the immunosensor could be integrated in a continuous flow device, which enables semi-automation of the system.[46,52] Such flow immunosensors have been applied successfully to the detection of trace levels of planktonic toxin in seafood (okadaic acid) and pesticide (2,4-D), with detection limits of $0.1\,\mu g\,L^{-1}$ and assay times of 20 min.

Other electro-assisted chemiluminescence enzyme immunoassays for 2,4,6-trinitrotoluene (TNT) and pentaerythritol tetranitrate (PETN) have been described.[53] Haptens corresponding to these explosives were covalently attached to high-affinity dextran-coated paramagnetic beads (Figure 14.6). The beads were mixed with the corresponding Fab antibodies fragments and the sample. After adding HRP-labelled anti-species specific antibody, the mixture was pumped into an electro-chemiluminometer where beads were magnetically concentrated on the working electrode. The amount of analyte in the sample was determined by measuring light emission when H_2O_2 was generated

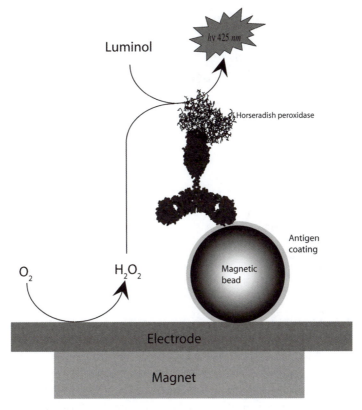

Figure 14.6 Electro-assisted chemiluminescence based immunosensor for TNT (2,4,6-trinitrotoluene) using horseradish peroxidase as label and antigen-coated, dextran-modified magnetic beads.

electrochemically in the presence of luminol and *p*-iodophenol. The detection limits obtained here for TNT and PETN were 0.11 and $19.8 \, \mu g \, L^{-1}$, respectively.

14.6.4 Electrochemiluminescent Immunosensors

Only few works have been published with such labels,[40,54] mainly because of the difficulties encountered when attempting to achieve their attachment through standard chemical reactions compatible with saving the protein integrity. Indeed, luminol possesses only an aromatic amine as available functional group, which is not easily covalently linked to biomolecules. Luminol derivatives having more reactive functions, such as *N*-(4-aminobutyl)-*N*-ethylisoluminol (ABEI), were then tested as label[55] but were found to exhibit lower light emission properties when grafted directly to proteins.

Luminol-labelled antibodies were prepared using glutaraldehyde as a crosslinking agent and used in a 2,4-dichlorophenoxyacetic acid (2,4-D)

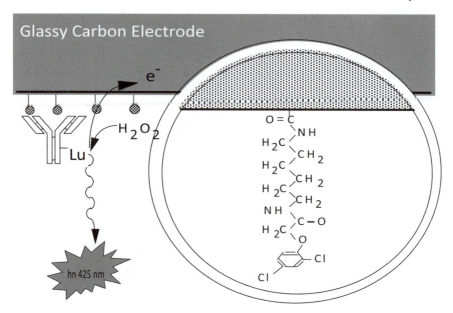

Figure 14.7 Electrochemiluminescence based immunosensors for 2,4-D using luminol labelled antibodies and direct immobilization of the antigen at a glassy carbon electrode surface.

competitive electro-chemiluminescent immunosensor (Figure 14.7). 2,4-D was covalently immobilized at a glassy carbon electrode surface, *via* a six-carbon spacer arm, by a procedure that allows the production of stable immobilized antigens that could be then stored dry, used and regenerated 50-times without loss of binding capacity. The luminol electrochemiluminescence detection was performed in a flow injection analysis system (FIA). The optimum conditions were found to be an oxidation potential of $+500\,mV$ *versus* a platinum pseudo-reference electrode, in the presence of $600\,\mu M$ H_2O_2. Under these conditions, luminol could be detected in the range 5.5 fmol–55 nmol. Luminol-labelled anti-2,4-D antibodies were tested for the 2,4-D immunodetection. The corresponding electrochemiluminescent immunoassay exhibits a detection limit of $0.2\,\mu g\,L^{-1}$ of free 2,4-D. The overall time of the experiment was 50 min and a linear range of 0.2–$200\,\mu g\,L^{-1}$ was obtained.

Numerous works by Wilson *et al.*[47,53,56] have been published on the use of the electrochemiluminescent reaction of luminol for the achievement of immunosensors. These works were based on glucose oxidase labelled antibodies used to locally generate the hydrogen peroxide required for the ECL reaction (Figure 14.8).

Thus, antibodies to atrazine were labelled with glucose oxidase and used in enzyme linked immunoassays.[56] Transparent, aminosilanized indium tin oxide (ITO) coated, glass electrodes were derivatized with aminodextran covalently modified with atrazine caproic acid. The labelled antibodies were used to

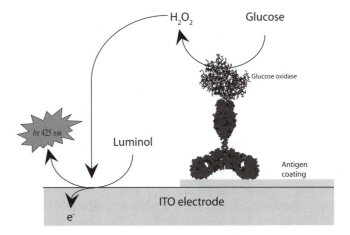

Figure 14.8 Electrochemiluminescence based immunosensor for TNT using glucose oxidase as label and antigen-coated, dextran-modified ITO (indium tin oxide).

investigate the derivatized electrodes in an electro-chemiluminescence flow injection analyser. Electrochemiluminescence immunoassay for atrazine in the range $0-1\,\mu g\,L^{-1}$ showed that it was possible to detect the target molecule at concentrations as low as $0.1\,\mu g\,L^{-1}$.

In a similar way, an electrochemiluminescence (ECL) enzyme immunoassay for TNT (2,4,6-trinitrotoluene) has been reported.[47] The deposition of a reusable immunosorbent dextran surface anchored to a gold surface in the flow cell by chemisorbed thiol groups has been described as sensing layer. Antibodies were here again labelled with the enzyme glucose oxidase and used in a competitive immunoassays in which the separation step was carried out by concentrating unbound antibodies on the immunosorbent surface. Hydrogen peroxide generated by the enzyme label when glucose was pumped through the flow cell was then subsequently detected using luminol ECL. The light intensities obtained were inversely proportional to the concentration of TNT in the sample in the range $2.3-100\,\mu g\,L^{-1}$.

14.6.5 Electrochemiluminescent DNA Sensors

Using a similar approach, a DNA sensor for sequence-specific DNA detection has been developed by Zhang *et al.*[57] In this design, a glucose oxidase labelled sandwich-type DNA sensor was built on a non-fouling surface made of mixed self-assembled monolayers (SAMs) incorporating thiolated oligonucleotides and oligo(ethylene glycol) (OEG) thiols (SH-DNA/OEG). The sequence-specific DNA sensing was accomplished by the electrochemiluminescent signal of luminol with the *in situ* generated H_2O_2. This sensor was able to detect 1 pM of target DNA in pure buffer matrix. In complex biological fluids such as human serum, this non-fouling platform-based sensor also revealed superior performance over conventional sandwich-type DNA sensors with mercapto-hexanol (MCH)-coated surfaces.

The luminol derivative N-(4-aminobutyl)-N-ethylisoluminol (ABEI) has also been used to label a known oligonucleotide sequence, subsequently used as a DNA probe for identifying a target single strand DNA.[58] The developed system consisted of a platinum-working electrode modified with electro-polymerized polypyrrole. This electrogenerated polymer was used to immobilize the probe nucleic acid sequence, subsequently involved in the specific hybridization reaction.

The hybridization events were evaluated by the ECL measurements of the ABEI labelled hybridized target sequence. The results showed that only a complementary sequence could form a double-stranded DNA with the DNA probe and gave a strong ECL response while a three-base mismatch sequence and non-complementary sequence gave no ECL signal. The intensity of the ECL was linearly related to the concentration of the complementary sequence in the range $9.6 \times 10^{-11} - 9.6 \times 10^{-8} \, \text{mol L}^{-1}$.

Figure 14.9 depicts the configuration of a DNA detection system[59] similar to the system described above for the immunodetection of explosives by Wilson *et al.* – *i.e.* based on electro-assisted chemiluminescence.

Nucleic acid probes were assembled on an Au-electrode using thiol derivates sequences. The resulting monolayer-functionalized electrode was then treated with the complementary target sequence, leading to the double strand DNA assembly on the electrode surface. This hybridized system was further treated with doxorubicin, a well known specific intercalator of double-stranded CG base-pair-containing DNA sequences.[60]

The electrochemical reduction of the intercalator led to the electrocatalysed reduction of O_2 to H_2O_2, which in the presence of luminol enabled catalysis of the chemiluminescent reaction by the free horseradish peroxidase. This electro-assisted chemiluminescent reaction enabled then the detection, down to the picomolar level of the target nucleic acid sequence.

14.6.6 Bioluminescent Biosensors

Bioluminescence is a very powerful analytical tool, since in addition to the direct measurement of ATP and NAD(P)H, any compound or enzyme involved in a reaction that generates or consumes these metabolites can be theoretically assayed by one of the appropriate light-emitting reactions. Some of these possibilities have been exploited for the development of fibre-optic biosensors, mainly with bacterial bioluminescence.

Bioluminescent determinations of ethanol, sorbitol, L-lactate and oxaloace-tate have been performed with coupled enzymatic systems involving the specific suitable enzymes (Figure 14.10). The ethanol, sorbitol and lactate assays involved the enzymatic oxidation of these substrates with the concomitant reduction of NAD^+ in NADH, which is in turn reoxidized by the

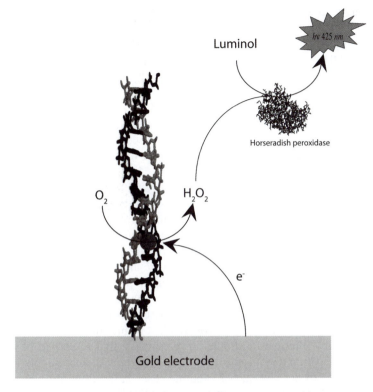

Figure 14.9 Schematic representation of the detection of a nucleic acid sequence through electro-assisted chemiluminescence. The sphere represents the doxorubicin intercalator.

$$\text{ethanol} + \text{NAD}^+ \xrightarrow{\;ADH\;} \text{acetaldehyde} + \text{NADH} + \text{H}^+$$

$$\text{D-sorbitol} + \text{NAD}^+ \xrightarrow{\;SDH\;} \text{D-fructose} + \text{NADH} + \text{H}^+$$

$$\text{L-lactate} + \text{NAD}^+ \xrightarrow{\;LDH\;} \text{pyruvate} + \text{NADH} + \text{H}^+$$

$$\text{oxaloacetate} + \text{NADH} + \text{H}^+ \xrightarrow{\;MDH\;} \text{NAD}^+ + \text{L-malate}$$

Figure 14.10 Example of dehydrogenase reactions that can be coupled with the bienzymatic bacterial bioluminescent system. ADH = alcohol dehydrogenase (EC 1.1.1.1), SDH = sorbitol dehydrogenase (EC 1.1.1.14), LDH = lactate dehydrogenase (EC 1.1.1.27), MDH = malate dehydrogenase (EC 1.1.1.37).

bioluminescence bacterial system. Thus, the assay of these compounds could be performed in a one-step procedure, in the presence of NAD^+ in excess. In contrast, the oxaloacetate measurement involved the simultaneous consumption of NADH by malate dehydrogenase and bacterial oxidoreductase and was therefore conducted in two steps.

As mentioned above, to extend the potentialities of the luminescence-based fibre-optic biosensors to other analytes, auxiliary enzymes can be used. The classical approaches consist either of the co-immobilization of all the necessary enzymes on the same membrane or of the use of microreactors including immobilized auxiliary enzymes and placed in a FIA system, upstream from the luminescence-based fibre-optic sensor (Figure 14.5a).

Another approach consists of compartmentalization of the sensing layer.[61–63] This concept, only applicable for multi-enzyme based sensors, consist of immobilizing the luminescence enzymes and the auxiliary enzymes on different membranes and then in stacking these membranes at the sensing tip of the fibre-optic sensor. This configuration results in an enhancement of the sensor response, compared with the case where all the enzymes are co-immobilized on the same membrane. This was due to a hyperconcentration of the common intermediate, *i.e.* the final product of the auxiliary enzymatic system, which is also the substrate of the luminescence reaction, in the microcompartment existing between the two stacked membranes.

Since, ideally, a biosensor should be reagentless, that is, should be able to specifically measure the concentration of an analyte without a supply of reactants, attempts to develop such bioluminescence-based fibre-optic biosensors have been made for the measurements of NADH.[64–66] For this purpose, the co-reactants FMN and decanal were entrapped either separately or together in a polymeric matrix placed between the fibre-optic surface and the bacterial oxidoreductase–luciferase membrane. In the best configuration, the period of autonomy was 1.5 h, during which about 20 reliable assays could be performed.

For luciferin, a firefly luciferase co-substrate, another method of retention has been evaluated that consisted of incorporating the substrate in acrylic microspheres during their formation; the microspheres were then confined in a polymeric matrix.[67] Using suitable co-immobilized enzymes (adenylate kinase and creatine kinase), the three adenylic nucleotides (ATP, ADP and AMP) could be assayed continuously and reproducibly with a self-containment working time of 3 h.

Table 14.3 summarizes the main performances of batchwise and flow luminescence-based fibre-optic sensors. As can be seen, the sensitivity achieved is generally better with bioluminescence-based sensors than with chemiluminescence-based sensors. This can be explained by considering the quantum yield of these light-emitting reactions. For the firefly luciferase reaction the quantum yield is close to 1 and for the bacterial bioluminescence reaction it is about 0.3, whereas it is only 0.01 for the luminol chemiluminescence reaction.

Table 14.3 Performances of batchwise and flow injection analysis (FIA) bioluminescence-based fibre-optic sensors.

Compound	Linearity of detection limit	Precision	System	Ref.
ADP[a]	1×10^{-11} mol	4.3%	FIA	67
AMP[a]	25×10^{-11} mol	6%	FIA	67
ATP	2.8×10^{-10}–1.4×10^{-6} M	///	Batch	70
ATP[b]	1×10^{-10}–1×10^{-6} M	6% at 9×10^{-9} M	Batch	71
ATP[b]	0.25×10^{-12} mol	4–4.5%	FIA	72
ATP[a]	2.5×10^{-12} mol	4%	FIA	67
Ethanol	4×10^{-7}–7×10^{-5} M	5.4% at 5×10^{-6} M	Batch	73
Lactate[c]	2×10^{-7} M	5.1%	Batch	62
LDH (lactate dehydrogenase)	5–250 IU L^{-1}		FIA	74
NADH	3×10^{-9}–3×10^{-6} M	///	Batch	70
NADH	1×10^{-9}–3×10^{-6} M	5% at 4×10^{-8} M	Batch	75
NADH	2×10^{-9} M^{d}	4.2% at $4\ 10^{-8}$ M	Batch	76
NADH	0.3×10^{-9} M^{e}	4.8% at 4×10^{-8} M	Batch	76
NADH	2×10^{-12}–1×10^{-9} mol	3.4% at 1×10^{-10} mol	FIA	77
NADH[a]	1×10^{-9}–1×10^{-6} M	6% at 4×10^{-8} M	Batch	71
NADH[a]	5×10^{-12} mol	4–4.5%	FIA	72
NADH[f]	5×10^{-12}–5×10^{-10} mol	<3%	FIA	78
Oxaloacetate	3×10^{-9}–2×10^{-6} M	5.1% at 5.5×10^{-8} M	Batch	73
Sorbitol	2×10^{-8}–2×10^{-5} M	6% at 4.4×10^{-7} M	Batch	73

[a]Reagentless biosensor with luciferin immobilized in microspheres included in a polymeric matrix.
[b]Firefly luciferase co-immobilized with the bacterial oxidoreductase–luciferase system.
[c]Compartmentalized system.
[d]Bacterial oxidoreductase–luciferase system from *V. harveyi*.
[e]Bacterial oxidoreductase–luciferase system from *V. fischeri*.
[f]Reagentless biosensor, *i.e.* with FMN and decanal entrapped in a polymeric matrix.

14.7 Conclusion

The field of biosensors is expanding continuously with a constant search for new transducing systems associated with stable biosensing elements. The ultra-sensitivity of bio- and chemiluminescence techniques together with the convenience of immobilized compounds in combination with fibre-optics constitutes an attractive opportunity for designing biosensors. In addition to the advantages of fibre-optic-based sensors, *i. e.* the possibilities of miniaturization and of remote sensing, bio- and chemiluminescence-based sensors require simpler instrumentation than those based on other spectroscopic techniques. The coupling of auxiliary enzymes allows extension of the range of compounds that can be monitored at the trace level, including enzyme activities.

References

1. U. Isacsson and G. Wettermark, *Anal. Chim. Acta*, 1974, **68**, 339.
2. D. H. Stedman and M. E. Fraser, *Analytical applications of gas phase chemiluminescence*, in *Chemi- and Bioluminescence.*, ed. J. G. Burr, Dekker, New York, 1985, p. 439.

3. L. J. Blum, *Bio-, Chemi-Luminescent Sensors*, World Scientific, Singapore, 1997.
4. D. F. Roswell and E. H. White, *The chemiluminescence of luminol and related hydrazides*, in *Methods in Enzymology*, ed. S. Fleischer and B. Fleischer, Academic Press, London, 1978, p. 409.
5. C. Dodeigne, L. Thunus and R. Lejeune, *Talanta*, 2000, **51**, 415.
6. A. Lundin and L. O. D. Hallander, *Mechanisms of horseradish peroxidase catalysed luminol reaction in presence of various enhancers*, in *Bioluminescence and Chemiluminescence: New Perspectives*, J. Schölmerich, R. Andreesen, A. Kapp, M. Ernst and W. G. Woods, John Wiley & Sons, Chichester, 1987, p. 555.
7. G. Merényi, J. Lind and T. E. Eriksen, *J. Biolum. Chemilum.*, 1990, **5**, 53.
8. M. Nakamura and S. Nakamura, *Free Radical Biol. Med.*, 1998, **24**, 537.
9. B. B. Haab, *Proteomics*, 2003, **3**, 2116.
10. A. N. P. Hiner, E. L. Raven, R. N. F. Thorneley, F. García-Canovas and J. N. Rodríguez-Lopez, *J. Inorg. Biochem.*, 2002, **91**, 27.
11. J. N. Rodríguez-Lopez, D. J. Lowe, J. Hernández-Ruiz, A. N. P. Hiner, F. García-Canovas and R. N. F. Thorneley, *J. Am. Chem. Soc.*, 2001, **123**, 11838.
12. S. Sakura, *Anal. Chim. Acta*, 1992, **262**, 49.
13. C. A. Marquette and L. J. Blum, *Anal. Chim. Acta*, 1999, **381**, 1.
14. W.-Y. Lee, *Microchim. Acta*, 1997, **127**, 19.
15. M. DeLuca, *Adv. Enzymol. Relat. Areas Mol. Biol.*, 1976, **44**, 37.
16. M. DeLuca and W. D. McElroy, *Methods Enzymol.*, 1978, **57**, 3.
17. W. D. McElroy and M. DeLuca, *Firefly bioluminescence*, in *Chemi- and Bioluminescence*, ed. J. G. Burr, Marcel Dekker, New York, 1985, p. 387.
18. L. Y. Brovko, O. A. Gandelman, T. E. Polenova and N. N. Ugarova, *Biochemistry (Moscow)*, 1994, **59**, 195.
19. N. N. Ugarova, *J. Biolumin. Chemilumin.*, 1989, **4**, 406.
20. O. A. Gandelman, L. Y. Brovko, A. Y. Chikishev, A. P. Shkurinov and N. N. Ugarova, *J. Photochem. Photobiol. B Biol.*, 1994, **22**, 203.
21. O. Shimomura, F. H. Johnson and Y. Saiga, *J. Cell. Comp. Physiol.*, 1962, **59**, 223.
22. H. H. Seliger and W. D. McElroy, *Arch. Biochem. Biophys.*, 1960, **88**, 136.
23. J. W. Hastings, T. O. Baldwin and M. Z. Nicoli, *Methods Enzymol.*, 1978, **57**, 135.
24. A. Gonzalus-Miguel, E. A. Meighen, M. M. Ziegler, M. Nicoli, K. H. Nealson and J. W. Hastings, *J. Biol. Chem.*, 1972, **247**, 398.
25. K. H. Nealson, *Methods Enzymol.*, 1978, **57**, 153.
26. E. Gerlo and J. Charlier, *Eur. J. Biochem.*, 1975, **57**, 461.
27. L. Clark and C. Lyons, *Ann. N. Y. Acad. Sci.*, 1962, **102**, 29.
28. S. J. Updike and G. P. Hicks, *Nature*, 1967, **914**, 986.
29. G. G. Guilbault and J. G. Montalvo, *J. Am. Chem. Soc.*, 1969, **91**, 2164.
30. E. Mallat, D. Barcelo, C. Barzen, G. Gaulitz and R. Abuknesha, *Trends Anal. Chem.*, 2001, **20**, 124.
31. F. Wu, Y. Huang and C. Huang, *Biosens. Bioelectron.*, 2005, **21**, 518.

32. F. Wu, Y. Huang and Q. Li, *Anal. Chim. Acta*, 2005, **536**, 107.
33. V. C. Tsafack, C. A. Marquette, F. Pizzolato and L. J. Blum, *Biosens. Bioelectron.*, 2000, **15**, 125.
34. P. Panoutsou and A. Economou, *Talanta*, 2005, **67**, 603.
35. Y. X. Li, L. D. Zhu, G. Y. Zhu and C. Zhao, *Chem. Res. Chin. Univ.*, 2002, **18**, 12.
36. N. Kiba, S. Ito, M. Tachibana, K. Tani and H. Koizumi, *Anal. Sci*, 2001, **17**, 929.
37. N. Kiba, T. Miwa, M. Tachibana, K. Tani and H. Koizumi, *Anal. Chem.*, 2002, **74**, 1269.
38. C. A. Marquette and L. J. Blum, Recent developments in luminol-H_2O_2 chemiluminescence-based flow injection assays, in *Recent Research Development in Pure Applied Analytical Chemistry*, Transworld Research Network, Kerala (India), 2001, p. 9.
39. D. J. Monk and D. R. Walt, *Anal. Bioanal. Chem.*, 2004, **379**, 931.
40. C. A. Marquette and L. J. Blum, *Sens. Actuators, B: Chem.*, 1998, **51**, 100.
41. M. F. Laespada, J. P. Pavon and B. M. Cordero, *Anal. Chim. Acta*, 1996, **327**, 253.
42. C. A. Marquette, S. Ravaud and L. J. Blum, *Anal. Lett.*, 2000, **33**, 1779.
43. V. C. Tsafack, C. A. Marquette, B. Leca, L. J. Blum and V. C. Tsafack, *Analyst*, 2000, **125**, 151.
44. C. A. Marquette, B. D. Leca and L. J. Blum, *Luminescence*, 2001, **16**, 159.
45. L. J. Blum, *Enzym. Microb. Technol.*, 1993, **15**, 407.
46. C. A. Marquette and L. J. Blum, *Talanta*, 2000, **51**, 395.
47. R. Wilson, C. Clavering and A. Hutchinson, *Analyst*, 2003, **128**, 480.
48. R. Zhang, K. Hirakawa, D. Seto, N. Soh, K. Nakano, T. Masadome, K. Nagata, K. Sakamoto and T. Imato, *Talanta*, 2005, **68**, 231.
49. R. S. Marks, A. Margalit, A. Bychenko, E. Bassis, N. Porat and P. Dagan, *Appl. Biochem. Biotechnol.*, 2000, **89**, 117.
50. J.-X. Luo and X.-C. Yang, *Anal. Chim. Acta*, 2003, **485**, 57.
51. M. Varshney, Y. Li, C. L. Griffis, R. Nanapanneni and M. G. Johnson, *J. Rapid Methods Auto. Microbiol.*, 2003, **11**, 111.
52. C. A. Marquette, P. R. Coulet and L. J. Blum, *Anal. Chim. Acta*, 1999, **398**, 173.
53. R. Wilson, C. Clavering and A. Hutchinson, *Anal. Chem.*, 2003, **75**, 4244.
54. L. S. Hersh, W. P. Vann and S. A. Wilheim, *Anal. Biochem.*, 1979, **93**, 267.
55. M. L. Calvo-Munoz, A. Dupont-Filliard, M. Billon, S. Guillerez, G. Bidan, C. Marquette and L. Blum, *Bioelectrochemistry*, 2005, **66**, 139.
56. R. Wilson, M. H. Barker, D. J. Schiffrin and R. Abuknesha, *Biosens. Bioelectron.*, 1997, **12**, 277.
57. L. Zhang, D. Li, W. Meng, Q. Huang, Y. Su, L. Wang, S. Song and C. Fan, *Biosens. Bioelectron.*, 2009, **25**, 368.
58. M. Yang, C. Liu, K. Qian, P. He and Y. Fang, *Analyst*, 2002, **127**, 1267.
59. E. K. I. W. Fernando Patolsky, *Angew. Chem., Int. Ed.*, 2002, **41**, 3398.
60. F. Arcamone, *Doxorubicin: Anticancer Antibiotics*, Academic Press, New York, 1983.

61. P. E. Michel, S. M. Gautier and L. J. Blum, *Enzyme. Microb. Technol.*, 1997, **21**, 108.
62. P. E. Michel, S. M. Gautier and L. J. Blum, *Anal. Lett.*, 1996, **29**, 1139.
63. A. Berger and L. J. Blum, *Enzyme. Microb. Technol.*, 1994, **16**, 979.
64. P. E. M. S. M. Gautier and L. J. Blum, *Anal. Lett.*, 1994, **27**(11), 2055.
65. S. M. Gautier, L. J. Blum and P. R. Coulet, *Anal. Chim. Acta*, 1991, **255**, 253.
66. S. M. Gautier, L. J. Blum and P. R. Coulet, *Anal. Chim. Acta*, 1991, **243**, 149.
67. P. E. Michel, S. M. Gautier and L. J. Blum, *Talanta*, 1998, **47**, 167.
68. N. Kessler, O. Ferraris, K. Palmer, W. Marsh and A. Steel, *J Clin. Microbiol.*, 2004, **42**, 2173.
69. C. A. Marquette and L. J. Blum, *Sens. Actuators B: Chem.*, 1998, **51**, 100.
70. S. M. G. L. J. Blum and P. R. Coulet, *Anal. Lett.*, 1988, **21**, 717.
71. S. M. Gautier, L. J. Blum and P. R. Coulet, *Sens. Actuators, B*, 1990, **1**.
72. S. M. Gautier, L. J. Blum and P. R. Coulet, *Anal. Chim. Acta*, 1990, **235**.
73. S. M. Gautier, L. J. Blum and P. R. Coulet, *J. Biolumin. Chemilum.*, 1990, **5**, 57.
74. S. M. Gautier, L. J. Blum and P. R. Coulet, *Anal. Chim. Acta*, 1992, **266**, 331.
75. S. M. Gautier, L. J. Blum and P. R. Coulet, *Biosensors*, 1989, **4**, 181.
76. L. J. Blum, S. M. Gautier and P. R. Coulet, *Anal. Lett.*, 1989, **22**, 2211.
77. L. J. Blum, S. M. Gautier and P. R. Coulet, *Anal. Chim. Acta*, 1989, **226**, 331.
78. S. M. Gautier, P. E. Michel and L. J. Blum, *Anal. Lett.*, 1994, **27**, 2055.

CHAPTER 15
Cell-based Bioluminescent Biosensors

KENDRICK TURNER,[a] NILESH RAUT,[a] PATRIZIA PASINI,[a] SYLVIA DAUNERT,[a] ELISA MICHELINI,[b] LUCA CEVENINI,[b] LAURA MEZZANOTTE[b] AND ALDO RODA[b]

[a] Department of Chemistry, University of Kentucky, Lexington, Kentucky KY, 40506-0055; [b] Department of Pharmaceutical Sciences, University of Bologna, 40126 Bologna, Italy

15.1 Introduction

A biosensor is defined as a device that employs biological components such as proteins, tissues, organelles, nucleic acids, or whole cells to detect a physico-chemical change and produce a measurable signal. Biosensors are typically composed of three parts: the biological sensing element, the signal transducing element, electronic and signal processing components and a display unit. The transducing element of a biosensor can produce various signal outputs such as optical, piezoelectric or electrochemical. Biosensors can be categorized as either molecular-based (binding proteins, enzymes, antibodies, *etc.*) or cell-based (whole cells, tissues, organisms, *etc.*). Biosensors can be designed with certain characteristics, which make them advantageous over traditional physico-chemical analysis methods. These characteristics include high specificity/selectivity, ease of use and the ability to provide relevant data related to the bioavailability of the target analyte in a given sample. Molecular-based biosensors offer the advantage of having generally faster response times than cell-based ones, although they are typically less rugged due to the often fragile nature of many isolated biomolecules and they fail to provide information on

Chemiluminescence and Bioluminescence: Past, Present and Future
Edited by Aldo Roda
© Royal Society of Chemistry 2011
Published by the Royal Society of Chemistry, www.rsc.org

the bioavailability of the compound of interest.[1] Additionally, the production and isolation of biomolecules can be expensive and time-consuming. In contrast, cell-based biosensors are usually more tolerant of extreme conditions, although, in many cases, they can require longer analysis times. Cell-based sensing systems also provide useful information on the bioavailability of the interrogated analyte and its ability to activate biochemical machinery, which can contribute to an increased understanding of the toxicity or physiological role of the compound of interest when this is, for example, an environmental pollutant or a biologically relevant molecule. As a result of these desirable properties, cell-based biosensing systems are finding increasing application in the fields of environmental and clinical analysis, drug discovery and toxicology, and are becoming the focus of much research to improve their characteristics and engineer them to respond to a greater variety of stimuli and analytes present in the environment.

15.2 Design and Construction of Bacterial Cell-based Bioluminescent Biosensors

Bacterial cell-based biosensors share a common basic design and include similar components. This basic design consists of an intact living cell containing a DNA sequence, in which the expression of a reporter gene/transducer is under the control of the promoter of a certain operon. Within this basic design, there can be much variation depending on the cells being used as a host, the type of promoter that is utilized (either constitutive or inducible reporter expression) and the detection strategy that is being employed.

As cells grow and are metabolically active, biosensing systems that employ a constitutive promoter to regulate the expression of the reporter protein often present a basal expression of reporter, which produces a high enough measurable signal even in the absence of the target analyte. This basal signal is reduced if the cells are subjected to stress as a result of exposure to toxic compounds or other adverse growing conditions. Thus, this kind of cell-based biosensor that employs constitutive promoters to regulate the production of a reporter protein can effectively provide information on the overall toxicity of a sample under investigation. However, notably, such a system does not provide any information on the specific nature of the compounds that cause the toxicity to the cells or the sample.

In contrast, the gene fusion of an inducible promoter to a reporter gene yields a cell-based biosensing system in which the expression of a reporter protein occurs only in the presence of a desired condition (Figures 15.1 and 15.2). Therefore, the presence of an analyte of interest activates the expression of a reporter protein resulting in an increase of the signal produced by the reporter. An inducible cell-based sensing system such as this is preferred when the goal is to detect and quantify a particular compound or condition in the sample. Furthermore, combinations of these genetic operons using different reporter genes under the control of different promoters can be achieved within a single

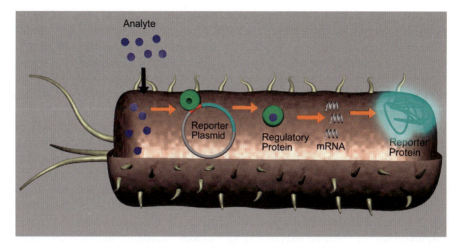

Figure 15.1 Schematic of an operon/regulatory protein cell-based biosensor featuring a negatively regulated operon. In the absence of the analyte, the regulatory protein is bound to the promoter on the reporter plasmid and prevents expression of the reporter protein. As the analyte concentration increases, it binds the regulatory protein, and the analyte–regulatory protein complex dissociates from the reporter plasmid. This triggers expression of the reporter protein, and a dose-dependent generation of the signal.

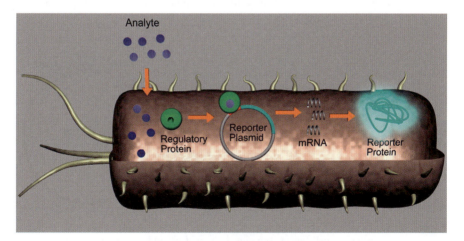

Figure 15.2 Schematic of an operon/regulatory protein cell-based biosensor featuring a positively regulated operon. When an analyte is present, it binds to the regulatory protein. This regulatory protein–analyte complex binds the promoter region on the reporter plasmid, thus triggering the expression of a reporter protein. This produces a luminescent signal that is dependent on the concentration of the analyte.

organism and yield a single cell-based biosensor capable of responding to multiple analytes, an advantage when developing multiplexed assays.[2,3] Inducible promoters are typically regulated by transcriptional regulatory proteins, a kind of protein that evolved as a response mechanism of an organism to stress. In general, when an extracellular stimulus/compound acts upon the organism, these proteins are able to specifically or selectively recognize and bind to this compound, and subsequently respond by triggering the production of other proteins involved in defense mechanisms. These include the production of protein pumps that control the cellular efflux of the toxic compound(s), metabolic and synthetic pathway enzymes that degrade them, proteins that can sequester them, receptor proteins, *etc.*[1] These inducible-promoter regulatory circuits are advantageous to the host organism in that they ensure that transcription of their gene products occurs only when required to increase their survival. In the absence of an external stimulus, the regulatory circuit is turned off, no proteins are expressed from the genes and the organism can conserve its resources and energy for other purposes.

In cell-based biosensing systems, recombinant DNA methods are employed to replace the gene products of the native operon with a reporter gene that produces a desired signal in response to the presence of a specific compound recognized by the regulatory protein. Thus, these systems combine the biospecific recognition afforded by the regulatory proteins with the signal generation stemming from the reporter gene. Figure 15.3 shows an example of the design of one such system, featuring the *hbp* operon from *Pseudomonas azelaica*. The genes *hbpC*, *hbpA* and *hbpD* of this operon are under the control of an operator/promoter that is regulated by the product of the gene, *hbpR*. The enzymes, *HbpC*, *HbpA* and *HbpD* are responsible for the degradation of

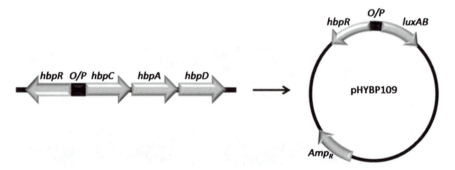

Figure 15.3   Design of the recombinant reporter plasmid component of a bioluminescent cell-based sensing system for hydroxylated polychlorinated biphenyls. The structural genes, *hbpC*, *hbpA* and *hbpD*, of the *hbp* operon from the chromosome of *Pseudomonas azelaica* were replaced with the *luxAB* reporter gene in the construction of reporter plasmid pHYBP109. Upon transformation of pHYBP109 into *E. coli*, a cell-based biosensing system to detect various hydroxylated biphenyls in both buffer and serum was constructed.[4]

2-hydroxybiphenyl. A bacterial cell-based biosensor was constructed by replacing these structural genes with the reporter gene *luxAB*. Further study of this construct demonstrated that it responded to various hydroxylated poly-chlorinated biphenyls (OH-PCBs), resulting in an increase in bioluminescence in a dose-dependent manner.[4]

15.2.1 Regulation of Reporter Gene Expression

Within the realm of inducible cell-based biosensing systems, the reporter gene expression may be positively or negatively regulated. In a negatively regulated operon (Figure 15.1), a regulatory protein binds to an operator DNA sequence and prevents the expression of the reporter protein, with this binding being dependent on the presence or absence of the effector/analyte. In one instance, the regulatory protein is initially bound to the operator region, thus repressing the expression of the reporter. As the effector is added, it binds to the regulatory protein causing it to dissociate from the operator DNA sequence. This results in an increased expression of the reporter protein. The alternative is that the effector is required for binding of the regulatory protein to the operator. In this case, as the effector is added, a regulatory protein–effector complex forms that binds to the operator, resulting in a decrease in reporter protein expression.

Alternatively, in a positively regulated system (Figure 15.2) the operon regulation is mediated by a DNA sequence that acts as an "enhancer" of the system. There are two different mechanisms by which this regulation occurs. In the first possible mechanism, the regulatory protein is bound to the enhancer and, consequently, produces a high level of initial expression of the reporter protein. As the concentration of the effector increases, it binds to the regulatory protein, causing its dissociation from the enhancer and, thus, reducing expression of the reporter gene. In the second case, the effector is required for the binding of the regulatory protein to the enhancer sequence. An increase in the effector concentration leads to the formation of a regulatory protein–effector complex that, in turn, binds to the enhancer DNA sequence and increases reporter protein expression.

From an analytical standpoint, it is desirable to have the lowest possible initial signal and an increasing dose-dependent response as the concentration of the analyte of interest increases. Thus, the most useful regulatory proteins are those from very tightly negatively regulated operons where the regulatory protein is initially bound to the operator and represses reporter protein synthesis. Accordingly, cell-based biosensing systems based upon these strategies, in general, afford the lowest limits of detection.

15.2.2 Reporter Genes and Their Attributes

While the sensing element of a cell-based biosensing system determines the selectivity of its response, the reporter gene largely determines the sensitivity of the system. In that regard, the variety of reporter genes available is steadily increasing and each of these reporter genes possesses unique attributes that must be considered in the design of a biosensing system. The reporter gene of a cell-based sensing system can yield a reporter protein that can generate various different measurable signals. This chapter will focus on those producing an optical/luminescent signal. Optical signals are easily measured, and an extensive array of instrumentation is available, ranging from very sophisticated, state-of-the art to less complex, cost-effective hardware depending on the needs of a particular application. Additionally, a great variety of luminescent proteins have been well-characterized.[1,5] Table 15.1 summarizes the attributes of various luminescent reporter proteins. Following is a discussion of each of the types of luminescent reporter proteins and enzymes capable of generating an optical signal when employing an appropriate substrate.

15.2.2.1 Luciferases

The term luciferase encompasses a class of enzymes that catalyze a reaction in which a substrate known generically as a luciferin reacts with oxygen to produce light. These enzymes offer excellent limits of detection, largely because there is no background luminescence from endogenous activity in other organisms or from the media and samples in which the measurements are being taken. The high quantum efficiency of the bioluminescent reactions catalyzed by these enzymes also contributes to the low detection limits that they afford. Given that luciferase emits light *via* a biochemical reaction, there is no need for an excitation source, making the instrumentation needed for measuring bioluminescence simple and cost-effective. These are clear advantages of luciferases over other luminescent systems such as those based on fluorescence and, therefore, make this family of enzymes great candidates for use as reporters.

Luciferases have been isolated from several organisms, including fireflies, bacteria, worms, fungi, *etc.* Commonly used as reporter genes in bacterial cell-based biosensing systems are those isolated from bacteria, including members of the genera *Vibrio*, *Xenorhabdus* and *Photobacterium*.[1] In these organisms, bioluminescence is conferred by the *lux* operon, which consists of the *luxCDABE* gene cassette. Of the five genes in the *lux* operon, only *lux A* and *lux B* (with the addition of a suitable substrate) are required for bioluminescence. The products of these two genes form the catalytically active dimer, which oxidizes a long-chain aliphatic aldehyde (luciferin) and a reduced flavin mononucleotide ($FMNH_2$) cofactor to the corresponding carboxylic acid and FMN, respectively. During this process, an unstable complex containing an excited oxide bond is formed, which relaxes to the ground state with concomitant emission of light at 490 nm with a quantum yield of about 0.1.[6]

While the *luxAB* genes alone, employed as a reporter, are enough to produce bioluminescence, the addition of a long-chain aldehyde substrate, usually decanal, is required unless the entire *luxCDABE* gene cassette is present. The enzyme products of the remaining genes in the cassette, *lux C*, *lux D* and *lux E* provide the aldehyde substrate needed for bioluminescence. However, it has

Table 15.1 Common reporter proteins and their characteristics.

1 Reporter protein 2 reaction catalyzed 3 detection method	Advantages	Disadvantages
1 Bacterial luciferase 2 $FMNH_2 + R\text{-}CHO + O_2 \rightarrow FMN + R\text{-}COOH + H_2O + h\nu$ (490 nm) 3 bioluminescence	High sensitivity; may not require substrate; no endogenous activity in mammalian or bacterial cells no light source needed.	Heat labile; narrow linear range; requires aerobic environment
1 Firefly luciferase 2 firefly luciferin $+ ATP + O_2 + Mg^{2+} \rightarrow$ oxyluciferin $+ AMP + P_i + h\nu$ (560 nm) 3 bioluminescence	High sensitivity; broad linear range; no endogenous activity in mammalian or bacterial cells; spectral variants available. no light source needed	Requires substrate; requires aerobic environment and ATP; requires solubilizers for substrate permeability into cells
1 *Renilla* luciferase 2 *Renilla* coelenterazine $+ O_2 \rightarrow$ coelenteramide $+ CO_2 + h\nu$ (480 nm 3 bioluminescence	No endogenous activity in bacterial or mammalian cells; no light source needed; required substrate is membrane permeable	Requires substrate and cell lysis
1 Aequorin 2 jellyfish coelenterazine $+ O_2 + Ca^{2+} \rightarrow$ coelenteramide $+ CO_2 + h\nu$ (465 nm) 3 bioluminescence	High sensitivity; no endogenous activity in mammalian or bacterial cells; no light source needed	Requires substrate and the presence of Ca^{2+}
1 Green fluorescent protein 2 formation of an internal chromophore, excitation/ emission: 395 nm/509 nm 3 fluorescence	Autofluorescent; no substrate or cofactors needed; spectral variants available; no endogenous activity in most systems; stable at physiological pH	Moderate sensitivity; background fluorescence from some samples may interfere; toxic to some cell types; formation of chromophore can be slow.
1 β-Galactosidase 2 hydrolysis of β-galactosides 3 chemiluminescence fluorescence colorimetric electrochemical	Sensitive and stable; moderate linear range; can be used in anaerobic environment	Endogenous activity; requires substrate

been noted that the use of *luxAB* alone results in increased bioluminescence and, consequently, improved sensitivity of measurement in a biosensing system.[7] This is likely the result of limited substrate production from the products of the *lux* cassette as opposed to the excess of substrate added when *luxAB* is used. The design of a cell-based biosensing system using the *luxCDABE* is advantageous in the development of simplified assays and applications to be used in real-time monitoring. For applications requiring increased sensitivity or improved control over when the generation of the bioluminescence signal occurs, the *luxAB* reporter gene is preferred.

In addition to bacterial luciferases, firefly luciferase has also been well-characterized and widely used as a reporter gene. Firefly luciferase is a 62 kDa monomer that catalyzes the oxidation of a benzothiazolyl-thiazole luciferin to oxyluciferin in the presence of ATP, oxygen and Mg^{2+}. The oxyluciferin produced is an excited molecule that subsequently relaxes to a ground state, resulting in the emission of light at a maximum wavelength of 550–575 nm. Firefly luciferase has approximately ten-fold higher quantum yield than bacterial luciferase, which endows it with a broad dynamic range (7–8 orders of magnitude).[8,9] Additionally, mutagenesis of firefly luciferase has resulted in enzymes that emit light at wavelengths in a wider range of the visible spectrum, allowing for the development of multiplexed analysis.[10] Examples of these include redshifted, thermostable variants of luciferase from *Photinus pyralis*.[11] However, the need for the addition of a substrate and the requirement for ATP can, in some cases, limit the application of firefly luciferase.

Another luciferase well-studied as a bioreporter has been isolated from the marine organism *Renilla reniformis*, a species of sea pansy. *Renilla* luciferase is a 31 kDa monomeric enzyme that catalyzes the oxidation of coelenterazine and results in the emission of light at 480 nm.[12] The sensitivity and dynamic range of *Renilla* luciferase is similar to that of firefly luciferase; however, *Renilla* luciferase is not as widely used as a bioreporter.[5,13] In applications where multiplexing or a dual-reporter based system is desirable this luciferase is sometimes used alongside firefly luciferase as they have distinct wavelengths of emission. However, the need for substrate addition and cell lysis make using either firefly or *Renilla* luciferases less appealing than bacterial luciferase since increased cost, time and error is introduced as a result.

Luciferases from other organisms have been successfully used as bioreporters in cell based sensing systems. In work by Wu *et al.*, a dual-reporter system was constructed by transforming NIH 3T3 fibroblast cells with a plasmid containing *Cypridina* luciferase (CLuc) fused to a target gene and a plasmid containing *Gaussia* luciferase (GLuc) as a control plasmid to monitor gene expression in these cells.[14] Both CLuc and GLuc are secreted into the growing medium upon expression and emit at different wavelengths (460 and 480 nm, respectively), allowing for simultaneous measurement without cell lysis.

The variety of luciferases is increasing as novel proteins are being identified and integrated as reporters into cell-based sensing systems. For example, the characterized luciferases from the organisms *Luciola italica*[15] and *Phrixothrix hirtus*[16] are expanding the palette of emission wavelengths available when selecting reporter proteins. Two other novel luciferases from the marine copepod *Metridia pacifica* are thermostable bioluminescent proteins with distinct emission kinetics and are efficiently secreted into culture medium upon expression as a result of the presence of an N-terminal signal peptide.[17] Because these are secreted into the culture medium, they could be incorporated into a cell-based sensing system and continuously monitored in a portion of culture medium independent of cell lysis.

15.2.2.2 Aequorin

Aequorin is a 22 kDa photoprotein native of the marine jellyfish *Aequorea victoria*. The emission of bioluminescence by aequorin is different from that of the luciferases, including *Renilla* luciferase. Aequorin needs an organic imidopyrazine substrate, coelenterazine and the presence of Ca^{2+} for emission of bioluminescence. Coelenterazine resides within a hydrophobic pocket within the structure of the protein, while Ca^{2+} binds to three conserved EF-hand regions of aequorin. It is the binding of aequorin to Ca^{2+} that causes the protein to undergo a conformational change, which causes coelenterazine, in the presence of molecular oxygen, to go through an excited state, from which it relaxes to form coelenteramide and emit bioluminescence at 460–470 nm.[1] The emission from aequorin follows flash-type kinetics with an emission from the native protein lasting about 3 s and a quantum yield of 0.15.[18] Mutagenesis of aequorin has led to the development and characterization of mutants with tuned emission lifetimes and altered wavelengths, allowing for multiplexing in both time and spatial (wavelength) domains.[19] While aequorin has found some application in cell-based biosensing systems, its use has been somewhat limited, due mostly to its sensitivity to the presence of calcium ions and the need for the addition of a substrate. Despite these limitations, aequorin has been employed in high-throughput screening assays where its sensitivity to calcium is imperative in the study of G-protein coupled receptors and in screening compounds that act as their agonists or antagonists.[20] In addition, the detection of specific pathogens, such as *Yersinia pestis* and *Bacillus anthracis*, has been achieved using a sensing system named CANARY (cellular analysis and notification of antigen risks and yields).[21] This system is constructed by engineering B cells that express both aequorin as a bioreporter and membrane-bound antibodies for the pathogen of interest. Even when exposed to low levels of pathogen, the antibodies are capable of recognizing their target. The resulting binding event triggers an increase in intracellular calcium concentration, which leads to the subsequent bioluminescence emission by aequorin within seconds.

Novel relatives of aequorin, such as the photoprotein clytin from *Clytia gregarium*, have been characterized recently and found to be less sensitive to calcium ions.[22] The availability of spectral variants for multiplexing, exceptional sensitivity (in the sub-attomole range) and lack of endogenous expression in other organisms warrant consideration of aequorin as a reporter gene.

15.2.2.3 Green Fluorescent Protein (GFP)

Like aequorin, the green fluorescent protein originates from the jellyfish *Aequorea victoria*. However, unlike the bioluminescence reporter genes mentioned thus far GFP emits fluorescence. Native GFP is a 238 amino acid protein possessing a β-barrel structure containing an internal fluorescent chromophore. This chromophore is formed from three amino acids (threonine-65, tyrosine-66 and glycine-67) on the interior of the protein by cyclization and oxidation of the tripeptide upon proper protein folding. The exceptional stability of GFP allows for the accumulation of reporter protein, which is particularly relevant when used in bacterial cell-based biosensors employing a weak promoter. GFP does not require a substrate to fluoresce; however, as with any fluorescent reporter, it requires irradiation at its excitation wavelength maximum of 395 nm, resulting in light emission at a wavelength of 509 nm with a quantum yield of 0.88.[6,23] Extensive research has generated numerous GFP mutants with altered excitation and emission maxima, stabilities and signal intensities.[8,24,25] GFP is also tolerated by various cell types, including mammalian and bacterial cells. However, because of interference from background fluorescence in samples, the detection limits afforded by GFP are not comparable with those of bioluminescent proteins such as the luciferases or aequorin.[1]

15.2.2.4 β-Galactosidase

The gene product of the *lacZ* gene from *Escherichia coli* is β-galactosidase, an enzyme whose biological function is to cleave lactose into galactose and glucose, although it has been shown that the enzyme may act on various substrates. Depending on the substrate employed, β-galactosidase can generate a fluorescent, chemiluminescent, colorimetric or electrochemical signal.[25] For the production of chemiluminescence signals, 1,2-dioxetane derivatives are typically used as a substrate. As a reporter gene, *lacZ* can afford detection limits as low as 2 fg with a dynamic range of 5–6 orders of magnitude.[1,5] Despite these advantages, the need for the addition of a substrate and the requirement for cell lysis to make the substrate available to the enzyme restrict the use of *lacZ* as a reporter gene for certain specialized applications.

15.2.3 Advantages and Limitations of Bacterial Cell-based Bioluminescent Biosensors

The continuous discovery of new reporter molecules and recognition elements by biologists along with advancement in the field of recombinant DNA broadens the range of types of sensors that can be developed with regard to selectivity, sensitivity and parallel analysis. Moreover, the identification and investigation of organisms that can be employed and survive in extreme environments, such as extremophiles and spore-forming microorganisms, should result in systems that present expanded storage and working conditions and are more resilient to extreme settings. This improved ruggedness and storage of cell-based biosensors should make them more amenable to field-portable environmental and clinical applications.[26]

Uniquely among biosensing systems, cell-based systems can provide significant information regarding the bioavailability of the compound being interrogated.[27] When employing an intact cell in the sensing system, the compound being detected must be transported into the cell and must activate certain cellular processes and pathways to produce a response. In this regard, a cell-based sensing system can identify those species, to be transported across the cell membrane. This information is especially useful in determining relevant toxicological characteristics of complex mixtures in which some components may be bioavailable while others may not. The bioavailable concentration of a species as detected using a cell-based biosensor is often related to the total concentration determined by standard physicochemical analysis to better characterize the sample being analyzed.[28]

While these advantages warrant further study and development of cell-based biosensing systems, there are several limitations that must be overcome before their full potential as an analytical method can be realized. Inherent to all biological systems is some degree of variability; cell-based biosensing systems are not immune to this. This variability can result from growth of the cells in non-ideal conditions, response to various components in complex samples or several other unidentified factors. This variation can contribute to inter- and intra-assay variability. To address this issue, cell-based systems have been developed that carry a secondary plasmid in which a unique reporter protein is under the control of a constitutive promoter.[29] This allows for the response from the analyte-inducible promoter to be normalized with respect to cell growth and metabolism.

As previously mentioned, cell-based biosensing systems provide bioavailability information. However, when bacterial whole-cell sensing systems are employed, this does not lead to a direct correlation to all relevant toxicological information as it applies to higher organisms such as humans. This can be addressed by the development of biosensing systems based on more complex cell types such as yeast or mammalian cells, thus giving data that is more applicable to these more complex organisms. In addition, further study of similarities in biochemical pathways between cell types used in biosensing systems and cells present in higher organisms may lead to more accurate extrapolation of relevant toxicological data.

Finally, there can be some degree of instability in the plasmid DNA within cell-based sensing systems, resulting in decreased reproducibility of measurement. This can occur as the cells carrying the exogenous genetic material reproduce and replicate the plasmid DNA contained within them. While the rate of error is very low, some mutations can occur. To negate this effect, plasmid DNA can be integrated into the chromosome of the cells being used, resulting in increased genetic stability.

15.3 Applications of Bacterial Cell-based Bioluminescent Biosensing Systems

The distinct properties and advantages discussed previously have allowed the application of cell-based biosensing systems in different fields. The use of bioluminescent reporters allows for compact, portable instrumentation due to the lack of need for an excitation source (a requirement in fluorescence measurements). In addition, bioluminescent bioreporters lack the background signal deriving from fluorescence generated by other components in the sample matrix when exciting fluorescent reporters, which contributes to their superior sensitivity. Many specific examples can be found throughout the current literature describing applications in the realm of on-site environmental monitoring, drug candidate screening, clinical testing, high-throughput screening, *etc.* Recent examples of these applications are highlighted here in more detail. Additionally, several bacterial cell-based biosensing systems are commercially available; Table 15.2 gives some examples.

Cell-based biosensors have been engineered that can determine factors such as general stress, oxidative stress and genotoxicity. In these constructs, a reporter gene is placed under the control of a promoter capable of responding in a dose-dependent manner to one of these stressors. Such systems have been developed to monitor oxidative stress,[2,30] protein damage[30,31] and DNA damage,[2] among others.

Bacterial cell-based biosensors are commonly used in environmental monitoring. Typically, soil and water samples concerning environmental contamination are complex in nature, containing both naturally occurring and foreign components. The specificity of the biological recognition element in a cell-based biosensing system is ideally suited for detecting a desired compound in a complex mixture. To that end, biosensors have been developed for various analytes, ranging from metals to organic pollutants, and representative examples of these are discussed below.

There are several reports in the current literature regarding cell-based assays for the detection of inorganic analytes, specifically metals. The design of these biosensors is based upon the use of regulatory elements from microorganisms, which natively regulate the expression of genes to confer metal resistance, to control the expression of a bioreporter instead. Sensing systems have been developed for various environmentally relevant toxic metals and metalloids, including mercury,[32–34] antimonite/arsenite/arsenate,[35–38] cadmium,[35,39,40] chromate[41] and aluminum[42] among others. In addition, inorganic compounds, such as nitrate, have been detected using cell-based sensing systems.[43] Detection limits as low as femtomolar with analysis times as short as 30 min have been reported for these species. Progress towards the development of a portable biosensing system for the detection of metals has been achieved by the engineering of a fiber-optic device consisting of mercury and arsenic sensing bacterial biosensors immobilized on optical fibers.[44] Environmentally relevant detection limits were obtained using this biosensing system: $2.6\,\mu g\,L^{-1}$ for mercury, $141\,\mu g\,L^{-1}$ for arsenic(v) and $18\,\mu g\,L^{-1}$ for arsenic(III).

Cell-based biosensing systems have been developed for several organic compounds, including endocrine disrupting compounds (EDCs), polychlorinated biphenyls (PCBs) and their metabolites, phenol, catechols, naphthalene/salicylic acid, benzene/toluene/ethylbenzene/xylene, *etc.*[5,45–51] Many of these compounds are nearly ubiquitous in the environment and pose negative health effects on many organisms, including humans. They share structural similarities and biological activities with naturally occurring compounds such as hormones.

Table 15.2 Commercially available bacterial cell-based biosensing systems.

1 Product name 2 responds to	Description	Manufacturer	Reference or web site
1 Microtox™ 2 general toxicity	Inhibition test based on freeze dried *V. fischeri*	Strategic Diagnostics, Inc.	http://www.sdix.com/
1 Mutatox™ 2 mutagenic toxicity	Engineered dark variant of *V. fischeri* recovers luminescence restored upon exposure to mutagenic compounds	Strategic Diagnostics, Inc.	http://www.sdix.com/
1 BIOMET™ 2 Zn, Cd, Cu, Ni, Pb, Cr or Hg	Engineered, metal-toler- ant *Ralstonia metalli- durans* produces luciferase in response to metals	Vito	http://wwwa.vito.be/ english/index.htm
1 Cellsense™ 2 chlorophenols, other organics	Engineered *E. coli* pro- duces amperometric response to analytes		Farre, *et al.*[105]
1 BioTox™ 2 general toxicity	Inhibition test based on freeze dried *V. fischeria* tailored for sediment samples	Aboatox	http:// www.aboatox.com/
1 LumiStox™ 2 general toxicity	Inhibition test based on frozen *V. fischeria*	Hach Lange	http:// www.drlange.com

Traditional quantification of these compounds from environmental samples involves extensive sample pretreatment, derivatization and extraction prior to detection by a suitable instrumental method. In addition to environmental samples, it has been demonstrated that several varied analytes can be detected in biological samples such as blood serum, making whole-cell sensing systems useful in the detection of biomarkers of exposure as well.[4]

Recently, cell-based biosensors have been developed for quorum sensing signaling molecules, the integral elements of the bacterial communication sys- tem. Quorum sensing is a phenomenon in which certain bacteria communicate by producing, secreting, sensing and responding to signaling molecules. The concentration of these molecules correlates to the density of the cells. This cell-to-cell communication allows the organisms to control the expression of specialized proteins, depending on the cell population size. Since bacterial processes, such as, production of virulence factors, formation of biofilms and ability to colonize a certain environment are regulated by quorum sensing, the detection and quantification of quorum sensing signaling molecules may be relevant in the investigation of the status of various diseases that have been linked to bacteria.[52] Cell-based biosensors have been developed by placing the expression of a bioreporter under the control of promoters and associated recognition/regulatory proteins from bacterial quorum sensing regulatory systems that respond to the presence of quorum sensing molecules.[53] Successful use of these sensors has also been demonstrated in biological samples such as saliva and stool.[54]

Cell-based biosensors also exist for several antibiotic compounds. As the use of antibiotics increases, antibiotic resistance mechanisms in microorganisms are becoming increasingly widespread. Antibiotics are also being found in environmental and food samples; for instance, they have been detected in chlorinated drinking water at trace levels (down to $\mu g\,L^{-1}$ levels).[55] Conventional methods to detect antibiotics rely on immunoassays, chromatographic methods and microbial growth inhibition tests. Because of their unique properties, especially the ability to characterize the bioavailability of an analyte, cell-based biosensors are well-suited for these applications. Such biosensors have been developed for the detection of antibiotics in various types of samples such as water and food products,[56] as well as blood and serum.[57] Cell-based biosensors have been used for the determination of antibiotic activity on several biochemical pathways in tandem with the screening of natural products for antibiotic activity. For example, a panel of five gene promoter regions from the soil bacterium *Bacillus subtilis*, which have altered mRNA expression profiles upon antibiotic exposure, were fused to a firefly luciferase reporter gene.[58] These genes, *yorB*, *yvgS*, *yheI*, *ypuA* and *fabHB*, participate in biosynthetic pathways such as the synthesis of DNA, RNA, proteins, cell wall and fatty acids, respectively. Biosensors based on the use of these genes along with luciferase have been used in a high-throughput screening mode to investigate the antibiotic activity of 14 000 natural products. Bacterial cell-based biosensors have also been developed to examine the microbicide activity of some antibacterial polymers. Luminescence produced by the *E. coli* strain O157:H7 modified to express bacterial luciferase *via* the *luxCDABE* gene has been monitored upon exposure to polymer compounds to determine bactericidal properties.[59] Bacterial biosensors also exist for the screening of antimicrobial activity of compounds in the gas phase that may find use in sterilization procedures. To that end, the bioluminescent bacterium *Pseudomonas fluorescens 5RL* was immobilized on a 0.2 μm membrane filter and exposed to varying concentrations of chlorine dioxide gas. In this system, a decrease in luminescence was correlated to increased antimicrobial activity of the gas.[60]

Other biologically relevant molecules, such as sugars, have also been detected using a cell-based biosensor approach. Cell-based biosensing systems have been developed for several sugars by placing the expression of a reporter gene under the control of a promoter and regulatory protein responding to the desired sugar. For example, cells have been engineered to detect arabinose,[3,61] glucose,[62] sucrose[62] and lactose[3,62]. Multi-analyte detection has been demonstrated with these compounds with the simultaneous detection of lactose and arabinose using a single biosensing organism in which two variants of GFP with distinct emissions were used as reporters for each sugar.[3] Cell-based biosensors are especially useful in the detection of sugars as they proved certain advantages over conventional detection methods. For example, detection of sugars using electrochemical methods often suffer from a lack of specificity and those based on spectroscopic methods require derivatization of sugar with a chromophore substrate.

Applications of bacterial cell-based biosensors can also be found in the field of molecular biology. The *luxCDABE* cassette from *Photorhabdus luminescens* was cloned into a pCRII vector and transformed into *E. coli*. The resulting cells were then grown with varying concentrations of lytic bacteriophage T4 at varying temperatures. The bioluminescence was monitored with respect to time, and as the bacteriophage lysed the bacteria a decrease in bioluminescent was observed. The results allowed the quantitation of the bacteriophage as well as determination of its thermal deactivation conditions.[63]

In the field of medicine, genetically engineered bacterial cell-based biosensors have found applications in imaging *in vivo* and *in vitro*. *E. coli* engineered to express GFP in response to quorum sensing molecules, *N*-acylhomoserine lactones (AHLs), were introduced into mice that had been infected with *Pseudomonas aeruginosa*, an opportunistic pathogen that uses these molecules for intercellular communication.[64] The lung tissue from the mice was examined by confocal scanning laser microscopy and the pathological damage observed was correlated to fluorescence measured as a result of increased AHL levels from bacterial presence. *E. coli* that migrate preferentially to tumor tissue, harboring a plasmid containing the *luxCDABE* cassette, were injected into mice with CT26 mouse colon cancer and the subsequent bacterial migration was imaged by detection of bioluminescence from the luciferase expressing bacteria.[65] The ability to image tumors *in vivo* in a non-invasive manner is a valuable tool when diagnosing and monitoring the spread of a different types of cancer, as well as when screening for novel antitumor drugs. Recently, Foucault *et al.*, engineered *E. coli* to express either bacterial luciferase or mutants of firefly luciferase that were employed for real-time *in vivo* monitoring of infection in mice.[66] The use of bioluminescent bacterial biosensors in these applications offers several advantages, perhaps the most important being the lower detection limits due to the lack of background emission associated with similar methods employing fluorescent reporters.

15.4 Bioluminescent Biosensors: the Eukaryotic Alternative

Whole-cell sensing systems based on genetically engineered yeast cells or mammalian cell lines have been widely used as biosensors, especially for the rapid detection of threats associated with food, environment and biosecurity as convenient alternatives to standard analytical chemical/biochemical techniques.[67] The assessment of hazard-induced physiological responses, like gene expression, apoptosis, receptor–ligand interactions and signal transduction, is paramount for understanding the molecular mechanisms of toxic substances. Among eukaryotic biosensors, yeast-based biosensors are the most employed.

15.4.1   Yeast-based Biosensors

Yeasts have been widely exploited thanks to their peculiar features: first they retain the typical "microbial advantages," *i.e.*, speed of growth, easy manipulation, low-costs for maintenance and growth on different carbon sources. Second they also keep the "eukaryotic advantage," meaning that when used for toxicity studies they are more predictive of potential hazards to other eukaryotes such as humans. Third, they have the so-called "yeast advantage," the fact they are very robust and tolerate a wide range of physicochemical conditions. This is of particular interest when such biosensors are used for real samples (*e.g.*, food and environmental samples) that show toxicity.[68]

15.4.2   Bioluminescent Cell-based Assays for Endocrine Disruptors Monitoring

Yeast-based biosensors have been widely exploited for the detection of endocrine disrupting compounds (EDCs). A lot of concern is arising about the presence of EDCs in the environment. In fact, many EDCs, which are molecules able to interact with hormone receptors with adverse health effects, are continuously introduced into the environment and need to be monitored. Such compounds are, in fact, able to mimic or antagonize the effects of endogenous hormones such as estrogens and androgens and they have shown involvement in sex differentiation and reproductive abnormalities in wildlife, human population and laboratory animals, also determining an increasing incidence of cancers. Another concern derives from unexpected sources of EDC; for example, bottled mineral water has been recently identified as potential source of the estrogenic compounds.[69]

Standard analytical methodologies for monitoring EDCs are currently based on HPLC or gas chromatography (GC) coupled with mass spectrometry (MS). These techniques are very sensitive, allowing the ultra-trace determination of target EDCs, but they are not suitable for screening purposes and do not take into consideration the biological effects of single EDCs or chemical mixtures. Their main limitation is that these methods are specific for one analyte or a limited class of structurally related compounds. Alternatively, *in vitro* bioassays, on the basis of the interaction between the EDCs and the hormone receptors, can determine the total hormonal activity of EDCs in complex mixtures. In particular, bioassays have been developed by engineering yeast or mammalian cells to express a reporter protein in the presence of compounds with androgenic activity.

Several cell-based assays have been developed to screen substances for their capacity to interfere with the endocrine system. Commercial bioassays, such as CALUX, are also available. Most bioassays for EDCs are based on the use of *Saccharomyces cerevisiae* cells genetically engineered to express a human nuclear receptor (*e.g.*, estrogen α or β or androgen receptors) and the corresponding hormone responsive elements (*e.g.*, estrogen or androgen responsive elements) driving the expression of a reporter gene such as luciferase from

P. pyralis. In the presence of compounds with pseudo-hormonal activity, the activated receptor moves into the nucleus and binds HRE sequences, resulting in luciferase expression, whose activity is measured by the BL emitted after addition of D-luciferin.[70–72]

Such assays are relatively simple to execute, are cost- and time-effective and can address gene transcription issues, are suitable for screening numerous samples simultaneously for their hormonal activities and they are easily performed in high-throughput formats such as 96- and 384-well microtiter plates. Thanks to their short assay time (usually less than 3 h) and possibility of automation, bioluminescent yeast-based assays have been proposed as bioavailability and activity screening method prior to more detailed chemical analysis.

In addition, much effort has been made recently in the direction of cell immobilization, to obtain more robust analytical tools that do not necessitate laboratories equipped with cell culture facilities and skilled personnel.[73] The use of different immobilization techniques to maintain cell viability cells for the development of high-throughput and chip-based biosensing systems is discussed elsewhere.[74–76]

Nevertheless yeast-based biosensors lack responsiveness to some estrogens and antiestrogens and to some chlorinated EDCs. Besides, permeability of compounds through the yeast cell wall differs from that of mammalian cell membranes.

For these reasons, mammalian cell-based reporter gene assays are becoming more common, not only in EDC and drug screening, but also in the analysis of bioactive levels of steroid hormones.[77,78] Various bioluminescent mammalian cell lines have been used in these assays based on different cell lines and most of them were developed in the 96-well microplate format.[79]

15.4.3 Bioluminescent Yeast and Mammalian Cell-based Assays for Genotoxic Compounds and ADMET Studies

Since many factors influencing a compound's ADMET properties are now understood at the molecular level, it is possible to study them with *in vitro* assays. Bioluminescent *in vitro* assays, which are usually high throughput, cost-effective, safe, fast and sensitive, have found many ADMET applications.

For example, an important issue is the assessment of potential genotoxicity and mutagenicity of drug candidates in the discovery phase of drug development. The most commonly used assays are based on bacterial microplates: Ames test and SOS response reporters; nevertheless, the complexity of DNA damage needs the analysis of several cellular pathways. A new high-throughput yeast-based assay has been proposed recently as an alternative to commercially available genotox assays.[80]

This assay is based on yeast cells co-transfected with two bioluminescent reporter genes: a *Renilla* luciferase under the regulation of a constitutive promoter (3-phosphoglycerate kinase, *PGK1*) used as internal cell number control and a firefly luciferase fused to the *RAD51* (bacterial *RecA* homolog) promoter to monitor variations in DNA repair activity. This assay was successfully employed to screen chemomutagenic potential of several compounds, but the authors reported lack of agreement with the traditional *in vivo* regulatory assays. This can be because a single model organism or cell line can offer only modest predictive power for genotoxicity in general. In addition, yeast cells have a cell wall and plasma membrane composition different from mammalian cells with ABC transporters that pump out many compounds.

Alternatively, an analogous dual- assay has been developed based on HepG2 cells.[81] The gene encoding *Renilla* luciferase was fused to the CMV promoter, and used as a control for cell numbers, and the firefly luciferase gene was fused to the GADD45beta promoter, and used to report an increase in DNA damage. A dual luciferase assay was then performed by measuring the firefly and *Renilla* luciferase activities in the same sample.

Many other ADMET assays have been obtained that rely on the use of whole-cell biosensors and luciferase as reporter gene, *e.g.*, for the screening of drugs acting as agonists of nuclear receptors such as pregnane X receptor (PXR), constitutive androstane receptor (CAR), aryl hydrocarbon receptor (AHR) and glucocorticoid receptor (GR). In addition, regulation of major drug-metabolizing cytochromes P450, *e.g.*, CYP3A4 and CYP1A1, has been studied with luciferase reporter assays.[82] These assays allow the testing of compounds in terms of fold effect relative to a vehicle treated control or relative to a prototypical inducer (*e.g.*, rifampicin for CYP3A4).

15.4.4 Bioluminescent Mammalian Cell-based Biosensors as Functional Detection Systems

Mammalian cell-based sensing systems that employ engineered mammalian cell lines as biorecognition elements are opening up a new area of functional diagnostics. In fact an urgent need is emerging for rapid functional detection systems for pathogens, toxins and bioactive compounds in clinical, pharmaceutical and environmental applications. More recently, the high threat of bioterrorism aimed at the water and food chain represents a challenge to current technologies. The routine monitoring of food and the environment for chemical and biological threat agents is not an easy task since available techniques usually require pure or clean samples and sophisticated equipment, thus being unsuitable for field use.

Alternatively, unconventional analytical tools that allow a rapid and low cost monitoring of complex matrices (*e.g.*, water, food, soil) are provided by whole-cell biosensors. One property common to all chemical or biological threat agents is that they damage mammalian cells; thus several threat detection and classification methods based on the effects of compounds on cells, genetically engineered or not, have been developed. Whole-cell biosensors have found application in various areas such as detection of biowarfare agents, waterborne and foodborne pathogens, and emerging infectious diseases.[83]

Considerable research has been carried out in the field of whole-cell sensing systems with the final goal of developing low-cost portable systems able to detect the presence of any toxic agent rapidly and with high sensitivity. To increase the robustness and portability of cell-based biosensors, several formats have been employed that combine the great potential of biochemiluminescence detection (high sensitivity, simple instrumentation) with recent advances in cell immobilization techniques to develop ready-to-use analytical devices.

For example, Fairey and colleagues have developed a reporter gene assay as a direct evolution of previously reported cytotoxicity assays for algal-derived toxins. *c-fos* was selected as biomarker for localizing the effects of toxins for its ability to be induced in neurons of mammals and fish as a result of neuronal stimulation. A mouse neuroblastoma cell line was stably transfected with a c-fos-luciferase reporter vector and brevetoxin-1 caused a concentration-dependent increase in luciferase activity with a half-maximal effect that occurred at a concentration comparable to that obtained by direct cytotoxicity assays.[84] A sensor that uses engineered B lymphocytes that emit light within seconds of exposure to specific bacteria and viruses has been developed by Rider *et al.* B cell lines were engineered to express cytosolic aequorin, a calcium-sensitive bioluminescent protein from the *Aequorea victoria* jellyfish, as well as membrane-bound antibodies specific for pathogens of interest. Crosslinking of the antibodies by even low levels of the appropriate pathogen elevated intracellular calcium concentrations within seconds, causing the aequorin to emit light. A feature of this biosensor is that the antibody expressed determines the cell specificity and can be tailored to a desired application, although antibody cross-reactivity can represent a problem, as in other antibody-based technologies.[85]

15.5 Multiplexed and Multicolor Assays

The main drawback encountered in the analytical application of cell-based biosensors is the high variability of their response, mainly caused by sample matrix aspecific effects on the BL signal. To overcome this problem and increase the robustness of the bioassay, an internal or external reference signal can be introduced to correct the analytical response and consequently separate the analytical signal from nonspecific interferences. The introduction of a second reporter gene in the same cell, for example, to correct the signal accordingly to cell vitality and sample matrix effects, needs the separation of the two BL signals. This can be achieved by using reporter proteins that require different substrates of BL reaction or that emit at a different wavelength or by using intracellular and secreted luciferases. The Promega Dual-Luciferase Reporter Assay System, which requires sequential measurement of both firefly and *Renilla* luciferases in one sample, is an example of a dual-reporter assay system based on bioluminescence detection. The weakness of this system is the necessity to add a reagent to stop one reaction before adding a second substrate and the need to express two dissimilar enzymes. More recently, reporter assay systems using red- and green-emitting luciferases from the Jamaican click beetles and *Phrixothrix* railroad worms have been reported.[86] Although these methods are promising, the sensitivity of these dual-color reporters might not be sufficient for some applications.

Very few works regarding multicolor reporter assay systems have been reported in literature.[87–89] The main bottleneck consists in the need for an efficient spectral unmixing procedure when using more than two luciferases requiring the same substrate. To perform such calculation, a Java plug-in for ImageJ has been written to deconvolute images composed of signals obtained with different filters.[90]

The development of new BL proteins by cDNA cloning and mutagenesis of wild-type genes is of particular relevance in this context because the use of reporter proteins emitting at different wavelengths facilitates the separation of analytical and control signals and expands the applicability of BL reporters to multiplexed cell-based assays.

For example, a cell-based BL assay for stabilization of IκBα has been developed in a B-cell lymphoma cell line. This assay employs green- and red-emitting beetle luciferases, with the green luciferase fused to IκBα (IκBα-CBG68) and the red luciferase (CBR) present in its native state. The IκBα-CBG68 reporter acts as a sensor of Iκ kinase (IKK) and proteasome activity, while CBR serves to normalize the BL signal for cell number and nonspecific effects.[91]

Nakajima *et al.* have proposed a novel reporter assay system in which three luciferases that emit at different wavelengths (green, orange and red) in the presence on the same substrate are used as reporter proteins.[88] The green-emitting luciferase of *Rhagophthalmus ohbai*, the orange-emitting luciferase (a point mutant of the green one: T226N) and the red-emitting luciferase of *Phrixothrix hirtus* were used as BL reporters. These enzymes are pH-insensitive, and thus do not change the emission color even if the intracellular pH changes.

By measuring the emission of the three BL reporters with long pass filters and applying a signal process algorithm, a monitoring system was developed for the simultaneous evaluation of the expression of three different genes within a cell, achieving a dynamic range of three orders of magnitude.

Alternatively, secreted BL reporter proteins, such as *Gaussia princeps* luciferase, that do not require cell lysis or special equipment (*e.g.*, filter-based luminometers) may be used although their expression has a higher variability than firefly luciferases.

We have previously reported a triple-color cell based assay that combines spectral unmixing of green- and red-emitting luciferases (the green-emitting wild-type *P. pyralis* luciferase and a red-emitting thermostable mutant of *L. italica* luciferase) with a secreted *Gaussia* luciferase requiring a different substrate, thus allowing the measurement of three separate signals with high sensitivity and rapidity. The green- and red-emitting luciferases were put under the regulation of CYP7A1 and CYP27A1 promoters, respectively, to monitor the two main bile acid biosynthesis pathways and the secreted *Gaussia* luciferase was used as vitality control under the regulation of a constitutive

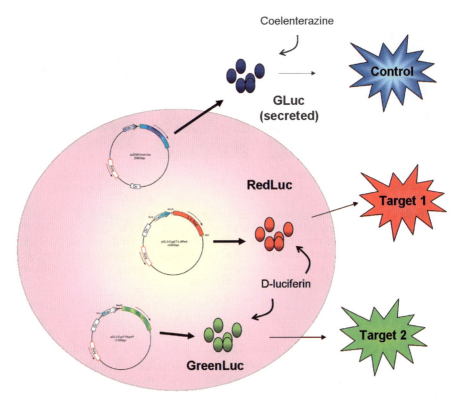

Figure 15.4 Schematic view of a triple-color whole cell-biosensor that combines green- and red-emitting luciferases, used as reporter proteins for two different molecular targets, and a constitutively expressed *Gaussia* luciferase as vitality control.

promoter (Figure 15.4).[92] The use of a secreted luciferase greatly simplified the measure of its activity because its expression was evaluated on small aliquots of cell culture medium. The association of secreted and intracellular reporters in the same cell-based assay has the advantage of complete absence of interference between the two signals; it is in fact possible to measure the two signals in separate wells of a high-throughput 96-well microtiter plate.

The assay was used to monitor *in vivo* the two main pathways of bile acid biosynthesis, showing good reproducibility and dose–response curves for bile acids consistent with previous published results.

15.6 Miniaturization of Cell-based Biosensing Systems

Cell-based bioluminescent biosensing systems have still to reach their fullest potential. Attractive technologies where these systems could find applications include rugged, compact portable sensing platforms and instrumentation for

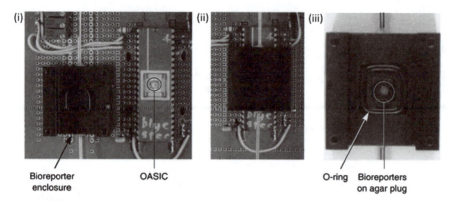

(i)   (ii)   (iii)

Bioreporter   OASIC   O-ring   Bioreporters
enclosure   on agar plug

Figure 15.5   Bioluminescent-bioreporter integrated circuit featuring (i) enclosure containing bacterial biosensing cells and optical application-specific integrated circuit (OASIC); (ii) enclosure mounted on the chip; and (iii) the enclosure showing bacterial biosensing cells on agar plug. A tight seal between chip and enclosure is maintained by using an O-ring. (Reproduced with permission from Simpson et al.[93])

on-site measurements of environmentally and clinically relevant analytes. To date, several important strides toward miniaturization have been achieved. An example of the progress made towards constructing miniaturized systems includes the whole-cell bioluminescent-bioreporter integrated circuit device developed by Simpson et al.[93] In this system, a toluene-selective genetically engineered *Pseudomonas putida* bioreporter strain was incorporated onto a chip provided with an optical application-specific integrated circuit (Figure 15.5). Upon interaction with toluene vapor, a bioluminescence signal was generated and measured by the integrated circuit, allowing detection of toluene concentrations down to 50–10 ppb, depending on the signal integration time. The main advantage of this system lies in the direct coupling of the bioluminescent bioreporter cells to an integrated circuit designed for detecting, processing and reporting of the light signal. This eliminates the need for large detection instrumentation and optical components for light collection and transfer, thus providing a self-contained portable device suitable for on-site applications.

Miniaturized cell-based biosensing systems have also found application in genome-wide transcription analysis. In a study performed by Van Dyk et al.,[94] sequenced random segments of *E. coli* DNA were inserted into plasmids as gene fusions with *Photorhabdus luminescens luxCDABE* gene cassette and transformed into host cells. A group of functional gene fusions known as LuxArray 1.0 was selected, which contained a total of 689 diverse reporter strains. These strains were printed on a porous nylon membrane (8 cm × 10 cm) at 16 spots cm^{-2} by means of a commercially available automated workstation. During and after this process the membrane was kept in contact with LB growth media in a culture dish. These reporter strains were employed for simultaneously evaluating gene expression in the presence of nalidixic acid, an antibiotic that

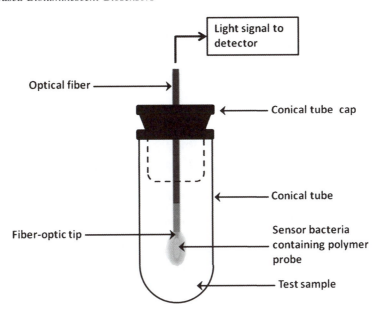

Figure 15.6 Fiber optic-based miniaturization. An optical fiber is placed in a conical tube containing the sample solution. The fiber optic tip is coated with an alginate matrix that incorporates the sensing bacterial cells. Bioluminescence is triggered when the analyte enters the cells and is recognized by the biosensing element. The bioluminescence emitted travels to the detector through the optical fiber. (Figure adapted from Polyak *et al.*[96])

induces DNA damage stress response by causing a change in gene regulation. Specifically, in the described reporter gene assay, upregulation of certain genes translated to increased bioluminescence signals. This system showed the feasibility of obtaining high-density bioluminescent reporter cell arrays and suggests their potential use for analytical purposes. Notably, a further increase in the density of the arrays may be limited due to cross-illumination from neighboring spots.

In another attempt towards miniaturization, fiber optic based systems have been designed and developed in which whole-cell biosensing bacteria were immobilized onto an exposed core of a fiber-optic.[95,96] In one case, reporter cells containing a gene fusion of the genotoxicant-inducible *recA* promoter of *E. coli* to the *P. luminescens luxCDABE* reporter were constructed.[96] These sensing cells emitted light in a dose-dependent manner in the presence of DNA damaging (genotoxic) agents, such as mitomycin C, which can react with the DNA structure, destabilize it and potentially cause deadly genetic mutations. The optical fiber was treated with acid for proper cleaning. The sensing cells were mixed with a polymeric solution, such as sodium alginate, and the fiber-optic tip was dipped into the mixture containing the cells (Figure 15.6). Further treatment with calcium chloride solution was performed to harden the cell-alginate matrix onto the fiber-optic core. When the cell-deposited fiber-optic tip

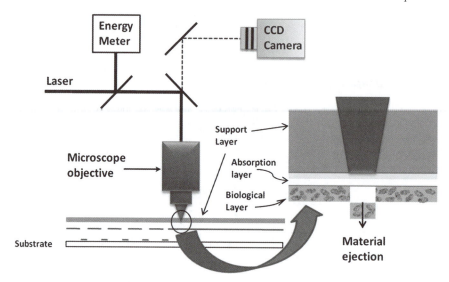

Figure 15.7 Schematic showing a BioLP™ device. A laser beam is focused on a spot at the interface of the support and absorption layers, causing ejection of the sensing cells containing material, which is then deposited on the substrate. (Figure adapted from Barron *et al.*[98])

was exposed to solutions of mitomycin C at various concentrations, a dose-dependent bioluminescence response was triggered, and then measured by the fiber optic system. Notably, this fiber-optic system was able to achieve the same detection limit, in a shorter period of time, as its larger scale counterpart. In another work by Gil *et al.*,[97] genetically engineered constitutively bioluminescent bacteria were deposited on an optical fiber, employing a solid matrix of glass beads and agar, to develop a biosensor for the detection of toxic gases. Specifically, the presence of toxic chemicals reduced the cells' bioluminescence intensity. Addition of glass beads increased both the porosity of the cell matrix, which facilitated the diffusion of vapors through the cell matrix layer, and the contact surface area of the cells with the gases, thus resulting in improved sensing ability of the bacterial sensor. This sensor is not as specific because it measures cell death, which can be caused not only by gases but also by other toxic compounds present in the sample.

Technologies such as biological laser printing (BioLP™) have been reported for the rapid deposition of biomolecules and live bacterial sensing cells onto various surfaces.[98] Forward transfer BioLP™ uses laser pulses to transfer material from a carrier support onto a receiving substrate (Figure 15.7). The carrier support is an absorption layer (mostly quartz coated with metal oxides) on which properly grown bacterial sensing cells are spread prior to printing. In the reported example, the sensing strain was *E. coli* harboring a plasmid-borne fusion of the *recA* gene promoter to the red fluorescent protein gene from *Discosoma*, capable of responding to genotoxicants like nalidixic acid. Then, the sensing cells on the carrier support were printed onto a receiving surface composed on a LB agar plate or a sterile glass slide with a thin film of LB agar. The laser pulse was focused on a spot in the absorption layer. The laser-material interaction produced photo-absorption, propelling a three-dimensional pixel of biomaterial towards the receiving substrate through photo-mechanical and/or photothermal effects. This method of printing was reported to be precise, with an average spot diameter of $70 \pm 6\,\mu m$ and an approximate volume of 5 pL. An alteration in the fluorescence emission was observed and attributed to the genotoxicity caused by nalidixic acid. This BioLP™ technique may be applied to several diverse sensing cells to produce miniaturized chip-based sensing systems that can be used in a laboratory setting or in the field.

During the last two decades, there has been considerable interest in and efforts made to miniaturize conventional bench-top analytical techniques and incorporate them into microfluidic chip-based platforms[99] as well as to integrate multiple analytical processes into a single chip.[100] The physical principles that govern mass transfer and fluid flow at the microscale level allow for rapid mass transfer and kinetics as well as high surface-to-volume ratio, which endow microfluidic systems with unique characteristics when compared to conventional volume analytical systems. Microfluidic devices such as micro-total-analysis systems (μTAS) and lab-on-a-chip platforms have been developed for several analytical tasks, including whole-cell-based biosensing. Generally, computer numerical control (CNC) machining and lithography techniques are used to fabricate these devices, employing polymeric materials such as poly(-methyl methacrylate) (PMMA), glass and silicon. Various microfluidic structures can be fabricated that incorporate features such as microreservoirs, microchannels, mixing devices, filtration, fractionation and separation devices and microvalves. The choice of structures incorporated into a microfluidic device depends on the specific application desired.[101] Propulsion of fluids on microfluidic platforms is accomplished by employing varied instruments such as syringe and peristaltic pumps, or by applying acoustic, magnetic or centrifugal forces.[102]

Microfluidic platforms employing centrifugal forces can be designed in the form and size of a compact disk (CD). This kind of centrifugal microfluidic device has been used for cell-based detection systems. Specifically, bacterial biosensing cells containing the gene for green fluorescent protein (GFP) under the transcriptional control of the promoter and regulatory genes of the *ars* operon were employed.[103] The biosensing system relies upon the recognition and binding of the target analytes arsenite/antimonite by the transcription regulatory protein ArsR, and the resulting expression of the reporter protein GFP inside the cytoplasm. The biosensing cells were incorporated into a CD microfluidic platform made of PMMA for detection of arsenite and antimonite. In this application, miniaturization significantly reduced the assay time (30 min *versus* <1 min) along with the volumes of reagents used, while retaining similar micromolar detection limits and dynamic ranges, when compared to the bench-top assay. The decrease in detection time is due to faster diffusion of the analyte

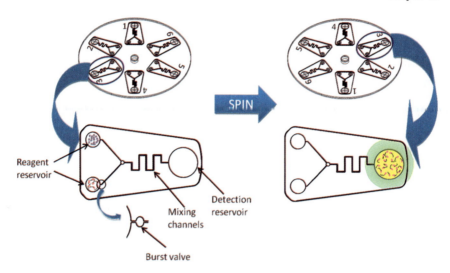

Figure 15.8 Schematic of a centrifugal microfluidic CD platform. Luminescent bio-sensing cells and respective analyte are added into the small reservoirs (17 µL capacity). When the CD is spun at high rpms, the liquids are forced through the microchannels into the detection chamber (34 µL capacity). Light emission can be measured by means of a suitable detector.

into the cells, thus increasing the reaction kinetics. The sensing system proved to be highly selective for arsenite and antimonite when incorporated into the microfluidic platform. Owing to precise manufacturing, all the structures have identical physical characteristics, making these platforms a very attractive solution for multiple parallel assays with potential for high-throughput screening as well as on-site monitoring. This type of sensing platform is suitable for the development of simple instrumentation based on readily-available, cost-effective hardware consisting of a drive motor, a power supply, a controller, lens optics and compact CCD cameras for the detection and quantification of emitted light. These components along with software for system control and data acquisition, processing and analysis can be easily integrated into a portable system. Figure 15.8 shows an example of centrifugal CD microfluidic platform. Multiple structures consisting of an arrangement of reagent reservoirs, burst valves and microchannels leading to detection reservoirs are shown. Fluid release from the reagent reservoirs is controlled by burst valves, located a very short distance from the reservoirs. Sufficient centrifugal force is needed to overcome the capillary force holding the liquids into the reservoirs and allow their flow to the detection chamber. Such force is generated by appropriate frequency of rotation (burst frequency) of the disk.

Whole-cell based biosensing systems employing firefly luciferase as a reporter for the detection of genotoxicants have been integrated into a chip-based three-dimensional microfluidic device, which was obtained by placing a silicon

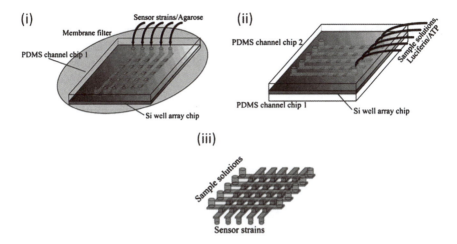

Figure 15.9 On-chip bioassay using a three-dimensional microfluidic network. (i) A mixture of agarose and bacterial sensor strain is added into the channels of PDMS chip 1 and cells are immobilized by gelation in the wells of the silicon chip. The chip assembly is turned over and, after peeling off the membrane filter, (ii) analyte solutions are introduced into the channels of PDMS chip 2. (iii) Schematic of the microfluidic network in the three-dimensional chip. (Reproduced with permission from Tani *et al.*[104])

substrate between two poly(dimethylsiloxane) (PDMS) layers.[104] Microchannels (volume 3 μL) in the two PDMS layers were connected *via* perforations in the silicon layer that served as micro-wells (volume 0.25 μL) to hold the sensing strains, thus forming a three-dimensional microfluidic network. The sensor strains were immobilized onto the micro-well array of the silicon chip by gelation upon injection of a cell/agarose mixture through the microchannels of one of the PDMS chips. Luciferase gene expression was then induced by passing sample genotoxicant solutions through the microchannels present on the second PDMS chip (Figure 15.9). Bioluminescence was triggered when a solution of luciferin/ATP was passed through the channels and detected by means of a CCD camera. The bioluminescence response obtained with this miniaturized microfluidic system (1 h) was significantly faster than the assay using test tubes (4 h). Low-cost materials were employed to make these platforms and low volumes of reagents were consumed, thus rendering the system very cost-effective. In addition, multi-analyte detection in multiple samples can be performed simultaneously by immobilizing different sensor bacteria on a single chip.

15.7 Conclusions and Future Potential

Continued work in the field of cell-based biosensors should contribute to further exploitation and enhancement of the unique advantages that they offer in the detection of various analytes in different types of samples using diverse analytical platforms. The identification of additional regulatory proteins and receptors should expand the ranks of chemical species that can be detected. Moreover, the identification, characterization and alteration of new light-emitting bioreporters should lead to advanced, multiplexed assays capable of measuring several analytes simultaneously. Advances in optical instrumentation and miniaturization should, undoubtedly, yield smaller, more rugged, less expensive methods along with the selectivity and sensitivity afforded by these genetically engineered cells. Miniaturization to array-based or microfluidic chip-based platforms reduces volumes of reagents and samples as well as wastes produced. There is no doubt that the future is bright for cell-based bioluminescent biosensors as they will find further application in the fields of environmental monitoring, toxicology, pharmacology, drug-screening and medical/clinical applications.

Acknowledgements

This work was supported in part by the National Institute of Environmental Health Sciences Superfund Research Program grant P42ES007380, the National Science Foundation grants CHE-0416553 and CHE-0718844, the National Institutes of Health, the Broad Foundation Broad Medical Research Program grant IBD-0198R. S. D. acknowledges support from a Gill Eminent Professorship. K. T. acknowledges a traineeship from the Superfund Research Program. K. T. and N. R. acknowledge support from the University of Kentucky Research Challenge Trust Fund.

References

1. S. Daunert, G. Barrett, J. S. Feliciano, R. S. Shetty, S. Shrestha and W. Smith-Spencer, *Chem. Rev.*, 2000, **100**, 2705.
2. R. J. Mitchell and M. B. Gu, *Biosens. Bioelectron.*, 2004, **19**, 977.
3. S. Shrestha, R. S. Shetty, S. Ramanathan and S. Daunert, *Anal. Chim. Acta*, 2001, **444**, 251.
4. K. Turner, S. Xu, P. Pasini, S. Deo, L. Bachas and S. Daunert, *Anal. Chem.*, 2007, **79**, 5740.
5. M. B. Gu, R. J. Mitchell and B. C. Kim, *Adv. Biochem. Eng. Biotechnol.*, 2004, **87**, 269.
6. S. Kohler, S. Belkin and R. D. Schmid, *Fresenius' J. Anal. Chem.*, 2000, **366**, 769.
7. E. A. Meighen, *Microbiol. Rev.*, 1991, **55**, 123.
8. P. Billard and M. S. DuBow, *Clin. Biochem.*, 1998, **31**, 1.
9. L. H. Naylor, *Biochem. Pharmacol.*, 1999, **58**, 749.
10. N. Kajiyama and E. Nakano, *Protein Eng.*, 1991, **4**, 691.
11. B. R. Branchini, D. M. Ablamsky, A. L. Davis, T. L. Southworth, B. Butler, F. Fan, A. P. Jathoul and M. A. Pule, *Anal. Biochem.*, 2009.

12. W. W. Lorenz, R. O. McCann, M. Longiaru and M. J. Cormier, *Proc. Natl. Acad. Sci. USA*, 1991, **88**, 4438.
13. J. Feliciano, S. Xu, X. Guan, H. J. Lehmler, L. G. Bachas and S. Daunert, *Anal. Bioanal. Chem.*, 2006, **385**, 807.
14. C. Wu, C. Suzuki-Ogoh and Y. Ohmiya, *BioTechniques*, 2007, **42**, 290.
15. B. R. Branchini, T. L. Southworth, J. P. DeAngelis, A. Roda and E. Michelini, *Comp. Biochem. Physiol. B Biochem. Mol. Biol.*, 2006, **145**, 159.
16. V. R. Viviani, E. J. Bechara and Y. Ohmiya, *Biochemistry*, 1999, **38**, 8271.
17. Y. Takenaka, H. Masuda, A. Yamaguchi, S. Nishikawa, Y. Shigeri, Y. Yoshida and H. Mizuno, *Gene*, 2008, **425**, 28.
18. J. C. Lewis, A. Feltus, C. M. Ensor, S. Ramanathan and S. Daunert, *Anal. Chem.*, 1998, **70**, 579A.
19. L. Rowe, K. Combs, S. Deo, C. Ensor, S. Daunert and X. Qu, *Anal. Chem.*, 2008, **80**, 8470.
20. M. A. Gilchrist 2nd, A. Cacace and D. G. Harden, *J. Biomol. Screen.*, 2008, **13**, 486.
21. T. H. Rider, M. S. Petrovick, F. E. Nargi, J. D. Harper, E. D. Schwoebel, R. H. Mathews, D. J. Blanchard, L. T. Bortolin, A. M. Young, J. Chen and M. A. Hollis, *Science*, 2003, **301**, 213.
22. S. Inouye and Y. Sahara, *Protein Expression Purif.*, 2007, **53**, 384.
23. S. Welsh and S. A. Kay, *Curr. Opin. Biotechnol.*, 1997, **8**, 617.
24. A. Muller-Taubenberger and K. I. Anderson, *Appl. Microbiol. Biotechnol.*, 2007, **77**, 1.
25. K. Yagi, *Appl. Microbiol. Biotechnol.*, 2007, **73**, 1251.
26. A. Date, P. Pasini and S. Daunert, *Anal. Chem.*, 2007, **79**, 9391.
27. S. Belkin, *Curr. Opin. Microbiol.*, 2003, **6**, 206.
28. S. Magrisso, Y. Erel and S. Belkin, *Microb. Biotechnol.*, 2008, **1**, 320.
29. M. Mirasoli, J. Feliciano, E. Michelini, S. Daunert and A. Roda, *Anal. Chem.*, 2002, **74**, 5948.
30. S. H. Choi and M. B. Gu, *Anal. Chim. Acta*, 2003, **481**, 229.
31. A. Molina, R. Carpeaux, J. A. Martial and M. Muller, *Toxicol. In Vitro*, 2002, **16**, 201.
32. L. Chu, D. Mukhopadhyay, H. Yu, K. S. Kim and T. K. Misra, *J. Bacteriol.*, 1992, **174**, 7044.
33. C. W. Condee and A. O. Summers, *J. Bacteriol.*, 1992, **174**, 8094.
34. T. Barkay, R. R. Turner, L. D. Rasmussen, C. A. Kelly and J. W. Rudd, *Methods Mol. Biol.*, 1998, **102**, 231–246.
35. P. Corbisier, G. Ji, G. Nuyts, M. Mergeay and S. Silver, *FEMS Microbiol Lett*, 1993, **110**, 231–238.
36. G. Ji and S. Silver, *J. Bacteriol.*, 1992, **174**, 3684–3694.
37. S. Ramanathan, W. Shi, B. P. Rosen and S. Daunert, *Anal. Chim. Acta*, 1998, **369**, 189–195.
38. D. L. Scott, S. Ramanathan, W. Shi, B. P. Rosen and S. Daunert, *Anal. Chem.*, 1997, **69**, 16–20.
39. K. P. Yoon, T. K. Misra and S. Silver, *J. Bacteriol.*, 1991, **173**, 7643–7649.

40. S. Tauriainen, M. Karp, W. Chang and M. Virta, *Biosens. Bioelectron.*, 1998, **13**, 931–938.
41. N. Peitzsch, G. Eberz and D. H. Nies, *Appl. Environ. Microbiol.*, 1998, **64**, 453–458.
42. J. Guzzo, A. Guzzo and M. S. DuBow, *Toxicol. Lett*, 1992, **64–65**(Spec No), 687–693.
43. A. G. Prest, M. K. Winson, J. R. Hammond and G. S. Stewart, *Lett. Appl. Microbiol.*, 1997, **24**, 355–360.
44. A. Ivask, T. Green, B. Polyak, A. Mor, A. Kahru, M. Virta and R. Marks, *Biosens. Bioelectron.*, 2007, **22**, 1396–1402.
45. M. B. Gu, J. Min and E. J. Kim, *Chemosphere*, 2002, **46**, 289–294.
46. H. Masuyama, Y. Hiramatsu, M. Kunitomi, T. Kudo and P. N. MacDonald, *Mol. Endocrinol.*, 2000, **14**, 421–428.
47. D. Abd-El-Haleem, S. Ripp, C. Scott and G. S. Sayler, *J. Ind. Microbiol. Biotechnol.*, 2002, **29**, 233–237.
48. X. Guan, S. Ramanathan, J. P. Garris, R. S. Shetty, M. Ensor, L. G. Bachas and S. Daunert, *Anal. Chem.*, 2000, **72**, 2423–2427.
49. X. Guan, E. D'Angelo, W. Luo and S. Daunert, *Anal. Bioanal. Chem.*, 2002, **374**, 841–847.
50. J. Trogl, G. Kuncova, L. Kubicova, P. Parik, J. Halova, K. Demnerova, S. Ripp and G. S. Sayler, *Folia Microbiol. (Praha)*, 2007, **52**, 3–14.
51. J. J. Dawson, C. O. Iroegbu, H. Maciel and G. I. Paton, *J. Appl. Microbiol.*, 2008, **104**, 141–151.
52. C. D. Sifri, *Clin. Infect. Dis.*, 2008, **47**, 1070–1076.
53. A. Kumari, P. Pasini, S. K. Deo, D. Flomenhoft, H. Shashidhar and S. Daunert, *Anal. Chem.*, 2006, **78**, 7603–7609.
54. A. Kumari, P. Pasini and S. Daunert, *Anal. Bioanal. Chem.*, 2008, **391**, 1619–1627.
55. Z. Ye, H. S. Weinberg and M. T. Meyer, *Anal. Chem.*, 2007, **79**, 1135–1144.
56. J. Scaria, S. Ramachandran, P. K. Jain and S. K. Verma, *Res. J. Microbiol.*, 2009, **4**, 104–111.
57. I. Vlasova, T. Asrieli, E. Gavrilova and V. Danilov, *Appl. Biochem. Microbiol.*, 2007, **43**, 422–428.
58. A. Urban, S. Eckermann, B. Fast, S. Metzger, M. Gehling, K. Ziegelbauer, H. Rubsamen-Waigmann and C. Freiberg, *Appl. Environ. Microbiol.*, 2007, **73**, 6436–6443.
59. T. R. Stratton, R. E. Garcia, B. M. Applegate and J. P. Youngblood, *Biomacromolecules*, 2009, **10**, 1173–1180.
60. M. del Busto-Ramos, M. Budzik, C. Corvalan, M. Morgan, R. Turco, D. Nivens and B. Applegate, *Appl. Microbiol. Biotechnol.*, 2008, **78**, 573–580.
61. R. S. Shetty, S. Ramanathan, I. H. Badr, J. L. Wolford and S. Daunert, *Anal. Chem.*, 1999, **71**, 763–768.
62. J. Svitel, O. Curilla and J. Tkac, *Biotechnol. Appl. Biochem.*, 1998, **27**(Pt 2), 153–158.

63. S. Kim, B. Schuler, A. Terekhov, J. Auer, L. J. Mauer, L. Perry and B. Applegate, *J. Microbiol. Methods*, 2009, **79**, 18–22.
64. H. Wu, Z. Song, M. Hentzer, J. B. Andersen, A. Heydorn, K. Mathee, C. Moser, L. Eberl, S. Molin, N. Hoiby and M. Givskov, *Microbiology*, 2000, **146**(Pt 10), 2481–2493.
65. J. J. Min, V. H. Nguyen, H. J. Kim, Y. Hong and H. E. Choy, *Nat. Protocols*, 2008, **3**, 629–636.
66. M. L. Foucault, L. Thomas, S. Goussard, B. R. Branchini and C. Grillot-Courvalin, *Appl. Environ. Microbiol.*, 2009.
67. F. Fan and K. V. Wood, *Assay Drug Dev. Technol.*, 2007, **5**, 127–136.
68. R. M. Walmsley and P. Keenan, *Biotechnol. Bioprocess. Eng.*, 2000, **5**, 387–394.
69. M. Wagner and J. Oehlmann, *Environ. Sci. Pollut. Res. Int.*, 2009, **16**, 278–286.
70. E. Michelini, P. Leskinen, M. Virta, M. Karp and A. Roda, *Biosens. Bioelectron.*, 2005, **20**, 2261–2267.
71. E. Michelini, L. Cevenini, L. Mezzanotte, P. Leskinen, M. Virta, M. Karp and A. Roda, *Nat. Protocols*, 2008, **3**, 1895–1902.
72. A. L. Välimaa, A. T. Kivistö, P. I. Leskinen and M. T. Karp, *J. Microbiol. Methods*, 2010, **80**, 44–448.
73. T. Fine, P. Leskinen, T. Isobe, H. Shiraishi, M. Morita, R. S. Marks and M. Virta, *Biosens. Bioelectron.*, 2006, **21**, 2263–2269.
74. M. B. Gu, R. J. Mitchell and B. C. Kim, *Adv. Biochem. Eng. Biotechnol.*, 2004, **87**, 269–305.
75. K. Yagi, *Appl. Microbiol. Biotechnol.*, 2007, **73**, 1251–1258.
76. K. Flampouri, S. Mavrikou, S. Kintzios and G. Miliadis Aplada-Sarlis, *Talanta*, 2010, **80**, 1799–1804.
77. P. Banerjee, B. Franz and A. K. Bhunia, *Adv. Biochem. Eng. Biotechnol.*, 2010, in press. DOI 10.1007/10_2009_21.
78. P. Roy, M. Alevizaki and I. Huhtaniemi, *Hum Reprod Update.*, 2008, **14**, 73–82.
79. P. J. Hofmann, L. Schomburg and J. Köhrle, *Toxicol. Sci.*, 2009, **110**, 125–137.
80. X. Liu, J. A. Kramer, J. C. Swaffield, Y. Hu, G. Chai and A. G. Wilson, *Mutat. Res.*, 2008, **653**, 63–69.
81. X. Liu, J. A. Kramer, Y. Hu, J. M. Schmidt, J. Jiang and A. G. Wilson, *Int. J. Toxicol.*, 2009, **28**, 162–176.
82. J. J. Cali, A. Niles, M. P. Valley, M. A. O'Brien, T. L. Riss and J. Shultz, *Expert Opin. Drug Metab. Toxicol.*, 2008, **4**, 103–120.
83. *Nano and Microsensors for Chemical and Biological Terrorism Surveillance*, ed. J. B. H. Tok, RSC Publishing, Cambridge, 2008, pp. 166–176.
84. E. R. Fairey, J. S. Edmunds and J. S. Ramsdell, *Anal. Biochem.*, 1997, **251**, 129.
85. T. H. Rider, M. S. Petrovick, F. E. Nargi, J. D. Harper, E. D. Schwoebel, R. H. Mathews, D. J. Blanchard, L. T. Bortolin, A. M. Young, J. Chen and M. A. Hollis, *Science*, 2003, **301**, 213–215.

86. Y. Kitayama, T. Kondo, Y. Nakahira, H. Nishimura, Y. Ohmiya and T. Oyama, *Plant Cell Physiol.*, 2004, **45**, 109–113.
87. E. Michelini, L. Cevenini, L. Mezzanotte, D. Ablamsky, T. Southworth, B. R. Branchini and A. Roda, *Photochem. Photobiol. Sci.*, 2008, **7**, 212–217.
88. Y. Nakajima, T. Kimura, K. Sugata, T. Enomoto, A. Asakawa, H. Kubota, M. Ikeda and Y. Ohmiya, Multicolor luciferase assay system: one-step monitoring of multiple gene expressions with a single substrate, *BioTechniques*, 2005, **38**, 891–894.
89. X. Li, Y. Nakajima, K. Niwa, V. R. Viviani and Y. Ohmiya, *Protein Sci.*, 2010, **19**, 26–33.
90. S. T. Gammon, W. M. Leevy, S. Gross, G. W. Gokel and D. Piwnica-Worms, *Anal. Chem.*, 2006, **78**, 1520–1527.
91. R. E. Davids, Y. Q. Zhang, N. Southall, L. M. Staudt, C. P. Austin, J. Inglese and D. S. Auld, *Assay Drug Dev. Technol.*, 2007, **5**, 85–103.
92. E. Michelini, L. Cevenini, L. Mezzanotte, D. Ablamsky, T. Southworth, B. Branchini and A. Roda, *Anal. Chem.*, 2008, **80**, 260–267.
93. M. L. Simpson, G. S. Sayler, B. M. Applegate, S. Ripp, D. E. Nivens, M. J. Paulus and G. E. Jellison, *Trends Biotechnol.*, 1998, **16**, 332–338.
94. T. K. Van Dyk, E. J. DeRose and G. E. Gonye, *J. Bacteriol.*, 2001, **183**, 5496–5505.
95. K. Hakkila, T. Green, P. Leskinen, A. Ivask, R. Marks and M. Virta, *J. Appl. Toxicol.*, 2004, **24**, 333–342.
96. B. Polyak, E. Bassis, A. Novodvorets, S. Belkin and R. S. Marks, *Sens. Actuators B: Chem.*, 2001, **74**, 18–26.
97. G. C. Gil, Y. J. Kim and M. B. Gu, *Biosens. Bioelectron.*, 2002, **17**, 427–432.
98. J. A. Barron, R. Rosen, J. Jones-Meehan, B. J. Spargo, S. Belkin and B. R. Ringeisen, *Biosens. Bioelectron.*, 2004, **20**, 246–252.
99. R. F. Ismagilov, *Angew. Chem. Int. Ed.*, 2003, **42**, 4130–4132.
100. U. Bilitewski, M. Genrich, S. Kadow and G. Mersal, *Anal. Bioanal. Chem.*, 2003, **377**, 556–569.
101. E. Dikici, L. Rowe, E. A. Moschou, A. Rothert, S. K. Deo and S. Daunert, in *Photoproteins in Bioanalysis*, ed. S. Daunert and S. K. Deo, Wiley-VCH Verlag GmbH, 2006, pp. 179–198.
102. M. Madou, J. Zoval, G. Y. Jia, H. Kido, J. Kim and N. Kim, *Annu. Rev. Biomed. Eng.*, 2006, **8**, 601–628.
103. A. Rothert, S. K. Deo, L. Millner, L. G. Puckett, M. J. Madou and S. Daunert, *Anal. Biochem.*, 2005, **342**, 11–19.
104. H. Tani, K. Maehana and T. Kamidate, *Anal. Chem.*, 2004, **76**, 6693–6697.
105. M. Farre, C. Goncalves, S. Lacorte, D. Barcelo and M. F. Alpendurada, *Anal. Bioanal. Chem.*, 2002, **373**, 696–703.

CHAPTER 16

Miniaturized Analytical Devices Based on Chemiluminescence, Bioluminescence and Electrochemiluminescence

LARRY J. KRICKA[a] AND JASON Y. PARK[b]

[a] Department of Pathology and Laboratory Medicine. University of Pennsylvania School of Medicine, 7.103 Founders Pavilion, 3400 Spruce Street, Philadelphia, Pennsylvania 19104, USA; [b] Department of Pathology, University of Texas Southwestern Medical Center, Children's Medical Center, 1935 Medical District Drive, Dallas, Texas 75235, USA

16.1 Introduction

A continuing trend in both chemical and biological analysis is the miniaturization of analytical procedures and the development of micro-miniature analyzers (microchips). The goal is to develop lab-on-a-chip or micro total analytical system (μTAS) devices in which all the sequential steps in an analytical procedure are performed within a single microchip.[1] In operation, the analyst would add a sample to the microchip, and then the microchip would automatically perform all of the necessary functions for analytical analysis: specimen processing; reagent additions; incubations; analyte extractions or other steps in the analysis; calculations; results display; communication of results to an information system. This type of miniaturization is viewed as a way in which assays can be simplified and made quicker so that these assays are suitable for extra-laboratory applications (*e.g.*, detecting bio-warfare agents, point-of-care tests in hospitals, clinics or in the home). Various types of microchips have been developed and these can be classified into microfluidic

Chemiluminescence and Bioluminescence: Past, Present and Future
Edited by Aldo Roda
© Royal Society of Chemistry 2011
Published by the Royal Society of Chemistry, www.rsc.org

microchips, microarrays, bioelectronic chips and finally micro total analytical system chips that combine and integrate microfluidic, microarray and micro-electronic components.

16.2 Fabrication Methods for Microchips

The different types of microchip fabrication methods can be classified into techniques designed to (a) produce physical structures, such as microchannels, microchambers, microposts, and (b) techniques that create microarrays of test areas by physical deposition or synthesis of reagents at defined locations on a planar surface. Materials used for microchip fabrication vary widely and include silicon, fused quartz, soda glass and a range of polymers, including poly(methyl methacrylate) (PMMA), polypropylene and poly(dimethylsiloxane) (PDMS). Specific physical structures are usually fabricated using photolithographic or reactive ion etching techniques adapted from the microelectronics industry.[2] Individual components can be bonded together to form multilayer structures and holes introduced by laser drilling.[3] Other fabrication methods include embossing, microinjection molding, micromilling and etching by X-ray or e-beam etching. Planar microarrays can be formed on a surface by ink jet printing or spotting pre-formed reagents, or light-directed combinatorial parallel synthesis of reagents at defined locations on a surface.[4]

Microchips for use with different type of luminescent reactions have been fabricated from the same range of materials that are diverse as those found in microchips developed for other applications. These materials include silicon,[5,6] glass,[7] PMMA[8,9] and acrylonitrile–butadiene–styrene (ABS) copolymer.[10]

16.3 Microchip Components

The types of micromachined components available for assembly into a microchip analyzer include: flow channels, reaction chambers, microposts, weir filters, strip-heaters, Peltier coolers, lenses, valves, pumps, diaphragms, motors, lasers, light emitting diodes, optical filters and ion selective electrodes.[11] Liquids can be manipulated and moved around a fluidic network inside a microchip by several techniques, including mechanical displacement or using electro- and magneto-kinetic forces.[12]

The successful manufacture of microchip-based analyzers will integrate the multiple components required to implement the different sequential steps in an analytical process. Although this field of component integration in analytical microchip technology is at an early stage, there have been numerous examples of successful integrations. Examples of microchip-based analyzer with multiple components include microchips that combine on-chip enzymatic digestion and delivery of the enzymatic digests to an electrospray mass spectrometer;[13] DNA arrays fabricated directly onto the surface of a CCD (charge coupled device) detector;[14] immunoassay reaction chambers coupled to capillary electrophoresis detectors;[15,16] microfilters for white blood cell isolation integrated into a PCR reaction chamber.[17,18]

In the context of microchips combined with chemiluminescence, bioluminescence and electrochemiluminescence, the scope of integration includes reaction channels and chambers, separation devices and detection devices. Microchips combining these integrated features are described in the following sections.

16.4   Advantages of Microchips for Analysis

Microchip-based analyzers have several potential advantages and benefits compared to analytical methods performed using conventional analytical procedures:

System integration – Total analytical system integration is, perhaps, the most compelling advantage. Current microfabrication technology permits the construction of relatively complex μm-sized and even nm-sized interconnecting structures with various processing functions (*e.g.*, valves, filters, heated reaction chambers),[19] and integrated electronic sensors and control circuitry. These features are all combined into a single microchip to form a micro total analytical system (μTAS).[1,20]

Ease of manufacture and design-For devices fabricated from silicon, the existing microelectronics industry manufacturing processes provide a high volume production of wafers with a high density of devices per wafer. A secondary benefit is that during the development stage many different designs can be simultaneously fabricated on the same wafer and then tested, thus permitting rapid design cycles and design iterations.

Lower reagent costs per microchip device – Many microchip devices have total volumes of $< 1 \,\mu L$; thus, the per test reagent consumption is greatly reduced. This decreased reagent consumption can lead to decreased overall cost per test.

Portability – The small size and hence portability of microchip devices renders them suitable for use in non-laboratory settings, such as point-of-care testing for medical, environmental or biowarfare detection applications.

Speed of operation – Shorter diffusion distances in miniaturized devices lead to faster reactions and hence faster analytical response times.

Multiplex analysis – Miniaturized arrays of different reagents on planar surfaces permits multiple simultaneous analyses of a given sample.

Reliability – On or within a microchip analyzer it is possible to fabricate multiple test sites for simultaneous multiplicate assays, and this built-in redundancy provides an analytical safeguard not easily achieved in a conventional singlet assay protocol on macroscale analyzers. In addition, encapsulated microchip analyzers may provide extended operation over a wider range of environmental conditions of humidity and temperature.

Sample size – Sample volumes for microchip analyzers are significantly less than that used by most conventional analyzers, and this is advantageous in many applications (*e.g.*, finger stick *versus* venipuncture in a clinical setting).

Disposal and safety – Reduced sample size minimizes exposure of operators to potentially hazardous samples, and the low capacity of devices minimizes the volume of waste fluids. Entombment of the contents of a microchip analyzer (unreacted sample and reagents and reaction mixtures) provides for safer disposal.

Quality control – By placing all of the aspects of analysis on a single platform, the consistency of testing will be improved. Indeed, interobserver and intraobserver variables should be decreased as the variables of sample handling and analytical manipulations by a user are decreased.

16.5   Applications of Microchip Analytical Devices

The overall scope of analytical applications for the different types of microchips is extensive and the reader is referred to the numerous reviews of such applications.[21,22] Detection of a fluorescence signal has been one the most popular detection modes for microchip devices. Additional detection methodologies that have been successfully utilized in microchip devices include the measurement of color or electrochemical signals. The application of chemiluminescence, bioluminescence and electrochemiluminescence has been less extensive and the following sections outline the scope and current status of these endpoints in microchip research.

16.5.1   Chemiluminescence

A range of chemiluminescent assays has been adapted to a microchip format, including antioxidant assays,[23] drug assays (*e.g.*, amphetamine, atrazine),[24,25] heme proteins,[26] glucose and ethanol from yeast cells immobilized in a microchip,[27,28] and inorganic anions such as sulfite (linear range $1.0–60\,\mu g\,mL^{-1}$, detection limit $0.5\,\mu g\,mL^{-1}$),[29] arsenate,[30] antibiotics[31] and nitrite (linear range $8–100\,\mu g\,L^{-1}$; detection limit $4\,\mu g\,L^{-1}$).[32] Various chemiluminescent analytical procedures have also been adapted to a microchip format, including nucleic acid hybridization,[7] RT-PCR,[33] small molecule–protein interaction studies,[5] and enzyme immunoassay (*e.g.*, atrazine).[34,35]

An early application of chemiluminescent detection of a reaction in a microchip was a glucose and lactate analyzer.[6] This analyzer was fabricated on a silicon chip capped with Pyrex glass ($15\,mm \times 20\,mm$; total internal volume of $15\,\mu L$). Inside the microchip, a microchamber was filled with 100-μm diameter enzyme immobilized beads that were reacted with analyte to generate hydrogen peroxide. The peroxide formed was transported to a mixing chamber where it was combined with a chemiluminescent reagent added *via* an inlet port and then pumped into a spiral flow cell. The light emission was detected using a photodiode. Another study investigated chemiluminescent light emission from alkaline phosphatase and horseradish peroxidase catalyzed reactions, and from the peroxyoxalate reaction contained in straight channels ($300\,\mu m$ wide $\times 20\,\mu m$

deep; volume 70.2 nL) and open chambers (812 μm wide, 400 μm deep, 5.2 mm long) linked by channels (100 μm wide, 20 μm deep). Light emission from the channels or chambers was detected using either a specially modified microplate holder with a microplate luminometer (photomultiplier tube detector) or a CCD camera.[36]

16.5.2 Bioluminescence

A significant development in bioluminescence in the context of microchips has been the "CANARY on a chip" concept. Harkening back to the days in which the viability of canary birds was used in coal mines to detect poisonous fumes, the "CANARY on a chip" concept is the utilization of living organisms on microchip sensors for the detection of pathogens.

The originally reported CANARY (cellular analysis and notification of antigen risks and yields) assay,[37] utilized B-lymphocytes as light emitting sensors that were activated in the presence of bacteria or other pathogens. The B-lymphocytes have two basic components: immunoglobulins expressed at the surface that are specific for a given bacteria or pathogen; calcium sensitive aequorin expressed in the cytoplasm (Figure 16.1). Aequorin is a bioluminescent protein that was cloned from the jellyfish *Aequorea victoria*. The presence of a specific bacteria or pathogen crosslinks the cell surface immunoglobulins, resulting in an increase of cytosolic calcium; the increase in calcium activates aequorin and results in light emission. The initial studies of these living sensors showed a sensitivity of detection comparable to polymerase chain reaction (PCR) accomplished in a fraction of the assay time. The limit of detection of the pathogen *Escherichia coli* strain O157:H7 was 500 colony forming units (CFU) per gram of inoculated lettuce, detected within 5 min; the report cited equivalent PCR studies that required 30 to 60 min to detect 10–10 000 CFU per gram.

The concept of living reporter systems has been applied to microchips. Bioluminescent reporter genes (*e.g.*, bacterial luciferase *lux* gene, firefly *luc* gene) have been used for the detection of chemical and biological agents.[38] In a sense, living bacteria are integrated into the analytical microchip. The bacteria act as analyte sensors and report the presence of analyte as a luminescent signal.

The combination of a microchip and bioluminescence has also been used to develop a marine bacterial *lux* bioluminescent bioreporter system for the detection and monitoring of pathogenic microbial species. Specific bioluminescent bioreporters were engineered to produce light in response to specific environmental inducers and then the light signal was measured with a 2.2 mm × 2.2 mm integrated circuit microluminometer (BBIC – bioluminescent bioreporter integrated circuit) to generate a quantitative assessment of inducer concentration. One motivation for this work was that the size and portability of BBIC biosensors could provide a deployable, interactive network sensing technology that could be used to monitor food, air and water for chemical and biological warfare agents.

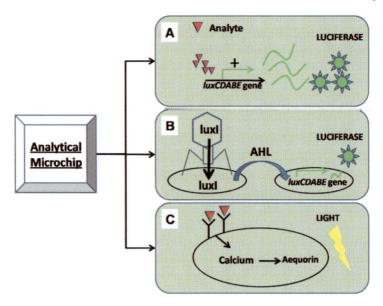

Figure 16.1   Three different bioluminescent systems. (A) Analyte activates a promoter
in a bioreporter cell that causes expression of a marine bacterial luciferase from a downstream *luxCDABE* gene. (B) Phage containing the
luxI gene specifically binds to a pathogen and the *luxI* gene is incorporated into the pathogen's genome where it produces an (AHL) acyl-
homoserine lactone autoinducer. This diffuses into bioreporter cells
where it interacts with the *luxR* gene and triggers the production of
marine bacterial luciferase by the *luxCDABE* gene. (C) Bioreporter cells
based on B lymphocytes that have surface expression of antibodies that
recognize a specific bacterial or viral agent and also have cytosolic
expression of the bioluminescent photoprotein aequorin. Binding of the
bacterial or viral agent to the surface antibody generates a calcium flux
that triggers light emission from the cytosolic aequorin.

In an initial feasibility study, bacteriophage carrying the portion of the *lux*
luciferase operon encoding autoinducer-N-acyl homoserine lactone was combined with pathogen bacteria in the presence of *Vibrio fisheri* as bioreporter
cells.[38] In theory, when pathogenic bacteria is present it becomes infected by
bacteriophage, which utilizes the pathogenic bacteria to produce autoinducer
N-acyl homoserine lactone. The autoinducer N-acyl homoserine lactone diffuses into the *Vibrio fisheri* bioreporter cells, resulting in the expression of high
levels of luciferase native to the *Vibrio fisheri* and consequent emission of light.

A microchip consisting of a three-dimensional microfluidic network constructed from a silicon perforated micro-well array chip and two poly(-
dimethylsiloxane) (PDMS) multi-microchannel chips has been developed for a
genotoxic bioassay using sensing *Escherichia coli* strains incorporating a firefly
luciferase *luc* reporter gene.[39] Mitomycin C was used as a representative
example of a genotoxic substance. The sensing strains were loaded into the

channels on one of the PDMS chips and immobilized into the silicon micro-wells. Next, samples containing genotoxic substances and substrates for the bioluminescent firefly luciferase reaction were added into the channels on the second PDMS chip. Light emission was detected using a CCD camera. Detection limits for the genotoxic agent (limit of detection for mitomycin: $0.2\,ng\,mL^{-1}$) comparable to those in the conventional method were obtained but with the advantage of a reduced analysis time. The concept of BBIC can be applied to a diverse range of analytes that have already been characterized in a non-microchip format.[40]

16.5.3 Electrochemiluminescence

One of the earliest applications of electrochemiluminescence was the use of a ruthenium chelate in the context of a microchamber designed for PCR analysis.[41] Electrochemiluminescence assays based on the tris(2,2'-bipyr-idyl)ruthenium(II) system have found several applications with microchip capillary electrophoresis, including dopamine, anisodamine, ofloxacin and lidocaine. In these assay methods tris(2,2'-bipyridyl)ruthenium(II), [Ru(bpy)$_3^{2+}$], is employed as an ECL reagent as well as a catalyst to generate tris(2,2'-bipyridyl)ruthenium(III), which then reacts with different analytes to produce an ECL emission.[42]

The combination of microchip capillary electrophoresis and electro-chemiluminescence detection has also been used for the rapid analysis of pro-line (limit of detection : $2\,\mu M$, linear range 25–$1000\,\mu M$) in a microchip device with PDMS electrophoresis microchannels integrated with a solid-state elec-trochemiluminescence detector,[43] and for assay of the antibiotic lincomycin (linear range 5–$100\,\mu M$, limit of detection in urine $9.0\,\mu M$).[44]

16.6 µTAS Chips

A total micro total analysis system (µTAS) has been fabricated that integrates three processes on a microchip: a competitive immunoassay (isoluminol label), electrophoresis and chemiluminescence detection for detection of human serum albumin or IAP (immunosuppressive acidic protein, a tumor marker). This µTAS chip contained two intersecting microchannels. The competitive immu-noassay was performed on antibody-immobilized glass bead. Electrophoresis was then used to move the immunological reaction mixture into the intersection of the two channels to form a sample plug. The sample plug was moved into a reservoir at the end of one of the channels that contained hydrogen peroxide. The resulting chemiluminescence reaction (microperoxidase-catalyzed oxida-tion of the isoluminol label) was then detected using a photomultiplier tube located under the reservoir.[45]

Integration of a detector with a microchip has been achieved using a planar silicon photodiode. The 10 mm wide by 20 mm long glass-capped silicon chip contained a fluidic network. Located on the backside of the chip was an array

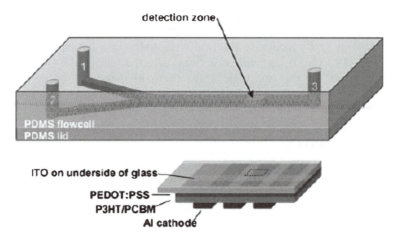

Figure 16.2   Microchip fabrication based on soft lithography. PDMS microchip
layout and photodiode for detection of chemiluminescence. "1" and "2"
are inlets for reagents and they lead to a common mixing channel
(800 μm wide and 5.2 cm long) that leads to an outlet "3". (Reproduced
with permission from ref. 47.)

of eight photodiodes. This chip was used to perform a flow injection assay for
hydrogen peroxide based on chemiluminescent oxidation of a luminol
reagent.[46] A detector integration strategy has also been used in the context of
arrays of oligonucleotides. The oligonucleotides were either immobilized
directly onto the face of the CCD, or onto a silicon chip that was then pressed
into contact with the CCD for measurement purposes.[14]

Thin-film photodiode detectors can also be integrated onto chips for detection purposes. In a proof-of-principle study an organic photodiode, consisting
of bilayers of copper phthalocyanine and fullerene, were deposited onto the
surface of a PDMS chip. This integrated device was used to detect hydrogen
peroxide in a chemiluminescent peroxylate reaction (Figure 16.2).[47]

16.7   Microarray Chips

A microarray is an ordered array of micron-sized test areas on a planar surface
of a small mm- to cm-sized chip. The test areas are created by spotting the
reagents or by *in situ* synthesis of the reagents (*e.g.*, oligonucleotides). Currently, fluorescence is the dominant detection technology used with microarrays. However, chemiluminescence has been effectively used as a detection
technique in combination with light imaging in a range of microarray applications for both protein[48–52] and nucleic acid[53,54] detection.

The microarray formats combined with chemiluminescence detection include
planar arrays,[55] arrays in flow-through microchips[7,56] and screen-printed
electrode networks.[57] Imaging of signals from the microarrays has been

achieved by utilizing a CCD detector.[58] Electrochemiluminescence has also been used as a means of quantitating assays performed on microarrays.[59] Particularly noteworthy has been the development of clinical analyzers for performing automated enzyme immunoassays (peroxidase label) based on arrays deposited onto microchips and detected using the luminol-based enhanced chemiluminescent reaction.[60] These analyzers (Evidence, Evidence Investigator and the Evidence MultiStat; Randox Laboratories, Inc.; http://www.randox.com) utilize Biochip Array Technology for simultaneous multiple analyte testing on a biochip that consists of a panel of 23 tests and two internal quality controls. Panels have been developed that include a wide range of analytes, including drugs of abuse, cytokines, cardiac tests and tumor markers.

Arrays of parallel channels in microchips can also be used for multiplex testing.[10] An advantage of the multiplicity of reaction channels in this study is that calibration and measurement are possible in a single experiment. A microchip design with multiple reaction channels has been utilized for the chemiluminescent detection of *Escherichia coli* O157:H7, *Salmonella typhimurium* and *Legionella pneumophila* using the appropriate capture antibodies immobilized in the channels and peroxidase labeled conjugates (luminol detection reagent). Signal was measured by a sensitive CCD camera, and detection limits in an 18-min assay were *E. coli* O157:H7 – 1.8×10^4 cells mL^{-1}, *S. typhimurium* – 2.0×10^7 cells mL^{-1} and *L. pneumophila* – 7.9×10^4 cells mL^{-1}.

16.8 Flow Injection Analysis (FIA) on Microchips

FIA is a popular analytical technique and the fluidic structures that are used to mix sample and reagents have been miniaturized into a microchip format.[61,62] For example, uric acid has been measured in human serum using a FIA system made from two transparent PMMA chips ($50 \times 40 \times 5$ mm) with CO_2 laser-etched microchannels (200 µm wide and 100 µm deep). The linear range of this assay for uric acid was 0.8–30 mg L^{-1} and the detection limit was 0.5 mg L^{-1}.[8]

16.9 Capillary Electrophoretic Analysis (CE) on Microchips

Microchip CE represents an early success for analytical microchips and the commercially available system uses fluorescence detection.[63] A basic CE microchip consists of two intersecting microchannels that terminate in reservoirs. Sample is added to one reservoir and is moved into the channel. The plug of sample at the intersection of the channels is then analyzed in the second channel (the separation channel). Application of an electric-potential difference across the separation channel causes electrophoretic separation of the components of the sample based on mass or charge differences, and these are detected near the outlet end of the second channel, usually by laser-induced fluorescence. Other luminescence-based detection methods can be employed as alternative to fluorescence detection.

Various groups have investigated combining capillary electrophoresis with either CL detection[42,45,64–66] or electrochemiluminescence detection.[43,44] It is also possible to conduct an immunoassay in a CE chip using chemilumines- cence detection.[67] Antibody immobilized on glass beads is located in one of the reservoirs of a CE chip and a competitive immunoassay is performed in the reservoir using analyte and fluorescein labeled analyte. The unbound fraction is then moved to the intersection of the two channels to form a sample plug. This was then moved to another reservoir and the free fluorescein labeled analyte reacted with bis[4-nitro-2-(3,6,9-trioxadecloxycarbonyl)phenyl] oxalate (TDPO) and hydrogen peroxide. Light emission from the TDPO reaction in the reservoir was measured using a photomultiplier tube. Assays for human serum albumin and immunosuppressive acidic protein both had detection limits (S/N = 3) of 1×10^{-7} M.

16.10   Point-of-care Applications for Microchips

Point-of-care or decentralized testing is a type of clinical testing that is per- formed in a clinic, or at the patient's bedside or even in the home. Devices for this type of testing need to be small and portable. Thus the small size of microchips leads to an obvious application in point-of-care testing. A multi- plexed test for myoglobin, cardiac troponin I, C-reactive protein and brain natriuretic peptide based on monoclonal antibodies immobilized onto a screen- printed electrode microarray has been developed based on CL imaging detec- tion of a peroxidase label. The device provided a 25-min assay for the four proteins in either buffer or diluted serum (dynamic ranges: 0.5–50, 0.1–120, 0.2–20 and 0.67–67 μg L^{-1} for C-reactive protein, myoglobin, cardiac troponin I and brain natriuretic peptide, respectively).[68] In these types of applications a significant enhancement of the on-chip chemiluminescence signal can be rea- lized using 0.2-mm^2 printed carbon electrodes onto which gold nanoparticles are electrodeposited. This has been demonstrated in the context of biomarkers for detecting prostate cancer. The gold nanoparticle surfaces of the electrodes were functionalized with, for example, prostate-specific antigen. The device was used to detect biotinylated monoclonal antibodies raised against prostate- specific antigen. Bound biotinylated antibody was detected using streptavidin labeled with a horseradish peroxidase. A burst of chemiluminescence was obtained from the oxidation of luminol catalyzed by the bound streptavidin peroxidase conjugate. Light intensity from the gold-modified screen-printed electrodes was enhanced by a factor of 229 (gold *versus* non-gold electrodes). This assay was applied successfully to assay p53 (assay range: 0.1–10 nM) and free prostate-specific antigen (detection limit: 5 ng mL^{-1}).[69]

A lab-on-a chip type device has been developed for point-of-care diagnosis of infection by HIV (human immunodeficiency virus). The one-hour RT-PCR assay for p24 and gp120 as markers of HIV was performed in a chip fabricated from cyclic olefin polymer (COP) with embedded silicone tubes as part of a series of pinch valves to control fluid flow. The chip architecture included

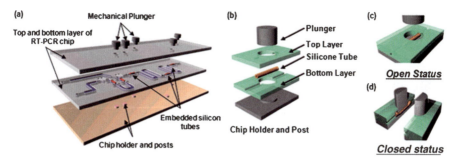

Figure 16.3 Polymer-based lab-on-a-chip for reverse transcription PCR (RT-PCR) and chip holder. (a) Schematic of a RT-PCR lab on a chip holder and mechanical plunger; (b) enlarged exploded view of embedded pinch valve that lies under the mechanical plungers shown on the top chip; and assembled valve in open (c) and closed (d) status. The two channels on the bottom layer of the chip shown in (a) lead to a room-temperature incubation chamber and then subsequently to a PCR chamber and then finally to a detection chamber and outlet. (Reproduced with permission from ref. 33.)

incubation, mixing, PCR and detection chambers (Figure 16.3).[33] Captured biotinylated probes were detected using streptavidinylated peroxidase and an enhanced luminol-based chemiluminescent assay for the bound peroxidase label.

Another area that requires decentralized testing is testing and monitoring for biological warfare agents. A chemiluminescent microchip immunoassay has been validated for *Bacillus globigii* spores as a surrogate for *Bacillus anthracis*. The assay was formatted into chambers placed in contact with an intensifier (from a night vision scope) and an integrated circuit photosensing array for light detection.[70] In this enzyme-linked immunosorbent assay, the alkaline phosphatase label was detected using a 1,2-dioxetane substrate and was effective in detecting the positive control that contained 5×10^5 *Bacillus globigii* spores.

16.11 Conclusions

The scope of application of chemiluminescence, bioluminescence and electro-chemiluminescence as detection technologies for microchip-based assays is now extensive. Over the past decade many new devices have been commercialized for medical and public health applications. Clearly, miniaturized analytical technology utilizing luminescent detection has matured into devices that not only replicate analytical processes that occur at the macro level but also demonstrate novel applications not available in the non-miniaturized format. A limiting factor to further integration and adoption of luminescent detection in miniaturized formats remains the discovery of commercially marketable applications.

References

1. A. Manz, N. Graber and H. M. Widmer, *Sens. Actuators B: Chem.*, 1990, **1**, 244.
2. K. E. Petersen, *Proc. IEEE*, 1982, **70**, 420.
3. S. Shoji and M. Esashi, *Bonding and Assembling Methods for Realizing a μTAS*, Kluwer Academic Publishers, Dordrecht, Boston, 1995.
4. M. Dufva, *DNA Microarrays for Biomedical Research: Methods and Protocols*, Springer, New York, 2009.
5. L. G. Hu, S. Y. Xu, C. S. Pan, H. F. Zou and G. B. Jiang, *Rapid Commun. Mass Spectrom.*, 2007, **21**, 1277.
6. M. Suda, T. Sakuhara and I. Karube, *Appl. Biochem. Biotechnol.*, 1993, **41**, 3.
7. B. J. Cheek, A. B. Steel, M. P. Torres, Y. Y. Yu and H. J. Yang, *Anal. Chem.*, 2001, **73**, 5777.
8. D. Y. He, Z. J. Zhang, Y. Huang, Y. F. Hu, H. J. Zhou and D. L. Chen, *Luminescence*, 2005, **20**, 271.
9. W. Liu, Z. J. Zhang and L. Yang, *Food Chem.*, 2006, **95**, 693.
10. X. Y. Karsunke, R. Niessner and M. Seidel, *Anal. Bioanal. Chem.*, 2009.
11. L. J. Kricka, O. Nozaki and P. Wilding, *J. Int. Fed. Clin. Chem.*, 1994, **6**, 52.
12. B. D. Iverson and S. V. Garimella, *Microfluidics Nanofluidics*, 2008, **5**, 145.
13. Q. F. Xue, Y. M. Dunayevskiy, F. Foret and B. L. Karger, *Rapid Commun. Mass Spectrom.*, 1997, **11**, 1253.
14. M. Eggers, M. Hogan, R. K. Reich, J. Lamture, D. Ehrlich, M. Hollis, B. Kosicki, T. Powdrill, K. Beattie, S. Smith, R. Varma, R. Gangadharan, A. Mallik, B. Burke and D. Wallace, *BioTechniques*, 1994, **17**, 516.
15. N. Chiem and D. J. Harrison, *Anal. Chem.*, 1997, **69**, 373.
16. L. B. Koutny, D. Schmalzing, T. A. Taylor and M. Fuchs, *Anal. Chem.*, 1996, **68**, 18.
17. L. J. Kricka and P. Wilding, *Anal. Bioanal. Chem.*, 2003, **377**, 820.
18. C. S. Zhang, J. L. Xu, W. L. Ma and W. L. Zheng, *Biotechnol. Adv.*, 2006, **24**, 243.
19. J. B. Angell, S. C. Terry and P. W. Barth, *Sci. Am.*, 1983, **248**, 44.
20. A. Manz, D. J. Harrison, E. M. J. Verpoorte, J. C. Fettinger, H. Ludi and H. M. Widmer, *Chimia*, 1991, **45**, 103.
21. K. Ohno, K. Tachikawa and A. Manz, *Electrophoresis*, 2008, **29**, 4443.
22. K. Sato, K. Mawatari and T. Kitamori, *Lab Chip*, 2008, **8**, 1992.
23. M. Amatatongchai, O. Hofmann, D. Nacapricha, O. Chailapakul and A. J. Demello, *Anal. Bioanal. Chem.*, 2007, **387**, 277.
24. H. R. Mobini Far, F. Torabi, B. Danielsson and M. Khayyami, *J. Anal. Toxicol.*, 2005, **29**, 790.
25. J. Yakovleva, R. Davidsson, M. Bengtsson, T. Laurell and J. Emneus, *Biosens. Bioelectron.*, 2003, **19**, 21.

26. X. Y. Huang and J. C. Ren, *Electrophoresis*, 2005, **26**, 3595.
27. R. Davidsson, F. Genin, M. Bengtsson, T. Laurell and J. Emneus, *Lab Chip*, 2004, **4**, 481.
28. R. Davidsson, B. Johansson, V. Passoth, M. Bengtsson, T. Laurell and J. Emneus, *Lab Chip*, 2004, **4**, 488.
29. D. Y. He, Z. J. Zhang and Y. Huang, *Anal. Lett.*, 2005, **38**, 563.
30. W. Som-Aum, H. Li, J. J. Liu and J. M. Lin, *Analyst*, 2008, **133**, 1169.
31. K. Kloth, M. Rye-Johnsen, A. Didier, R. Dietrich, E. Martlbauer, R. Niessner and M. Seidel, *Analyst*, 2009, **134**, 1433.
32. D. Y. He, Z. J. Zhang, Y. Huang and Y. F. Hu, *Food Chem.*, 2007, **101**, 667.
33. S. H. Lee, S. W. Kim, J. Y. Kang and C. H. Ahn, *Lab Chip*, 2008, **8**, 2121.
34. I. H. Cho, E. H. Paek, Y. K. Kim, J. H. Kim and S. H. Paek, *Anal. Chim. Acta*, 2009, **632**, 247.
35. J. Yakovleva, R. Davidsson, A. Lobanova, M. Bengtsson, S. Eremin, T. Laurell and J. Emneus, *Anal. Chem.*, 2002, **74**, 2994.
36. L. J. Kricka, X. Y. Ji, O. Nozaki and P. Wilding, *J. Biolumin. Chemilumin.*, 1994, **9**, 135.
37. T. H. Rider, M. S. Petrovick, F. E. Nargi, J. D. Harper, E. D. Schwoebel, R. H. Mathews, D. J. Blanchard, L. T. Bortolin, A. M. Young, J. Z. Chen and M. A. Hollis, *Science*, 2003, **301**, 213.
38. S. Ripp, J. Young, A. Ozen, P. Jegier, C. Johnson, K. Daumer, J. Garland and G. Sayler, *Phage-amplified Bioluminescent Bioreporters for the Detection of Foodborne Pathogens*, Proc. SPIE vol. 5329, SPIE, Bellingham, WA, 2004.
39. K. Maehana, H. Tani and T. Kamidate, *Anal. Chim. Acta*, 2006, **560**, 24.
40. D. E. Nivens, T. E. McKnight, S. A. Moser, S. J. Osbourn, M. L. Simpson and G. S. Sayler, *J. Appl. Microbiol.*, 2004, **96**, 33.
41. M. A. Northrup, C. Gonzales, S. Lehew and R. Hills, *Development of a PCR Microreactor*, Kluwer Academic Publishers, Dordrecht, Boston, 1995.
42. H. B. Qiu, X. B. Yin, J. L. Yan, X. C. Zhao, X. R. Yang and E. K. Wang, *Electrophoresis*, 2005, **26**, 687.
43. Y. Du, H. Wei, J. Z. Kang, J. L. Yan, X. B. Yin, X. R. Yang and E. K. Wang, *Anal. Chem.*, 2005, **77**, 7993.
44. X. C. Zhao, T. Y. You, H. B. Qiu, J. L. Yan, X. R. Yang and E. K. Wang, *J. Chromatogr. B*, 2004, **810**, 137.
45. K. Tsukagoshi, N. Jinno and R. Nakajima, *Anal. Chem.*, 2005, **77**, 1684.
46. A. M. Jorgensen, K. B. Mogensen, J. P. Kutter and O. Geschke, *Sens. Actuators B: Chem.*, 2003, **90**, 15.
47. X. H. Wang, O. Hofmann, R. Das, E. M. Barrett, A. J. Demello, J. C. Demello and D. D. C. Bradley, *Lab Chip*, 2007, **7**, 58.
48. B. I. Fall, B. Eberlein-Konig, H. Behrendt, R. Niessner, J. Ring and M. G. Weller, *Anal. Chem.*, 2003, **75**, 556.
49. K. A. Heyries, M. G. Loughran, D. Hoffmann, A. Homsy, L. J. Blum and C. A. Marquette, *Biosens. Bioelectron.*, 2008, **23**, 1812.

50. R. P. Huang, R. C. Huang, Y. Fan and Y. Lin, *Anal. Biochem.*, 2001, **294**, 55.
51. J. W. Pickering, J. D. Hoopes, M. C. Groll, H. K. Romero, D. Wall, H. Sant, M. E. Astill and H. R. Hill, *Am. J. Clin. Pathol.*, 2007, **128**, 23.
52. T. Urbanowska, S. Mangialaio, C. Zickler, S. Cheevapruk, P. Hasler, S. Regenass and F. Legay, *J. Immunol. Methods*, 2006, **316**, 1.
53. M. T. Beck, L. Holle and W. Y. Chen, *BioTechniques*, 2001, **31**, 782.
54. M. T. Coiras, M. R. Lopez-Huertas, G. Lopez-Campos, J. C. Aguilar and P. Perez-Brena, *J. Med. Virol.*, 2005, **76**, 256.
55. A. Roda, M. Guardigli, C. Russo, P. Pasini and M. Baraldini, *BioTechniques*, 2000, **28**, 492.
56. A. Wolter, R. Niessner and M. Seidel, *Anal. Chem.*, 2008, **80**, 5854.
57. C. A. Marquette, M. F. Lawrence and L. J. Blum, *Anal. Chem.*, 2006, **78**, 959.
58. S. J. Brignac, R. Gangadharan, M. McMahon, J. Denman, R. Gonzales, L. G. Mendoza and M. Eggers, *IEEE Eng. Med. Biol. Mag.*, 1999, **18**, 120.
59. E. G. Hvastkovs, M. So, S. Krishnan, B. Bajrami, M. Tarun, I. Jansson, J. B. Schenkman and J. F. Rusling, *Anal. Chem.*, 2007, **79**, 1897.
60. S. P. FitzGerald, J. V. Lamont, R. I. McConnell and E. O. Benchikh, *Clin. Chem.*, 2005, **51**, 1165.
61. T. Kamidate, T. Kaide, H. Tani, E. Makino and T. Shibata, *Anal. Sci.*, 2001, **17**, 951.
62. Z. J. Zhang, D. Y. He, W. Liu and Y. Lv, *Luminescence*, 2005, **20**, 377.
63. S. F. Y. Li and L. J. Kricka, *Clin. Chem.*, 2006, **52**, 37.
64. M. Hashimoto, K. Tsukagoshi, R. Nakajima, K. Kondo and A. Arai, *J. Chromatogr. A*, 2000, **867**, 271.
65. S. D. Mangru and D. J. Harrison, *Electrophoresis*, 1998, **19**, 2301.
66. T. Nogami, M. Hashimoto and K. Tsukagoshi, *J. Sep. Sci.*, 2009, **32**, 408.
67. K. Tsukagoshi, K. Tsuge and R. Nakajima, *Anal. Sci.*, 2007, **23**, 739.
68. C. A. Marquette, F. Bouteille, B. P. Corgier, A. Degiuli and L. J. Blum, *Anal. Bioanal. Chem.*, 2009, **393**, 1191.
69. B. P. Corgier, F. Li, L. J. Blum and C. A. Marquette, *Langmuir*, 2007, **23**, 8619.
70. D. N. Stratis-Cullum, G. D. Griffin, J. Mobley and T. Vo-Dinh, *Anal. Bioanal. Chem.*, 2008, **391**, 1655.

CHAPTER 17

Recent Analytical Application Areas of Chemiluminescence and Bioluminescence

MASSIMO GUARDIGLI, MARA MIRASOLI, ELISA MICHELINI, LUISA STELLA DOLCI AND ALDO RODA

Department of Pharmaceutical Sciences, University of Bologna, Via Belmeloro, 6, 40126 Bologna, Italy

17.1 Introduction

Chemiluminescence (CL) and bioluminescence (BL) have found application in each sector of analytical chemistry. In the other chapters many of these applications have been reviewed, with particular emphasis on the uses of CL and BL techniques in life sciences and related fields. This chapter deals with CL and BL applications not described in previous chapters, in particular as concerns forensic sciences (*i.e.*, the well-known luminol test for bloodstain detection), detection of explosives (which nowadays is a topic of particular relevance and requires rapid, sensitive and selective analytical techniques) and the study and conservation of cultural heritage.

17.2 Forensic Sciences

Forensic sciences have taken advantage of many recent developments in CL and BL, *e.g.*, the new sensitive detection systems for separative analytical techniques. However, the best-known application of CL in forensics is undoubtedly the luminol test for bloodstain detection.

Chemiluminescence and Bioluminescence: Past, Present and Future
Edited by Aldo Roda
© Royal Society of Chemistry 2011
Published by the Royal Society of Chemistry, www.rsc.org

17.2.1 Luminol Test for Bloodstain Detection

This test allows detection of faint and/or hidden blood traces with high sensitivity and has been utilized by forensic scientists in investigations involving violent crime for more than 40 years. Even though it should be considered a presumptive test for blood (*i.e.*, further confirmation by other analytical techniques is usually required) it is nevertheless of substantial importance in forensics. The principles of this test and its use and limitations have been reviewed recently.[1]

Briefly, the luminol test relies on the CL oxidation reaction of luminol (5-amino-2,3-dihydro-1,4-phthalazine-dione) that takes place when an alkaline solution containing luminol and an oxidizing agent (*e.g.*, hydrogen peroxide) is sprayed on dried bloodstains. Bloodstains contain several degradation products of haemoglobin, in particular hematin (the heme prosthetic group of haemoglobin in which the Fe^{2+} ion has been replaced by a Fe^{3+} ion coordinated by an hydroxyl group). When in contact with a luminol formulation, hematin catalyses both the decomposition of hydrogen peroxide and the oxidation of luminol, leading to light emission (Figure 17.1).[2,3] Several luminol preparations have been proposed over the years in an attempt to improve sensitivity, specificity and duration of light emission. However, the most used forensic luminol formulations are still those described in the 1950s and 1960s by Grodsky *et al.*[4] and Weber,[5] based on luminol/sodium perborate in sodium carbonate and luminol/hydrogen peroxide in sodium hydroxide, respectively. Indeed, these formulations combine good performance with easy preparation, low cost and ready availability of the reagents. In forensic applications, luminol solutions are usually directly sprayed on the surfaces to be examined in a dark environment. The light obtained can be photographed or filmed or the luminescent areas can be marked to allow their identification once the light emission disappeared (use of high-sensitivity intensified cameras for recording light emission has been proposed, but the light emission is usually quite intense and these devices are not of general use at crime scenes). In addition to detecting blood traces, the test allows the study of bloodstain pattern, which could help investigators to reconstruct the events of a crime.

17.2.2 Factors Affecting the Luminol Test

The first factor affecting the luminol test is the physical nature of the surface.[6] Materials with irregular porous surfaces (wood-finish panelling, walls, interstitial spaces between tiles, cracks on wood objects, *etc.*) or with high absorbing properties (leather, fabric, *etc.*) retain significant amounts of blood. Moreover, they protect blood from chemical and physical degrading agents (solar rays, moisture) and even from attempts at cleaning up after the crime has been committed. Therefore, these surfaces usually give strong signals with the luminol test. In contrast, non-absorbent smooth surfaces (linoleum, glass, metal) retain very small amounts of blood and can be easily cleaned, leading to weak or absent reaction with luminol.

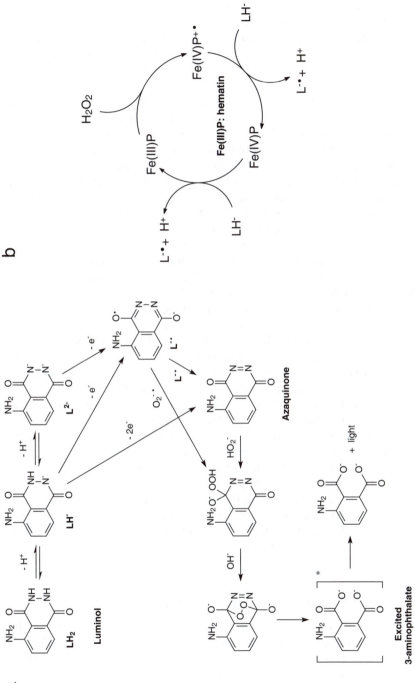

Figure 17.1 (a) Currently most accepted mechanism of the luminol CL reaction in the presence of oxidizing agents. (b) Mechanism of luminol oxidation by hydrogen peroxide catalysed by hematin.

A wide range of environmental, domestic and industrial substances can affect the blood-induced luminol CL. For example, ligands with a high affinity for a specific oxidation state of iron (*e.g.*, sulfide and cyanide) or antioxidant compounds are able to suppress the luminol CL emission.[7] Nevertheless, most of these substances are unlikely to come into contact with blood, so that false-negative results are uncommon in the forensic literature. The most problematic chemicals for a correct interpretation of luminol test results are those producing intensification or generation of a CL signal even in the absence of blood, thus affording false-positive results. For example, minerals containing Fe^{2+} and Fe^{3+} ions such as hematite (Fe_2O_3), magnetite (Fe_2O_3), siderite ($FeCO_3$) and pyrite (FeS_2), as well as rust and other hydroxide and oxide forms present in humid or aquatic aerobic environments, are substantially insoluble but can catalyse the CL luminol reaction on their surfaces. Other metal ions (cobalt, copper, nickel, chromium and manganese) are also capable of producing CL when exposed to luminol solutions. In addition, many biological molecules, including redox active prosthetic groups containing ferric or ferrous ions, heme-containing proteins such as peroxidases, catalase and cytochromes, efficiently catalyze the luminol CL reaction (horseradish peroxidase is widely used in bioanalytical chemistry as an enzymatic label suitable for CL detection). In particular, peroxidases are abundant in fruit and vegetables, as well as in photosynthetic microorganisms, and in their presence the luminol test can give false positive indications.[6,8] Hypochlorite (OCl^-) is another chemical responsible for misinterpretation of the results of a luminol test, because it is capable of amplifying the CL emission resulting from the oxidation of luminol by hydrogen peroxide.[9] It is a common component of industrial and domestic bleaches; thus its presence on the crime scene may be either fortuitous or the result of an attempt to remove blood evidence. As a consequence, hypochlorite is one of the major chemicals responsible for false-positive results in the forensic literature. Despite the limitations described above, the luminol test is still a fundamental tool in forensic science. In two recent studies, the luminol test has been compared with other presumptive tests for blood commonly used in crime investigation.[10,11] Even though the reported absolute detectability of bloodstains were quite different, both studies concluded that the performance of the luminol test was comparable or better than those of the other tests. Moreover, they demonstrated that DNA can be recovered and amplified from bloodstains previously treated with luminol forensic reagents.

17.2.3 Improvement of the Luminol Test

To minimize interferences and increase the CL emission intensity, many different approaches have been investigated, including pretreatment of the substrate to be tested with suitable reactives and addition to the luminol reagent of chemical substances able to destroy interfering species.

In 1975, it was reported that pretreatment of experimental samples with strong bases (*e.g.*, NaOH) reduced interferences due to isolated Fe^{2+} ions, whereas the

CL signal from bloodstain was much less affected.[12] This effect was attributed to the dissociation of the hematin prosthetic group from the proteins, while free iron ions were converted into poorly soluble $Fe(OH)_3$. However, since the study primarily aimed at the quantitative determination of hematin compounds in environmental chemistry, this approach was never applied to the forensic field. In the following years, investigations on the chemical species interfering with the luminol reaction focused mainly on hypochlorite. Kent *et al.* reported that primary and secondary amines could inhibit the luminol CL due to hypochlorite in alkaline conditions, presumably due to the reaction of amines with hypochlorous acid to form chloramines.[13] The highest inhibition efficiency under conditions similar to those of common forensic luminol sprays (*i.e.*, at pH > 10) was observed with strongly basic amines such as 1,2-diaminoethane. Amines did not interfere with the haemoglobin-catalysed CL reaction of luminol and, thus, their use has been proposed in forensic applications. Notably, recent studies have shown that hypochloride rapidly decomposes upon exposition to air;[14,15] therefore, the interference effect by bleach with standard luminol reagents disappeared if the bloodstain was left to air for a period of 1–2 days before measurement. Recently, in 2006, a new luminol-based formulation (Bluestar® Forensic, by ROC Import Group) has been commercialized for forensic applications. According to the producer, this formulation should give stronger and longer emission than common luminol formulations. Indeed, a comparative study between Bluestar® Forensic and other luminol formulations demonstrated that in the presence of bloodstain Bluestar® Forensic produced an intense and long-lived CL emission that could be visualized even in the presence of some ambient light.[16] Moreover, this formulation was easy to prepare and quite stable (it could be used for several day after mixing of reagents).

17.2.4   Other Applications of the Luminol Test

The forensic luminol test has found application in other fields, such as for the control of patient-to-patient transmission of hepatitis C virus (HCV). As blood and blood products are routinely screened for the presence of such virus, attention has been focused on its nosocomial transmission pathways, especially in hemodialysis units in which contaminated dialysis machines and environmental contamination represent risk factors for the spread of HCV. The forensic luminol test proved suitable for identifying blood contamination in the hospital environment and, in perspective, for assessing the efficacy of cleaning and disinfection procedures.[17] Even though this mode of transmission of HCV has never been demonstrated, diffusion of the luminol test would reduce the risks related to contamination of the environment with blood.

17.3   Detection of Explosives

Detection of explosives (Table 17.1) and related compounds is important in both environmental and forensic applications, *i.e.*, for the study of

Table 17.1   Main classes of explosives and representative examples[a]

Structural class	Representative examples	Abbreviation
Nitroaliphatic	Nitromethane	NM
	2,3-Dimethyl-2,3-dinitrobutane	DMNB
Nitroaromatic	2,4,6-Trinitrotoluene	TNT
	2,4,6-Trinitrophenol	Picric acid
Nitramine	1,3,5-Trinitro-1,3,5-triazacycloexane	RDX
	1,3,5,7-Tetranitro-1,3,5,7-tetraazacyclooctane	HMX
Nitrate ester	Nitroglycerin	NG
	Pentaerithrytol tetranitrate	PETN
Peroxide	Triacetone triperoxide	TATP
	Hexamethylene triperoxide diamine	HMTD
Inorganic	Ammonium nitrate	AN

[a]Explosives for military and civilian use often have complex compositions; for example, the military explosive C4 contains RDX, polyisobutylene, di(2-ethylhexyl) sebacate and fuel oil.

environmental and toxicological effects of explosives, to detect land mines and to improve aviation security and prevent terrorist attacks. The analytical techniques employed for the detection of explosives can be roughly classified in two categories: (a) techniques for bulk detection, used to identify a macroscopic mass of explosive and based on the employment of X-ray scanners or similar equipment, and (b) techniques for trace detection, which allow the identification of microscopic residues of explosive compounds either in the form of vapour or particulate. The latter are the most demanding in terms of analytical performance, because the amount of explosive to detect is very low (explosive residues due to their manipulation, as well as in the environment and in post-blast debris, are very faint) and complex matrices, such as soils and solid surfaces, have to be analysed.

Among the various explosives, nitrated compounds, such as TNT, nitramines (*e.g.*, RDX), and nitrate esters (*e.g.*, PETN), are the most studied. It should be considered, however, that explosive compounds for military or civilian applications are often composite materials consisting of different explosives, fuels and other excipients. Therefore, analytical methods for the detection of such non-explosive compounds could also be of interest for specific applications. More recently, liquid and peroxide-based explosives (*e.g.*, TATP and HMTD) have become one of the major concerns in explosive detection. Even though these new explosives, mainly due to their low stability, are not suitable for military applications, they can be easily synthesized from readily available materials and have therefore been used in terrorist attacks. Unfortunately, since these materials are significantly different from the other classes of explosives, the best methods for their determination have yet to be identified.

A wide range of analytical methods have been used for detection of explosives, including – among others – chromatographic methods (gas chromatography, high-performance liquid chromatography and ion chromatography), capillary electrophoresis and X-ray systems. Thanks to their high sensitivity, luminescence-based detection methods, employing either fluorescence or chemiluminescence, offer many benefits over other common techniques, especially when combined with separative techniques for the analysis of complex matrices. Luminescence-based methods for explosives have been reviewed in several recent papers, such as those by Meaney and McGuffin[18] (mainly dealing with fluorescence-based methods) and Jimenez and Navas[19] (focused on chromatographic analytical techniques).

17.3.1 Analysers Based on the NO–Ozone CL Reaction

Most explosive compounds contain either nitro (NO_2) or nitrate (NO_3) groups and their vapours can be decomposed by high-temperature pyrolysis (up to 675–1000 K) in thermal energy analysers (TEA) to produce nitrogen oxide (NO). The NO radical can be then detected through its reaction with ozone, which produces electronically excited nitrogen dioxide (NO_2^*) that emits in the near-infrared.[20] Portable explosive analysers based on this technology are commercially available and have been used for the detection of nitrated compounds in various matrices (air samples, post-blast debris, solvent extracts of hand-swabs and biological samples) with detection limits of the order of ppb or below. On-field applications of TEA-based explosive detectors employing the NO–ozone CL reaction have been reviewed:[21] concerning the range of samples that can be analysed and the detection limits achievable, these detectors were clearly superior to those based on other NO detection principles, such as laser-induced fluorescence.

17.3.2 Analysers Based on the Luminol CL Reaction

Nitrated explosives can be also detected by CL by employing the well-known luminol CL reaction. In a method described by Nguyen *et al.*[22] the nitrite ions generated by pyrolysis of nitrated explosives were detected through the CL oxidation of luminol in the presence of sodium sulfite. This system has been used to identify several nitrated explosives, including NG, DMNB, TNT, RDX and PETN. Interestingly, a non-nitrated explosive (triacetone triperoxide, TATP) can be also detected because its pyrolysis produces oxidizing species able to trigger the luminol CL reaction. Most recently, a sensor for detection of nitrated explosives employing laser photofragmentation instead of thermal pyrolysis has been described (Figure 17.2).[23] In this device, the sample is heated to generate vapour explosive that is irradiated by a 193-nm ArF laser to produce NO and NO_2. Nitric oxide is oxidized to NO_2 in an oxidation reactor containing chromium trioxide (CrO_3), and then mixed with an aerosol of alkaline luminol solution. Finally, the luminescence emission resulting from the CL reaction between luminol and NO_2 gas is measured by a cooled photomultiplier tube. Thanks to the large interface area between luminol and NO_2 gas, development of the CL signal is very efficient, and detection limits of a few ppb were achieved for different nitrated explosives. The device also proved suitable for the direct analysis of trace explosives on surfaces and in soil

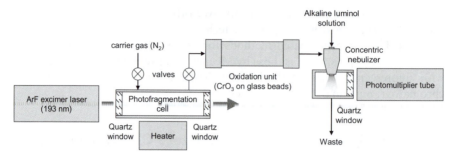

Figure 17.2   Scheme of the sensor for detection of nitrated explosives employing laser photofragmentation and luminol-based CL detection of NO$_2$. (Drawn according to the description reported in ref. 23.)

samples, with analytical performance comparable to other currently available analytical techniques.

17.3.3   Direct Detection of Degradation Products of Nitrated Explosives

A few CL methods have been reported for the direct detection of degradation products of nitrated explosives. An electrochemiluminescent (ECL) assay for the determination of 2,4- and 3,4-diaminotoluene (DAT) isomers has been developed by Bruno *et al.*[24] The assay is based on the formation of DAT complexes with Au$^+$ and Cu^{2+} followed by the measurement of their ECL emission. The apparent specificity of Au$^+$ for 2,4-DAT and Cu^{2+} for 3,4-DAT was, at least in part, attributed to the different size of the metal ions (Au$^+$ has nearly twice the ionic diameter of Cu^{2+} and, thus, it may form a coordination complex with the meta, but not the ortho, diaminotoluene). Sensitivities were in the ppm range, making this method potentially suitable for the determination of some aminoaromatics deriving from degradation of explosives (*e.g.*, TNT). In 2005, a flow CL method for aminonitroaromatics in environmental aqueous samples based on ion-pair formation and on-line liquid–liquid extraction was described.[25] The analytes were protonated at low pH, then extracted in dichloromethane as neutral ion-pairs with the tetrachloroaurate(III) ion. Subsequently, the tetrachloroaurate(III) ion was detected by reverse micelle-mediated CL (RMM-CL). Reverse micelle-mediated CL is an interesting approach for producing the CL signal within the organic phase: the solvent containing the ion-paired tetrachloroaurate(III) ion was mixed with a reversed micellar luminol reagent (luminol/cetyltrimethylammonium water micelles in dichloromethane–cyclohexane) to catalyse the luminol CL reaction into the organic medium. As with the previously reported method, the detection limits for different aminonitroaromatics were in the ppm range, which may limit the practical application of these methods for the detection of degradation products of nitrated explosives in environmental samples.

17.3.4 Detection of Explosives by Gas Chromatography

Since several explosives have a quite high vapour pressure, gas chromatography (GC) separative techniques have been widely used for the analysis of explosives in conjunction with TEA-based CL detection. Applications of GC-TEA for the detection of explosives have been extensively reviewed in ref. 19 and are not described in detail here. However, notably, despite the expected specificity of TEA detectors for nitro- and nitroso-compounds, in some cases unidentified peaks appear in the chromatogram and therefore confirmation by another method is required. This, in conjunction with the wide availability of mass spectrometric detectors suitable for confirmation analyses, may explain at least in part the limited number of GC-TEA applications reported in the literature in the recent years.

Crowson *et al.* have reported an alternative CL method for the detection of explosives that is potentially suitable for application in GC analyis.[26] Upon heating the vapours of a range of explosive compounds (*i.e.*, nitrate esters and nitramines such as simple alkyl nitrates, RDX, NG and PETN) at sub-atmospheric pressures they observed that the decomposition process originated a blue-green CL emission. In addition, the emission intensity was enhanced by using helium or nitrogen as a carrier gas to assist the transport of the vapour into the reaction vessel. According to the authors, this detection method could offer a better selectivity over current detection methods for explosive compounds of forensic interest and samples might be directly analysed or detected after gas chromatographic separation. Bowerbank *et al.* have described a solvating gas chromatography (SGC) method for the analysis of nitrogen-containing explosives employing a TEA detector.[27] Solvating gas chromatography is a chromatographic technique that can be considered a hybrid between gas and liquid chromatography.[28] It utilizes packed capillary columns and a carbon dioxide mobile phase and combines the solvating power of supercritical carbon dioxide with rapid analyte elution. Since the eluate of the SGC column is gaseous, this separative system can be easily coupled with TEA detectors. An additional benefit is that carbon dioxide does not give rise to any additional background noise in the detector; therefore, the TEA detection limits are not increased. This method has been applied for the detection of nitroaromatic explosives, such as TNT, NG and PETN, and of the TNT degradation products 2,4-dinitrotoluene and 2,6-dinitrotoluene. Detection limits in terms of absolute amounts of injected analytes were in the low picogram range, thus rendering this method suitable for the analysis of explosive residues in the environment.

17.3.5 Detection of Explosives by Liquid Chromatography

Thermal energy CL analysers have also been used in combination with HPLC, even though the applicability of this detection technique in HPLC is reduced since it is not compatible with aqueous mobile phases. In addition, the high temperatures required for pyrolysis of explosives increase the background noise and reduce sensitivity and selectivity. Owing to these disadvantages, only a few examples of HPLC-TEA for the analysis of explosives have been reported.

Lafleur *et al.* have described an HPLC-TEA method for the identification and determination of explosives and other related compounds possessing thermally labile nitro or nitroxy groups.[29] They reported analytical results for a wide range of different explosives (NG, PETN, RDX and others), with detectabilities in the ng range. In addition, they compared the retention times on two different columns (with silica or NH_2-bonded phase) to facilitate identification of the compounds. The authors reported that standard explosive preparations could be analysed without purification, because no interferences due to the presence of ancillary components such as plasticizers and stabilizers were observed. Several years later, Selavska *et al.* incorporated a post-column, on-line UV photolysis unit in an HPLC-TEA system to improve the detectability of mono-, di- and tri-nitrotoluenes.[30] Following an investigation of the mechanism of signal enhancement, the authors proposed that it involves a photochemically induced isomerization leading to homolytic cleavage of the C–NO_2 bond, followed by hydrogen abstraction. They obtained a significant increment in the TEA signals (more than one order of magnitude) and, most interestingly, were able to detect mononitrotoluene, which is not responsive under conventional TEA conditions. The detectability of other nitro-based high explosives was not influenced by the presence of the UV photolysis units.

A quite different detection technique has been used by Tsaplev for the HPLC analysis of nitrate esters such as NG and PETN.[31] Nitrate esters have low absorption in the spectral region above 210 nm and, therefore, conventional UV–Vis spectrophotometry is not suitable for these analytes. The detection method is based on the post-column CL reaction of a luminol analogue (4-dimethylaminophthalhydrazide) with the labile products (mainly peroxynitrite ions) of the alkaline hydrolysis of nitrate esters: the eluate from the HPLC column is mixed with a strongly alkaline solution of the luminol analogue and the resulting CL signal is recorded using a flow-through CL cell made of a coiled polytetrafluoroethylene tube and a photomultiplier tube. The detection limits (0.01 and 1 ng for NG and PETN, respectively) were several orders of magnitude lower than those attained in HPLC with spectrophotometric detection. Notably, PETN and NG can hardly be detected using gas analysers; therefore, this method is of particular relevance for the determination of these explosives.

17.3.6   Detection of Explosives by Immunoassays

Luminescence immunoassays have been widely studied for the detection of explosives, especially of nitrated ones such as TNT, RDX and PETN. The high detectability of luminescence signals and the specific recognition of the target analyte by the antibody allowed for the development of sensitive and rapid assays, suitable for the analysis of, for example, water samples and soil extracts at low cost and with minimal interference. However, most of the immunoassays for explosives described so far are based on fluorescence detection and only a few examples of CL-based assays have been reported in the literature.

Wilson *et al.* have developed competitive electrochemiluminescence enzyme immunoassays for TNT and PETN.[32] These assays rely on the competition between the analyte and a suitable hapten, which is covalently attached to dextran-coated paramagnetic beads, for the binding to a limited amount of anti-analyte antibody. The antibody bound to the beads is then revealed, upon addition of a second horseradish peroxidase (HRP)-labelled antispecies-specific antibody, by magnetically concentrating the beads on the working electrode of an electrochemiluminometer and by triggering the CL signal of HRP with electrochemically generated H_2O_2 in the presence of luminol and an enhancer. The reported limits of detection were 0.11 and 19.8 ppb for TNT and PETN, respectively. A portable flow-injection immunosensor for field analysis has also been described.[33] The most interesting features of this device were the use of an exchangeable single-use chip containing the analyte-specific antibody immobilized on a gold surface and the enzyme tracer and the sample reservoirs, as well as the possibility to operate on field for quite a long time (up to 6 h) without external power supply. The immunoassay format was a competitive one and the HRP-labelled tracer was detected using a luminol/enhancer CL substrate. As a proof of principle, the device was employed for the detection of explosives (TNT) and pesticides (diuron and atrazine), achieving a detection limit of $0.1 \,\mu g \, L^{-1}$ for TNT. More recently, another example of CL immunoassay for TNT (probably the most studied explosive due to both its extensive use in the twentieth century and its well-known toxicity) has been reported.[34] The assay was performed in 96-well microtiter plates as an indirect competitive format: the analyte contained in the sample competed with an immobilized hapten–protein conjugate for binding to a limited amount of anti-TNT monoclonal antibody. After incubation, the antibody bound to the hapten–protein conjugate was detected by CL using a HRP-labelled anti-species antibody and a luminol/enhancer CL HRP substrate. The limit of detection of the CL immunoassay varied between 0.2 and $0.6 \, ng \, mL^{-1}$ of TNT depending on the residual amount of methanol (used for the extraction of TNT from cotton swabs and tissue and plastic post-blast debris) in the solution analysed. Such detection limits were comparable with those reported for other ELISA formats based on colour reactions and spectrophotometric detection, but the possibility of detecting lower amounts of HRP by CL enabled reduction of the optimum antibody and conjugate concentrations. In 2009, an ultrasensitive immunoassay for TNT was described.[35] This heterogeneous sandwich-type immunoassay employs monoclonal anti-TNT antibodies immobilized on magnetic beads for capturing the analyte. Then, the captured TNT is detected using a second anti-TNT monoclonal antibody labelled with polystyrene beads heavily loaded with an ECL label [a hydrophobic ruthenium(II) tris(2,2'-bipyridine) complex]. The sandwich immunocomplex is separated from the reaction medium, then the polystyrene beads are dissolved with acetonitrile and the released ruthenium(II) complex is detected by ECL. Thanks to the extremely high amplification signal factor achievable with this technique (each polystyrene bead contains about 7 billion ruthenium(II) complex molecules), the limit of detection is of the order of 0.10 ppt, *i.e.*, much lower than those of other TNT detection methods reported in the literature. This immunoassay has been used for the measurement of TNT in contaminated soil and water samples, employing the standard addition method to eliminate the possible matrix effect.

17.4   Study and Conservation of Cultural Heritage

The study and conservation of cultural heritage is of fundamental importance for the entire international community. While in the past these studies were addressed in a quasi-empirical and often poorly documented manner, nowadays much attention from scientific community is aimed at the study and preservation of historical and artistic heritage. Almost all the current analytical techniques, from X-ray analysis to chromatographic techniques, have been used in this field. As artwork samples are generally very small and the compounds to be determined are usually in low concentration, sensitive and selective analytical techniques are often required. Nevertheless, only a few examples of application of CL and BL have been reported in the literature.

17.4.1   Detection of Microbial Contamination by BL Imaging

It is well known that biodeteriogen agents, such as bacteria, fungi, yeast, algae and lichens, can cause physical–chemical damage to cultural heritage.[36] Suitable analytical methods are thus required for early detection of the presence of biodeteriogen agents and to monitor the efficacy of biocide treatments aimed at removing such agents from the surface of artworks. The bioluminescent (BL) detection of adenosine triphosphate (ATP) through the firefly luciferin–luciferase reaction is a powerful technique for the measurement of ATP. Detection limits are of the order of attomoles of ATP (*i.e.*, the amount of ATP in a single bacterial cell), thus enabling direct enumeration of bacterial cells without cultivation (see Chapter 5). Indeed, ATP measurements are widely employed as a reliable, non-specific bioindicator of total microbial presence in several fields, such as surface hygiene monitoring and quality control in foods, pharmaceutics and wastewater treatments.

The ATP-dependent firefly luciferin–luciferase reaction has been exploited for the development of a BL low-light imaging technique for the evaluation of biodeteriogens' spatial distribution in artworks.[37] Optimization of experimental conditions taking into account sample geometry and surface properties (*e.g.*, porosity) was crucial to obtain accurate quantitative evaluation of ATP distribution on the artwork surface. When compared with a batch-type ATP assay performed by employing a photomultiplier-based luminometer the proposed BL low-light imaging technique was found to be less sensitive (ATP detection limits were about 1 and 10 pg, respectively). However, it presented significant advantages over batch-type ATP assays because it allowed a very accurate assessment of different contamination levels on the sample surface, rather than an overall measurement of total microorganisms contamination. Moreover, this technique could be used as a non-destructive rapid diagnostic tool for *in situ* applications employing a suitable low-light portable imaging device.

17.4.2 Detection and Localization of Proteins in Artworks by CL Imaging Microscopy

Throughout history, artists such as painters have experimented with a number of techniques and materials, but their origins and details are often unknown to modern restorers. Paintings have a complex and heterogeneous structure and correct identification and localization of their components is essential to obtain information on painting techniques, to evaluate the state of deterioration and to undertake conservation and restoration policies. The identification and localization of organic materials (pigment binders, adhesives and varnishes) is particularly challenging due to their low amount and susceptibility to deterioration. In addition, many techniques for protein detection (*e.g.*, chromatography) require complete destruction of the sample, thus losing the spatial information, or do not allow discrimination between different proteins (*e.g.*, infrared reflectance spectroscopy). The localization of proteins in a painting cross section by immunological techniques has been proposed since 1971.[38] By taking advantage of the high specificity of the antigen–antibody reaction, these techniques allow us to distinguish between different proteins (or the same protein from different species) and permit highly sensitive detection and localization of the target antigens in tissues and cells with submicrometer resolution.[39] Immunoassays are widely employed in bioanalytical and clinical chemistry but few data have been reported in the literature concerning its use for the localization and characterization of organic compounds in paintings.[38,40–43] In addition, all these applications were based on fluorescence detection and in many cases significant interferences due to sample autofluorescence (*e.g.*, from pigments and/or binding media) were observed. Since CL does not require an excitation source, it thus represents a promising alternative detection technique.

In 2008, an immunochemical method based on CL imaging microscopy was proposed for the localization of proteins in paintings.[44] In this method, the protein ovalbumin (chicken egg white albumin) was detected by employing a specific anti-ovalbumin primary antibody followed by a HRP-labelled secondary antibody and a CL enzyme substrate. Spatial resolution on the order of micrometres was achieved, thus allowing discrimination of the protein content among the different painting layers (Figure 17.3). Possible interferences from pigments or other materials commonly used in paintings (which could act either as catalysts of the CL reaction, thus leading to non-specific signals, or as inhibitors, causing false negative results) were investigated. Ovalbumin could be detected in cross-sections of samples taken from a Renaissance wood painting, demonstrating the suitability of the method for the study of ancient paintings. In perspective, multiplexed methods allowing the co-localization of

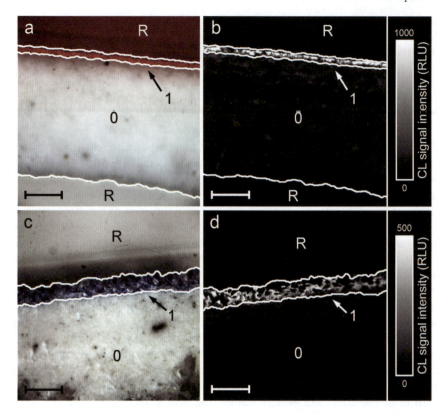

Figure 17.3   Immunolocalization of ovalbumin by CL imaging microscopy in resin-embedded cross-sections of standard painting samples with a layer of whole-egg tempera containing cinnabar (top) or smalt (bottom) pigments. Panels (a) and (c) show the cross-sections, while panels (b) and (d) show the corresponding CL images, which confirm the localization of the CL signal in the egg tempera layer corresponding to the binding medium. The different parts of the cross-sections are indicated (R = resin, 0 = ground layer, 1 = egg tempera layer with pigments). Bars represent 200 μm. (Reproduced with permission from ref. 44.)

several organic components could be developed for the characterization of artworks.

17.4.3   Dating of Ancient Bones

Various methodological approaches have been reported in the literature for the assessment of the post-mortem interval (PMI) of skeletal human remains.[45–48] At present, only isotope-based dating techniques appear to provide valid results, but they are expensive and have only been tested on small samples. Thus, reliable dating of skeletonized remains still poses a scientific challenge. A specific question is the possibility of differentiating, in the absence of an historical context, between historical (*i.e.*, PMI > 100 years) and recently ske-letonised human remains. Among the numerous morphological, chemical, physical and histological dating techniques, the measurement of the CL of bone finds by means of the luminol reaction can represent an inexpensive, quick and simple dating method.

In 1969, the monitoring of the decay of haemoglobin, assessed using a col-orimetric benzidine-based test, was proposed by Knight *et al.* for dating human and animal bones.[49] In subsequent years, there were a few attempts to employ the luminol CL reaction for detecting haemoglobin in bones. Experiments performed on powdered bone samples demonstrated that the intensity of the CL emission decreases with increasing sample age.[50] However, a significant number of false-negative results was reported: while all the very young samples (PMI < 3 years) showed an intense CL and none of the bone samples more than 80 years old resulted in a positive test, a high number (33%) of forensically relevant samples with PMI < 35 years did not give any CL emission. Use of the luminol test for dating bones has been re-examined recently. Two recent stu-dies[51,52] concluded that this test was reproducible and repeatable and repre-sented a relatively easy and economical method for distinguishing between remains of medico-legal (PMI < 100 years) and historical interest. However, due to its limitations, it is not valid as an only method and a reliable dating of skeletonized remains requires a combination of different techniques, including the analysis of environmental conditions and interpretation of the anthro-pological profile, as well as absolute dating methods using radionuclide techniques.

References

1. F. Barni, S. W. Lewis, A. Berti, G. M. Miskelly and G. Lago, *Talanta*, 2007, **72**, 896.
2. D. A. Svistunenko, *Biochim. Biophys. Acta*, 2005, **1707**, 127.
3. S. Baj and T. Krawczyk, *J. Photochem. Photobiol. A.*, 2006, **183**, 111.
4. M. Grodsky, K. Wright, P. L. Kirk and J. Crimin, *Law Criminol. Police Sci.*, 1951, **42**, 95.
5. K. Weber, *Dtsch. Z. Gesamte Gerichtl. Med.*, 1966, **57**, 410.
6. D. L. Laux, in *Principles of Bloodstain Pattern Analysis: Theory and Practice*, ed. P. E. Kish and T. P. Sutton, CRC Press, Boca Raton, FL, 2005, p. 369.
7. H. Cui, R. Meng, H. Jiang, Y. Sun and X. Lin, *Luminescence*, 1999, **14**, 175.
8. T. I. Quickenden and J. I. Creamer, *Luminescence*, 2001, **16**, 295.
9. J. Arnhold, S. Mueller, K. Arnhold and E. Grimm, *J. Biolumin., Chemi-lumin.*, 1991, **6**, 189.
10. J. L. Webb, J. I. Creamer and T. I. Quickenden, *Luminescence*, 2006, **21**, 214.
11. S. S. Tobe, N. Watson and N. N Daeid, *J. Forensic Sci.*, 2007, **52**, 102.

12. L. Ewetz and A. Thore, *Anal. Biochem.*, 1975, **71**, 564.
13. E. J. M. Kent, D. A. Elliot and G. M. Miskelly, *J. Forensic Sci.*, 2003, **48**, 64.
14. J. I. Creamer, T. I. Quickenden, L. B. Crichton, P. Robertson and R. A. Ruhayel, *Luminescence*, 2005, **20**, 411.
15. A. Castello, F. Frances and F. Verdù, *Talanta*, 2009, **77**, 1555.
16. L. J. Blum, P. Esperanca and S. Rocquefelte, *Can. Soc. Forensic Sci. J.*, 2006, **39**, 81.
17. P. W. M. Bergervoet, N. van Riessen, F. W. Sebens and W. C. van der Zwet, *J. Hosp. Infect.*, 2008, **68**, 329.
18. M. S. Meaney and V. L. McGuffin, *Anal. Bioanal. Chem.*, 2008, **391**, 2557.
19. A. M. Jiménez and M. J. Navas, *J. Hazard. Mater.*, 2004, **106**, 1.
20. D. H. Fine, D. Lieb and F. Rufeh, *J. Chromatogr.*, 1975, **107**, 351.
21. D. S. Moore, *Rev. Sci. Instrum.*, 2004, **75**, 2499.
22. D. H. Nguyen, S. Locquiao, P. Huynh, Q. Zhong, W. He, D. Christensen, L. Zhang and B. Bilkhu, in *Electronic Noses and Sensors for the Detection of Explosives*, eds. J. W. Gardner and Y. Yinon, Kluwer, Norwell, MA, 2004, p. 71.
23. M. P. P. Monterola, B. W. Smith, N. Omenetto and J. D. Winefordner, *Anal. Bional. Chem.*, 2008, **391**, 2617.
24. J. G. Bruno and J. C. Cornette, *Microchem. J.*, 1997, **56**, 305.
25. I. U. Mohammadzai, T. Ashiuchi, S. Tsukahara, Y. Okamoto and T. Fujiwara, *J. Chin. Chem. Soc.*, 2005, **52**, 1037.
26. A. Crowson, R. W. Hiley, T. Ingham, T. McCreedy, A. J. Pilgrim and A. Townshend, *Anal. Commun.*, 1997, **34**, 213.
27. C. R. Bowerbank, P. A. Smith, D. D. Fetterolf and M. L. Lee, *J. Chromatogr. A*, 2000, **902**, 413.
28. Y. Shen and M. L. Lee, *Anal. Chem.*, 1997, **69**, 2541.
29. A. L. Lafleur and B. D. Morriseau, *Anal. Chem.*, 1980, **52**, 1313.
30. C. M. Selavka, R. E. Tontarski and R. A. Strobel, *J. Forensic. Sci.*, 1987, **32**, 941.
31. Yu. B. Tsaplev, *J. Anal. Chem.*, 2008, **64**, 299.
32. R. Wilson, C. Clavering and A. Hutchinson, *Anal. Chem.*, 2003, **75**, 4244.
33. I. M. Ciumasu, P. M. Krämer, C. M. Weber, G. Kolb, D. Tiemann, S. Windisch, I. Frese and A. A. Kettrup, *Biosens. Bioelectron.*, 2005, **21**, 354.
34. S. Girotti, S. Eremin, A. Montoya, M. J. Moreno, P. Caputo, M. D'Elia, L. Ripani, F. S. Romolo and E. Maiolini, *Anal. Bioanal. Chem.*, 2010, **396**, 687.
35. T. L. Pittman, B. Thompson and W. Miao, *Anal. Chim. Acta*, 2009, **632**, 197.
36. P. Fernandes, *Appl. Microbiol. Biot.*, 2006, **73**, 291.
37. G. Ranalli, E. Zanardini, P. Pasini and A. Roda, *Ann. Microbiol.*, 2003, **53**, 1.
38. M. Johnson and E. Packard, *Stud. Conserv.*, 1971, **16**, 145.
39. D. Wild, *The Immunoassay Handbook*, 3rd edn, Elsevier, Amsterdam, 2005.
40. L. Krockaert, P. Gausset and M. Dubi-Rucquoy, *Stud. Conserv.*, 1988, **34**, 183.

41. B. Ramirez-Barat and S. de la Vina, *Stud. Conserv.*, 2001, **46**, 282.
42. A. Heginbotham, V. Millay and M. Quick, *J. Am. Inst. Conserv.*, 2006, **45**, 89.
43. M. Vagnini, L. Pitzurra, L. Cartechini, C. Miliani, B. G. Brunetti and A. Sgamellotti, *Anal. Bioanal. Chem.*, 2008, **392**, 57.
44. L. S. Dolci, G. Sciutto, M. Guardigli, M. Rizzoli, S. Prati, R. Mazzeo and A. Roda, *Anal. Bioanal. Chem.*, 2008, **392**, 29.
45. B. Knight, *Med. Sci. Law*, 1969, **9**, 247.
46. F. Facchini and D. Pettener, *Am. J. Phys. Anthropol.*, 1977, **47**, 65.
47. D. L. Sparks, P. R. Oeltgen, R. J. Kryscio and J. C. Hunsaker III, *J. Forensic Sci.*, 1989, **34**, 197.
48. M. Verhoff and K. Kreutz, in *Forensic Pathology Reviews,* Vol. 3, ed. M. Tsokos, Humana Press, Totowa. NJ, 2005, p. 239.
49. B. Knight and I. Lauder, *Hum. Biol.*, 1969, **41**, 322.
50. F. Introna Jr., G. Di Vella and C. P. Campobasso, *J. Forensic Sci.*, 1999, **44**, 535.
51. F. Ramsthaler, K. Kreutz, K. Zipp and M. A. Verhoff, *Forensic Sci. Int.*, 2009, **187**, 47.
52. J. I. Creamer and A. M. Buck, *Luminescence*, 2009, **24**, 311.

Subject Index